SEISMIC HAZARD IN MEDITERRANEAN REGIONS

MODERN APPROACHES IN GEOPHYSICS

formerly Seismology and Exploration Geophysics

Seismic Hazard in Mediterranean Regions

Proceedings of the Summer School organized in Strasbourg, France, July 15 - August 1, 1986 by European Mediterranean Seismological Centre and Institut de Physique du Globe de Strasbourg

edited by

J. BONNIN

M. CARA

and

A. CISTERNAS

Institut de Physique du Globe de Strasbourg, Université Louis Pasteur, Strasbourg, France

and

R. FANTECHI

Commission of the European Communities, Brussels, Belgium

KLUWER ACADEMIC PUBLISHERS
DORDRECHT / BOSTON / LONDON

for the Commission of the European Communities

Library of Congress Cataloging in Publication Data

Seismic hazard in Mediterranean regions : proceedings of the Summer School organized in Strasbourg, France July 15-August 1, 1986 / by European Mediterranean Seismological Centre and Institut de Physique du Globe de Strasbourg ; edited by J. Bonnin ... [et. al].
p. cm.
"Summer School on Seismic Hazard in the Mediterranean Regions, held in Strasbourg, France during July 1986"--Pref.
ISBN 90-277-2779-1
1. Earthquakes--Mediterranean Region--Congresses. 2. Seismology--Mediterranean Region--Congresses. 3. Earthquake engineering--Mediterranean Region--Congresses. I. Bonnin, J. II. Summer School on Seismic Hazard in the Mediterranean Regions (1986 : Strasbourg, France) III. European Mediterranean Seismological Centre. IV. Université Louis Pasteur de Strasbourg. Institut de physique du globe. V. Commission of the European Communities.
QE536.2.M43S45 1988
551.2'2'091822--dc19 88-12997
CIP

ISBN 90-277-2779-1

Publication arrangements by
Commission of the European Communities
Directorate-General Telecommunications, Information Industries and Innovation,
Scientific and Technical Communications Serrvice, Luxembourg

EUR 10706

Published by Kluwer Academic Publishers,
P.O. Box 17, 3300 AA Dordrecht, The Netherlands.

Kluwer Academic Publishers incorporates the publishing programmes of D. Reidel, Martinus Nijhoff, Dr W. Junk and MTP Press.

Sold and distributed in the U.S.A. and Canada
by Kluwer Academic Publishers,
101 Philip Drive, Norwell, MA 02061, U.S.A.

In all other countries, sold and distributed
by Kluwer Academic Publishers Group,
P.O. Box 322, 3300 AH Dordrecht, The Netherlands.

Printed in The Netherlands

TABLE OF CONTENTS

Summer School : "Seismic Hazard in Mediterranean Regions",

Strasbourg July 15 - August 1, 1986

Scientific Committee

A. Cisternas	IPG of Strasbourg	Chairman
K. Aki	USC, Los Angeles	
C. Allen	Caltech, Pasadena	
D. Aubry	Ecole Centrale, Paris	
J. Bonnin	CSEM, Strasbourg	
M. Bouchon	LGI, Grenoble	
M. Cara	IPG of Strasbourg	
R. Gaulon	IPG of Paris	
G. King	University of Cambridge	
J. Mercier	University of Orsay	
St Mueller	ETH, Zurich	
B. Papazachos	University of Thessaloniki	
H. Philip	University of Montpellier	
A. Udias	University of Madrid	

Financial support provided by

Délégation aux Risques Majeurs, Paris
Ministère des Affaires Etrangères, Paris
Commissariat à l'Energie Atomique, Paris
Commission of the European Communities, Brussels
Conseil de l'Europe, Strasbourg
Centre National de la Recherche Scientifique, Paris
Institut National des Sciences de l'Univers, Paris
UNESCO, Paris
City of Strasbourg
Université Louis Pasteur, Strasbourg
Banque Populaire, Strasbourg
Crédit Mutuel, Strasbourg

List of participants

Najib Abou Karaki, Marcus Antonini, Marco Arellano, Katriina Arhe, Luciana Astiz, Naz Aybey, Vladimir Bartak, Belen Benito Oterino, Djillali Benouar, Andrée Bergeron, Mourad Bezzeghoud, Panceva Biserka, Nadia Boumbar, Abdallah Bounif, Pierre Briole, Thierry Camelbeeck, Yves Caristan, Licio Cernobori, Jean-Guy Chabellard, M'Hamed Chadi, Patrick Compte, Nicholas Deichmann, Borislav Georgiev Dimitrov, George Drakatos, Jose Carlos F. Falcao Lucas, Susanna Falsaperla, Arezou Farahmand-Modaressi, Lea Feldman, Jean-Christophe Gariel,, Jean-Luc Got, Heidi Houston, Eleni Ioannidou, Myung Soon Jun, Ioannis Kalogeras, Basil Karacostas, Fekadu Kebede Alamneh, Asma Kerris, Beji Khedaier,, Despina Kondopoulou, Xavier Lana Pons, Gideon Leonard, Hormos Modaressi-Esfeh, Antoine Mocquet, Dolores Munoz Sobrino, Suleyman Pampal, Daniela Pantosi, Panayotis Papadimitriou, Elftheria Papadimitriou, Christos A. Papaioannou, Geneviève Patau, Mohamed Ramdani, Kaspar Renggli, Luis Rivera, Philippe Roth, Stephen L. Salyards, Jean B. Savy, Edelvays Nicolaev Spasov, Peter Suhadolc, M.Teresa Susagna Vidal, Benaïssa Tadili, Bart W. Tichelaar, Helen Toytoydakis, Gianluca Valensise, Aristoteles Vergara Munoz, Jan Zednik, Aldo Zollo.

PREFACE

This book contains a selection of the papers presented at the Summer School on Seismic Hazard in the Mediterranean Regions, held in Strasbourg, France during July 1986. A previous school having the same philosophy, and with special emphasis on the Caribbean region, had been organized in Pointe-à-Pitre (Guadaloupe) in July 1983.

These summer schools responded to the feeling that meetings on earthquake hazard are often too specialized, dealing mainly with current research, concentrating on the immediate interest of the speakers, but not addressing the need of a broad audience of young scientists. We had in mind a school, rather than a research meeting, that would be open to students interested in understanding seismic sources and their effects. We wanted to provide a framework in which to discuss basic problems rather than more technical aspects of the subject. After several years of research on earthquake prediction and the physics of seismic sources, of development of exciting new methods in the study of neotectonics and microtectonics, and of advances in modeling near field seismic wave propagation, we were convinced that such a project could be effective. The very enthusiastic reaction of the participants of the first school in Pointe-à-Pitre encouraged us to repeat the experience for the Euro-Mediterranean region. The response to this initiative was strong; we received some 80 participants from 26 countries. For two weeks formal lectures, seminars, and countless spontaneous discussions broadened our understanding of the physical problems related to earthquake hazard.

For this school we selected five main topics: 1) Physics of earthquake sources. 2) Ground motion in the near field. 3) Engineering seismology. 4) Seismotectonics. 5) Historical seismicity.

Professor K. Aki, one of the world leaders in source theory, delivered the keynote address, as he had done for the previous school. He singled out many of the oustanding problems remaining to be solved in the five topics. He initiated the first lecture series by discussing the complementary views on the nature of seismic sources (barriers and asperities) and reviewing recent trends in the search for a model that could explain departures from self-similarity that come either from very short periods (f_{max} , departures from Gutenberg-Richter law) or long periods (complex wave forms). F. Lund presented a very elegant application of the physics of strings to the calculation of the wavefield generated by a propagating rupture. Classical kinematic source theory and its application to the determination of the source parameters was next presented by A. Udias. P. Bernard and A. Zollo completed the first lecture series with an application of the previous methods, including analysis of wave polarization, to the retrieval of a source model for the 1980 Campania-Lucania earthquake.

The section on ground motion in the near field was intended to show some of the new results concerning the media encountered in seismic engineering: resonant effects in basins, topographic irregularities, stratification, etc. An introduction to the theory of Gaussian beams (A. Cisternas) was followed by a paper on response of basin-like geometries by F. Sanchez Sesma. The simulation of near field seismograms by the superposition of high frequency normal modes (P. Suhaldoc et al.) and a study of a microearthquake swarm thought to be associated with volcanic hazard in a densely populated zone next to Naples (the Campi Flegrei crisis; A. Zollo and G. de Natale), were two examples of applications.

The section on engineering seismology illustrated the variety of subjects of this vast domain: the dynamic behaviour of soils (D. Aubry), the definition of reference motion for antiseismic design (B. Mohammadioun), a spectral study of the seismic response of Strasbourg's medieval Gothic cathedral (M. Cara), a quantitative definition of isoseimals (P. Suhaldoc et al.), and the probabilistic estimation of seismic hazard in Turkey (P. Gulkan).

A most important section concerned the interaction of seismology and neotectonics. After the El Asnam earthquake, one of the clearest lessons was that team work between seismologists and specialists in neotectonics is fundamental for the good understanding of the mechanics of faulting. A general introduction to the geodynamics of the Mediterranean by J. Bonnin, was followed by H. Philip, who explained the methods of neotectonics such as they are applied in the Mediterranean and the form in which stress tensor analysis may lead to a classification of different modes of deformation. One of the few cases in which paleoseismology has been used in the Mediterranean region is that of the El Asnam fault: the study of the lake formed by reverse faulting after each major earthquake (M. Meghraoui) constitutes an oustanding example of the possibilities opened to seismic hazard assessment. B. C. Papazachos presented the seismotectonics of Greece, one of the most active areas of the region, where subduction coexists with back-arc extension. H. Haessler and P. Hoang, and J.L. Plantet and Y. Cansi, illustrated another interesting situation, that of the Rhine Graben, with their analysis of the Remiremont earthquake of December 1986. Here, the structures work in left-lateral strike slip due to the northward push of the Alps.

Finally, the section on historical seismicity contains a methodological paper by J. Vogt, whose long experience may be appreciated here, that shows the renewed interest in this subject and its lasting importance. A. Levret, C. Loup and X. Goula provided new insights on the Lambesc (1909) earthquake, the most important seismic event in France during this century.

Most of the funding for the school was provided by the "Délégation aux Risques Majeurs". Major financial support also came from the "Ministère des Affaires Etrangères", the Council of Europe, the CEE, the "Commissariat à l'Energie Atomique", the "Centre National de la Recherche Scientifique (CNRS), and UNESCO. We thank these organizations for helping to sponsor the summer school. We also wish to graciously thank the authors of the papers for the high level of interest and participation as reflected in this volume. Finally we would also like to thank Soizic Cara, Jane Hartley and Monique Martiny for their help in organizing the Summer School and/or editing the proceedings.

A. Cisternas.

CHAPTER I

EARTHQUAKE SOURCES

PHYSICAL THEORY OF EARTHQUAKES

Keiiti Aki
Department of Geological Sciences
University of Southern California
Los Angeles, CA 90089-0741

ABSTRACT. If earthquake phenomena are controlled by the heterogeneity scale length of a fault zone such as the barrier interval, asperity size and cohesive zone size, there should be a departure from self-similarity. Evidence supporting such a departure from self-similarity is presented. We find that the ω-square model with a constant stress drop cannot explain observed scaling law of seismic spectra for the entire seismic frequency range. We also find that the upper limit frequency f_{max} depends only weakly with seismic moment for a broad range of magnitude, and that the often observed constant corner frequency for earthquakes smaller than M=3 agrees well with f_{max} for earthquakes greater than 3. Another new evidence supporting the above departure from self-similarity is the kink in the magnitude-frequency relation observed at about M=3. We also speculate on the possible link between the heterogeneity of lithosphere and the heterogeneity of fault zone.

INTRODUCTION

To understand the earthquake phenomena, studies have been made on the seismic motions generated by earthquakes, the recurrence behavior of earthquakes in space, time and magnitude, and the tectonic structure of the earth in a seismic region. These studies in the past were made, however, more or less separately by the use of different approaches.

In the present paper, we shall review recent developments in the above three areas of study and call attention to several observations which suggest a unifying new theory of earthquakes. More specifically, for example, we find that the upper limit frequency of acceleration spectrum of major earthquakes, called f_{max}, is related to the constant corner frequency of small earthquakes, to the kink in frequency-magnitude relation, to the fault zone width observed on the surface, and to the recurrence behavior of earthquakes in space, time and magnitude.

We believe that the above observations on various aspects of earthquake phenomena, which have been considered so far unrelated, can all be explained in terms of the slip weakening instability model in

J. Bonnin et al. (eds.), Seismic Hazard in Mediterranean Regions, 3–33.

which the cohesive zone is identified as the fault zone containing branches, subfaults, and other sources of non elastic deformation.

Another example is that the extra bump in the acceleration spectra of earthquakes found by Gusev (1983) may be related to the peak in the attenuation spectra of S waves and coda waves observed for the earth's crust in seismically active regions. We suggest that the heterogeneity of a fault plane which is responsible for Gusev's bump, may be a 2-D cross-section of 3-D heterogeneity of the seismic region which causes scattering loss responsible for the peak in attenuation spectra.

All these observations mentioned above are, however, not yet widely accepted by seismologists. The f_{max} of large earthquakes and the constant corner frequency of small earthquakes have sometimes been considered due not to the source effect but to the recording site effect (e.g. Hanks, 1982). The Gusev bump is also a controversial matter (e.g. Houston and Kanamori, 1986). The interpretation of attenuation spectra obtained by the coda method in terms of the strong heterogeneity of seismic region was also questioned by Frankel and Clayton (1986). Thus, further work is needed to substantiate all of these observations for which we are building the unifying theory. We shall present, therefore, a theory of earthquakes, which is intended to unify observations on many different aspects of the phenomena, as a working hypothesis which should be tested by further observations.

Let us begin with a review of results obtained by the study of earthquakes using observed seismograms.

ASPERITIES, BARRIERS AND CHARACTERISTIC EARTHQUAKE

For many years, seismologists have modeled earthquake fault motion by a uniform slip or a uniform stress drop over the entire fault plane. Recently, they recognized the need for an irregular slip motion over a heterogeneous fault plane in order to explain high-frequency seismic radiation, in particular, strong motion from large earthquakes. Various

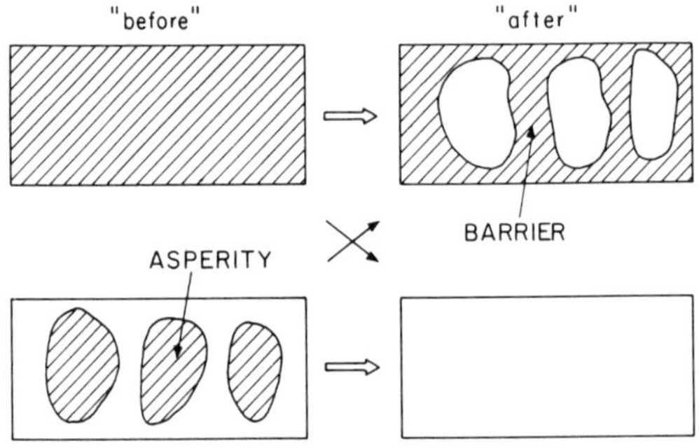

Fig. 1. The barrier model and asperity models, respectively, for the aftershock and foreshock processes.

terms have been proposed to describe such heterogeneity in a fault plane, but "asperities" and "barriers" appear to be the ones most frequently used in recent literature. Both terms refer to strong patches of the fault plane that are resistive to breaking, but they are used for modeling distinctly different roles of strong patches in the process of earthquake faulting.

Figure 1 compares the state of a fault plane before and after an earthquake. The shaded region is stressed, and the blank region is slipped. The completely shaded fault plane at the upper left corner corresponds to a uniformly stressed fault, while the completely blank one at the lower right corner represents a smoothly slipped fault without any unbroken patches. On the upper right of Figure 1 we show the fault plane containing unbroken strong patches after an earthquake. Numerical experiments by Das and Aki (1977) and Mikumo and Miyatake (1978) showed that a rupture can propagate leaving behind unbroken patches. These patches are called "barriers" and are used for modeling strong motion observed for various earthquakes (Aki et al., 1977; Aki, 1979; Papageorgiou and Aki, 1983a, b). This model explains the occurrence of aftershocks as the release of stress concentration around barriers through static fatigue (Mogi, 1962; Otsuka, 1976; Das and Scholz, 1981) and offers a physical mechanism for a stress-roughening process to explain the observed stationary frequency-magnitude relation (Andrews, 1978).

On the lower left of Figure 1, we show the fault plane containing strong patches under stress surrounded by a region where stress has already been released by preslips and foreshocks. These strong patches are called "asperities" by Kanamori and Stewart (1978), and they considered the breaking of these asperities as the model of faulting during the Guatemala earthquake of 1976. In their model the stress is heterogeneous before the main shock and becomes homogeneous afterward. Thus this model represents the main shock as a stress-smoothing process in contrast to the barrier model in which the main shock is considered as a stress-roughening process.

Because both aftershocks and foreshocks exist, we expect that some of the strong patches on a fault plane behave as asperities and others as barriers. It is, however, true that aftershocks are more abundantly observed than foreshocks, suggesting that barriers may be dominant over asperities. On the other hand, barriers along a plate boundary must eventually be broken. Asperities may be more important for great earthquakes along a plate boundary. This view is consistent with the observations that small aftershocks are still being observed for the Nobi earthquake of 1891, a large midplate earthquake with segmented faults but not for the great California earthquakes of 1857 and 1906.

Asperities also play an important role in rupture initiation as in the case of the San Fernando earthquake in which the initial high stress drop event was unequivocally identified by Hanks (1974) and others.

Despite the difference of stress roughening versus smoothing, the barrier model (as described by Papageorgiou and Aki (1983a, b)) and the asperity model (as described by Rudnicki and Kanamori (1981)) cannot be distinguished by the high-frequency seismic radiation, because, according to Madariaga (1983), both may give rise to a flat high-frequency spectrum of the type found by Hanks and McGuire (1981) in most accelerograms.

Recent progress in paleoseismology (Wallace, 1981; Sieh, 1981) has led Schwartz et al. (1981), Coppersmith and Schwartz (1983), and Schwartz and Coppersmith (1984) to propose an earthquake recurrence model called the "characteristic earthquake model." Earthquake recurrence has usually been modeled by the Gutenberg-Richter frequency-magnitude relation which applies very well to earthquakes in a given region. The data from paleoseismology, however, suggest that a specific fault segment generates characteristic earthquakes having a very narrow range of magnitudes and that the Gutenberg-Richter relation may not apply to earthquakes from a given individual fault segment. A similar conclusion was obtained by Wesnousky et al. (1982), who compared the 400-year historical record of seismicity with geologically determined slips on Quaternary faults in Japan.

The idea of characteristic earthquakes was interpreted in terms of the stability of asperities and barriers on faults by Aki (1984), who welcomed it as a promising way of reducing uncertainty in earthquake hazard estimation by a quantitative prediction of strong ground motion directly from geologic observations of fault behavior.

QUANTITATIVE CHARACTERIZATION OF HETEROGENEOUS FAULT PLANES

Papageorgiou and Aki (1983a,b) constructed a specific barrier model having an idealized geometry. Their model consists of circular cracks of equal diameter filling up a rectangular fault nicknamed as "cookie box". As the rupture front sweeps the fault plane, a stress drop takes place in each crack starting from its center and spreading with a fixed velocity. The slip stops suddenly and is frozen over the whole crack as soon as the crack radius reaches its final size. (The seismic far-field waveform radiated from such a circular crack was given by Sato and Hirasawa (1973) in a compact form and has essential features of the waveform from the more realistic models studied by Madariaga (1976, 1977)). The region between circular cracks represents barriers left unbroken after the passage of the rupture front. The ruptures of individual cracks are assumed to take place statistically independently. Thus this model is a hybrid of deterministic and stochastic ones and is described by five parameters, namely fault length, width, rupture velocity, maximum slips, and barrier interval (diameter of the circular crack).

Papageorgiou and Aki (1983a, b) chose California earthquakes for which all the above parameters except the barrier interval are already known from geological and seismological observations other than strong motion. Then they estimated the acceleration power spectra over the time interval in which the seismic motion can be considered as a sample of stationary time series and determined the barrier interval by fitting the predicted spectra to those observed. In this process they had to introduce a sixth parameter to define a cutoff frequency (called "f_{max}" by Hanks (1982)) beyond which the acceleration spectrum decays sharply

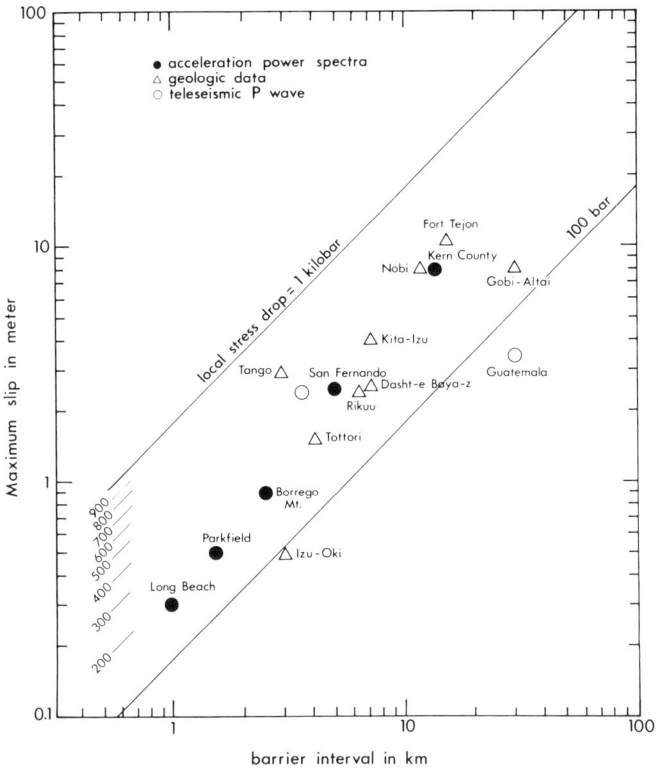

Fig. 2. Relation between the maximum slip and the barrier interval. Different symbols correspond to different data from which the barrier intervals are estimated.

with increasing frequency. Thus they obtained from the observed accelerograms the barrier interval and f_{max}.

The resultant barrier intervals are plotted against the maximum slip (known from earlier studies) in Figure 2. The figure also shows the amount of static stress drop within the circular crack estimated from the maximum slip and barrier interval, which Papageorgiou and Aki (1983a, b) called "local stress drop" as opposed to the usual "global stress drop" estimated from the total fault area assuming a uniform stress drop over the entire fault plane. Although the global stress drop varies greatly among these California earthquakes, the local stress drop appears to be very stable, in the range from 20 to 40 MPa, increasing slightly with earthquake size.

The difference between large and small earthquakes studied here is quite interesting. For example, the Kern County earthquake of 1952 has a barrier interval of about 12 km over a fault plane of 70 x 20 km^2 with a maximum slip of 7-9 m, while the Parkfield earthquake of 1966 has a barrier interval of 1-2 km on a fault plane of 35 x 15 km^2 with a maximum slip of only about 0.5 m. Thus both have comparable fault areas but differ by more than an order of magnitude in slip and in global stress drop. The local stress drop, however, is similar because the barrier interval roughly scales with the amount of slip.

The above result has an interesting implication on why a certain fault segment generates a certain characteristic earthquake. According to Sibson (1982), an earthquake with a large characteristic slip occurs on a segment where the temperature increase with depth is slower and the brittle-ductile transition occurs at greater depth. Since the strength in the brittle region obeys the Coulomb-Byerlee law, it will increase with the confining pressure or depth. Thus, the great earthquake occurs when a deeper and stronger asperity is broken. This explanation is consistent with the slight increase in stress drop with the increase in fault slip obtained by Papageorgiou and Aki (1983), but does not obviously lead to the order of magnitude difference in barrier interval (asperity size or sub-event size) between, say, the Parkfield earthquake and the 1857 earthquake.

The main parameter which determines a characteristic earthquake may not be the strength but the scale length of heterogeneity along the fault plane.

The same problem can be looked at from the point of failure criterion in terms of the slip-weakening model which has been used by many seismologists since Ida (1973). As described in more detail later in our discussion of f_{max}, the slip-weakening model may be characterized by the average cohesive stress σ_c and the critical slip-weakening displacement D. Papageorgiou and Aki (1983) showed that the difference between the small-slip characteristic earthquake (such as the Parkfield) and large-slip characteristic earthquake (such as 1857) may be attributed mainly to the difference in D rather than in σ_c. In other words, we may say that the scale length of fault plane heterogeneity (such as barrier interval) is directly related to the critical slip-weakening displacement.

SLIP-WEAKENING MODEL OF FAILURE AND f_{max}

f_{max} and the critical slip-weakening displacement can be related through a cohesive zone crack tip model as illustrated in Fig. 3 reproduced from Papageorgiou and Aki (1983a). The upper part of Fig. 3 shows the distribution of shear stress and slip as a function of distance along a fault plane in the direction of rupture propagation. The stress at the infinity is σ_o, and it drops to the dynamic frictional stress σ_f after the rupture. The stress rises to σ_u ahead of the crack tip where non-zero slip is occurring over the cohesive zone of length d. The stress drop to σ_f occurs when the slip reaches the critical slip-weakening displacement D.

In the lower part of Fig. 3, we show the constitutive relation between the slip and stress for the cohesive zone. The area of shaded region gives the apparent Griffith energy (the energy needed to create a

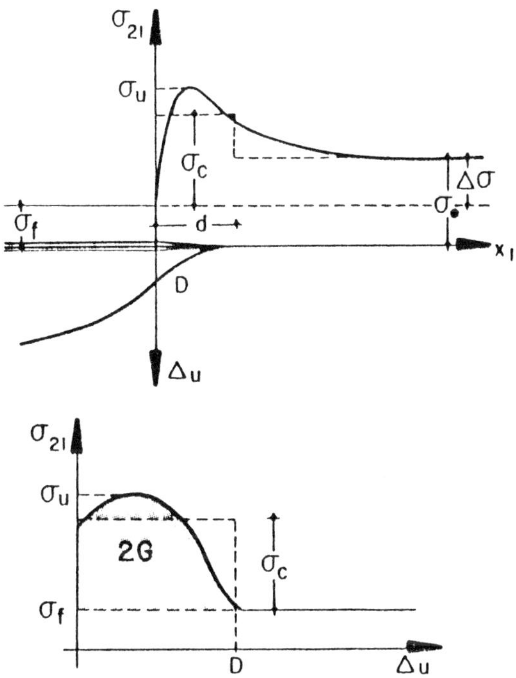

FIGURE 3

unit area of the ruptured fault plane). As summarized by Aki (1979) (see also Aki and Richards (1980) chapter 15), the Griffith energy can be written in two different ways using the parameters introduced above.

$$g = 2\sigma_c{}^2 d/\mu\pi C \qquad (1)$$

$$g = \sigma_c D/2 \qquad (2)$$

where σ_c is the average cohesive stress as indicated in Fig. 3, μ is rigidity and C is a constant of order 1 which depends on the velocity of rupture propagation. From Equations (1) and (2), we obtain

$$d = (\mu\pi C/4\sigma_c) \;\; D \qquad (3)$$

As can be seen from the sketch of slip function in Fig. 3, the cohesive zone acts as a low pass filter with the spatial constant d and the corresponding time constant d/v, where v is the velocity of rupture propagation. This time constant would correspond to the reciprocal of f_{max}

$$f_{max} \simeq v/d \simeq 4\sigma_c' v/\mu\pi CD. \tag{4}$$

Thus, we can express f_{max} in terms of either d or D. The observed f_{max} by Papageorgiou and Aki (1983) is shown in Fig. 4 as a function of seismic moment, together with the observed f_{max} for aftershocks of the Central Japan Sea earthquake of 1983 obtained by Irikura and Yokoi (1984) using the data from a single station located roughly equidistant to all the earthquakes. The observed corner frequencies of the same earthquakes are also shown in Fig. 4 with the corner frequency-moment

FIGURE 4

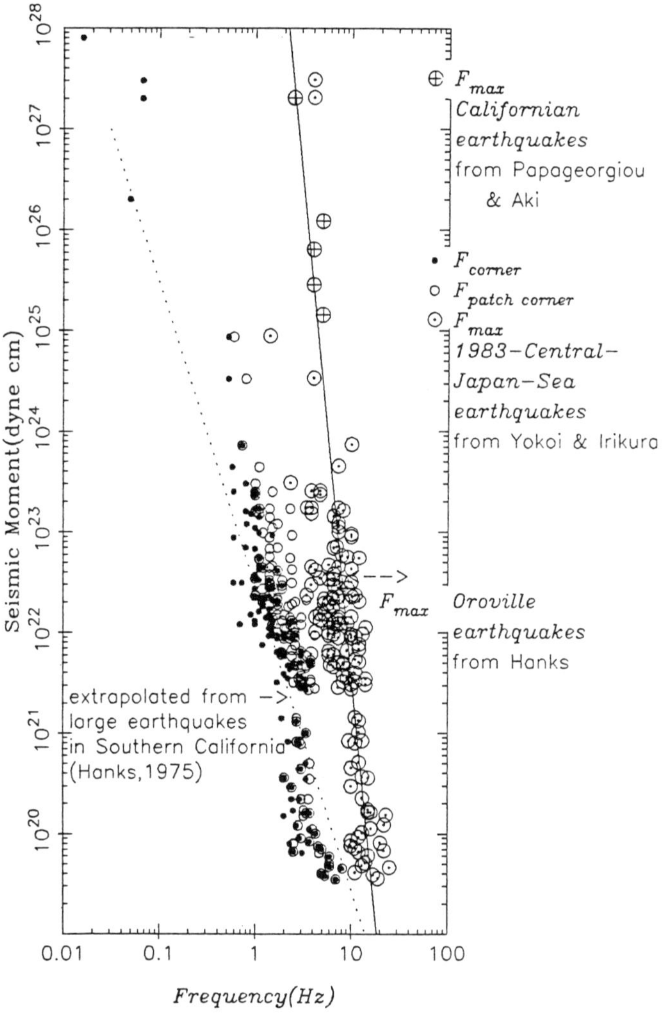

relation obtained by Aki (1967) for large earthquakes. As shown in Fig. 4, the observed f_{max} shows a slight dependence on seismic moment, and varies from a few Hz to a few tens of Hz for the moment range of 10^{28} to 10^{20} dyne cm. This range of f_{max} roughly corresponds to the range of d (cohesive zone size) from 100 m to 1 km, and the range D (critical slip-weakening displacement) from 10 cm to 1 m.

The critical slip-weakening displacement of about 10 cm was used by Day (1982) for 3-dimensional simulation of earthquake faulting as a spontaneous rupture. One of his justifications for the above choice of the critical displacement was that shear stress may not exceed the maximum strength of the slip weakening model.

Recently, several numerical experiments have been made on the recurrence behavior of an earthquake fault using the slip-weakening model.

$$\tau = S \exp\left[-\frac{(u-u_o)^2}{a^2}\right] \cdots\cdots \qquad (5)$$

Where τ is the frictional stress, u is the fault slope, and S, u_o, and a are constants. For example, according to Stuart (1985), the simulation of major earthquakes along the San Andreas fault from Parkfield to Salton Sea required the critical slip-weakening displacement D to be from about 50 cm to 1 m. If a smaller D is chosen, only small earthquakes will occur and great ones such as the 1857 earthquake cannot be produced. This value of D is 4 orders of magnitude greater than the value found in Dieterich's (1981) experiments on a large rock sample. A similar large value of D was used by Mavko (1983), but he did not explain why such a large value was used. Cao and Aki (1984), on the other hand, showed that the critical displacements of 10 cm to 1 m are required for producing widely observed "quiescence" before a major earthquake.

The above agreement of the estimates of absolute value of D from the strong motion data and numerical experiments on recurrence behavior is encouraging, because it suggests a possibility to unify the seismicity data and the strong motion data hitherto considered unrelated. Obviously, we need further studies to establish the reality of this apparent agreement. The agreement also supports the assertion that the absolute value of the order of magnitude of f_{max} is determined by the source effect. The weak but significant dependence of f_{max} on seismic moment can be attributed to the interaction between the individual earthquake fault length and the length d of the fault cohesive zone. As discussed later, we interpret the cohesive zone length d as the outer fractal limit of a fault zone which is essentially fault zone width.

DEPARTURE FROM THE ω-SQUARE SCALING LAW AND IN-PLANE HETEROGENEITY

The first scaling law of seismic spectrum was constructed by Aki (1967) based on the assumption of self-similarity for two types of dislocation models called "ω-square" and "ω-cube", using the spectral ratio data obtained by Berekhemer (1962). The ω-cube model was rejected by Aki (1967) because of its failure in explaining the M_s-M_b relation empirically determined by Gutenberg and Richter (1956). The ω-square model was further tested by Aki (1972) and shown to be valid for frequency less than 0.1 Hz and M>6.

Recently, the ω-square model was found to explain many measures of ground motion important to earthquake engineering, including peak acceleration and velocity by McGuire and Hanks (1980), Hanks (1979) and McGuire et al. (1984). Boore (1986) and Houston and Kanamori (1986) showed its validity for great earthquakes up to moment magnitude 9.5 using teleseismic P waves.

It is important to note a significant difference between the above two ω-square models. The relations between the corner frequency and seismic moment are significantly different. The ω-square model of Aki (1967, 1972) corresponds to the stress drop of 0.2 to 0.3 bar when interpreted by the circular crack model of Brune (1970, 1971), while the ω-square models used for explaining strong motion data have the stress drops ranging from 30 to 100 bars. The difference is more than 100, and is beyond the error of observations.

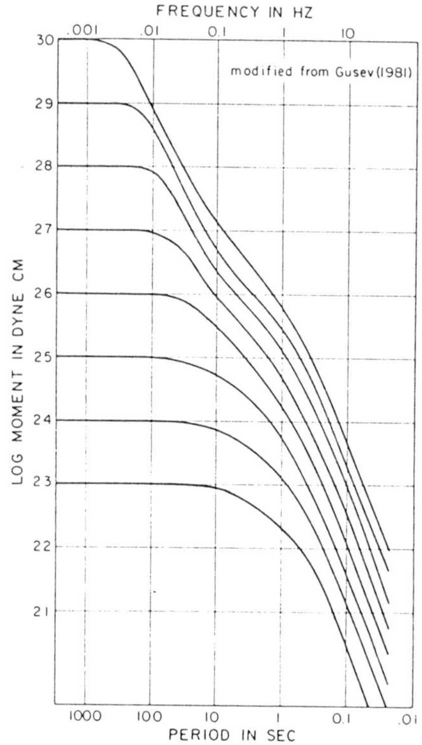

Fig. 5. Scaling law of displacement source spectra based on global data of various magnitude scales. Adapted from GUSEV (1983).

In fact, Boore (1986) concludes that his ω-square model fails to explain the observed M_S-moment relation for large earthquakes. Thus, we must conclude that an ω-square model with a constant stress drop cannot cover the whole range of seismic spectrum.

Gusev (1983) modified the ω-square model to satisfy available data on the relations among different magnitudes. The moment-corner frequency relation in Gusev's scaling law is close to that for Aki's ω-square model, and agrees well with the observed.

In order to agree with the observed short-period data, Aki's ω-square model had to be modified in such a way that a bump appears for frequencies from .1 to 10 Hz as shown in Fig. 5 for displacement spectra and in Fig. 6 for acceleration spectra.

The Gusev scaling law implies a departure from the self similarity among large and small earthquakes. It means that the fault length is not the only length parameter controlling the earthquake phenomena. It is natural to suspect that the cause of Gusev's bump may be the heterogeneities of fault plane which was recognized as barriers and asperities by the study of individual earthquakes.

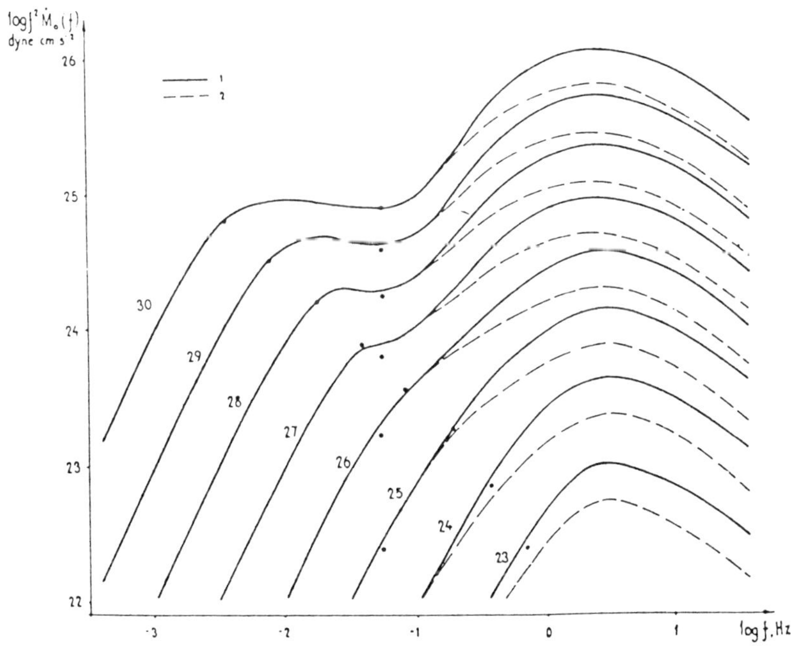

Fig. 6 Scaling law of acceleration source spectra based on global data of various magnitude scales. Reproduced from GUSEV (1983). The solid line corresponds to the observed spectra at the earth's surface while the broken line to the ones reduced to the source. Circles lined up vertically indicate spectral levels derived from M_{IH} (M_0) curves given by GUSEV (1983) and points lined up along a slanted line represent spectral levels at corner frequencies according to the model $\dot{M}_0(f) = M_0[1 + (f/f_0)\gamma]^{-1}$.

The observed barrier intervals shown in Fig. 2 ranges from 1 to a few tens of km. The corresponding time constant may be estimated by dividing them by the rupture velocity, which would correspond to the characteristic frequency from 0.1 to a few Hz where Gusev's bump lies (Papageorgiou and Aki, 1985). Thus, we have a reasonable basis for

FIGURE 7

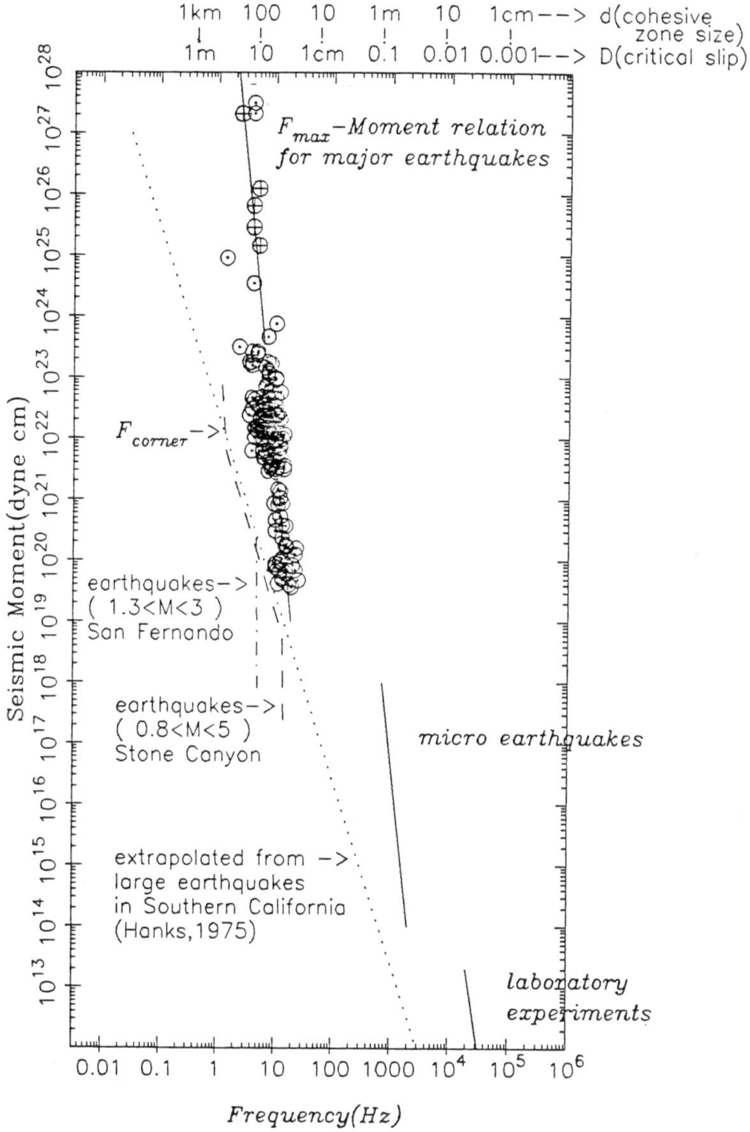

relating Gusev's bump to the heterogeneity along the fault plane, which we shall call "in-plane heterogeneity".

Let us now return to f_{max} and present evidence for its causal relation with the heterogeneity along the direction normal to the fault plane, which we shall refer to "out-of-plane heterogeneity".

f_{max} AND OUT-OF-PLANE HETEROGENEITY

If the earthquake phenomena are strictly self similar, f_{max} should be proportional to the fault length and to the cube root of seismic moment like the corner frequency. The observed f_{max} shown in Fig. 4, however, is more weakly dependent on seismic moment. Since, by definition, the f_{max} curve cannot cross the corner frequency curve, the f_{max}-moment relation for a wider range of moment may be broken into *en echelon* segments as sketched in Fig. 7.

The right-lower corner of Fig. 7 corresponds to laboratory experiments on rock samples. For example, in Okubo and Dieterich's (1984) experiment, the fault length is 2 m (seismic moment on the order of 10^{13} dyne cm) and the critical displacements are 5 to 25 x 10^{-4} cm. Between these two extreme segments of f_{max}-moment relation, an additional segment was arbitrarily drawn in Fig. 7. This figure is intended to show that uniquely different rock structure may be responsible for different branches of f_{max}-moment relation. For the Okubo-Dieterich experiment, the roughness of fault plane appears to be the controlling factor. The relation for major earthquakes, on the other hand, may be controlled by the complex secondary features of an earthquake fault such as branching, offsets, bifurcation, bend, non-elastic deformation, etc.

Whatever the factors controlling the f_{max} of major earthquakes may be, they appear to affect the corner frequency-moment relation for small earthquakes as indicated by a dashed line and a dash-dot line deviated from the self-similar relation with the slope of -3. An apparently constant corner frequency for a wide range of seismic moment for earthquakes with moment less than about 10^{21} dyne cm (corresponding to magnitude about 3) has been widely observed and attributed to the source effect by Chouet *et al.* (1978) and Rautian *et al.* (1978) among others in which source, path and site effects have been separated by the use of the simple nature of coda waves.

Fig. 7 shows that between the major earthquake and the laboratory experiment, f_{max} appears to be grossly proportional to the reciprocal of fault length. If the self similarity holds strictly, the critical slip-weakening displacement should scale with the fault length and the Griffith energy given in equation (2) should also increase proportionally to the fault length. On the other hand, Wong (1982) compared the Griffith energy for earthquakes and laboratory results and suggested that the disagreement may be due to uncertainty in the seismological measurement. We maintain that the branching and other secondary features of a fault may be responsible for the large apparent Griffith energy of an earthquake fault. A similar explanation was suggested by Friedman et al. (1972) for the one to two orders of magnitude higher apparent Griffith energy for various rocks than for the constituent single crystals of these rocks.

An example for illustrating how a secondary feature of a fault may create a large apparent Griffith energy is the famous offset of the San Andreas fault between Parkfield and Cholame. Fig. 8 is reproduced from

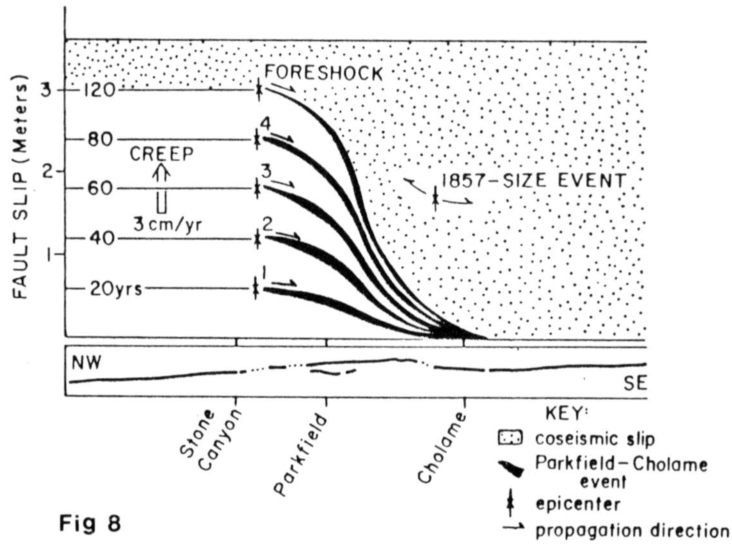

Fig 8

Sieh (1978) who showed schematically the distribution of accumulated slip along the fault near the offset due to creep and earthquakes on the segment to the northwest and blocked segment to the southeast. The curve drawn by Sieh is strikingly similar to the slip function given in Fig. 3, suggesting that the offset and associated region of non-elastic deformation may be the geological expression of the cohesive zone.

The size of the cohesive zone is a few kilometers according to Fig. 8, in an order-of-magnitude agreement with the estimation from the acceleration spectra and the recurrence simulation experiments discussed earlier.

If the secondary features of a fault are an important factor in controlling the seismic radiation and recurrence behavior, we would like to describe them quantitatively and objectively. One such attempt has been made by Okubo and Aki (1984) who measured the fractal dimension of a fault zone (Mandelbrot, 1977) using published fault maps of the San Andreas fault zone. Their preliminary results show that the fractal dimension of the part of the San Andreas fault broken in 1857 is distinctly different for different scale lengths. The fractal dimension of the fault trace is 1.2 for the scale length shorter than 300 m to 1 km (depending on particular segments) and 1.0 for the longer scale length. This means that the fault trace is fractal for scale lengths shorter than a certain critical value, the outer fractal limit. In other words, the fault zone becomes a smooth line for scales greater than this limit. If we define a fault zone as a zone containing secondary branches of fractal nature, the outer fractal limit is precisely the fault zone width. The magnitude of these critical values agrees with our earlier estimates of the cohesive zone size from observations on strong motion spectra as well as recurrence behavior. The agreement supports the idea that the fracture energy barrier for major earthquakes is related to the fractal nature of the fault zone (King, 1983).

f_{max} AND A KINK IN THE FREQUENCY-MAGNITUDE RELATION

If the above idea on f_{max} is correct, there should be a kink in the magnitude-frequency relation at magnitude around 3, reflecting a departure from self-similarity due to the effect of fault zone width. In order to examine if there is such a kink in the magnitude-frequency relation, we used seismograms obtained at a borehole station in the Newport-Inglewood fault zone operated by the University of Southern California. It is shown in the following that we indeed found the kink.

In order to examine the departure from self-similarity for the frequency-magnitude relation, it is essential to develop an accurate procedure to estimate the seismic moment from observed seismograms. We cannot use the magnitude reported routinely for small earthquakes, because they are based on an arbitrary extrapolation of some empirical relation toward small magnitudes.

It is also important to know the sampled area or volume and time period for each magnitude so that the frequency for all magnitudes refers to the common spatial and temporal extent. In order to meet the above requirements, we applied the coda method (Aki, 1969; Aki and Chouet, 1975) to single station data in the following manner.

First, we developed a set of calibration curves which relate the coda amplitude (A) at a lapse time (t, measured from the origin time) with the moment magnitude. Second, we selected a time period over which a continuous visible recording is available. For all seismic events visible on the record, the moment magnitude and t_{s-p} were determined. Third, examining seismic signals on the record, we determined the maximum value of t_{s-p} for each moment magnitude for which signals were unambiguously detected on the record. Fourth, we counted the number of earthquakes for each moment magnitude interval and normalized it to unit area using the maximum t_{s-p} in estimating the total sampled area (or volume).

Because of the use of single station data described above, it is possible to underestimate the frequency of occurrence for small earthquakes if we choose our station in a stable aseismic block. In order to avoid this possibility (which may introduce the kink for a different reason), we chose a station within the Newport-Inglewood fault zone, where the activity of microearthquakes has been known (Teng *et al.*, 1973).

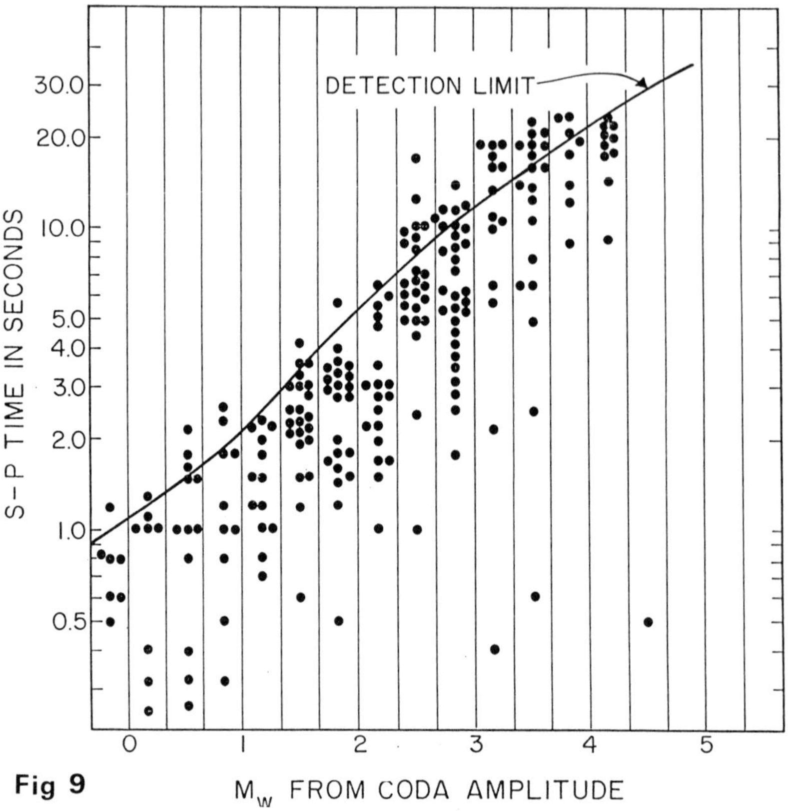

Fig 9

Figure 9 shows t_{s-p} and M_W determined from coda amplitude for 225 earthquakes. In this diagram, we draw a line showing detection limit below which signals are strong and unambiguously identified as earthquakes on the record. We are confident that earthquakes located below the line in this diagram were not missed.

We, then, count the number of earthquakes N falling in the 1/3 magnitude unit interval below the detection line.

Since the hypocentral distance in km is about 8 times t_{s-p} in sec, and focal depths of our earthquakes are in the range from 0 to about 15 km, the sampled volume of seismic region for a given magnitude may be proportional to $(\bar{t}_{s-p})^2$ for $\bar{t}_{s-p} > 2$ sec, and to $(\bar{t}_{s-p})^3$ for $\bar{t}_{s-p} < 2$ sec. First, we calculated the areal density of earthquakes by dividing the number of earthquakes below $\bar{t}_{s-p}$ by $(\bar{t}_{s-p})^2$. Then, we calculated cumulative density as shown in Figure 10 by the curve labeled "AREA". We then corrected the curve for $M < 1$ for which $\bar{t}_{s-p}$ is less than 2 sec and the volume density is proportional to $(\bar{t}_{s-p})^3$. The corrected curve is labeled "VOLUME" in Figure 10.

The observed curves shown in Figure 10 correspond to the cumulative number of earthquakes in a time period effectively about 1 year long and in an area of about 200 km^2; the area of a circle with radius 8 km corresponding to $\bar{t}_{s-p}=1$ sec.

frequency relation from self-similarity; the corner frequency tends to become constant for earthquakes with magnitude smaller than about 3. As mentioned earlier, this constant corner frequency agrees roughly with the extrapolation of the moment-f_{max} relation from large earthquakes. We also suggested that the absolute values of f_{max} determined from actual strong motion acceleration spectra are consistent with the slip-weakening failure criterion in which the critical slip is a few tenths of a meter to a few meters and the cohesive zone size is a few hundred meters to a few kilometers, roughly comparable to the width of a fault zone.

Relevant to this discussion is the prediction of minimum earthquake for a slip-weakening failure by Dieterich (1979). Since the stress intensity factor decreases with the crack size for a constant stress drop, there is a minimum earthquake for the failure mechanism with a given critical weakening slip. He estimates the minimum magnitude to be 1 for the critical slip of 1 mm. If we extrapolate his prediction, the critical slip corresponding to the minimum magnitude of 3 would be 1 cm. Since the model parameters used in his prediction are not strictly constrained by observations, his result may be considered to be consistent with our conclusion.

Some of the observations which supported the constant corner frequency for $M<3$ as the source effect might be unfounded, but others in increasing numbers have been well founded. If the suggested equivalence between the constant corner frequency of small earthquakes and the f_{max} of large earthquakes is well established, we shall have a direct means to estimate one of the most important parameters of strong ground motion from the study of weak motion of small earthquakes which are more easily observable.

HETEROGENEITY OF LITHOSPHERE INFERRED FROM CODA WAVES

As evidenced abundantly from aftershocks located outside the fault plane of a mainshock, the rupture along a fault plane can create fractures in the volume surrounding the fault plane. Such fractures may become scatterers of seismic waves and cause attenuation of seismic waves propagating through the volume.

On the other hand, a seismically active region may be intrinsically heterogeneous due to various tectonic processes and the fault planes which cut across such a region may be characterized by the 3-D heterogeneity. Either way, we may expect some relation between the heterogeneity in the fault zone discussed earlier and the heterogeneity in the lithosphere surrounding the fault zone. We shall discuss about the lithospheric heterogeneity using the results obtained from studies of scattering and attenuation of short period seismic waves.

Most of our knowledge of small-scale heterogeneity in the lithosphere comes from the study of coda waves. If we did not discover coda waves, the enormous difficulty in deciphering numerous factors affecting primary waves prevented us from even recognizing the existence of small scale heterogeneity.

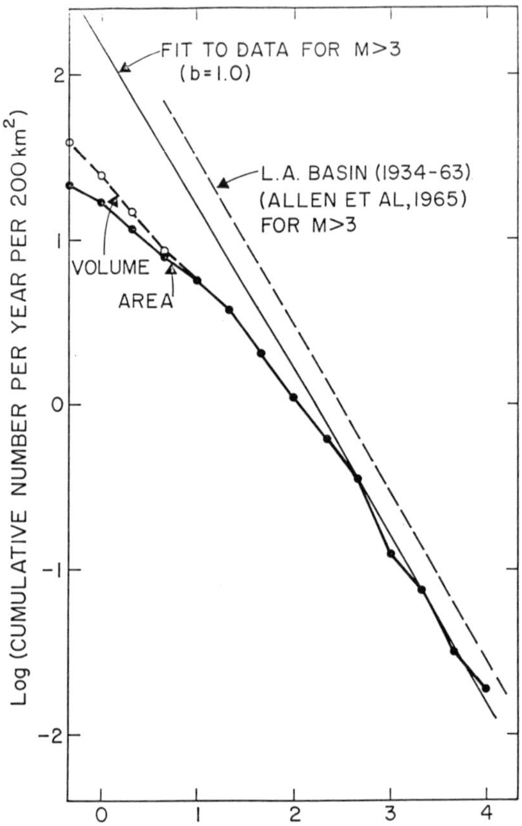

Fig 10 M_W FROM CODA AMPLITUDE

For comparison, we show an extrapolation from the empirical result for earthquakes with magnitude greater than 3 occurring in the Los Angeles basin in the 29 year period from 1934 to 1963 (Allen et al.. 1965). Since the result of Allen et al. was given in terms of the number of earthquakes per year per 1000 km^2, we divided their result by a factor of 5. The slope of the line, so called b-value, is 1.0 for the Los Angeles basin in agreement with our result for M>3. We see a clear departure of the observed curve for M<3 from the line extrapolated from the data for M>3. The observed frequency of earthquakes with M≃-0.5 is almost 10 times less than that expected from the extrapolation of empirical relation for earthquakes with magnitude larger than 3.

Our conclusion is very simple. The observed frequency magnitude relation departs from self-similarity for earthquakes with magnitude smaller than about 3. Although we have not yet confirmed with earthquakes in the Newport-Inglewood fault zone, this departure coincides with the widely observed departure of the moment-corner

Coda waves of local earthquakes recorded at a station and passed through a band-pass filter with center frequency f have a remarkable property that the envelope $A(f|t)$ of filter output plotted as a function of lapse time t measured from the earthquake-origin time can be expressed as

$$A(f|t) = S(f)C(f|t) \tag{6}$$

for t greater than about twice the shear wave travel time (Rautian and Khalturin, 1978), where S(f) depends on source alone, and $C(f|t)$ is a function of f and t alone and is independent of locations of the source and receiver.

The above property of coda waves expressed by Equation (6) has been confirmed for local earthquakes in many parts of the world. It was first recognized by Aki (1969) for aftershocks of the Parkfield, California earthquake of 1966. Later, many investigators demonstrated the validity of Equation (6) for earthquakes in various areas such as California, Hawaii, Japan, Mexico, Central Asia, the USSR, and Norway (Aki and Chouet (1975), Chouet (1976), Rautian and Khalturin (1978), Tsujiura (1978), and others). An example of such demonstration due to Tsujiura (1978) is given in Fig. 11, which shows the envelope of coda

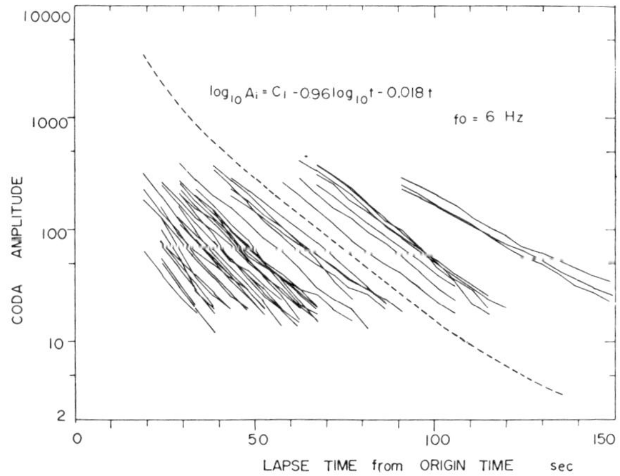

FIGURE 11

amplitude measured on the output of a band-pass filter with center frequency at 6 Hz for local earthquakes observed at Tsukuba, Japan. The curves for individual earthquakes do not intersect each other and suggest the existence of a common decay curve for all earthquakes as implied in Equation (6). The dashed curve in Fig. 11 is a common curve which is determined by a least-squares fitting of the form shown in the figure. The factor C_i represents the strength of source at 6 Hz for the ith earthquake.

The above remarkable property of coda waves can be explained if they are backscattered waves from heterogeneities distributed more or less uniformly in the Earth. Consider for simplicity, an unbounded medium containing a seismic source and a receiver. The primary waves spreading from the source and passing the receiver will meet heterogeneities and send the scattered waves back to the receiver. Let t be the time measured from the time when the waves left the source. Then, the heterogeneities that send the backscattered waves to the receiver within the range $(t, t+\Delta t)$ lie within a spheroidal shell having the receiver and source as foci.

Consider a second source at a location different from the first. Then, for a given $(t, t+\Delta t)$, the spheroidal shell for the first source and that for the second have the same size partially overlapping with each other. As t increases, the spheroid becomes closer to a sphere, and the overlapped part will increase. Even for the non-overlapping part, the properties of scatterers within the two shells may be quite similar statistically, and the sum (average) of contributions from numerous scatterers will produce similar coda waves. Since the contributions from different scatterers are incoherent, the similarity is not in deterministic wave form (wriggle to wriggle correspondence), but in statistical property (power spectrum and envelope shape).

Since the observed coda wave amplitude is independent of distance from the source to receiver, we can simplify our problem by placing the source and receiver at the same point. Let us write the Fourier transform of displacement due to primary S waves as $\phi_o(\omega|r)$, where r is the distance from the source point. Assume that the scatterers responsible for coda waves are distributed randomly in space, and let $N(r)$ be the number of scatterers within r radius of the station. The number of scatterers in the zone bounded by $(r, r+\Delta r)$ will be $(dN/dr)\Delta r$. We shall assume that both primary and scattered waves are S waves with velocity β. The back-scattering waves from scatterers in $(r, r+\Delta r)$ will arrive at the receiver in the time interval $(t, t+\Delta t)$, where $t = 2r/\beta$ and $\Delta t = 2\Delta r/\beta$. For a Δr long enough so that the corresponding Δt is greater than the duration of an individual backscattering wavelet, and for a random distribution of scatterers, the total energy of coda contained in the interval $(t, t+\Delta t)$ will be approximately equal to the sum of energy carried by the individual wavelets arriving in the same interval. On the other hand, for an arbitrary stationary time series $f(t)$, the autocorrelation function estimated from a finite sample of $f(t)$ for a time interval T may be written as

$$\phi(\tau) = \langle f(t)f(t+\tau)\rangle \sim \frac{1}{T} \int_o^T f(t)f(t+\tau)dt \qquad (7)$$

Putting the power spectra of $f(t)$ as $P(\omega)$, we have

$$P(\omega) = \int_{-\infty}^{\infty} \phi(\tau)e^{i\omega\tau}d\tau \sim \frac{1}{T}|F(\omega)|^2 \tag{8}$$

where $F(\omega)$ is the Fourier transform of the finite sample of $f(t)$. Applying the above relation to our case, the power spectral density $P(\omega|t)$ of the coda waves can be written as

$$P(\omega|t) = \frac{1}{\Delta_t} \sum_{r<r_n<r+\Delta r} |\phi_n(\omega)|^2 \tag{9}$$

where $\phi_n(\omega)$ is the Fourier transform of backscattering wavelet from the nth scatterer located at r_n. Assuming that a scatterer at distance r generates a backscattering wavelet with the Fourier transform $\phi(\omega|r)$, we obtain

$$P(\omega|t) = \frac{\Delta r}{\Delta t}\frac{dN}{dr}|\phi(\omega|r)|^2 \tag{10}$$

Let us now separate the effect of seismic source and that of scattering in $\phi(\omega|r)$. In general, consider a plane wave $A_o\, e^{i\omega(t-x/\beta)}$ propagating in the x-direction and incident upon a heterogeneous region of volume V. At a long distance R from V, we measure the scattered waves which may be expressed as $A_1\, e^{i\omega(t-R/\beta)}$. We introduce the directional scattering coefficient $g(\theta)$ which is 4π times the fractional loss of energy by scattering per unit travel distance of primary waves and per unit solid angle at the radiation direction θ measured from the x-direction. Then, we can write

$$|A_1|^2 = |A_o|^2 \frac{Vg(\theta)}{4\pi R^2} \tag{11}$$

In our case, putting the Fourier transform of primary wave as $\phi_o(\omega|r)$, we consider that they are locally plane waves for large r. Then, we can write $A_o = \phi_o(\omega|r)$. The volume of the heterogeneous regions is $4\pi r^2\Delta r$ and the number of scatterers is $\frac{dN}{dr}\Delta r$. Assuming random contribution from each scatterer, we have $|A_1|^2 = (dN/dr)\Delta r|\phi(\omega|r)|^2$. Putting these quantities into equation (11), we obtain

$$\frac{dN}{dr} |\phi(\omega|r)|^2 = g(\pi)|\phi_0(\omega|r)|^2 \qquad (12)$$

where we put $\theta=\pi$, assuming a literal backscattering. Putting (12) into (10), we obtain the formula for coda power spectrum in terms of backscattering coefficient $g(\pi)$ and primary wave spectrum $\phi_0(\omega|r)$.

$$P(\omega|t) = \frac{\beta}{2} g(\pi)|\phi_0(\omega|\beta t/2)|^2 . \qquad (13)$$

This equation was first derived by Aki and Chouet (1975).

Equation (13) was derived under the assumption of single scattering or the first Born approximation. Multiple scatterings are neglected. Under this assumption, the loss of energy from primary waves is of the second order and should be neglected. In fact, it was neglected in deriving equation (13). On the other hand, the primary waves before entering the heterogeneous volume under consideration have lost some energy by absorption (change to heat) and by scattering. We shall write down $|\phi_0(\omega|r)|$ explicitly as $|\phi_0(\omega|r)| = |S(\omega)|r^{-1}e^{-\omega r/(2Q_c\beta)}$. Including also the attenuation for propagation from the scatterers to the receiver, we obtain the equation (14).

$$P(\omega|t) = \frac{\beta}{2} g(\pi)| S(\omega)| \left(\frac{\beta t}{2}\right)^{-2} e^{-\omega t/Q_c} \qquad (14)$$

We denote the effective quality factor as Q_c, which can be obtained by fitting equation (14) to the observed envelope of coda waves. If our S to S single-scattering theory is correct, Q_c should be equal to Q_β determined directly from S waves.

Numerous studies on coda waves have been done using the above formula for the purpose of determining the scattering, attenuation and source effects on high frequency seismic waves (see e.g. Sato (1984) and Wu (1984) for review). The general validity of the formula is supported by the agreement between the quality factor Q_c determined from coda waves and the quality factor Q_β for shear waves determined by independent methods for a broad frequency range, although the validity of the assumption of single scattering has been questioned by Gao et al. (1983), Frankel and Clayton (1986) and others.

Introducing multiple scattering, however, make the coda analysis very unstable, because of non-uniqueness in simultaneously determining the scattering loss and the intrinsic absorption loss. Under the assumption of single scattering, we can determine the apparent Q uniquely from the data although we cannot distinguish the contributions from scattering and absorption. For this reason, we shall continue to use the formula (14), but we shall have to always explicitly describe the time interval over which the coda waves are sampled.

Figure 12 shows the value of Q^{-1} as a functon of frequency for various regions of the earth. The dependences of Q^{-1} on frequency and region are very systematic. For old stable continent, Q is high and frequency dependence is weak. For young active regions, Q is low and shows a strong frequency dependence. If Q values determined from

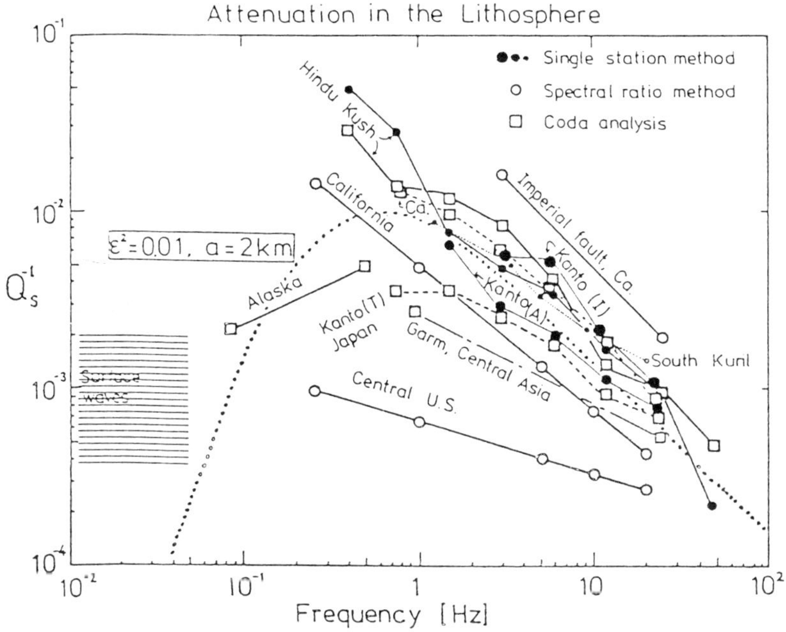

Fig. **12** Observed Q^{-1} of S and coda waves in the lithosphere and the curve theoretically predicted by SATO (1984) assuming that the attenuation is due to scattering by random heterogeneity of the exponential autocorrelation function with correlation distance a and RMS velocity fluctuation ε.

long-period surface waves are combined with these results, we find that the Q^{-1} for active regions shows a strong peak around about 1 Hz.

After rejecting various attenuation mechanisms, the only viable mechanism that can explain observed frequency dependent Q was found to be the loss by scattering due to heterogeneity. The scattering loss is needed also for explaining the very existence of coda waves as back-scattered waves. In fact, the theoretical curve given by Sato (1984) as shown in Fig. 12 explains observed excitation level of coda waves satisfactorily.

CODA Q AND LONG-TERM SEISMICITY

As shown in Fig. 12, the value of Q^{-1} at about 1 Hz correlates well with the degree of tectonic activity. It means that the peaked spectral characteristic of Q^{-1} around 1 Hz is enhanced in a seismically active region. Since this is the frequency range where the Gusev bump lies, it is natural to consider that these two may be related. The Gusev bump was attributed to the in-plane heterogeneity of fault plane. The peak in Q^{-1} was attributed to the lithosphereic heterogeneity in a seismic region. The in-plane heterogeneity of fault plane may be 2-D cross-section of 3-D heterogeneity in a seismic region. It may be that seismic activity creates the heterogeneity in the surrounding lithosphere. If so, coda Q may be an indicator of long term seismicity.

An important question in regard to application to earthquake hazard estimation is how the above relation between coda Q and seismic activity holds in different scales of time and space. For example, if we find that Q is low in the Charleston area, but high elsewhere in the eastern U.S., we may be able to conclude that there will be no events of similar size to the Charleston earthquake of 1886 in other parts of the east.

In order to test this possibility, Jin and Aki (1986) made a preliminary survey of coda Q at 1 Hz using Herrmann's (1980) method at various locations in China where historic data on major earthquakes are relatively complete.

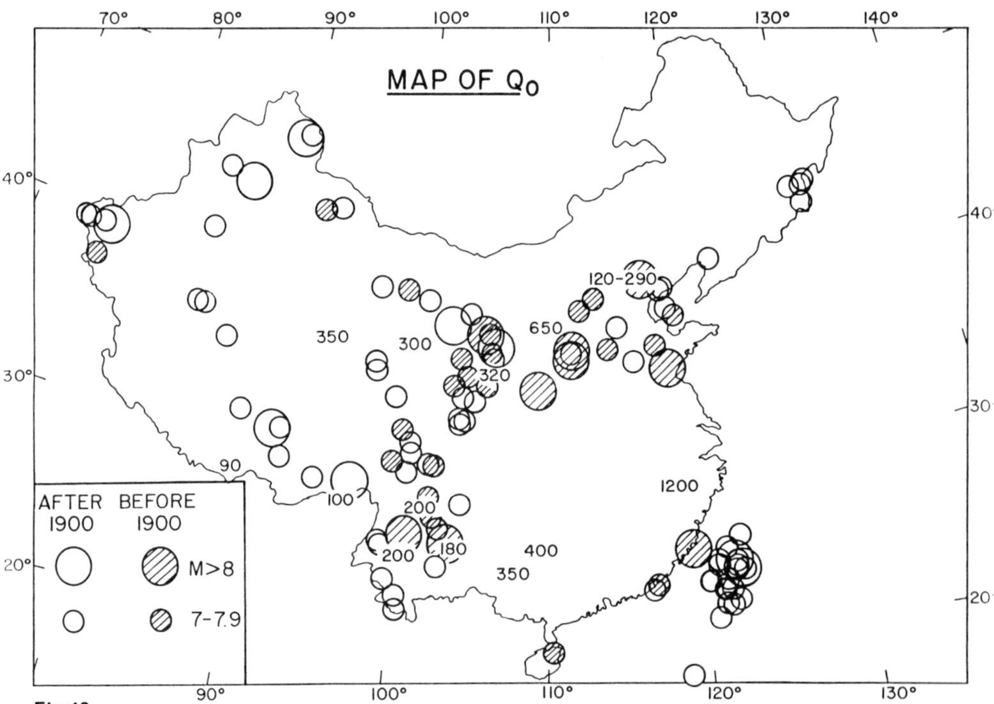

Fig 13

Figure 13 shows our preliminary map of Q_o in China together with major earthquake epicenters before (shaded circles) and after (open circles) 1900. The historic records on seismicity goes back 3000 years for the northeastern part, 2000 years for the southeastern part, 500 years for the southcentral part, and 100 years for the western part.

It appears that there is a critical Q_o value around 300-350. For regions with Q_o above 300-350, there has not been a major earthquake. The value of 650 obtained for the region surrounded by the Yellow River, in fact, refers to a historically aseismic area in the generally seismic region. Active regions in the northeastern China as well as in the southwestern China show Q_o less than 300.

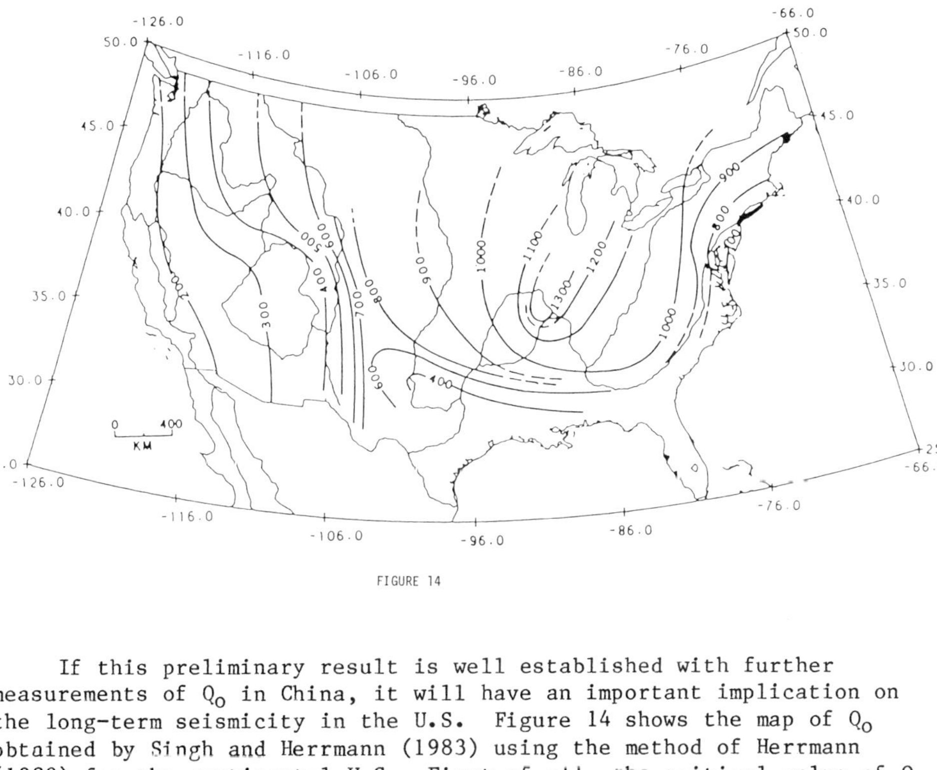

FIGURE 14

If this preliminary result is well established with further measurements of Q_o in China, it will have an important implication on the long-term seismicity in the U.S. Figure 14 shows the map of Q_o obtained by Singh and Herrmann (1983) using the method of Herrmann (1980) for the continental U.S. First of all, the critical value of Q_o (300-350) obtained from China roughly separates the seismically active part of the western U.S. from the rest of the country. Q_o is lower in the active region and higher in the stable region.

We have some problems in the eastern U.S. The observed Q_o shows great variability. For example, the map of Singh and Herrmann shows the Q_o values of 700-800 for the Charleston area using earthquakes at regional distance but Reah (1984) obtained the value around 200 using local earthquakes.

The high Q_o obtained from distant earthquakes may be representative of the average of the eastern U.S., and the low Q_o obtained from local earthquakes may apply to the Charleston area. Clearly, further study is needed to resolve the discrepancy. More detailed map of Q_o using local events such as quarry blasts will give a firm basis for identifying and grading seismic source regions in the eastern and central U.S.

SUMMARY AND CONCLUSION

"Asperities" and "barriers" along a fault, which express the heterogeneity of strength, have been recognized by seismologists from the behavior of fault slip during the rupture. They are apparently stable over many earthquakes and may be considered as the physical basis for the idea of "characteristic earthquake" proposed by geologists.

Quantitative modeling of observed strong ground motion by the specific barrier model gave estimates of the barrier interval which roughly agree with geologists' estimates of length of segmented fault observed on the surface.

The modeling result also gave a clue about what controls the characteristic slip for a given fault segment (0.5 m for the Parkfield segment, and 10 m for the Carrizo plain segment). The scale length of heterogeneity (such as the barrier interval) rather than strength appears to be the dominant factor. In terms of the slip-weakening failure model, the critical slip-weakening displacement rather than cohesive stress controls the characteristic slip.

The upper limit frequency f_{max} that can be generated by fault rupture is related to the cohesive zone length d as well as the critical slip-weakening displacement D. The observed values of f_{max} for major California earthquakes are consistent with the values of D inferred from the numerical simulation of recurrence behavior of the San Andreas fault and with the values of d inferred from a geological interpretation of cohesive zone. The cohesive zone may be interpreted as the zone which contains secondary features of fault such as branching, bifurcation, stepping, etc., and may be quantified as the upper fractal limit.

The existence of unique heterogeneity scale length in a fault zone such as the barrier interval, critical slip-weakening displacement, and the length of cohesive zone suggests a departure of earthquake phenomena from self-similarity in which total fault length is the only physical parameter controlling the phenomena. In fact, we find that the ω-square model with a constant stress drop cannot cover the entire seismic frequency range. A modified scaling law proposed by Gusev (1983) shows a bump in the frequency range from 0.1 to 10 Hz, which probably corresponds to the range of barrier intervals obtained by Papageorgiou and Aki (1983), which we now refer to as "in-plane heterogeneity".

Another departure from self-similarity is related to f_{max}, and is caused by the existence of fault-zone width containing the secondary features of a fault, which we now refer to as "out-of-plane heterogeneity". This departure shows up as a very weak dependence of f_{max} of large earthquakes on seismic moment and also as a nearly constant corner frequency of small earthquakes observed widely for magnitude less than about 3. The constant corner frequency for small earthquakes and f_{max} for large earthquakes are very close to each other, suggesting a common origin. Another new evidence supporting the above departure from similarity is the kink in the frequency-magnitude relation observed at magnitude about 3 for the Newport-Inglewood fault zone.

If all these are due to a common cause, namely, the finite fault zone width, we shall have a basis to predict f_{max} of large earthquakes using the corner frequency of small earthquakes or the frequency-magnitude relation for small earthquakes.

Finally, we find a strong tie between the heterogeneities of fault plane described above and the heterogeneities of lithosphere surrounding earthquake faults as estimated from seismic scattering and attenuation. It appears that the in-plane heterogeneity of fault plane is a 2-D cross-section of lithospheric heterogeneity, or the latter may be created by the earthquake faulting process.

If so, long-term seismicity may be reflected in the scattering-attenuation property of the lithosphere in the seismic region. In fact, we find a strong correlation between the quality factor Q (at 1 Hz) and the degree of seismic activity both in U.S. and China. The quality factor may be used to infer the long term seismicity in a region where historic data are lacking.

ACKNOWLEDGEMENT

This work was supported by U.S. National Science Foundation Grant ECE-8408227.

REFERENCES

Aki, K., Scaling law of seismic spectrum, J. Geophys. Res., 72, 1217-1231, 1967.

Aki, K., Analysis of the seismic coda of the local earthquakes as scattered waves, J. Geophys. Res., 74, 615-631, 1969.

Aki, K., Scaling law of earthquake source time function, Geophys. J. R. astr. Soc. 31, 3-25, 1972.

Aki, K., Characterization of barriers on an earthquake fault, J. Geophys. Res., 84, 6140-6148, 1979.

Aki, K., Asperities, barriers, characteristic earthquakes and strong motion prediction, J. Geophys. Res., 89, 5867-5872, 1984.

Aki, K. and B. Chouet, Origin of coda waves: Source, attenuation, and scattering effects, J. Geophys. Res., 80, 3322-3343, 1975.

Aki, K., M. Bouchon, B. Chouet, and S. Das, Quantitative prediction of strong motion for a potential earthquake fault, Ann. Geofis., 30, 341-368, 1977.

Aki, K. and P. G. Richards, Quantitative Seismology: Theory and Methods, Vol. I and II, W. H. Freeman and Co., 1980.

Allen, C. R., P. St. Amand, C. F. Richter and J. M. Nordquist, Relationship between seismicity and geologic structure in the southern California region, Bull. Seis. Soc. Am., 55, 753-798, 1965.

Andrews, D. J., Coupling of energy between tectonic processes and earthquakes, J. Geophys. Res., 83, 2259-2264, 1978.

Berckhemer, H., Die Ausdehnung der Bruchflache in Erdbebenherd und ihr Einfluss auf das seismiche wellen spectrum, Gerlands Beitr. Geophys., 71, 5-26, 1962.

Boore, D. M., Short-period P-and S-wave radiaton from large earthquakes: Implications for spectral scaling relations, Bull. Seis. Soc. Am., 76, 43-64, 1986.

Boore, D., Stochastic simulation of high frequency ground motion based on seismological models of the radiated spectra, Bull. Seis. Soc. Am., 73, 1865-1894, 1983.

Brune, J. N., Tectonic stress and the spectra of seismic shear waves from earthquakes, J. Geophys. Res., 75, 4997-5009, 1970 (Correction, J. Geophys. Res., 76, 5002, 1971).

Cao, T., and K. Aki, Seismicity simulation with a mass-spring model and a displacement hardening-softening friction law, PAGEOPH, 122, 10-24, 1984/1985.

Chouet, B., Source, scattering and attenuation effects on high frequency seismic waves, Ph.D. Thesis, M.I.T., 1976.

Chouet, B., K. Aki, and M. Tsujiura, Regional variation of the scaling law of earthquake source spectra, Bull. Seis. Soc. Am., 68, 49-80, 1978.

Coppersmith, K. J., and D. P. Schwartz, The characteristic earthquake model: Implications to recurrence on the San Andreas fault (abstract), Earthquake Notes, 54, 61, 1983.

Das, S., and K. Aki, Fault planes with barriers: A versatile earthquake model, J. Geophys. Res., 82, 5648-5670, 1977.

Das, S., and C. H. Scholz, Theory of time dependent rupture in the earth, J. Geophys. Res., 86, 6039-6051, 1981.

Day, S. M., Three-dimensional simulation of spontaneous rupture: The effect of non-uniform prestress, Bull. Seis. Soc. Am., 72, 1881-1902, 1982.

Dieterich, J. H., Modeling of rock friction, 1. Experimental results and constitutive equations, J. Geophys. Res., 84, 2161-2168, 1979.

Dieterich, J., Constitutive properties of faults with simulated gouge, mechanical behavior of crustal rocks, Monograph 24, eds. N. L. Carter, N. Friedman, J. M. Logan, and D. W. Stemns, American Geophysical Union, 103-120, 1981.

Frankel, A., and R. W. Clayton, Finite difference simulations of seismic scattering: implications for the propagation of short-period seismic waves in the crust and models of crustal heterogeneity, J. Geophys. Res., 91, 6465-6490, 1986.

Friedman, M., J. Handin and G. Alani, Fracture-surface energy of rocks, Int. J. Rock Mech. Min. Sci., 9, 757-766, 1972.

Gao, L. S., L. C. Lee, N. N. Biswas and K. Aki, Comparison of the effects between single and multiple scattering on coda waves for local earthquakes, Bull. Seis. Soc. Am., 73, 377-389, 1983.

Gusev, A. A., A descriptive statistical model of earthquake source radiation and its application to an estimation of short-period strong motion, Geophys. J. R. astron. Soc., 74, 787-808, 1983.

Gutenberg, B., and C. F. Richter, Earthquake magnitude, intensity, energy and acceleration, part 2, Bull. Seis. Soc. Am., 46, 105-145, 1956.

Hanks, T. C., The faulting mechanism of the San Fernando earthquake, J. Geophys. Res., 79, 1215-1229, 1974.

Hanks, T. C., b values and $\omega^{-\gamma}$ seismic source models: implications for tectonic stress variations along active crustal fault zones and the estimation of high-frequency strong ground motion, J. Geophys. Res., 84, 2235-2242, 1979.

Hanks, T. C., and R. K. McGuire, The character of high frequency strong ground motion, Bull. Seismol. Soc. Am., 71, 2071-1096, 1981.

Hanks, T. C., f_{max}, Bull. Seismol. Soc. Am. 72, 1867-1880, 1982.

Hanks, T. C. and D. M. Boore, Moment-magnitude relations in theory and practice, J. Geophys. Res., 89, 6229-6235, 1985.

Herrmann, R. B., Q estimates using the coda of local earthquakes, Bull. Seis. Soc. Am., 70, 447-468, 1980.

Houston, H., and H. Kanamori, Source spectra of great earthquakes: Teleseismic constraints on rupture process and strong motion, Bull. Seis. Soc. Am., 76, 19-42, 1986.

Ida, Y., The maximum acceleration of seismic ground motion, Bull. Seis. Soc. Am., 63, 959-968, 1973.

Irikura, K., and T. Yokoi, Scaling law of seismic source spectra for the aftershocks of 1983 Central-Japan-Sea earthquake, Abstracts of Seismological Society of Japan, 1984, No. 1.

Jin, A., and K. Aki, Temporal change in coda Q before the Tangshan earthquake of 1976 and the Haicheng earthquake of 1975, J. Geophys. Res., 91, 665-673, 1986.

Kanamori, H., and G. S. Stewart, Seismological aspects of the Guatemala earthquake of February 4, 1976, J. Geophys. Res., 83, 3427-3434, 1978.

King, G., The accommodation of large strains in the upper lithosphere of earth and other solids by self-similar fault systems: The geometrical origin of b-value. PAGEOPH, 121, 761-816, 1983.

Madariaga, R., Dynamics of an expanding circular fault, Bull. Seismol. Soc. Am., 66, 639-666, 1976.

Madariaga, R., High frequency radiaton from crack (stress drop) models of earthquake faulting, Geophys., J. R. astron. Soc., 51, 625-651, 1977.

Madariaga, R., High frequency radiation from dynamic earthquake fault models, Ann. Geofis., 1, 17-24, 1983.

Mandelbrot, B. B., Fractals, W. H. Freeman and Co., San Francisco, pp. 365, 1977.

Mavko, G., Large-scale earthquakes from a laboratory friction law, submitted to J. Geophys. Res., Oct., 1983.

McGuire, R. K., and T. C. Hanks, RMS accelerations and spectral amplitudes of strong ground motion during the San Fernando, California Earthquake, Bull. Seismol. Soc. Am., 70, 1907-1919, 1980.

McGuire, R. K., A. M. Becker, and N. C. Donovan, Spectral estimates of seismic shear waves, Bull. Seis. Soc. Am., 74, 1427-1440, 1984.
Mikumo, T., and T. Miyatake, Dynamics rupture process on a three-dimensional fault with non-uniform frictions and near-field seismic waves, Geophys J. R. Astron. Soc., 54, 417-438, 1978.
Mogi, K., Study of elastic shocks caused by the fracture of heterogeneous material and its relation to earthquake phenomena, Bull. Earthquake Res. Inst. Univ. Tokyo, 40, 125-173, 1962.
Okubo, P. G., and K. Aki, Fractal geometry in the San Andreas fault system, EOS, 64, 766, 1984.
Okubo, P. G., and J. H. Dieterich, Effects of physical fault properties on frictional instabilities produced on simulated faults, J. Geophys. Res., 89, 5815-5827, 1984.
Otsuka, M., A simulation of earthquake occurrences, 5, An interpretation of aftershock phenomena (in Japanese), Jishin, 29, 137-146, 1976.
Papageorgiou, A., and K. Aki, A specific barrier model for the quantitative description of inhomogeneous faulting and the prediction of strong ground motion, I, Description of the model, Bull. Seismol. Soc. Am., 73, 693-722, 1983a.
Papageorgiou, A., and K. Aki, A specific barrier model for the quantitative description of inhomogeneous faulting and the prediction of strong ground motion, II, Applications of the model, Bull. Seismol. Soc. Am., 73, 953-978, 1983b.
Papageorgiou, A. S. and K. Aki, Scaling law of far-field spectra based on observed parameters of the specific barrier model, PAGEOPH, 123, 353-374, 1985.
Rautian, T. G., and V. I. Khalturin, The use of the coda for determination of the earthquake source spectrum, Bull. Seismol. Soc. Am., 68, 923-948, 1978.
Rautian, T. G., V. I. Khalturin, V. C. Martynov and P. Molnar, Preliminary analysis of the spectral content of P and S waves from local earthquakes in the Garm, Tadjikistan region, Bull. Seis. Soc. Am., 68, 949-971, 1978.
Rhea, S., Q determined from local earthquakes in the South Carolina coastal plain, Bull. Seis. Soc. Am., 74, 2257-2268, 1984.
Rudnicki, J., and H. Kanamori, Effects of fault interaction on moment, stress drop, and strain energy release, J. Geophys. Res., 86, 1785-1793, 1981.
Sato, H., Attenuation and envelope formation of three-component seismograms of small local earthquakes in randomly inhomogeneous lithosphere, J. Geophys. Res., 89, 1221-1241.
Sato, T., and T. Hirasawa, Body wave spectra from propagating shear cracks, J. Phys. Earth, 21, 415-431, 1973.
Schwartz, D. P., and K. J. Coppersmith, Fault behavior and characteristic earthquakes: Examples from the Wasatch and San Andreas fault zone, J. Geophys. Res., 89, 5681-5698, 1984.
Schwartz, D. P., K. J. Coppersmith, F. H., Swan III, P. Somerville, and W. U. Savage, Characteristic earthquakes on intraplate normal faults (abstract), Earthquake Notes, 51, 71, 1981.

Sibson, R. H., Fault zone models, heat flow, and the depth distribution of earthquakes in the continental crust of the United States, Bull. Seis. Soc. Am., 72, 151-163, 1982.

Sieh, K. E., Central California foreshocks of the great 1857 earthquake, Bull. Seis. Soc. Am., 68, 1731-1749, 1978.

Sieh, K. E., A review of geological evidence for recurrence times of large earthuakes, In Earthquake Prediction: An international Review, Maurice Ewing Ser., vol. 4, edited by D. W. Simpson and P. G. Richards, pp. 181-207, AGU, Washington, D.C., 1981.

Singh, S., and R. B. Herrmann, Regionalization of crustal coda Q in the continental United States, J. Geophys. Res., 88, 527-538, 1983.

Stuart, W. D., Instability model for recurring large and great earthquakes in southern California, PAGEOPH, 1985.

Teng, T.-L., C. R. Real, and T. L. Henyey, Microearthquakes and water flooding in Los Angeles, Bull. Seis. Soc. Am., 63, 859-876, 1973.

Tsujiura, M., Spectral analysis of the coda waves from local earthquakes, Bull. Earthq. Res. Inst., Univ. of Tokyo, 53, 1-48, 1978.

Wallace, R. E., Active faults, paleoseismology, and earthquake hazards in the western United States, in Earthquake Prediction: An International Review, Maurice Ewing Ser., vol. 4, edited by D. W. Simpson and P. G. Richards, pp. 279-289, AGU, Washington, D. C., 1981.

Wesnousky, S. G., C. H. Scholz, K. Shimazaki, and T. Matsuda, The occurrence rate of earthquakes on Quaternary faults in interplate Japan, paper presented at the Chapman Conference on Fault Behavior and the Earthquake Generation Process, AGU, Snowbird, Utah, Oct. 11-15, 1982.

Wong, T. F., Shear fracture energy of Westerly granite from post-failure behavior, J. Geophys. Res., 87, 990-1000, 1982.

Wu, R., Scismic wave scattering and the small scale inhomogeneities in the lithosphre, Ph.D. thesis, MIT, 1984.

DISLOCATION DYNAMICS : APPLICATIONS TO NEAR FIELD SEISMOLOGY

Fernando Lund
Departamento de Fisica
Facultad de Ciencias Fisicas y Matematicas
Universidad de Chile
Casilla 487-3, Santiago, Chile

ABSTRACT. This lecture presents a new look at the formalism used in studying the generation of high frequency seismic waves, the idea being to spend some human time trying to extract as much as possible from the formalism in order to be in as strong a position as possible at the moment of starting a given computation and thus make the most of available computer time. The question adressed is that of wave generation by a rupturing surface that is evolving in a prescribed way.

INTRODUCTION

Recent work on strong motion seismology has provided a detailed close-up picture of the seismic source (for a review, see the contributions of Aki and Bouchon in these proceedings). The information thus acquired appears to be rather complex, reflecting the fact that much of it comes from the frequency band 1-10 Hz of particular interest in seismic engineering, and this high frequency information is able to look at small distance scales with sufficient sensitivity for the records to reflect the heterogeneities in the fault plane that are responsible for the complexity of the rupture process that generates earthquake phenomena. In spite of the complexity of strong motion records, it has been shown (Rudnicki and Kanamori, 1981 ; Papageorgiou and Aki, 1983, a,b ; Madariaga, 1983 ; Luco and Anderson, 1985), that at least some of their salient features can be understood in simple physical terms. This basic understanding is interesting not only for its own sake, but also because in practical applications one is limited by available computer time and the better the starting point of any given computation the better the information one will be able to extract from it. The question that will be adressed in this paper is that of wave generation by a rupturing surface that is evolving in a prescribed way. The important problem of why a given rupture does what it does will not be adressed, at least directly, here.

THE GENERAL THEORETICAL FRAMEWORK

The mathematical framework is continuum mechanics and the source will be taken as a dislocation loop, alias the rupture front, which is a closed curve $\mathbf{X}(\sigma,t)$ bounding a surface of uniform slip (Figure 1). The curve is represented by three functions of two coordinates :

J. Bonnin et al. (eds.), Seismic Hazard in Mediterranean Regions, 35–44.

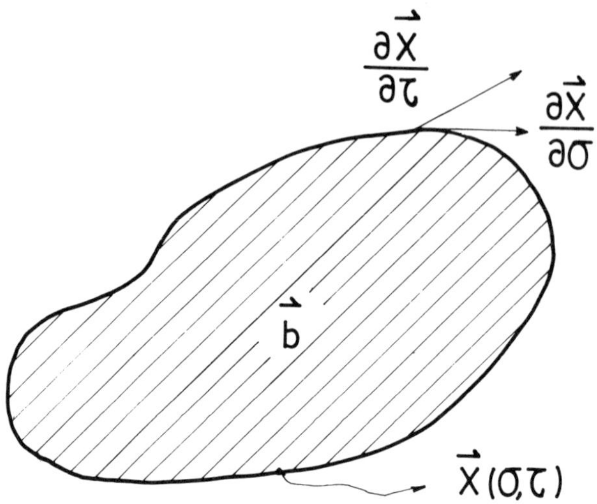

Figure 1 : A dislocation loop **X** (σ,τ) : it consists of three functions of two variables, the time τ and a (time independent) label σ. ∂**X**/∂ σ is the tangent and ∂**X**/∂ τ the velocity at each point. In general they are arbitrary functions of σ and τ. The loop **X**(σ, τ) is the boundary of a surface of constant slip **b**.

time, and a Lagrangean parameter σ labelling each point. Only flat surfaces and shear slip will be considered here, although the formalism is by no means restricted to those cases.

The well-known representation theorem for the displacement **u** (**x**,t) at point **x** and time t generated by a fault plane with a slip function Δ**u** (ξ, t) on it is (Aki and Richards, 1980).

(1) $$u_n(\mathbf{y},t) = \int d\tau \iint \Delta u_i(\xi,\tau)\, c_{ijpq} \frac{\partial G_{np}(\mathbf{y},\xi\,;t-\tau)}{\xi_p}\, \nu_j\, d\Sigma$$

where $\nu_j d\Sigma$ is the surface element of the fault plane, c_{ijpq} are elastic constants and G_{np} is the impulse response function of the elastic medium.

When the slip on the fault is uniform, Δ**u** (ξ,τ) = **b**, it was shown by Nabarro (1951) that an alternative representation was

(2) $$u_n(\mathbf{y},t) = \int d\tau \oint d\sigma\, b_i c_{ijpq} \left(\frac{\partial \mathbf{X}}{\partial \tau} \wedge \frac{\partial \mathbf{X}}{\partial \sigma}\right)_j \frac{\partial G_{np}(\mathbf{y},\xi\,;t-\tau)}{\partial \xi_q}\Bigg|_{\xi=\mathbf{X}}$$

where $\tilde{G}$ is the response of the medium to a force that is localized at one point in space but is a step function in time. In other words, $\partial\tilde{G}/\partial t = G$. $\partial \mathbf{X}/\partial\sigma$ is the tangent to the rupture front. The line integral over σ involves only the rupture front and not the whole fault plane. In the sense of this representation, the seismic source is localized at the rupture front only, which is simpler that having a source over a whole plane. On the other hand, it features the step function response $\tilde{G}$ which has an infinitely long tail in time and is thus less manageable than the impulse response G which (at least for infinite homogeneous isotropic media) consists of two wave fronts travelling at the P and S wave velocities plus something in between. A nice representation in terms of the impulse response and a source localized at the rupture front can be obtained for particle velocity simply differentiating (2) and using $\partial\tilde{G}/\partial t = G$:

$$v_n(\mathbf{y},t) = \frac{\partial u_n}{\partial t} = \int d\tau \oint d\sigma\, b_i\, c_{ijpq} \left(\frac{\partial \mathbf{X}}{\partial \tau} \wedge \frac{\partial \mathbf{X}}{\partial \sigma} \right)_j \left. \frac{\partial G_{np}(\mathbf{y}, x\, ;\, t-\tau)}{\partial \xi_q} \right|_{\xi = \mathbf{X}} \tag{3}$$

which was found by Mura (1963). This representation is valid for all frequencies and for any

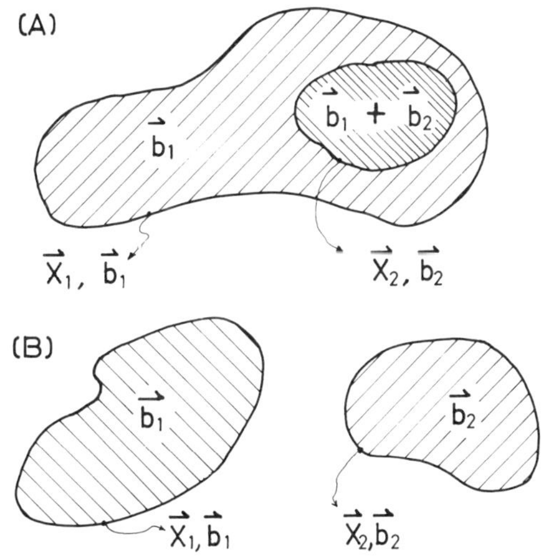

Figure 2 : (a) Fault of variable slip. Inside the dislocation $\mathbf{X}_2$ the slip is $\mathbf{b}_1 + \mathbf{b}_2$. Outside $\mathbf{X}_2$ but inside $\mathbf{X}_1$ it is $\mathbf{b}_1$. (b) Fault consisting of two disconnected patches, one with slip $\mathbf{b}_1$, the other with slip $\mathbf{b}_2$. On both cases the particle velocity is obtained adding the contributions of $\mathbf{X}_1$ and $\mathbf{X}_2$ in the representation (4).

medium whose impulse response can be obtained, either with an exact algebraic expression as in an idealized model problem, or approximately, as in a concrete application. It displays explicitely an allowance for a non uniform propagation of the rupture front, as $(\partial\mathbf{X}/\partial t)\wedge(\partial\mathbf{X}/\partial\sigma)$ may be variable both in time (acceleration of one point of the loop) or along σ (different velocities at different points).

The case of a surface with variable slip (including cases of finite rise time) is obtained simply by superposition, as the system is supposed to be linear everywhere outside the rupture front. For instance, in Figure 2a a fault is shown where the slip vector takes two values, and the evolution of each region may be independent of the other. This is simply achieved by taking two dislocation loops $\mathbf{X}_1$ and $\mathbf{X}_2$ with slip vectors $\mathbf{b}_1$ and $\mathbf{b}_2$ appropriately superposing them. If they don't overlap, two independent rupture patches are described, as in Figure 2b.

Dislocation loops, having uniform slip, need infinite gradients for the displacement in their vicinity and are thus surfaces of infinite stress drop. Surfaces whose stress drop is finite are called cracks and, if the medium is linearly elastic up to the crack tip, there still is a (square root) singularity, this time in slip, which is smoothed out by the introduction of a "cohesive zone". In any case, the crack surface can still be thought of as a (possibly continuous) distribution of dislocation loops (Bilby and Eshelby, 1968) and the representation (3) can be used. The most important practical limitation of this method will appear when very many dislocation loops are needed to model a surface of complicated slip (Figure 3). Representing the crack as a number of dislocations has the advantage that each one may have its own time evolution, and the whole crack need not stop simultaneously along its edge. Such "quasi-dynamical" models have been used by Boatwright (1980), Archuleta and Hartzell (1981), Campillo (1983), and Bernard and Madariaga (1984a).

For the case of shear faulting along a plane we choose coordinates as in Figure 4a. In this case $b_2 = 0$, $\mathbf{X}(\sigma, \tau) = (X_1, 0, X_3)$ and $(\partial\mathbf{X}/\partial t)\wedge(\partial\mathbf{X}/\partial\sigma) = (0, \dot{X}_1 X'_3 - \dot{X}_3 X'_1, 0)$, where an overdot (resp. prime) means τ (resp. σ) differentiation. If a rupture proceeds along a more complex surface such as a hinged plane, superposition is used again (Figure 4b). In the case of a homogeneous isotropic medium, representation (3) becomes

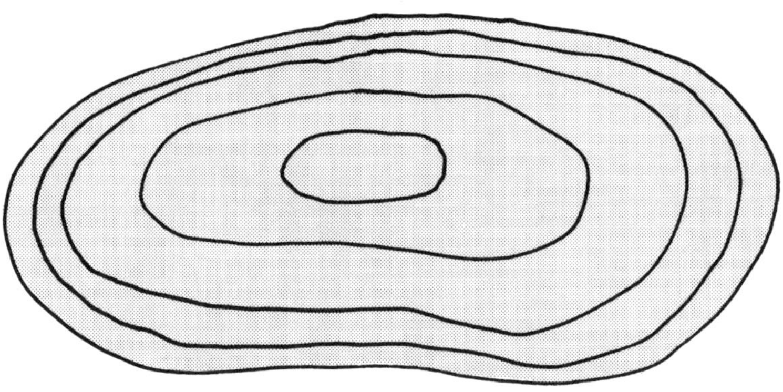

Figure 3 : A crack may be considered as a superposition of several dislocation loops.

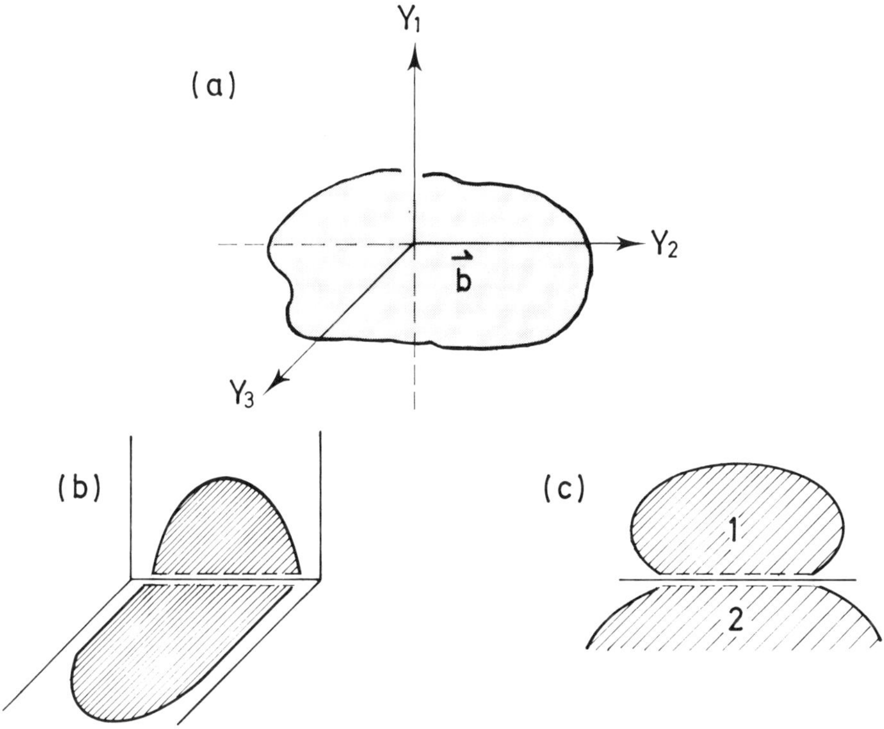

Figure 4 : (a) Coordinates for a dislocation loop representing shear faulting in the y_1 - y_3 plane. (b) Faulting of a hinged plane is represented as two loops joined along the hinge. (c) Faulting in a layered medium is obtained by superposing two loops joined along the interface.

$$v_n\,(\mathbf{y},t) = \mu b_a \int d\tau \oint ds\, (\dot{X}_1 X_3' - X_1' \dot{X}_3) \left(\frac{\partial G_{na}}{\partial y_2} + \frac{\partial G_{n2}}{\partial y_a} \right) \tag{4}$$

with a = 1,3. Rupture in a layered medium is, once more, obtained by superposition (Figure 4c).

EXAMPLE: THE TWO DIMENSIONAL PROBLEM

The simplest example one can image to illustrate the above representation is the two dimensional problem of a fault of infinite width (Figure 5a). In this case ($X'_1 = 0$, $X'_3 = 1$, and for a rupture front moving with piecewise constant velocity the integrals in (4) can be done analytically (Lund, 1986). A fault of finite length L and constant rupture velocity V_1 generates a particle velocity that comes out decomposed as

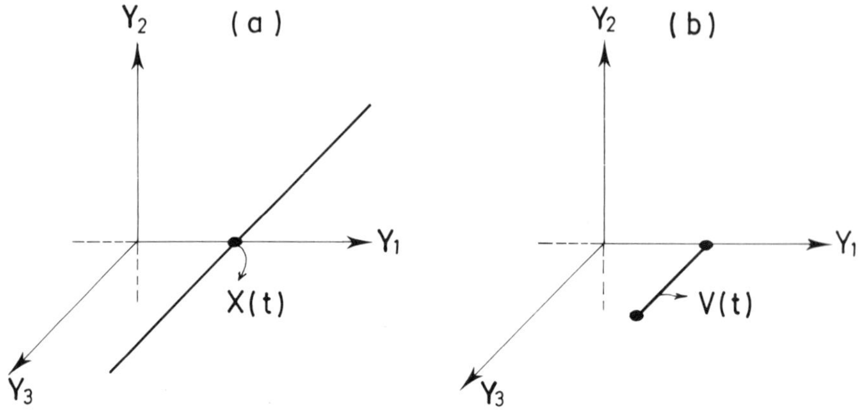

Figure 5 : (a) An infinitely wide fault is represented as a moving straight line. (b) A fault of finite width is represented as a moving dislocation line segment.

(5) $\qquad \mathbf{v}$ = (starting phase) + (stopping phase) + (passage of rupture phase)

The starting and stopping phases have the well-known inverse square root singularities inherent to a two dimensional situation with an infinitely short rise time. They are smoothed by finite rise times and are a source of high frequency radiation. The passage of rupture phase is smoother, of importance near the fault, where it scales as the inverse of the distance d to the fault, and is visible within a forward cone of aperture α with $\cos \alpha = V_1/c$ where c is either the P or S wave velocity. This phase is only slightly modified by a finite rise time. These results were first obtained by Luco and Sotiropoulos (1985) using Cagniard-de-Hoop methods. The particle acceleration generated by the passage of rupture phase has a frequency content with a maximum at V_1/d. In other words, this phase generates higher frequencies in accelerographs placed closer to the fault. Phases quite similar to the starting and stopping ones will be obtained when a rupture front abruptly changes its velocity, say from V_1 to V_2 along a given plane, in which case the amplitude of the pulse will be proportional to $|\mathbf{V}_1\text{-}\mathbf{V}_2|$, or by changing the direction of the fault plane by an angle ϕ, and the amplitude of the pulse will be proportional to ϕ (for small ϕ). Indeed, a sudden change in velocity is nothing but an abrupt stop plus an immediate abrupt departure.

A second example of the use of the representation (4) is a fault of finite width (Figure 5b), in which case again the integrals can be done analytically (Madariaga, 1978 ; Lund, 1986). A noteworthy feature of the solution, which can be easily read off when using the representation (4) is the existence of a spherical passage of rupture front (SPR) phase, of importance near the edges of the fault, generated by the end points of the rupture front as they move. It is illustrated in Figure 6a, together with two (P and S) starting phases. The amplitude of the SPR phase is inversely proportional to the distance of the observer to the edge. It is this type of qualitative behaviour that would be very hard (and expensive) to learn from a purely numerical approach. As an example of how one would mimic the smoothing influence of a finite rise time in our present approach, the velocity generated by two dislocations each with slip (Burgers) vector b/2 and starting 0.2 seconds apart, has been plotted in Figure 6b. Clearly, the starting phases are smoothed out (although they keep having a strong high frequency content) but the SPR phase is not much modified, as a quick thought about what would happen with five dislocations, each with slip b/5 and starting at intervals of 0.05 seconds, indicates.

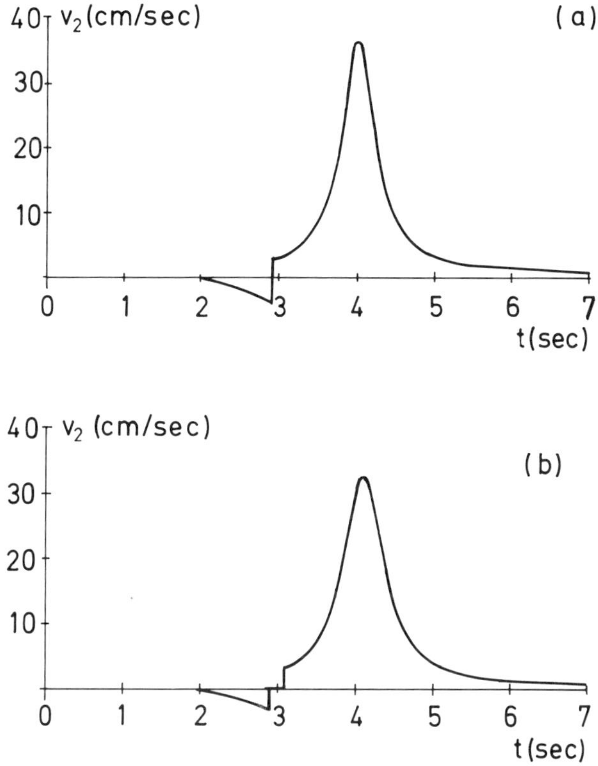

Figure 6 : (a) Spherical passage of rupture phase (SPR) generated by the end points of a fault of finite width. The discontinuities are starting phases. For details see Lund (1986). (b) Smoothing of starting phases due to a finite rise time modeled as two dislocations, each with half the total slip, starting 0.2 sec apart.

The main lesson to be learned from these analytic calculations is that it is considerably simpler to use the time domain representation (3) than Cagniard-de Hoop methods based on the representation (1). The natural conjecture that comes to mind is that this might also be true of numerics in more realistic models.

HIGH FREQUENCY SEISMIC WAVE GENERATION

Recently, interesting advances in the understanding of the high-frequency seismic wave generation mechanism have been made by Bernard and Madariaga (1984a,b) and Spudich and Frazer (1984). The time domain representation (3) affords a new insight into these results. Indeed, (3) gives particle velocity as a convolution of a source localized at the rupture front with a Green function :

$$v = G * (\text{source localized at rupture front})$$

But $G = G^P + G^S$ where (Aki and Richards 1980)

$$(6) \qquad G^c(\mathbf{r}, \omega) = \frac{e^{i\omega r/c}}{rc^2} \left(0(1) + 0\left(\frac{c}{\omega r}\right) + 0\left(\frac{c}{\omega r}\right)^2\right)$$

The 0(1) term in (6) is the one that, in the time domain, gives a sharp wave front for G and it is tempting to consider situations in which it dominates, since use of ray theory would allow treatment of signals propagating in heterogeneous media. This happens whenever c/ω is much greater than the distance of closest approach between station and rupture front, which gives precise meaning to the phase "high frequency approximation". In this range of frequencies (3) becomes $\mathbf{v} = \mathbf{v}^S + \mathbf{v}^P$ with

$$(7) \qquad v^c_m(\mathbf{y},t) = \int d\tau \oint d\sigma\, (\dot{\mathbf{X}} \Lambda \mathbf{X}')_j\, (\tau,\sigma)\, b_i\, c_{ijkl} \frac{\partial}{\partial y_l} \left[\frac{R^c_{mk}\, \delta(t - \tau - c^{-1} |\mathbf{y} - \mathbf{X}|)}{4\pi \rho c^2 |\mathbf{y} - \mathbf{X}|} \right]$$

where ρ is the density and $R^c{}_{mk}$ is the radiation pattern. Using

$$\int dx\, g(x)\, \delta'(f(x)) = -\sum_{x_o} \left(\frac{g}{f'}\right)' \frac{1}{|f'|}\Bigg|_{x=x_o}$$

where $f(x_o) = 0$, the τ integration of (7) yields an expression of the form

$$(8)\ v^c = \int d\sigma \left(\frac{\text{constants and}}{\text{radiation pattern}}\right) \frac{1}{|y-X|} \left\{ 0\left(\frac{\ddot{X}}{c}\right) + 0\left(\frac{\dot{X}}{|y-X|}\right) + 0\left(\frac{(\dot{X})^2}{c\,|y-X|}\right) \right\}$$

in which the integrand is evaluated at the retarded time τ_R, the solution of

$$c\,(t - \tau_R) - |\mathbf{y} - \mathbf{X}(\tau_R, \sigma)| = 0$$

The curves that contribute to v^c (**y**.t) at a given **y** and t are the isochrones of Spudich and Frazer (1984) and Bernard and Madariaga (1984b). An interesting feature of (8) is that it clearly shows that for abrupt changes in rupture front velocity, the term $0(\ddot{X}/c)$ will dominate. That is, most of the high frequency radiation comes from those points of the rupture front that suffer sharp discontinuities due to heterogeneities of the fault plane such as barriers and asperity boundaries. As an example, take a circular dislocation loop that grows with constant rupture velocity until it meets a straight barrier at a distance d from the nucleation point (Figure 7) ; the dominant high frequency contributions come only from two

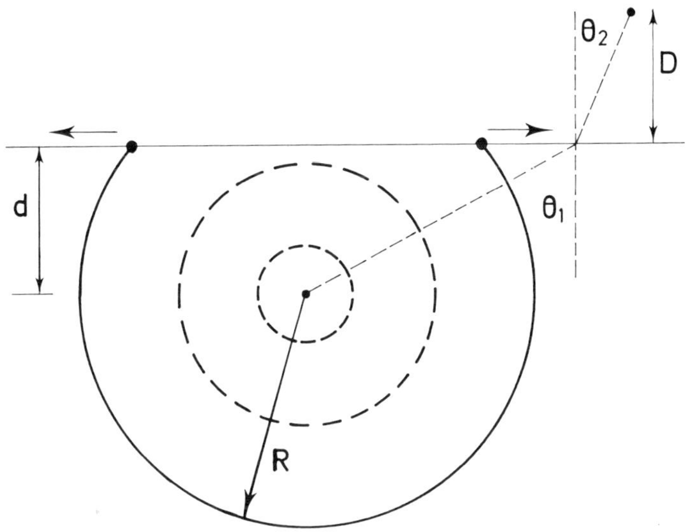

Figure 7 : Most of the high frequency waves generated by a circular dislocation loop meeting a straight barrier comes from the two points of intersection shown as dark dots, which act as point sources.

points moving along the barrier with speed $\dot{R}R/(R^2 - d^2)^{1/2}$ which is very large immediately after the rupture front touches the barrier ($R \geq d$) and decreases to the asymptotic value $\dot{R}$ when $R \rightarrow \infty$. The first arrival from this moving point source say, at a point located a distance D from the barrier, comes from a point on the barrier satisfying $\sin \theta_1 = (\dot{R}/c) \sin \theta_2$, where c is the P-or S-wave velocity. This is the critical ray approximation of Bernard and Madariaga (1984b).

ACKNOWLEDGEMENTS

This work was supported by DIB Grant E-2489-8612 and UNDP-UNESCO Program CHI/005/84.

REFERENCES

Aki, K., and P.G. Richards (1980). *Quantitative Seismology : Theory and Methods*, Freeman.

Archuleta, R.J., and S.M. Hartzel (1981). Effects of Fault Finiteness on near-source ground motion, *Bull. Seism. Soc. Am.* ,*71*, 939-957.

Bernard, P., and R. Madariaga (1984a). High Frequency seismic radiation from a buried circular crack, *Geophys. J.R. Astr. Soc.*,*78*, 1-18.

Bernard, P., and R. Madariaga (1984b). A new asymptotic method for the modelling of near-field accelerograms, *Bull. Seism. Soc. Am.*, *74*, 539-557.

Bilby, B.A., and J.D. Eshelby (1968). Dislocations and the theory of fracture, in *Fracture, an Advanced Treatise*, Vol. I, H. Liebowitz, Editor, Academic Press, 99-182.

Boatwrigth, J. (1980). A spectral theory for circular seismic source ; simple estimates of source dimensions, dynamic stress drop and radiated energy, *Bull. Seism. Soc. Am.*,*70*, 1-77.

Campillo, M. (1984). Numerical evaluation of the near field high-frequency radiations from quasi-dynamic circular faults, *Bull. Seism. Soc. Am.*, *73*, 723-734.

Luco, J.E., and J.G. Anderson (1985), Near Source ground motion from kinematic fault models, in *Proc. EERI Workshop on Strong Ground Motion Simulation and Earthquake Engineering Applications*, April 30 to May 3, 1984, J. King, Editor, Electric Power Research Institute, Palo Alto, California.

Luco, J.E., and D.A. Sotiropoulos (1985). On a decomposition of near-source earthquake ground motion (submitted for publication).

Lund, F. (1986). A note on the synthesis of near-field ground motion, *Bull. Seism. Soc. Am.*, *76*, 1790-1800.

Madariaga, R. (1978). The dynamic field of Haskell's rectangular dislocation fault model, *Bull. Seism. Soc. Am.*, *68*, 869-887.

Madariaga, R. (1983). High frequency radiation from dynamic earthquake fault models, *Ann. Geophys.*, *1*, 17-23.

Mura, T. (1963). Continuons distributions of moving dislocations, *Phil. Mag.*, *8*, 843-857.

Nabarro, F.R.N. (1951). The synthesis of elastic dislocations fields, *Phil. Mag.*, *42*, 1224-1251.

Papageorgiou, A., and K. Aki (1983a). A specific barrier model for the quantitative description of inhomogeneous faulting and the prediction of strong ground motion, I, Description of the model, *Bull. Seism. Soc. Am.*, *73*, 693-722.

Papageorgiou, A; and K. Aki (1983b). A specific barrier model for the quentitative description of inhomogeneous faulting and the prediction of strong ground motion, II, Applications of the model, *Bull. Seism. Soc. Am.*, *73*, 953-978.

Rudnicki, J., and Kanamori (1981). Effects of fault interaction on moment, stress drop and strain energy release, *J. Geophys. Res.*, *86*, 1785-1795.

Spudich, P. and L.N. Frazer (1984). Use of ray theory to calculate high frequency radiation from earthquake sources having spatially variable rupture velocity and stress drop, *Bull. Seism. Soc. Am.*, *74*, 2061-2082.

METHODS OF DETERMINATION OF SOURCE PARAMETERS

A. UDIAS
Catedra de Geofisica
Universidad Complutense
28040 Madrid
Spain

ABSTRACT. Since the 1930's source mechanism studies based on the analysis of the recorded seismic waves have developed very rapidly. Early methods use very small information of the seismograms and only give the orientation of the source. Modern developments allow the determination of more parameters of the source and the use of more realistic models. A short review is presented of some of these methods.

1. REPRESENTATION OF SEISMIC SOURCES

Under certain conditions, the seismic source can be represented by a shear dislocation on a plane. Using the representation theorem, in terms of the Green function, the displacement field due to a dislocation $\Delta u_j(\xi_i,\tau)$ on a plane Σ with normal $n_k(\xi_i)$ (Fig. 1) can be written as (Aki and Richards, 1980)

(1) $$U_n(X_1, t) = \int_{-\infty}^{\infty} d\tau \int_{\Sigma} C_{jklm}\, \Delta U_j(\xi_i, \tau)\, G_{ln,m}(\xi_i, t, x_i, t)\, n_k(\xi_i)ds$$

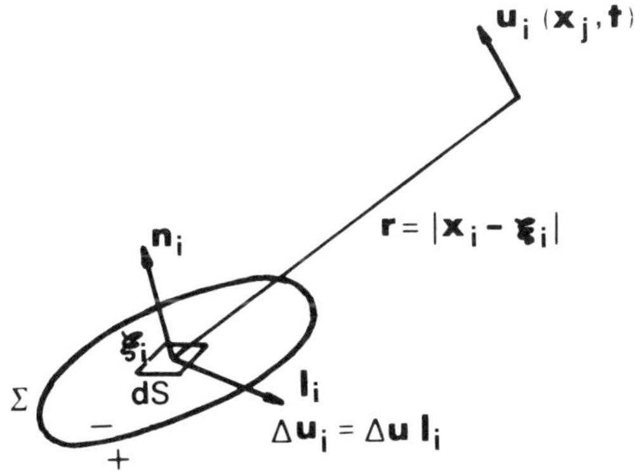

Figure 1. Elements of a shear fracture and displacements produced at a distance r.

J. Bonnin et al. (eds.), Seismic Hazard in Mediterranean Regions, 45–58.

In terms of the seismic moment tensor M_{ij}, for a point source at the far field the displacement are given by the convolution of M_{ij} with the derivatives of the Green function $G_{in,j}$.

(2) $u_n = M_{ij} * G_{in,j}$

For a shear dislocation on a plane of area S, normal n_i and displacement Δu in the direction of l_i the moment tensor has the form,

(3) $M_{ij} = \mu \, \Delta u \, S \, (l_i n_j + l_j n_i) = M_o \, (l_i n_j + l_j n_i)$

This type of mechanism is equivalent to a double couple force system, with couples in the direction of l_i and n_i.

1.1. Orientation of point sources

For a point source shear dislocation, the parameters which define the orientation of the source are n_i (normal to plane), l_i (direction of slip). Using spherical coordinate these are defined by the angles (Φ_n, θ_n) and (Φ_l, θ_l) (Figure 2A). Since n_i and l_i are orthogonal,

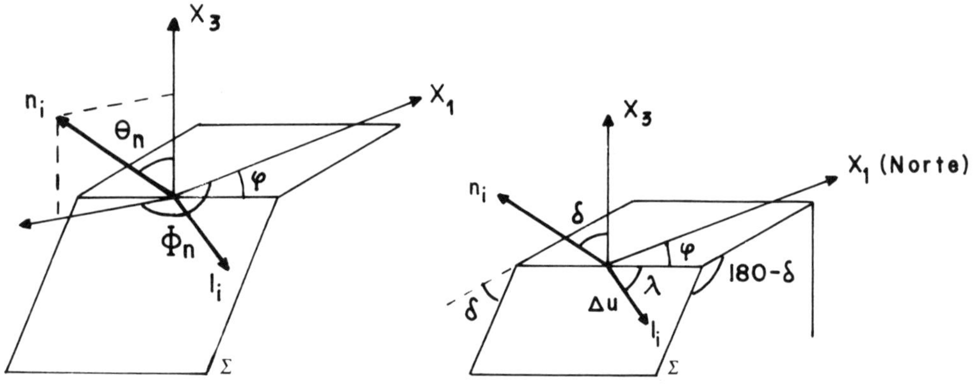

Figure 2. A) Orientation of the normal to the fault plane n_i with respect to geographical axes X_1, (North), X_3 (Zenit), B) Relation of the angles φ, δ, λ of the fault plane, normal (n_i) and direction of slip (l_i) and geographical axes (X_1, X_3).

only 3 angles are independent. On the fault plane, one can define the angles φ, δ, λ (strike, dip, slip) related to n_i and l_i by (Figure 2B) (Udias et al., 1985)

(4) $$\varphi = \Phi_n - \frac{\pi}{2}$$

(5) $$\delta = \theta_n$$

$$\lambda = \sin^{-1}\left[\frac{\cos\theta_1}{\sin\theta_n}\right] \tag{6}$$

Using DC (double couple) terminology n_i and l_i correspond to axes X and Y. In the equivalent system P and T are at 45° to X and Y (Figure 3).

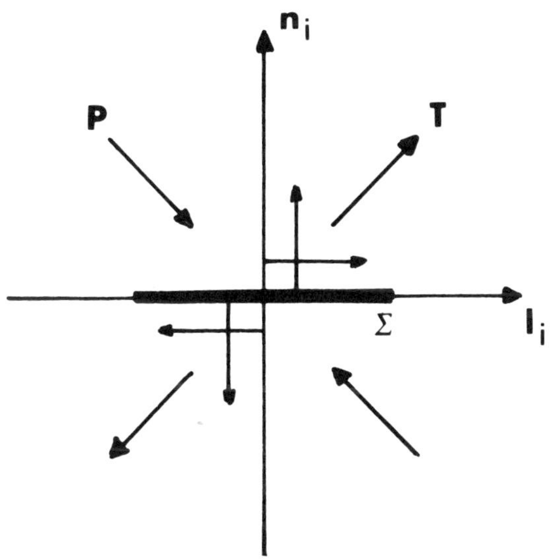

Figure 3. Relation between the shear dislocation and the equivalent double-couple force system

1.2. Seismic moment, source time function and dimensions

Orientation of the point source is also defined by the 6 different components of the seismic moment tensor M_{ij}. If no net-change in volume at the source is allowed $M_{11}+M_{22}+M_{33}=0$, and for a pure shear dislocation or DC source Det $(M_{ij}) = 0$. If only the first condition is imposed, the mechanism can be split into a DC (double couple) and a CLVD (Compensated Linear Vector Dipole). This separation allows the determination of the amount that a particular source with no volume change separates from a pure DC. In the case that M_{ij} are obtained by an inversion method from observational data of the seismic displacements u_i, one may or maynot impose the about mentioned conditions. In all cases the scalar seismic moment $M_0 = \mu \, \Delta u \, S$ gives an estimated of the size of the source.

The slip $\Delta u(\xi_i,t)$ is a function of time and the coordinates on the fault. The dependence of time defines the rise time τ, that takes Δu in getting its maximum value. As the source is limited in dimension, a total time T can be defined for total duration of the faulting process, $T = L / V_R$, (L length, V_R rupture velocity).

The dimensions of the source are given by the area of the fault plane S. If the shape of the source is circular, the linear dimension is its radius r. In case of a rectangular fault, linear dimension are its length L and width D.

2. MECHANISM FROM FIRST MOTION OF P-WAVES

The first method used to determine the orientation of a point source is that based on polarities of first motion of P waves or Byerly's method. Basically, the method consists in separating by two orthogonal planes the four quadrants of alternating compressions and dilatations, predicted in an homogeneous medium by a double couple source. Using spherical coordinates r, θ, φ, defined with respect to X_1, X_2, X_3 if n (0,0,1) and 1 (1,0,0), the radial (P) displacement is given by

$$u_r^P = \frac{M_o}{4\pi\rho\alpha^3 r} \sin 2\theta \cos\varphi \quad f(t - \frac{r}{\alpha}) \tag{7}$$

Nodal planes are $\varphi = \frac{\pi}{2}$ and q = 0, or planes (X_1, X_2) (X_2, X_3) (Figure 4)

2.1. Focal sphere

To reduce the observational data to the conditions of an homogeneous medium, a unit radius sphere of homogeneous material (the focal sphere) is constructed around the source. Observations are reduced to points on the surface of the focal sphere by back tracing the

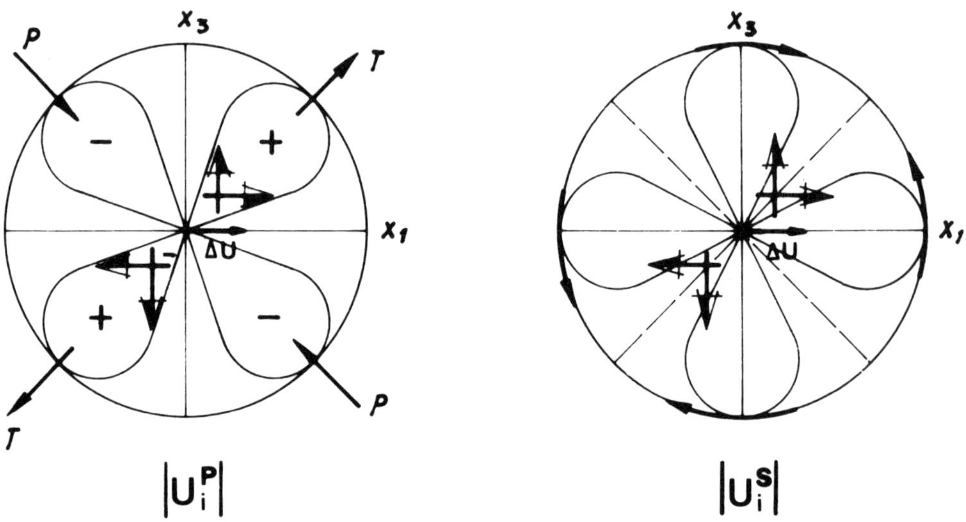

Figure 4. Radiation patterns for P and S waves from a double couple source

rays from the observation point to the source. A point at distance Δ and azimuth Φ, is reduced to a point on the sphere with coordinated (i_h, Φ) (Fig.5). For teleseismic distances, the take off angle in terms of travel time is given by

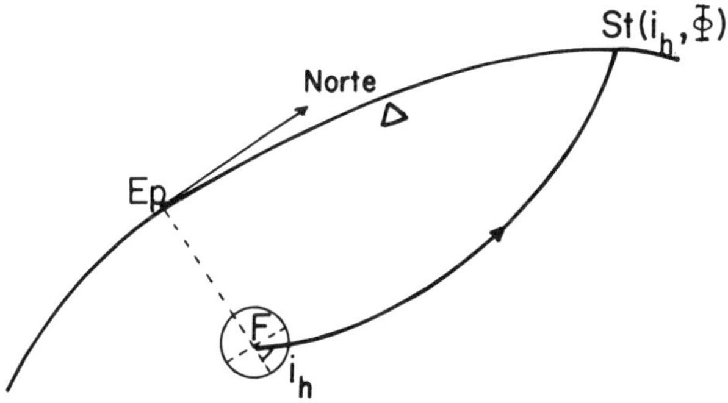

Figure 5. The focal sphere and the seismic ray to the station

(8) $$\text{Sin } i_h = v_h \frac{dt}{d\Delta}$$

For near stations, rays are traced back through known heterogeneous models of the crust and upper mantle and i_h is consequently determined.

The source is defined on the focal sphere by the orientation of the two orthogonal nodal planes A and B (motion on each plane is defined by the angles φ, δ, λ) of the principal axes of stress P and T, or of the axes of the double couple X and Y (Figure 6).

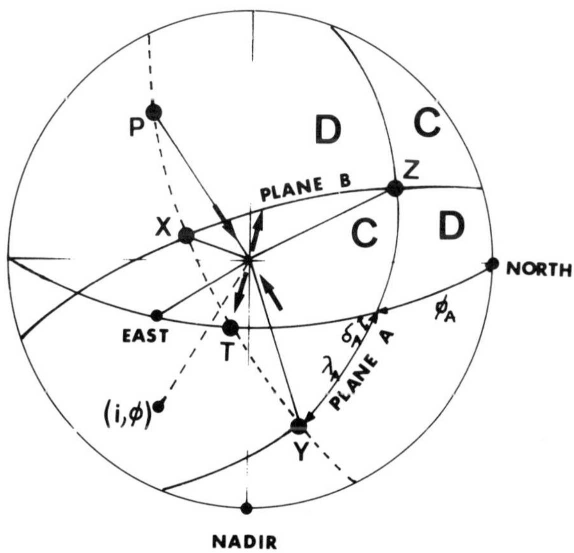

Figure 6. Focal sphere ; orientation of nodal planes (A and B), principal stresses (P and T), direction of the DC source (X and Y) and observation point (i, φ)

2.2. Graphical method

Graphical methods for fault plane solutions are based on performing the separation of compressions and dilatations on a suitable projection of the focal sphere upon a plane. Several projections have been used for this purpose such as Byerly's, central, Wulff and Schmidt. The last one, Schmidt-Lambert or equal area projection is most generally use at present. On the projection, observation points are located at azimuth φ and distance $b(i_h)$ from the center of the projection. Two orthogonal great circles A and B are used to separate compressions and dilatations in four quadrants of alternating polarity. From the projection, one determines the orientation of the normals to the planes A and B or X and Y axes given by the angles (θ_x, Φ_x) and (θ_y, Φ_y). On a plane containing X and Y at 45° of both axes, T and P axes are situated ; T (θ_T, Φ_T) on the compressional quadrant and P (θ_p, Φ_p) on the dilatational one (figure 6). The two planes A and B are given by their respectives angles φ, σ and λ. An example of such a representation and solution is given in figure 7.

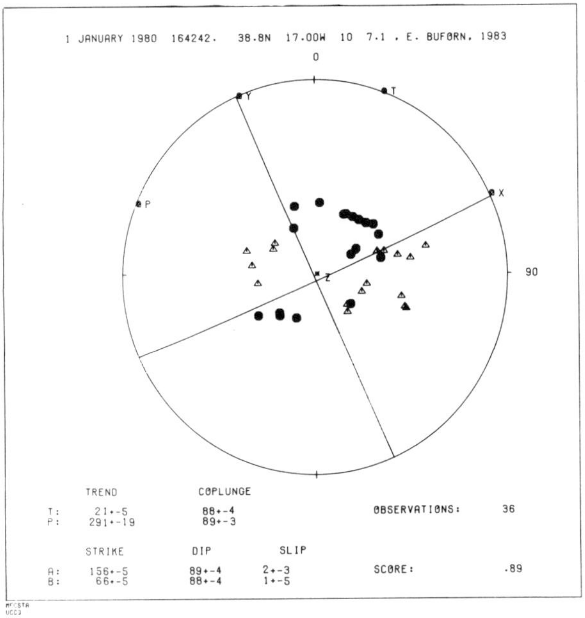

Figure 7. Fault plane solution for the earthquake of 1 January 1980 (Azores Islands) determined by the algorithm of Brillinger, Udias and Bolt (1980)

2.3. Numerical method

Several numerical methods to estimate the best fitting planes separating compressions and dilatations and their error margins have been proposed. The presented by Brillinger, Udias and Bolt (1980) is based on defining the probability of reading a compression at a station i by

(9) $$\pi_i = \text{prob}\,\{Y = 1\} = \gamma + (1 - 2\gamma)\ \Phi\,(\rho\ A_i)$$

where γ accounts for reading errors, Φ is the Gaussian cumulative function, ρ is a constant inverse proportional to the variance of the noise and A_i the expected amplitude of P. A set of N observations of first motions Y_i for an earthquake has a probability P of corresponding to a source orientation given by the three angles that define the orientation source

$$P = \Pi_{i=1}^{N} \pi_i^{1/2\,(1+\gamma_i)} + (1 - \pi_i)^{(1-y_i)/2} \tag{10}$$

The logarithmic likekihood function L = log P is maximized to obtain the best estimates of the 3 angles. Since L is a continuous function of these angles, first and second derivatives can be calculated to determine the covariance matrix. Error margins of the parameters are,then, estimated. Total score of solution is given by

$$P = \frac{1}{N} \sum_{i=1}^{N} \frac{1}{2} \left(1 + \frac{A_i Y_i}{|A_i|}\right) \tag{11}$$

This method allows also joint solutions of M earthquakes using the parameter ρ_j as a variables which are related to the probability that the observations of a given event j belong to the joint solution. The probability at each observation point is now given by

$$\pi_{ij} = \gamma + (1 + 2\gamma)\,\Phi\,(\rho_j A_{ij}) \qquad \begin{array}{l} i = 1, N_j \text{ stations} \\ j = 1, M \text{ events} \end{array} \tag{12}$$

Maximization process includes derivates with respect to the three angles of the source and ρ_j.

3. SEISMIC MOMENT AND FAULT DIMENSIONS FROM AMPLITUDE SPECTRA

Practically independently of the source model, estimates of the seismic moment and fault dimensions can be obtained from the amplitude spectra of seismic waves. The scalar seismic moment M_o appears in the expression for a point source as a proportionality constant of the amplitude of the displacement in the far field. Since the dimensions do not influence long wave lengths ($\lambda \gg L$), spectral amplitude at low frequencies are used to obtain M_o, once corrected by anelastic attenuation and geometrical spreading .
For Rayleigh waves, the expression is

$$M_o = \frac{U_o \sin \sqrt{\frac{\Delta}{a}}\, e^{\gamma_r \Delta}}{R_R(\varphi,h)} \tag{13}$$

where U_o is the ground motion spectral amplitude at low frequencies, Δ is the epicentral distance and R_R (φ, h) the radiation pattern at azimuth φ, for a focus of depth h.

A good estimation of the source dimensions can be obtained from the width of the S wave pulse. This width is related to that of the spectra, or corner frequency f_c (Savage, 1972) (Figure 8). According to Brune's model for a circular fault the radius is given by

$$\hat{r} = \frac{2.34\ \beta}{2\ \pi\ f_c} \tag{14}$$

where β is the S wave velocity at the focus.

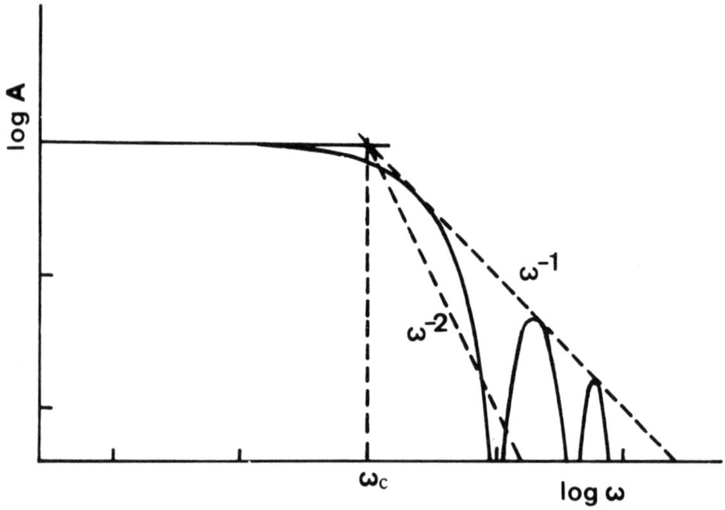

Figure 8. Shape of the spectrum of seismic waves, corner frequency (ω_c) and low-frequency amplitude

4. INVERSION OF SEISMIC TENSOR

The far field displacements for a point source can be expressed in the time domain as a convolution of the derivatives of the Green function with the seismic moment tensor.

$$U_i\ (X_l, t) = \int_{-\infty}^{\infty} G_{ij,k}\ (X_l, t - \tau)\ M_{jk}\ (\tau)\ d\tau \tag{15}$$

and in the frequency domain as their product

$$U_i\ (X_l, \omega) = G_{ij,k}\ (X_l, \omega)\ M_{jk}\ (\omega) \tag{16}$$

It is, then, possible to obtain by linear inversion the 6 components of M_{ij} from the values of the displacements in the time or frequency domain. If no restrictions are placed, the source is general and may have a component of volume change. Generally, the condition $M_{11} + M_{22} + M_{33} = 0$ is imposed. The condition for a DC source, Det $M_{ij} = 0$ is not linear and is not used. The 5 independent components of M_{ij} are interpreted as a sum of a DC and CLVD source.

In general, the derivatives of the Green function $G_{ki,j}$ have 27 different components, however for a source without change of volume only 8 selected combinations are needed. For example, for the SH component at azimuth φ

$$u_n^{SH} = \frac{-1}{2} (G_{n1,2} + G_{n2,1}) \sin 2\varphi (M_{11} + M_{22}) + (G_{n1,2} + G_{n2,1}) \cos 2\varphi M_{12} - (G_{n2,3} + G_{n3,1}) \sin\varphi M_{13} + (G_{n2,3} + G_{n3,1}) \cos\varphi M_{23} \tag{17}$$

Knowing the derivatives of the Green function or excitation function for a determined media and wave component, M_{ij} can be obtained from the observed vales of the displacement u_i.

An eigenvalue-eigenvector analysis of M_{ij} will provide three real eigenvalues σ_1, σ_2, σ_3.

If $\sigma_2 = 0$, the source is DC, $\sigma_3 = - \sigma_1$ and $M_o = (\sigma_1 + \sigma_3) / 2$. If $\sigma_2 \neq 0$ the source can be interpreted as DC + CLVD. The relative value of σ_2 with respect to σ_1 and σ_3 give the deviation of the source from a DC.

Formally the inversion can be performed as

$$U_i = G_{ij} M_j , \tag{18}$$

U_i, $i = 1, N$ Observed seismograms,
M_j, $j = 1, 6$ Components of M_{ij},
G_{ij} Components of excitation functions.

Using the generalized inverse method

$$M = G^{-1} U \tag{19}$$

and G^{-1} , the generalized inverse of G, is given by

$$G^{-1} = V \Lambda^{-1} W^T \tag{20}$$

where V is ghe eigenvector matrix of G^TG, W is the eigenvector matrix of GG^T and Λ the eigen value diagonal matrix of G^TG.

4.1. Inversion of Rayleigh waves

According to Mendiguren (1977) and Romanowicz (1982) inversion of spectral amplitudes of Rayleigh waves can be formulated as follows : The vertical component of the spectral amplitudes of Rayleigh waves recorded at distance r and azimuth φ is given by

$$U (r, \varphi, \omega) = S(\omega)\, I(\omega)\, e^{ir\omega/c}\, e^{-\gamma r}\, A_o (\omega, \varphi)\, e^{\Phi_o(\omega, \varphi)} \tag{21}$$

S and I correct for propagation and instrument effects and A_o and Φ_o are the amplitude and phase at the source.

The real and imaginary parts of the spectra at the source are functions of the components of M_{ij}. With the condition of zero trace these are

(22) $$R_e(\omega, \varphi) = M_{33}(G_1 - G_2) - (M_{11} - M_{22}) G_2 \operatorname{Cos} 2\varphi$$

$$2 M_{12} G_1 \sin 2\varphi$$

(23) $$\operatorname{Im}(\omega, \varphi) = M_{13} G_3 \cos\varphi + M_{13} G_3 \sin\varphi$$

where $G_i(\omega, h)$ are the excitation functions. At each station (value of φ) there are M values of frequency ω. The problem is linear and the unknowns are 5 ; $M_{11} - M_{22}$, M_{33}, M_{12}, M_{13}, M_{23}.

Since a large number of observations is used, the problem is solved by least squares techniques for the 2N (stations) by M (frequencies) equations. Excitation functions depend on frequency w and depth h. Hypocentral depth is usually a badly defined parameter. For this reason, to avoid the errors that it may introduce in the determination of M_{ij} ; it is determined in the same process. This is done by representing the total error as a function of h, and selecting the value of h that corresponds to minimum error.

Romanowicz (1982) used a two step method. The first, independent of depth and azimuth, solves for 5xM unknowns from 2MxN equations. The second uses 5M equations with 5 unknowns which are solved for different values of depth. Minimum error selects the value of h.

For the spectral amplitude A, the problem is not linear. If P_i, i = 1,5 are the unknowns (components of M_{ij}), linearization gives

(24) $$A(P_i) = A(P_i^o) + \sum_{i=1}^{5} \left[\frac{\partial A}{\partial P_i}\right]_{P_i^o} \delta Pi$$

where P^o_i are initial values of the 5 unknowns and $\delta\rho_i$ the corrections. The number of equations is NxM. An iterative converging process solves for ρ_i.

4.2. Centroid seismic moment

Dziewonski et al. (1981) consider that the values obtained by hypocentral determination from first arrival times may not be adequate for the determination of M_{ij}. They propose that the coordinates of the of the focus must be determined at the same time that M_{ij}. This focus will represent the centroid of the source and not the initial point of rupture. In spherical coordinates the 10 unknowns are M_{rr}, $M_{\theta\theta}$, $M_{\varphi\varphi}$, $M_{r\theta}$, $M_{r\theta}$, $M_{\theta\varphi}$ (f_i, i = 1,6) and r, θ, φ, t, coordinates and time of the source. The centroid ($\mathbf{x}_s$, t_s) is defined as that which minimize the second spatial and time moments ; which may be written in the form

(25) $$\Lambda_{ij} = \int_V (\mathbf{X} - \mathbf{X}_s)\, m_{ij}\, dV$$

$$(26) \qquad H_{ij} = \int_V (t - t_s)\, m_{ij}\, dV$$

The problem is solved in two steps :

In the first step the 6 components f_i of M_{ij} are found corresponding to a fixed value of epicentral coordinates x_s, and origin time t_s, by linear inversion of

$$(27) \qquad U_k\,(X,t) = \sum_{i=1}^{6} \Psi_{ik}\,(X, X_s, t, t_s) * f_i\,(X_s, t_s)$$

In the second step, the result f^0_i from the first are modified with the variation of the **x** and t_s and f_i according to the equations

$$(28) \qquad U_k = U_k^o + \frac{\partial \psi_{ik}}{\partial r} * f_i\,\delta r + \frac{\partial \psi_{ik} \partial q}{\partial q} * f_i\,\delta q$$

$$+ \frac{\partial \psi_{ik}}{\partial \varphi} * f_i\,\delta\varphi + \frac{\partial \psi_{ik}}{\partial t} * f_i\,\delta t + \psi_{ik} * \delta f_i$$

where ψ_{ik} are the excitation functions.

The problem is solved in the time domain. This method is actually used in a routine basis for all large earthquakes using data from digital stations.

5. MODELIZATION OF POINT SOURCES

This method (Langston and Helmberger, 1975 ; Deschamps et al., 1980, Shudofsky, (1985) is based on the calculation of theoretical seismograms produced by a determined model of a point source defined by its orientation, given by φ, δ, λ ; size, M_0 and source time function f(t) (generally trapezoidal). It is used generally for long period P and S waves at large distances $\Delta > 30°$. Theoretical seismograms are compared with observations at different azimuths. Parameters of the source are changed, to bring theoretical wave forms to fit observed ones. The adjusted parameters are : orientation of point source φ, δ, λ, size M_0, depth h, and source time function f(t).

For an non-attenuating homogeneous medium, displacements of P waves can be written as

$$(29) \qquad U^P = \frac{M_o}{4\,\pi\rho\,\alpha^3 r} R(\varphi, i_h)\, f\,(t - \frac{r}{\alpha})$$

where R is the radiation pattern, i_h the take-off angle and φ the azimuth.

Geometrical spreading of the wave front along distance is given by g (Δ), a function of i_h at the source, i_o at the station and second derivatives of the travel time curve t (Δ). Influence of free surface is given by C_z which depends on i_o, α and β at the station.

For the reflected phase pP and sP that arrive shortly after the direct P waves, expressions for the reflection coeficients V_{pp} and V_{ps} are functions of i_o, j_o, α and β.

The delay of the pP an sP waves with respect to P for depth h can be approximated by

$$\Delta t_{pP} = t_{pP} - t_P = \frac{2 \text{ h cos in}}{\alpha_n} \tag{30}$$

$$\Delta t_{sP} = t_{sP} - t_p = h \left(\frac{\text{cos in}}{\alpha_n} + \frac{\text{cos in}}{\beta_n} \right) \tag{31}$$

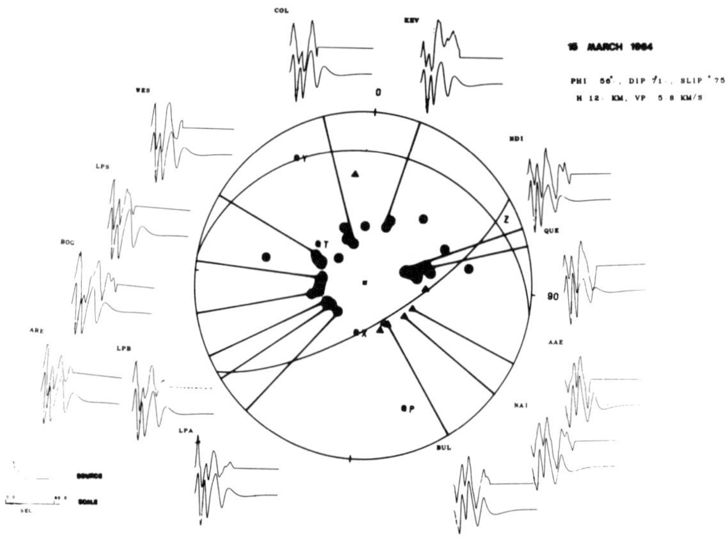

Figure 9. Modelization of P waves for the earthquake of 15 March 1964 (Gulf of Cadiz).

The vertical component of displacements of P waves are, then, finally given by

$$U^P = \frac{M_o}{4 \pi \rho_h \alpha_h a^3} \, g(\Delta) \, C_z(i_o) \{ R^P(\varphi, i_h) \, F(t - t_p) \tag{32}$$

$$+ R^P(\varphi, \pi - i_h) \, V_{pP} \, F(t - t_P - \Delta t_{pP})$$

$$+ R^S(\varphi, \pi - j_h) \, V_{sP} \, F(t - t_P - \Delta t_{sP})$$

Computations is carried out in the frequency domain, where the effect of anelastic attenuation G(T/Q,ω) and instrumental reponse I (ω) must be introduced. Since i_h depends on Δ

$$(33) \qquad U^p(\Delta,\varphi,\omega) = \tilde{U}^p(\Delta,\varphi,\omega)\, G(T/Q,\omega)\, I(\omega)$$

To find the displacements as function of time the inverse Fourier transform is taken. In a sufficiently good approximation for nearly vertical incident rays the structure at the station is not taken into account.

The observed complexity of the wave form may come from complexities in the reflected phases near the source or complexities of the source time function.

This method allows the determination of the orientation of the point source, its seismic moment, depth and time function by visual comparison of theoretical and observed wave-forms. It has become a standard procedure in source mechanism studies, an example is given in figure 9.

CONCLUSIONS

Some of the methods for source mechanism determination have been reviewed. Fault-plane solution based on polarities of P waves are still a very simple method with a minimum requirements on data. Seismic moment and dimensions can be also easily determined from amplitude spectra of body and surface waves. Wave form analysis is also relatively simple procedure, although restricted to earthquakes of moderate magnitude recorded worldwide on the LP seismographs of the WWSSN stations or digital data. Moment tensor inversion techniques are better applied with digital data.

ACKNOWLEDGEMENTS

The author thanks Dr. E. Buforn for her help in drafting the text and using some of her mechanism results. Partial support is acknowledge from the CICYT, project PB86-0431-C05-01 and the Comité Conjunto Hispano-Norteamericano, N°. CCA-8309/009.

Contribution n° 300 of the Catedra de Geofisica, Universidad Complutense de Madrid.

REFERENCES

Aki, K. and P.G. Richards (1980). Quantitative Seismology. Theory and methods. W.H. Freeman. San Francisco.

Brillinger, D., A. Udias and B. Bolt (1980). A probability model for regional focal mechanism solutions. *Bull. Seism. Soc. Am. 70*, 149 - 170.

Deschamps, A., H. Lyon-Caen and R. Madariaga (1980). Mise au point sur les méthodes de calcul de seismogrammes synthétiques de long période. *Ann. Geophys. 36*, 167 - 178.

Dziewonski, A.M., T.A. Chou and J.H. Woodhouse (1981). Determination of earthquake sourceparameters from wave form data studies of global and regional seismicity. *J.*

Geophys. Res. 86, 2825 - 2852.

Langston, C.A. and D.V. Helmberger (1975). A procedure for modelling shallow dislocation sources. *Geophys. J. R. astr. Soc. 42*, 117 - 130.

Mendiguren, J. (1977). Inversion of surface wave data in source mechanism studies. *J. Geophys. Res. 82*, 889 - 894.

Romanowicz, B. (1982). Moment tensor inversion of long period Rayleigh waves. A new approach. *J. Geophys. Res. 87*, 5395 - 5407.

Savage, J.C. (1972). Relation of corner frequency to fault dimensions. *J. Geophys. Res. 77*, 3788 - 3795.

Shudofsky, G.N. (1985). Source mechanisms and focal depths of east African earthquakes using Rayleigh wave inversion and body wave modelling. *Geophys. J. R. astr. Soc. 83*, 563 - 614.

Udias, A., D. Munoz and E. Buforn (1985). Mecanismo de los terremotos y tectonica. Ed. Universidad Complutense de Madrid. 232 pp.

INVERSION OF S POLARIZATION FROM NEAR-SOURCE ACCELEROGRAMS APPLICATION TO THE RECORDS OF 1980 IRPINIA EARTHQUAKE (ITALY)

Pascal BERNARD and Aldo ZOLLO
Laboratoire de Sismologie
I.P.G. PARIS, 4, Place Jussieu
75252 PARIS CEDEX 05 FRANCE

ABSTRACT. We propose a statistical method to invert near-source accelerograms, which may provide the focal mechanism, location and time parameters of local subevents which have source dimensions of a few kilometers. The distance source-station should range between 10km and 50km, so that realistic synthetics of the S acceleration can be generated by dynamic ray tracing of direct S waves. The procedure can be described as follows. The accelerograms are filtered 1-2Hz, and the peak vectors are selected. The synthetics are generated by a set of point sources with a given mechanism, and similarly only the peaks are retained. A norm, mainly associated with the fit of the polarisation angles, and secondary to the arrival times and amplitudes, is used to correlate the two sets of peak vectors. A large number of fault geometries, focal mechanisms and time histories can be tested with reduced computational expenses. Preliminary results of the application of this method to the strong motion records of the Irpinia (Italy) 1980 earthquake are very satisfactory. In particular, it could constrain the normal mechanism of the main shock to have no important strike slip component, with a slip angle in the range -90° ± 25°. It also shows that the northwestern rupture presents a mechanism significantly different from that of the central main fault.

1. INTRODUCTION

Due to the usual complexity of seismic sources for large events and the generally poor accelerometric station coverage in their vicinity, the use of near-source accelerograms in order to constrain the source parameters is difficult. Usual inversion methods concern the low-frequency slip amplitude distribution on the fault, when the plane geometry, the slip vector and sometimes even the rupture velocity is specified (e.g., Archuletta,1984; Hartzell & Heaton,1986). Unfortunately, the numerous earthquakes which provide good quality near-source acceleration or velocity records are generally poorly constrained by other data sets (levelling, teleseismic records, aftershock studies,...). In these cases, the accelerograms are mainly used for simple time or spectral analysis. The aim of the present paper is to show that important information on the fault parameters can be extracted by polarisation analysis of the direct S phase. The inversion method is applied to the records of the Irpinia 1980 earthquake.

The magnitude Ms=6.9 Irpinia earthquake of November,23 1980 was the most damaging seismic event to have occured in Italy over the last 50 years. It is also the best documented European earthquake, and numerous studies have been published on the available geophysical data, which mainly concern the epicentral location (e.g., Del Pezzo et

J. Bonnin et al. (eds.), Seismic Hazard in Mediterranean Regions, 59–69.

al,1983; Westaway,1985), the aftershock distribution and mechanisms (Deschamps & King, 1983,1984), the focal mechanisms from teleseismic body waves (Gasparini et al,1982; Martini & Scarpa,1983; Del Pezzo, 1983; Deschamps & King,1984), long period focal mechanisms (Boschi et al,1981; Kanamori & Given, 1982; Brustle & Muller,1983), levelling data (Arca et al,1983; Crosson et al,1986), and reports of surface breakage (Westaway & Jackson,1984). The strong motion analog records of the SMA1 network (ENEL) were first processed and analysed by Berardi et al (1981). The locations of the accelerometers and the aftershock sequence is presented in Figure 1.

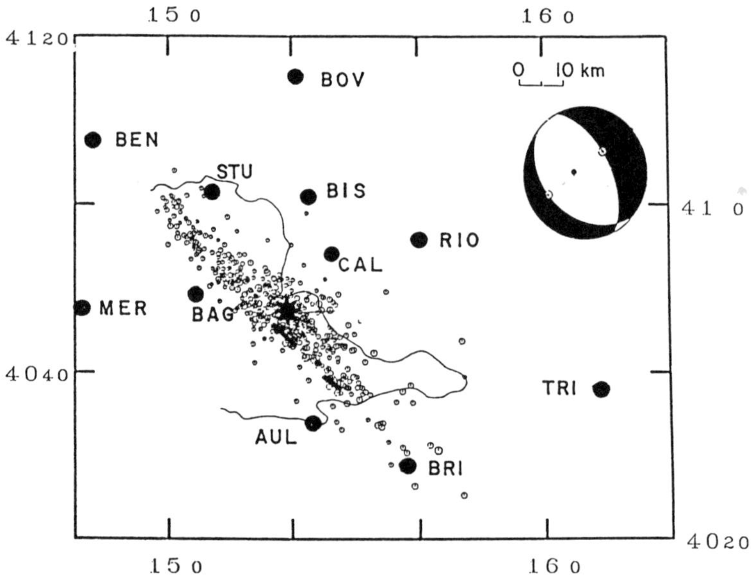

Figure 1. Aftershock location from Deschamps & King (1984). The star represents the main epicenter. The solid dots are the closest accelerometric stations. The continuous line is the levelling path. The two thick segments are the surface breakage. The focal mechanism is inverted from long period teleseismic records.

The earthquake consisted of the rupture of a complex sequence of normal faults striking mainly towards the NW, nearly parallel to the axis of the southern Apenninic chain. Its location coincides with the well defined area of historical and instrumental seismicity associated with this chain (Martini & Scarpa, 1983; ENEL, 1977).

The complex character of this event makes the different information sets difficult to analyze. An interesting faulting pattern with two orthogonal faults has been proposed by Crosson et al (1986) in order to integrate various data sets in a common model. Nevertheless, these authors did not make a detailed use of the strong motion records to constrain the late subevent at 40s, as they only considered qualitatively the acceleration peak amplitudes at various stations. Even for the first seconds of the rupture, the focal mechanisms from regional and teleseismic short-period records are not correctly constrained : although the strike (-45° N) and the dip (60° NE) are well defined, the slip angle varies from -45° , i.e. a normal fault with left-lateral strike-slip component, to +115°, i.e. a normal fault with right-lateral strike-slip component, according to different authors and different methods. Westaway and Jackson (1984) observed a small left-lateral component (slip≃-75°) on the central fault scarp at the surface. It will be shown in the following that the filtered 1-2 Hz S phases provide constraining information on this mechanism.

2. OBSERVATION AND THEORETICAL MODELING OF S POLARIZATION.

The horizontal polarization of the band-pass filtered acceleration (1-2 Hz) recorded at station STU is shown in Figure 2. Our choice of the representation of a 2D vector depending on time is an alternative to the classical particle motion diagram. It provides more information than the latter, because the polarization in the plane (South,East), which is the plane of the figure, is immediatly accessible at each time step along the time axis.

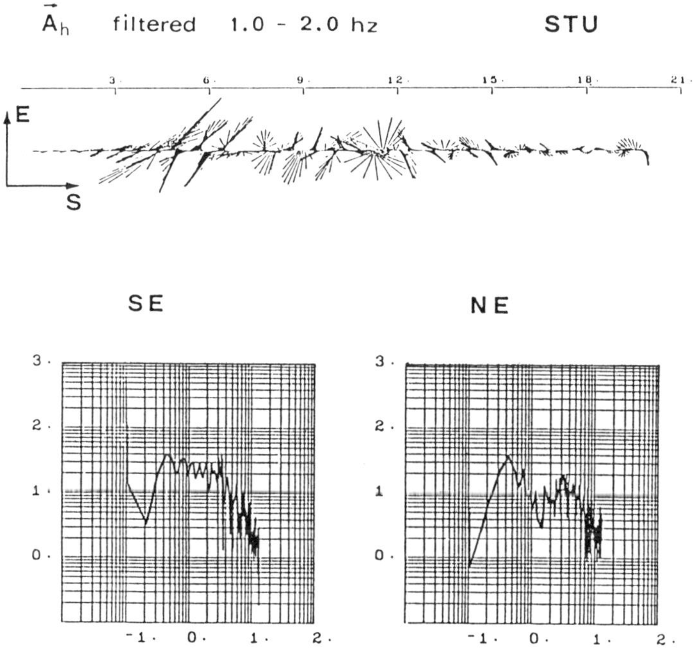

Figure 2. Horizontal polarization of the 1-2 Hz acceleration at STU and spectra of the SE and NE components.

The stability is striking, even in the 3D space, which can be deduced by similar figures on orthogonal planes. This appears to be a general property for all the near-source records, and if reliable Green's functions were calculated, the direction of polarization would constrain the ratio of the SV to SH wave amplitude at the source, and therefore the mechanism. It is important to note that the polarization characteristics disappear for frequencies greater than 5 Hz, as is observed on the acceleration spectra at STU where the energy is equally distributed on the NE and the SE components above this frequency. For all the accelerograms, we selected the frequency band 1-2 Hz in order to deal with polarized S acceleration associated with moderate-size subevents (a few kilometers), and to apply the far-field approximation of the Green's function, as the closest station to the source is at about 10km, i.e., at a distance of several wave-lengths.

In a homogeneous medium, the far field approximation of the double couple point source can be expressed as :

$\mathbf{u} = \Delta u\, \mathbf{A}\, \mathbf{s}$ (e.g., Aki & Richards,1980)

where **u** is the observed displacement vector at a given site, **A** a matrix depending on the radiation pattern, Δu the amplitude of the slip velocity at the source, and **s** the unit slip vector. Any inversion procedure working on the three components of **u** has therefore to deal with both the mechanism (through **A** and **s**) and the amplitude at the source (through Δu), unless one of these parameters is a priori constrained. Similarly, the acceleration can be expressed as :

$\mathbf{a} = S\, \mathbf{A}\, \mathbf{s}$

where S is a scalar function at the source, depending on time. Let one set :

$\mathbf{a} = a \cdot \mathbf{p}$

where **p** is the polarisation unit vector of the acceleration which does not depend on time for a point source. Here, a is a function of both the mechanism and the amplitude, and **p** is a function of the mechanism only. When the amplitude and mechanism are both poorly constrained, it seems more suitable to first of all invert for the mechanism with **p** , and afterwards for the source amplitude, rather than to simultaneously analyze the three components of **a**. The norm to minimize in the first step of the inversion has to be an increasing function of the angle between the observed acceleration peak and the synthetic **p**. Such an inversion is clearly non linear, because several mechanisms give identical polarisation at the same station.

For more realistic media with varying elastic parameters, ray theory can be applied in the selected frequency band. The greatest difficulty arises from the free surface effect, where the reflected P and S waves interact with the incident S to produce the observed motion. In particular, for incidence angles around the critical angle (from about 35^0 to 45^0), the polarization of a given peak is unstable (elliptical motion) and very sensitive to small changes in incidence angle or frequency (Booth & Crampin, 1985). Nevertheless, for the closest Irpinia records, the first P waves are nearly vertical, with incidence angle generally smaller than 20^0, because of the presence of a thick low-velocity upper layer. The incident S waves are therefore expected to be subcritical. In this case, the use of the classical plane wave complex coefficients at the surface gives a very good approximation of the free surface effect on the incident wave.

Acceleration peaks generated by a source can be interpreted as having been produced by localized discontinuities of the rupture velocity or of the slip on the fault plane. An important consequence of the far-field approximation is that the acceleration pulse has the same polarization as the displacement which would be generated by a point source at the same place as the rupture dicontinuity and with the same focal mechanism. Of course, the amplitude of the acceleration will present a much more complicated scalar wave-form, depending on the local dynamics and geometry of the rupture, which we will not attempt to resolve. The observation of the stability of **p** for the recorded acceleration peaks is consistent with the description of the source in terms of a distribution of a few localized areas of the fault acting as isolated point-like sources (Bernard & Madariaga, 1984). An observed peak direction **p** may be explained by different subsources located at various places on the fault plane with specific mechanisms. If the location is constrained, the mechanism will be the main inverted parameter.

3. THE CORRELATION METHOD

In order to present the correlation method, we will consider the modeling of the focal parameter for the central part of the 1980 Irpinia fault. We selected the first seconds of the S phases on the three closest accelerometric stations, CAL, RIO and BIS, which correspond to a rupture propagation of 10km at most away from the hypocenter. We considered the hypocentral location by Westaway (1985), and assumed a bilateral rupture propagation. We modelled the source as a set of point sources located at the same depth, spaced by 2 km on a 20km long segment striking -45°. Each point source radiates at a time defined by the arrival of the rupture front propagating at the fixed rupture velocity of 2.7 $km.s^{-1}$. The direct S phases are calculated for the same mechanism and the same amplitude at the different points. The scalar time function at the source is not important, as only the peaks are selected. We used a computing program developped by Farra (Farra et al,1986) for calculating the synthetics at the free surface. The selection of the peak vectors gives a set ASx_{ij} ,ASy_{ij} ,ASz_{ij} , T_{ij} where i refers to the point source and j to the station. The polarization of the vector AS at a given time T_{ij} is then interpreted as the acceleration polarization associated with a localized rupture discontinuity at point *i*. A constant source amplitude for all the points along the segment means that the classical radiation pattern of the double-couple is assumed not to be affected by geometrical parameters, such as the directivity of the rupture. Moreover, for a given direction of AS, we will not distinguish between the two opposite vectors, because they are related to the sign of the rupture velocity gradient, or to the spatial gradient of the dislocation. These functions may reverse at a scale which remains unresolved by our analysis. The accelerograms are band-pass filtered 1-2 Hz, and the S time window corresponding to the synthetic source is selected for each station. In this time interval, the acceleration peaks are selected, giving a set of k data peak vectors ADx_{kj} ,ADy_{kj} ,ADz_{kj} ,T_{kj} where ADx ,ADy and ADz are the three components of the acceleration vector at time T_{kj} observed at station *j*.

The misfit between data and synthetic peaks is defined in three successive steps, A,B and C, by a norm in a three dimensional space, related to the polarisation, the time and the amplitude.

A: The distance d_{kij} between one data peak k recorded at a station j and one synthetic peak i is defined by :

$$d_{kij}^{\ 2} = d_p^{\ 2} + d_a^{\ 2} + d_t^{\ 2} ,$$

where d_p is related to the polarisation, d_a to the amplitude and d_t to the time.

$$d_p^{\ 2} = \epsilon \ . \sin \Theta^{\ 2}$$

where Θ is the angle between the two vectors, ϵ an adjustable parameter.

$$d_a^{\ 2} = \frac{A_j}{A_{ij}} - 1$$

where $A_j = \max_i A_{ij}$ and A_{ij} is the synthetic peak amplitude.

$$d_t^{\ 2} = \frac{t^2}{q.TT^2}$$

where $t = T_{kj} - T_{ij}$, $TT = T_{22j} - T_{11j}$, q an adjustable parameter.

The functions d_p , d_a and d_t were chosen in order that they all take the value 1 for the expected error of the corresponding parameter. The distance is the same for Θ and $\pi+\Theta$,which means that the sign of the spatial gradients of the source parameters is not resolved. One takes ϵ=5 and q=0.75. This gives an error in angle of about 25°, an acceptable amplitude ratio of 0.5 and an uncertainty in time of about half the duration of the synthetic. This choice gives a higher weight to the polarization than to the time or the peak amplitude by setting a greater maximal distance for d_p than for d_a or d_t.

B: The distance d_{kj} between one data peak k and the set of synthetic peaks i=1,I corresponding to the entire segment, is defined as :

$$d_{kj} = \min_i d_{kji} = d_{kji_o}$$

This means that a data peak is estimated to be well represented by the synthetic set if at

least one of the synthetic peaks is close enough to it. The greater distances associated with the other peaks have no influence on d_{kj}. By doing so, some synthetic peaks may not be represented by any data peak.
Another definition could be :

$$d_{kj} = \sum_i A_{ij}{}^2 \, d_{kij}{}^2 \; ,$$

which gives a large weight to all the peaks with large amplitude. This formulation implicitely means that all the point sources on one segment are expected to have equally important amplitudes. We prefer the physical meaning of the minimal distance rather than that of the root mean square distance because it assumes less a priori constraint on the source amplitude distribution.
C: The distance between the set of data peaks and the set of synthetic peaks for a site j is defined as d_j :

$$d_j{}^2 = \sum_k A_{kj}{}^2 \, d_{kj}{}^2$$

This means that all the significant data peaks have to be close enough to at least one synthetic peak. The distance d_j gives an estimate of the fit for each station j, and characterizes the quality of the current model. A weight may be given to all the stations, W_j, depending on the distance to the segment and site characteristics. The global fit between all the synthetics and records for one segment is defined by the distance d :

$$d^2 = \frac{\sum_j W_j \, d_j{}^2}{\sum_j W_j}$$

Finally, the uncertainty on the model parameters can be estimated by the distance d_o :

$$d_o{}^2 = \frac{d^2}{3}$$

where the number 3 is the data space dimension.

The best model for a set of point sources is defined as the one which minimizes d_o. The problem of finding the minimum value and the associated parameters is highly nonlinear, and the existence of several local minima are expected in the model space (in our example : depth, dip and slip). The simplest way is to explore this space in the entire domain of interest. This space is discretized by a grid, and the step in parameters is choosen in order to make a compromise between reduction in computing time and resolution in the model. The step in dip and slip is 5^0 , and the step in depth is 3 km. These steps produce rotations of the polarization vector of the order of 5^0, which is small enough with respect to the 25^0 of assumed uncertainty. The plot of the function d_o around the selected minimal values may provide a good representation of the resolution of the model.

An important advantage of this method is that the correlation between data and synthetics is realized with a reduced number of data points, with respect to a correlation of the whole signal. The selection of peaks reduces the number of parameters by a factor of about 5 (only 2 peaks for 10 points by period), and therefore the computation time of the correlation by a factor of about 25.

We have now to define the range of parameter values for depth, slip and dip. The slip is assumed to range between -45⁰ and -135⁰, i.e., from a normal fault with left lateral strike slip component of equal amplitude to a normal fault with right lateral component. The dip is assumed to range between 50⁰ and 90⁰ NE. The trial depth are 6, 9 and 12 km. The strike to the NW is well constrained by short and long period focal mechanisms, and by the report of surface breakages. A weight of 3 is given to RIO and BIS, and only 1 to CAL because of non planar local velocity heterogeneities.

4. RESULTS AND DISCUSSION.

Two minima of d_0 are found, 0.69 and 0.65, with very similar parameters. The resolution is given in Figure 3 for the best fit. The dip is found to be 55⁰ at the minimum. It is consistent with the teleseismic result of 60⁰ dip, but the resolution is poor. The depth is preferentially 12km. The best resolved parameter is the slip, which is found to be -90⁰, i.e., a pure normal fault, with 25⁰ of uncertainty.

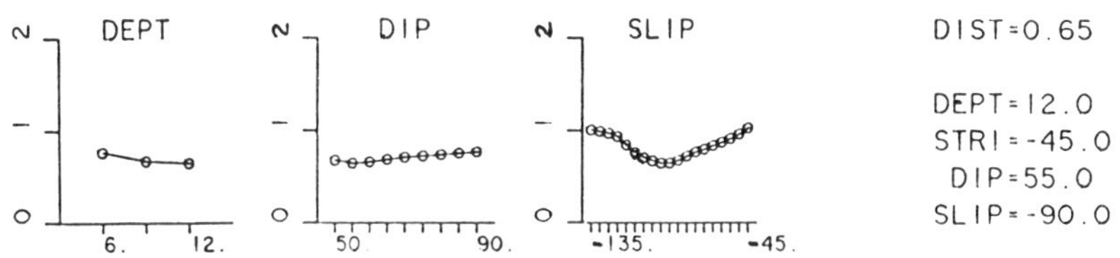

Figure 3. Misfit of the polarization vector, parametrized by the distance d_0, versus the fault parameters depth (in km), dip and slip angles (in degrees).

Figure 4 shows the horizontal polarisation of the synthetic peaks at the three stations, generated by the line of point sources for the three slip values -135⁰, -90⁰ and -45⁰, the other parameters being fixed at the minimum. The fit to the data is obviously better for -90⁰, in terms of polarisation.

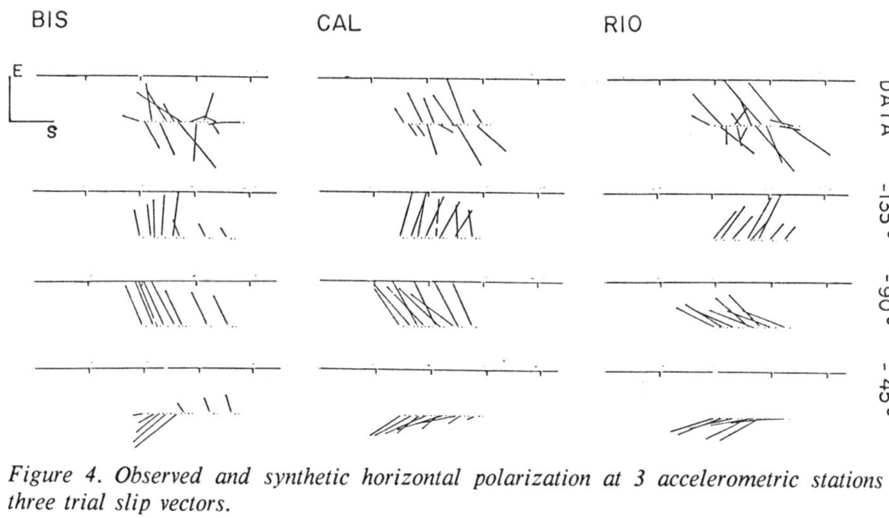

Figure 4. Observed and synthetic horizontal polarization at 3 accelerometric stations for three trial slip vectors.

Figure 5 shows the horizontal and vertical polarisation of the synthetic peaks at RIO for the best values of the parameters. They are in good visual agreement with the data, which means that the synthetics succeeded in reproducing not only the horizontal polarization, but also the three components of the vector **p**.

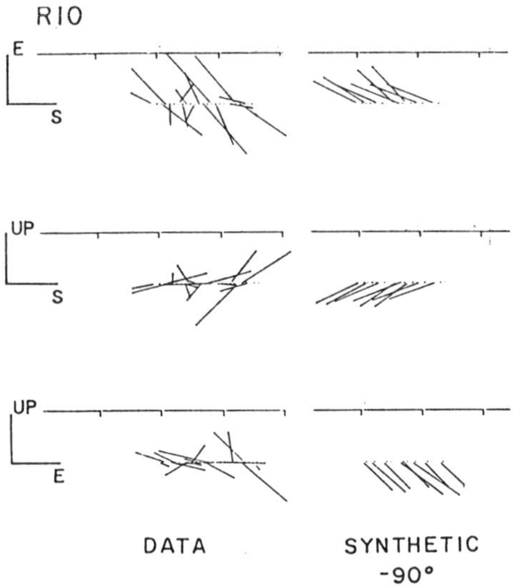

Figure 5. Observed and synthetic polarization of the acceleration in three orthogonal planes at station RIO for the best fitting source parameters.

A similar analysis could be performed for the northern part of the source. Nevertheless, the strike and dip angles are unconstrained, and the selection of the time window on the accelerograms will strongly dependent on the assumed rupture velocity. The resulting size of the parameter space becomes quite large. We do not have yet complete results, but it may already have been noted that the mechanism of the central part is not suitable for the northern rupture. A strike -45°N, dip 60° NE and slip -90° gives acceleration vectors polarized NE-SW at station STU, which is perpendicular to the observation. A better fit is obtained for subfaults striking towards the North instead of the NW.

The main difficulty for the proposed method concerns the generation of realistic synthetics. Although the far field approximation is probably the simplest approach to the problem, the complexity of the real structures could alter the adequacy of this choice. The geological structure near the faulted area is strongly laterally heterogeneous, and the existence of dipping interfaces with important velocity contrast (contact between limestones and clays) may produce large multiples which we do not model properly as well as rotations of S polarisation by angles greater than our assumed error of 25°. The method proposed here could be used as a first step before a more refined inversion procedure. It may sufficiently reduce the space of model parameters to allow the use of more expensive techniques with complete synthetics, which could again deal with polarisation only, or or which could attempt the inversion for the amplitude.

5. CONCLUSION

We propose a statistical method for correlating strong motion records with synthetics generated by a trial fault model. The source-station distance should range between 10 and 50km, so that realistic synthetics of the S acceleration can be generated by dynamic ray tracing of direct S body waves. The accelerograms are filtered between 1 and 2Hz, and the peak vector of the S phases are selected. The synthetics are generated by a set of point sources with a given mechanism, and similarly only the peak vectors of the direct S are retained. The two sets of peaks are correlated with the help of a norm, mainly associated with the fit of the polarisation angles, and secondary to the arrival times and to the amplitudes. The simplicity of the method makes possible the correlation computation for a very large number of models, which present varying geometries, focal mechanisms or time histories.

The application of this method to the strong motion records of the Irpinia 1980 earthquake has provided new information on this complex normal faulting event. It could resolve the slip angle in the central part ($-90^{0} \pm 25^{0}$), and suggests a rotation of the strike of the northern subfaults. Because of the large uncertainty in the angle (20^{0} to 30^{0}) due to the usual complexity of the geological structure in the vicinity of a fault, the proposed method is not suitable for refined imaging of faulting details. It could be useful when the fault geometry is poorly constrained, as in the case of the Irpinia 1980 earthquake, and might therefore provide important results concerning the local stress field and active geological accidents.

These results are part of the preliminary study of a more complete analysis of the accelerograms together with the other data sets, which will be presented in Bernard & Zollo (1987a & b).

AKNOWLEDGMENTS. We would like to thank V. Farra, who kindly provided us with her computing program for the generation of synthetics. This work was supported by ENEA in Italy, the Institut des Sciences de l'Univers (INSU) and the Bureau de Recherches Géologiques et Minières (BRGM) in France.

REFERENCES.

Aki, K., and Richards,P., 1980. Quantitative seismology, theory and methods (W.H. Freeman and Co., S. Francisco, 2 vol.)

Arca, S., Marchioni, A.,1983. I movimienti verticali del suolo nellezone della Campania e della Basilitica interessate dal sisma del Novembre 1980, Boll. Geod. e Sc.Affini,42,125-135.

Arca, S., Bonasia, V., Gaulon, R., Pingue, F.,Ruegg,J.C., and Scarpa,R., 1983. Ground movements and faulting mechanism associated to the November 23, 1980 Southern Italy earthquake. Boll. geod. sci. affini XLII, 137-147.

Archuleta, R.J.,1984. A faulting model for the 1979 Imperial valley earthquake. J. Geophys. Res.,89,4559-45859.

Berardi,R., Berenzi,A., and Capozza,F.,1981. Campania-Lucania earthquake on 23 november 1980. Accelerometric recordings of the main quake and relating processing. ENEA-ENEL report.

Bernard,P. and Madariaga,R.,1984. A new asymptotic method for the modelling of near-field accelerograms. Bul. Seism. Soc. Am.,74,539-557.

Bernard,P. and Zollo, A.,1987a. The Irpinia (Italy) 1980 earthquake : Detailed analysis of a complex normal fault. In preparation.

Bernard,P. and Zollo,A.,1987b. Inversion of S wave polarization from near-source accelerograms of the 1980 Irpinia earthquake. In preparation.

Booth,D.C. and Crampin,S., 1985. Shearwave polarizations on a curved wavefront at an isotropic free surface. Geophys.J.R.astr.Soc.,83,31-45.

Boschi, E., Mulargia, F., Mantovani, E., Bonafede, M., Dziewonski, A.M. and Woodhouse,J.H.,1981. The Irpinia earthquake of november 23, 1980. (abstr.) EOS, Trans.
Am. Geophys. Union 62,330.

Brüstel,W. and Muller,G.,1983. Moment and duration of shallow earthquakes from Love wave modelling for regional distances, Phys. Earth Planet. Int.,32,312-324.

Crosson, R.S., Martini, M., Scarpa, R. and Key, S.C.,1986. The southern Italy earthquake of 23rd November 1980: an unusual pattern of faulting. Bull. seism. Soc. Am. 76, 381-394

Del Pezzo,E.,Iannacone,G.,Martini,M.,Scarpa,R., 1983. The 23rd November 1980 Southern Italy earthquake, Bull. Seism. Soc.,73,187-200.

Deschamps,A. and King,G.C.P., 1983. The Campania-Lucania (southern Italy) earthquake of 23 November 1980. Earth Planet. Sci. Lett. 62 296-304.

Deschamps,A. and King,G.C.P.,1984. Aftershocks of the Campania-Lucania (Italy) earthquake of 23 november 1980. Bull. Seism. Soc. Am. 74,2483-2517.

ENEL,1977. Catalogo dei terremoti italiani dall'anno 1000 al 1975. Delivered to the CNR. Progetto Finalizzado Geodynamica.

Farra,V., Bernard,P. and Madariaga,R.,1986. Fast near source evaluation of strong ground motion for complex source models. 5th Ewing Symposium proceedings, A.G.U., Washington, .

Fels, A., Pugliese, A. and Muzzi, F.,1981. Campano-Lucano earthquake, November 1980, Italy: Strong motion data related to local site conditions. Contributo alla caratterizzazione della sismicita del territorio Italiano, 105-124. contribution to the annual convention of the Italian national research project on Italian seismicity. Udine,12-14 May 1981.

Hartzell, S.H. and Heaton, T.H.,1986. Rupture history of the 1984 Morgan Hill, California, earthquake from the inversion of strong motion records. Bull. Seism. Soc. Am.,76,649-674.

Kanamori ,H. and Given,J.W.,1982. Use of long period surface waves for rapid determination of earthquake source parameters, 2. Preliminary determination of source mechanisms of large earthquakes (Ms>6.5) in 1980. Phys. Earth Planet. Inter. 30, 260-268.

Martini,M., Scarpa,R.,1983. Earthquakes in Italy in the last century. In earthquakes : observation, theory and interpretation, H. Kanamori and E. Boschi, eds, LXXXV corso,479-47m92. (Soc.Italiana di Fisica, Bologna)

Westaway,R.W.C. and Jackson,J.,1984. Surface faulting in the southern Italian Campania -Basilitica earthquake of 23 November 1980. Nature Vol 312.

Westaway,R.W.C,1985. Active tectonics of Campania, southern Italy. Ph.D. thesis, University of Cambridge.

CHAPTER II

GROUND MOTION

Classical theory of Gaussian beams.

A. Cisternas
Institut de Physique du Globe de Strasbourg
5 rue René Descartes. 67084 Strasbourg. France.

Abstract

A compact formulation of the theory of Gaussian beams is obtained by the study of the characteristics of the equations of elastodynamics. The exact Hamiltonian of the system is replaced by its parabolic approximation and then trajectories and wave fronts are easily obtained. Gaussian beams are obtained as the analytic continuation of this approximation into the complex plane. The approximate solution may be interpreted as consisting of harmonic oscillators around the central ray. The wave field from a cylindrical source is given as example.

1. INTRODUCTION.

The method of Gaussian Beams is one among the many asymptotic methods of solving the wave equation, and hence it corresponds to the high frequency approximation to the solution. The advantage over other known methods comes from the fact that it may be applied to laterally heterogeneous media, and that it gives synthetic seismograms without large, time consuming computations.

The idea is to continue into the complex plane the parabolic approximation to the wave equation originally due to Fock (1965, the first paper being published in 1946) and largely employed in the study of laser signal propagation in plasma during the sixties. The applications to seismology began in the seventies with a long series of papers of the School of Leningrad (see for instance, Babic and Popov, 1981; Babic and Buldyrev, 1972). Cerveny and his collaborators in Prague, became very active contributors and introduced the theory in western countries by means of an interesting collection of papers, a periodic meeting on the subject, and a popular but unpublished volume of lectures (1981). K. Aki realized the practical possibilities opened by the method for the applications to laterally heterogeneous media, and conducted a seminar on Gaussian Beams at M.I.T. and in Paris in 1982.

Many advantages favor the application of Gaussian Beams to seismological problems: 1) It is fast. Once the central rays are computed, the algoritms are simple. 2) The passage through caustics and shadow boundaries is automatic. 3) It may be adapted either for 2 or 3-dimensional laterally heterogeneous media. 4) It may be applied not only for body wave, but also for surface waves (Yomogida and Aki, 1985).

There are nevertheless some aspects that should be handled with care. 1) Since there is an arbitrary selection of some parameters that determine the geometry of the beams, the resulting seismogram is affected accordingly. 2) Some smoothing of the solution comes from the discretization involved in selecting a finite number of beams. 3) It is not well suited

J. Bonnin et al. (eds.), Seismic Hazard in Mediterranean Regions, 73–83.

for dealing with a large number of rays interfering within a layer, and other similar cases, by the same reasons that restrict other asympthotic methods.

This paper, that is based on an introduction to Gaussian beams published in spanish (Cisternas et al., 1983), will make a short review of some of the main steps in the method from the point of view of the classical theory, and will include a simple application in order to illustrate the way in wich the approximation works, and the properties of the results.

2. ABOUT THE WAVE EQUATION.

Wave Theory is well known and well structured. We may think of it as a building with three stages where we can move from one to the other in a natural way. The first stage is that of the wave equation, that represents physical wave theory. The second one is that of the equation of wave fronts, or eikonal equation, or Hamilton-Jacobi equation, that is the equation satisfied by the phase of the solutions of the wave equation. The third stage is that of the equations of the rays, or Hamilton's equation, in other words, the equation of the trajectories orthogonal to the wave fronts. The following equations illustrate these statements:

WAVE EQUATION $$\nabla^2 \Phi = \frac{1}{v^2}\frac{\partial^2 \Phi}{\partial t^2} \qquad \Phi = A(q,\omega)\, e^{iS(q,\omega)}$$

HAMILTON-JACOBI $$(\nabla S)^2 = \frac{\omega^2}{v^2} \qquad S = \omega \int L\, ds = \omega\tau$$

RAYS: HAMILTON'S EQ. $$\dot{q} = \frac{\partial H}{\partial p} \qquad H = \sum p\dot{q} - L \;;\; p = \frac{\partial L}{\partial \dot{q}}$$

$$\dot{p} = -\frac{\partial H}{\partial q}$$

Here Φ is the wave function, v the wave velocity that is assumed to vary little within a wavelength; A and S, the amplitude and the phase respectively, ω the frequency, L and H are the Lagrangian and Hamiltonian, τ the travel time along the ray, **p** and **q** are the momentum and coordinate vectors respectively. The wave fronts are the characteristics and the rays the bicharacteristics of the wave equation. Three scalar functions are associated to these equations: Φ to the wave equation, S (or L) to the Hamilton-Jacobi equation, and H to the equations of the rays.

3. ELASTODYNAMICS.

The equation of motion of an elastic, isotropic, heterogeneous medium is

$$\rho\ddot{\mathbf{u}} = (\lambda+\mu)\,\nabla\,(\,\nabla.\mathbf{u}\,) + \mu\,\nabla^2\,\mathbf{u} + \\ + \nabla\lambda\,(\,\nabla.\,\mathbf{u}\,) + \nabla\mu\wedge(\,\nabla\wedge\mathbf{u}\,) + 2\,(\,\nabla\mu.\nabla\,)\,\mathbf{u}$$

where λ and μ are the Lamé parameters, ρ the density and **u** the displacement vector.

The characteristics are obtained in the high frequency approximation, that is to say, as the lowest order term in the asymptotic development of the wave equation in power series of ω^{-1}. We try a solution of the form

$$\mathbf{u} = \begin{bmatrix} u_1 \\ u_2 \\ u_3 \end{bmatrix} e^{iS}$$

where each u_i is developed in power series of ω^{-1}. $S = \mathbf{k.x}$ in the homogeneous case, where the wave fronts are planes. In the heterogeneous case S is a general function of the coordinates. The series must satisfy the equation of motion and hence each coefficient of ω^{-n} must vanish. The lowest order term gives three homogeneous linear equations in the u_i. The determinant of the coefficients must be zero for non trivial solutions. Therefore

$$\begin{vmatrix} (\lambda+\mu)S_1^2 + \mu(\nabla S)^2 - \rho\omega^2 & (\lambda+\mu)S_1S_2 & (\lambda+\mu)S_1S_3 \\ (\lambda+\mu)S_1S_2 & (\lambda+\mu)S_2^2 + \mu(\nabla S)^2 - \rho\omega^2 & (\lambda+\mu)S_2S_3 \\ (\lambda+\mu)S_3S_1 & (\lambda+\mu)S_3S_2 & (\lambda+\mu)S_3^2 + \mu(\nabla S)^2 - \rho\omega^2 \end{vmatrix} = 0$$

where $S_i = \partial_i S$. This expression may be factorized as follows

$$[\mu\,(\nabla S)^2 - \rho\omega^2]^2\,[\,(\lambda+2\mu)\,(\nabla S)^2 - \rho\omega^2\,] = 0$$

the first double factor corresponding to the S, and the second one to the P wave front. P and S phases are thus separated in the lowest order approximation.

4. COORDINATE SYSTEM.

The first thing to be shown is the construction of a metric around a central ray. Let us suppose that a ray Γ_o has been calculated. A point $\mathbf{r}_o(s)$ of the central ray is characterized by its distance s, taken along the ray, from an initial point. Points on neighbouring rays are situated with respect to a plane passing through $\mathbf{r}_o(s)$ and orthogonal to Γ_o. The unit vectors on this plane are $(\mathbf{e}_1,\mathbf{e}_2)$ and the coordinates are (q_1,q_2). Hence, a point **r** on the vicinity of the central ray has coordinates (s,q_1,q_2). We have then

$$\mathbf{r} = \mathbf{r}_o(s) + q_1\hat{e}_1(s) + q_2\hat{e}_2(s)$$

After differentiation we obtain

$$dr^2 = h^2 ds^2 + dq_1^2 + dq_2^2$$

with $h = 1 - C\,(q_1\cos\phi + q_2\sin\phi)$; where C is the curvature of the ray and ϕ is the angle between $\mathbf{e}_1$ and the normal to the ray.

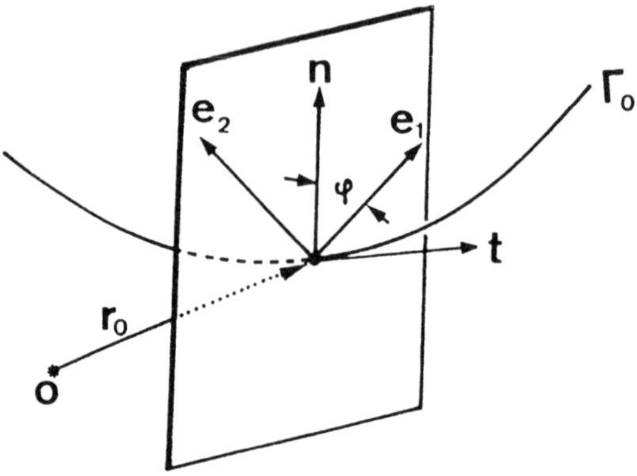

Figure 1: Geometry of the local coordinates around the central ray of the beam. The coordinates q_1 and q_2 are measured on a plane normal to the ray.

The element of length is expressed as a sum of squares after imposing the condition that the crossed terms vanish. This condition is equivalent to

$$T = \frac{d\phi}{ds}$$

where T is the torsion of the ray. This expression allow us to construct ϕ as a function of s in such a way as to eliminate the cross products.

5. VARIATIONAL PRINCIPLE.

The metric that has been chosen simplifies the formulation of the variational principle. In fact, the travel time along a paraxial ray (a ray in the vicinity of the central ray) must be minimized, what is equivalent to say that the phase function S must be an extreme. The travel time is equal to the sum of the arc elements divided by the velocity. Thus

$$S = \omega\tau = \omega \int_{s_o}^{s} \frac{\sqrt{h^2 + \dot{q}_1^2 + \dot{q}_2^2}}{v(s,q_1,q_2)}\, ds = \omega \int_{s_o}^{s} L(q_1,q_2,\dot{q}_1,\dot{q}_2,s)\, ds = \text{extreme}$$

The function L is the Lagrangian of the system, and the dots indicate derivatives with respect to s. The momenta are easily calculated from the Lagrangian:

$$p_i = \frac{\partial L}{\partial \dot{q}_i} = \frac{\dot{q}_i}{v\sqrt{h^2 + \dot{q}_1^2 + \dot{q}_2^2}}$$

Now we are in a position to obtain the value of the Hamiltonian:

$$H = \sum p_i \dot{q}_i - L = -\frac{h}{v}\sqrt{1 - v^2(p_1^2 + p_2^2)}$$

and we have all the elements that are needed to formulate the problem from the point of view of the Classical Theory.

6. PARABOLIC APPROXIMATION.

The coordinates q_i are zero all along the central ray. Therefore, along the central ray we have

$$q_i = \dot{q}_i = p_i = \dot{p}_i = 0$$

hence, from Hamiltons equations, we obtain

$$\frac{\partial H}{\partial p_i} = \frac{\partial H}{\partial q_i} = 0$$

on the central ray. This means that there are no first order terms in the Taylor development of the Hamiltonian around the central ray. If we carry on the development up to second order terms, we obtain

$$H(\mathbf{q},\mathbf{p}) \approx -\frac{1}{v_o} + \frac{1}{2v_o^2}\mathbf{q}^T \mathbf{V}\,\mathbf{q} + \frac{v_o}{2}\mathbf{p}^T\mathbf{p}$$

where $\mathbf{V} = [\,\partial^2 v/\partial q_i \partial q_j\,]$ is the matrix of second derivatives of the wave velocity.

Therefore, the parabolic approximation is obtained by using a truncated Taylor series development of the Hamiltonian that gives a quadratic function of the coordinates and momenta. It has the same form as the Hamiltonian of the harmonic oscillator.

7. RAY PATHS.

Paraxial rays satisfy Hamilton's equations in the parabolic approximation

$$\dot{\mathbf{q}} = \frac{\partial H}{\partial \mathbf{p}} = v_o \mathbf{p}$$

$$\dot{\mathbf{p}} = -\frac{\partial H}{\partial \mathbf{q}} = -\frac{1}{v_o^2} \mathbf{V}\,\mathbf{q}$$

This set of equations may be solved if wanted, but it is more practical to use the properties of paraxial rays in terms of quantities evaluated at the central ray.

8. WAVE FRONTS.

The phase $S = \omega\tau$ satisfies the Hamilton-Jacobi equation as shown in classical mechanics (Landau and Lifshitz, 1967)

$$\frac{\partial\tau}{\partial s} = H(q_i, \frac{\partial\tau}{\partial q_i}, s) = \frac{1}{v_o} - \frac{1}{2v_o^2}\mathbf{q}^T\,\mathbf{V}\,\mathbf{q} - \frac{v_o}{2}\left[\left(\frac{\partial\tau}{\partial q_1}\right)^2 + \left(\frac{\partial\tau}{\partial q_2}\right)^2\right]$$

where we have replaced $\mathbf{p}$ by $\partial\tau/\partial\mathbf{q}$. We may try a solution of the form

$$\tau = \tau(s,0,0) + \frac{1}{2}\mathbf{q}^T\,\mathbf{M}\,\mathbf{q}$$

where we do not keep a linear term in $\mathbf{q}$ since $\mathbf{p} = \partial\tau/\partial\mathbf{q}$ must be zero for $\mathbf{q} = 0$. If we replace this expression into the Hamilton-Jacobi equation and we identify terms of equal order in $\mathbf{q}$ we obtain

$$\mathbf{p} = \partial\tau/\partial\mathbf{q} = \mathbf{M}\mathbf{q} \quad ; \quad \partial\tau/\partial s = 1/v_o \quad ; \quad d\mathbf{M}/ds + v_o\,\mathbf{M}^T\mathbf{M} + v_o^{-2}\,\mathbf{V} = 0$$

where $\mathbf{M} = [\,\partial^2\tau/\partial q_i \partial q_j\,]$ is the matrix of curvature of the wave front, and it is obviously symmetric. The momentum $\mathbf{p}$ is the gradient of the wave front τ. $\mathbf{M}$ satisfies Ricatti's differential equation. Finally, we have

$$\tau = \int_{s_o}^{s} \frac{ds}{v_o} + \frac{1}{2}\mathbf{q}^T\mathbf{M}\mathbf{q}$$

9. GAUSSIAN BEAMS.

We remember that the displacement $\mathbf{u} = \mathbf{u}_o\, e^{iS} = \mathbf{u}_o\, e^{i\omega\tau}$, where

$$\tau = \tau(s,0,0) + \frac{1}{2}\mathbf{q}^T\mathbf{M}\mathbf{q}$$

is the travel time. The Gaussian Beams are obtained when we make an analytic continuation of **M** into the complex plane. In fact, let us change **M** by **M** + i **m**, where **m** is a positive definite real matrix. Then, the displacement becomes

$$\mathbf{u} = \mathbf{u}_o\, e^{i\omega[\tau_o + \frac{1}{2}\mathbf{q}^T\mathbf{M}\mathbf{q}] - \frac{\omega}{2}\mathbf{q}^T\mathbf{m}\mathbf{q}}$$

and we see that the phase has the same form as that of the parabolic approximation, but the amplitudes decrease exponentially from the central ray. The beam is concentrated around the central ray with a gaussian profile. The approximation generates parabolic wave fronts and gaussian amplitude decay.

10. AMPLITUDES.

The next term in the asymptotic development of the equation of motion gives what is known as the transport equation

$$\nabla.(\rho u^2 v\mathbf{t}) = 0$$

where the term inside the divergence is the energy flux density along the ray.

Moreover, if J is the Jacobian of the mapping (x,y,z) --> (s,q_1,q_2) , the cross-sections of a ray tube are proportional to J, and we have a second conservation law:

$$\nabla.\left(\frac{\mathbf{t}}{J}\right) = 0$$

but since $\rho u^2 v\mathbf{t} = \rho u^2 vJ.(\mathbf{t}/J)$, from these two equations we easily deduce that

$$\rho u^2 vJ = \text{constant}$$

along the ray. Thus, we obtain the conservation of the energy flux across the section of a tube of rays. Then we may obtain the amplitudes by the formula

$$u = \frac{\text{constant}}{\sqrt{\rho vJ}}$$

which may be generalized to a vector field (u_1,u_2,u_3).

11. THE CYLINDRICAL SOURCE: AN EXAMPLE.

a. We will discuss a simple example in order to illustrate the concept of a Gaussian beam. A two dimentional source function (Cerveny et al. 1982) in a homogeneous medium, representing the wave field from a line source with cylindrical symmetry source, may be written as

$$u = \frac{H(t - \frac{r}{v})}{\sqrt{t^2 - \frac{r^2}{v^2}}}$$

where $\tau = r/v$, r is the distance to the source, v the wave velocity, t the time and H(t) the Heaviside step function. The displacement is zero for $t<\tau$, it grows suddenly at the time of arrival of the wave and then it decreases with time. The Fourier transform of this source function is (look for the Hankel function $H_0^{(2)}(\omega\tau)$ and its asymptotic behaviour, Abramowitz and Stegun, 1972).

$$u(\omega) = \int_\tau^\infty \frac{e^{-i\omega t}}{\sqrt{t^2 - \tau^2}} dt = \int_0^\infty e^{-i\omega\tau\cosh\phi} d\phi \approx \sqrt{\frac{\pi}{2i\omega\tau}} e^{-i\omega\tau}$$

The last term is the high frequency asymptotic approximation of the integral obtained by steepest descent.

b. The asymptotic approximation of the spectrum shows a product of two functions of ω (a convolution in the time domain). The first one, $1/\sqrt{(i\omega)}$, is nothing but a source time function. The second one, $\sqrt{(\pi/2\tau)}\exp(-i\omega\tau)$, contains the propagation effects: an amplitude with a geometrical spreading varying as $1/\sqrt{r}$, and a phase $\omega\tau$. We will now concentrate our study on this second factor.

c. First, we select the central rays as straight lines passing through the source at equal intervals $\Delta\phi$ covering the interval $[0,\pi]$. Second, the local coordinates (s,q) for a given beam are obtained by passing the perpendicular to the central ray from the observation point. Third, we will give equal weight to each beam in order to approximate a cylindrically symmetrical line source. This is the analog of summing plane waves in all possible directions with equal weigths in order to construct a cylindrical wave. The advantage of the Gaussian beams over plane waves comes from the fact that they are concentrated around the central ray instead of being spread over the space.

d. We proceed now to perform the parabolic approximation. The wave fronts of the cylindrical source are circles, and the phase is given by

$$\tau = r/v = \sqrt{(s^2+q^2)}/v$$

where q is the distance of the observation point to the central ray, and s is the projection of r onto the central ray. Therefore the parabolic approximation gives

$$\tau = s/v \sqrt{(1+q^2/s^2)} \approx s/v + q^2/(2sv)$$

This approximation is good in the neighbourhood of the central ray.

e. The question now is how to make an analytic continuation of this expression into the complex plane. The answer is obtained by taking complex values of the curvature, that is to say, complex solutions of Ricatti's equation (§8), which for constant velocity becomes:

$$dM/ds + v\,M^2 = 0$$

with solution $$M = 1/[v(s + \varepsilon)]$$

where ε is a complex integration constant. Then, the complex phase function for the Gaussian beam becomes:

$$\tau = s/v + q^2/[2v(s + \varepsilon)]$$

f) Passage to the time domain: The Fourier inversion formula for a real time function may be written as

$$f(t) = \frac{1}{\pi}\,\mathrm{Re}\int_0^\infty \bar{f}(\omega)\,e^{i\omega t}\,d\omega$$

thus we obtain the following expression for the jth beam

$$f_j(t) = \frac{1}{\pi}\,\mathrm{Re}\sqrt{\frac{\pi v}{2s_j}}\int_0^\infty e^{i\omega(t-\tau_j)}\,d\omega = \frac{1}{\pi}\,\mathrm{Re}\sqrt{\frac{\pi v}{2s_j}}\;\frac{i}{t-\tau_j}$$

where the geometrical spreading factor has been evaluated at the central ray ($q=0, \tau=s/v$). Notice that the convergence at $\omega = +\infty$ is guaranteed by the choice $\mathrm{Im}\varepsilon > 0$.

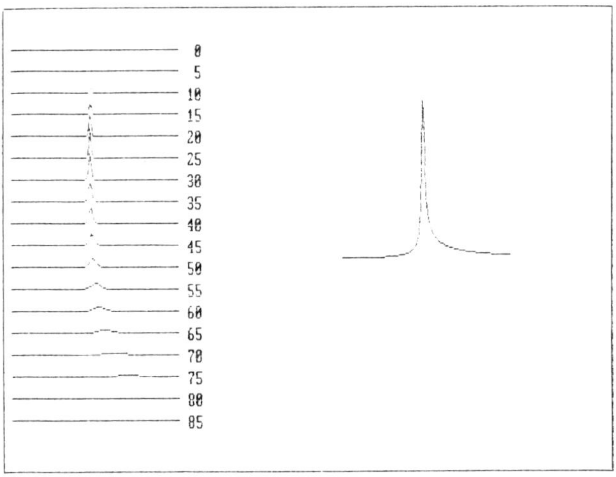

Figure 2: Gaussian beam summation for a cylindrical source. The elementary seismograms (left) are the contributions of each beam. The total seismogram (right) is the sum of all contributions. The angle of each beam is measured with respect to the source-observer direction.

g. For a given observation point we have to sum over all of the beams with equal weights. Thus we obtain

$$f(t) = \sqrt{\frac{v}{2\pi}} \sum_j \mathrm{Re} \frac{1}{\sqrt{s_j}} \frac{i}{t - s_j/v - q_j^2/[2v(s_j + \varepsilon)]}$$

Figure 2 shows the contribution of each beam and the seismogram that results from the sum of all the contributions. Half of the beams are shown, the first one in the figure being the one that is closest to the observer.

A further convolution with the function corresponding to $1/\sqrt{(i\omega)}$, namely $H(t)/\sqrt{(\pi t)}$, would be necessary in order to obtain the right approximation for the original source time function.

A short Fortran program that calculates the sum over the beams is included in Table I.

Table I: Fortran program to compute the Gaussian beam approximation of a cylindrical source wave function.

```
C M=N° OF CENTRAL RAYS. DPHI=ANGLE BETWEEN 2 BEAMS.
C N=N° OF POINTS IN SEISMOGRAM. DT=TIME INTERVAL.
C R=SOURCE-OBSERVER DISTANCE. EPSIL=COMPLEX BEAM PARAMETER.
      DIMENSION F(200),FJ(200)
      COMPLEX S1,TAU,FFJ,EPSIL
      DATA V,PI,M,DPHI,N,DT,R,ER,EI/6.,3.1415926,18,5.,100,.35,
     *100.,10.,10./
      OPEN (6,FILE='GAUSS.DAT',STATUS='NEW')
      DPHI=DPHI*PI/180.
      A0=SQRT(V/2./PI)
      EPSIL=CMPLX(ER,EI)
         DO 30 J=1,M
         PHI=(J-1)*DPHI
         S=R*COS(PHI)
         Q=R*SIN(PHI)
         S1=S+EPSIL
         TAU=S/V + Q*Q/2./V/S1
            DO 40 I=1,N
            T=(I-1)*DT
            FFJ=CMPLX(0.,1.)/(T-TAU)/SQRT(S)
            FJ(I)=A0*REAL(FFJ)
            F(I)=F(I)+FJ(I)
40          CONTINUE
         WRITE(6,10) (FJ(I),I=1,N)
30       CONTINUE
      WRITE(6,10) (F(I),I=1,N)
10    FORMAT(10F8.4)
      CLOSE(6)
      STOP
      END
```

12. CONCLUSIONS.

We have shown in this paper that the examination of Gaussian beams from the point of view of classical Hamilton-Jacobi theory gives a simpler and more unified frame than when theory is developed from first principles. The equations governing the propagation of the beams are inmediatly obtained from the Hamiltonian of the system. The parabolic approximation is made only once, over the Hamiltonian, without the effort needed when elementary methods are used.

The example that illustrates the concept of a Gaussian beam concerns the source rather than the propagation of the beams. Nevertheless, its study leads to a fast understanding of several important features of the method. In particular, the concept of paraxial rays (discussed in detail by Madariaga,1984), is illustrated in this simple case by the elementary solution of Ricatti's equation.

REFERENCES.

Abramowitz, M. and Stegun, I. A. (1972), Handbook of Mathematical Functions. Dover Publications, Inc.

Babich, V.N. and Kirpichnikova, N.J. (1974), Boundary Layer Method in Diffraction Problems. Leningrad University Press (in russian, English translation by Springer-Verlag, 1980)

Babic, V. N. and Buldyrev, N. J., (1972) Asymptotic Methods in Problems of Diffraction of Short Waves. Nauka, Moscow (in Russian).

Babic, V. N. and Popov, M. M., (1981) Propagation of concentrated acoustical beams in a three-dimensional inhomogeneous medium. Akust. Zh., 27, 823-835 (in Russian).

Cerveny V. (1981), Seismic Wave Fields in Structurally Complicated Media. (Ray and Gaussian beam aproaches). Unpublished Lectures Notes.

Cerveny V. (1983), Synthetic body wave seismograms for laterally varying layered structures by Gaussian beam method. Geophys. J. R. astr. Soc., 73, PP. 389-426.

Cerveny, V., Popov, M. M. and Psencik, I. (1982), Computation of wave fields in inhomogeneous media, Gaussian beam approach. Geophys. J. R. astr. Soc., 70, pp. 109-128.

Cisternas, A., G. Jobert, P. Compte. (1983), Teoria Clasica de haces Gausianos. Rev. de Geofisica, Madrid, 40, pp 27-32.

Fock, V. A. (1965), Electromagnetic Diffraction and Propagation Problems. Pergamon Press.

Landau, L. D. and Lifshitz E. M. (1969), Mecanique, Editions MIR, Moscou.

Madariaga, R. (1984), Gaussian beam synthetic seismograms in a vertically varying medium. Geophys. J. R. astr. Soc., 79, pp. 589-612.

Psensik, I. (1983), Gaussian beams in two-dimensional elastic inhomogeneous media., Geophys. J. R. astr. Soc., 72, pp. 417-433.

Popov, M. M. (1981) A new method of computation of wave fields in the high frequency approximation., LOMI, A. N. SSSR, E-I-81, Leningrad.

Yomogida, K. and Aki, K. (1985), Waveform Synthesis of Surface Waves in a Laterally Heterogeneous Earth by the Gaussian Beam Method. J. Geophys. Res., 90, pp. 7665-7688.

ON THE SEISMIC RESPONSE OF ALLUVIAL VALLEYS

F.J. Sánchez-Sesma
Instituto de Ingeniería, UNAM
Cd. Universitaria 04510 DF
Mexico, and
Fundación J. Barros Sierra A.C.
Carretera al Ajusco 203, Tlalpan 14200 DF
Mexico

ABSTRACT. Paper reviews some of the methods available for studying the seismic response of alluvial valleys. The current formulation of the problem is discussed in brief. Some examples are given of results obtained using analytical and numerical approaches. The need for a unified treatment of source, path, and site effects on seismic risk assessment is pointed out.

1. INTRODUCTION

Local soil conditions can significantly affect the characteristics of ground motion during earthquakes giving rise to large amplifications and spatial variations of shaking. These effects can not be ignored in the assessment of seismic risk, in studies of microzonation, and in the seismic design of important facilities (Esteva, 1977; Ruiz, 1977). The seismic behavior of local irregularities is relevant in the response of life-line systems.

Effects of soil conditions on ground motion have been observed in well documented earthquakes (*e.g.* Sozen *et al.*, 1968; Jennings, 1971; Singh *et al.* 1987) and in regression analyses of strong motion data (see *e.g.* Campbell, 1985). In addition there is evidence that localized damage distribution is related to lateral irregularities in subsurface topography.

These effects were observed and documented for the July 26, 1963 Skopje, Yugoslavia earthquake (Poceski, 1969). It has been suggested that focusing of the wave energy by irregular interfaces greatly amplified ground motions in parts of the city (Jackson, 1971). A recent case history was offered by the September 19, 1985 Michoacán, Mexico earthquake which caused unprecedented destruction in Mexico City. In addition to important source and path effects, the lacustrine formations of the valley originated observed spectral amplifications of ten to 50 times compared with what was observed on nearby firm ground in the frequency band of 0.2 to 1.0 Hz (Singh *et al.*, 1987).

J. Bonnin et al. (eds.), Seismic Hazard in Mediterranean Regions, 85–104.

When the soil layers are horizontal, with few or no lateral irregularities, use of one dimensional models (*e.g.* Haskell, 1962) may suffice to account for site effects. However, lateral irregularities can generate focusing and local surface waves. This can produce significant discrepancies between observed ground motion and the values predicted when assuming horizontal stratification. This is precisely the case in some Mexico City's locations during the 1985 earthquake. There is conclusive evidence of important generation of local surface waves (*e.g.* Sánchez-Sesma and Singh, 1986).

Theoretical studies (*e.g.* Trifunac, 1971; Wong and Trifunac, 1974; Sánchez-Sesma and Esquivel, 1979; Bard and Bouchon, 1980; Dravinski, 1982; Sánchez-Sesma, 1983; Bard and Gariel, 1986) have shown the importance of lateral irregularities. Results for a wide variety of shapes of alluvial deposits have been obtained and in some cases compared with observations (*e.g.* Bard and Tucker, 1985). The different approaches to the problem have emphasized the physical understanding of site effects so that quantitative predictions can be made. However, there is still lack of practical criteria for dealing with irregular local soil conditions. It is desirable that such criteria take into account the characteristics of the source and path of seismic waves. The recent progress on strong motion prediction using mathematical modeling techniques (Aki, 1982) is encouraging. Much of the research is concentrated on the physical description of the fault rupture process and wave propagation in the Earth.

The aim of this work is to review the problem of evaluating the seismic response of alluvial valleys. For this purpose the current formulation of the problem, the known analytical solutions, and some of the available numerical methods are briefly described. Paper points out the need for a unified treatment of the source, path, and site conditions in the assessment of seismic risk.

2. FORMULATION OF THE PROBLEM

There is no doubt that the seismic source mechanism governs the way in which the released energy is radiated in space and time. Seismic waves, once emitted by the source, are dependent on the mechanical properties of earth materials and the heterogeneities encountered in their path. Moderate changes in mechanical impedances or irregularities with size comparable to incident wave lengths can generate significant spatial variations of seismic wave fields. Surficial soil formations with mechanical impedances lower than that of overlaying materials are efficient in trapping the energy carried by seismic waves.

It is generally accepted in seismic wave modeling that Earth materials are elastic, linear, homogeneous, and isotropic. This is not accurate. However, the success of the classical elastic model in the quantitative description of most seismic phenomena is impressive (see *e.g.* Aki and Richards, 1980). The assumption of linearity has been very useful as it allows use of well established mathematical tools and the development of new ones. However, in some situations nonlinear behavior has to be invoqued to satisfactorily explain observations. This is the case for most soils when subjected to large strains. As the mathematical difficulties to deal with nonlinear problems are formidable the bulk of the work in this area is based on the elastic model.

In most studies of seismic response of alluvial valleys and other surface irregularities incidence of some kind of elastic waves is assumed, usually plane waves, which propagate from a distant source in a hypotetical half-space. Elastic plane waves are reflected back and refracted forward as they arrive at a plane interface. The amounts of reflected and transmitted energy depend on the mechanical properties of the media involved. Reflection and refraction in elastic wave propagation can be well described by geometrical means. We will call diffraction every change in the wave path that can not be described as reflection. Irregular interfaces produce diffraction; to describe the elastic wave fields it is necessary to solve a boundary value problem for the governing equations of linear elasticity (*e.g.* Achenbach, 1973; Aki and Richards, 1980).

For the sake of illustration, consider an elastic homogeneous, isotropic half-space containing an irregular alluvial valley, as shown in Figure 1. This model is, of course, an oversimplification, as real geological configurations present very complicated features, which are rarely known in detail. Mathematical treatment of local irregularities requires care in the selection of the model in order to properly represent the salient characteristics of their seismic response. Under incident elastic waves, our ideal irregular alluvial deposit will be excited by refracted waves. Diffracted waves will also be generated (diffraction is frequently called scattering). Diffracted and refracted waves must satisfy the governing equations (Navier equations) and, together with incident waves, the boundary conditions (zero tractions at free surfaces and continuity of displacements and tractions at interfaces). Additionally, the diffracted wave fields in the half-space must fulfill the Sommerfeld (1949) radiation condition at infinity, which means that they must scatter to infinity; *i.e.* no energy may be radiated from infinity into the irregular region. The Sommerfeld radiation condition has been extended to elastic wave fields by Kupradze (1965).

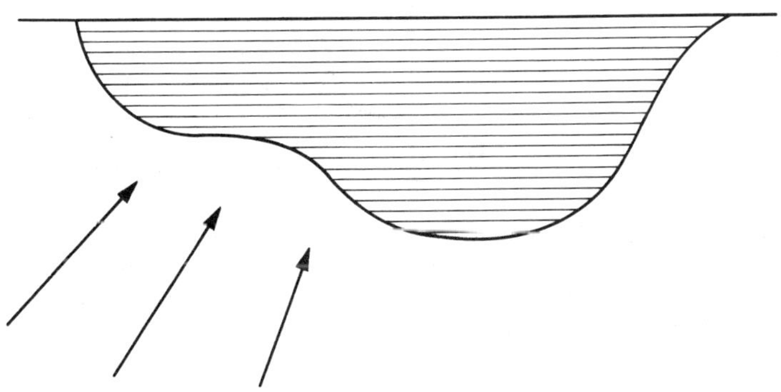

Figure 1. Model of a stratified alluvial valley in an elastic half-space under incidence of elastic waves.

In this formulation of the problem one can recognize an approach similar to that of most frequency domain studies. It is sufficiently general and useful for reference.

3. ANALYTICAL SOLUTIONS

The simplest problems in elastic wave diffraction are the two-dimensional SH-wave problems because they can be analyzed separately from other body waves. The governing equation for this case is the scalar wave equation. Analytical solutions can be obtained for geometries of the scatterer which allow separation of variables (Mow and Pao, 1971). Using this method, exact solutions have been obtained for the diffraction of SH-waves by canyons and alluvial valleys with semi-circular (Trifunac, 1971, 1973) or semielliptical (Wong and Trifunac, 1974a,b) shapes. Even with these simple models of local irregularities, complicated interference patterns were found and the obtained surface displacements vary pronouncedly in space and frequency (Figure 2). They are very sensitive to the incidence angle. Results for alluvial valleys show the importance of the two-dimensional behavior which gives much larger amplifications than those obtained from unidimensional calculations. These analytical solutions have shown the importance of the phenomenon and they provide a check for numerical procedures.

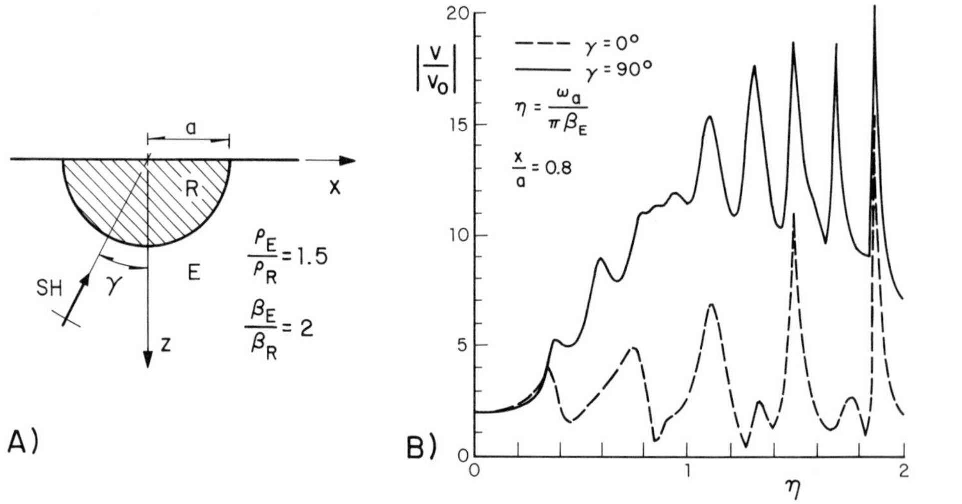

Figure 2. Incidence of SH waves upon a semicircular alluvial valley. (A) Problem configuration and (B) normalized displacement amplitudes on the surface at $x/a = 0.8$ for two incidence angles plotted versus dimensionless frequency η. Here ω = frequency, ρ = mass density, and β = shear-wave velocity. (After Trifunac, 1971).

Several analytical solutions can be easily obtained if a moving rigid basement is assumed. This would be a useful approximation for very soft soils overlaying rock formations. For example, very simple results have been recently obtained for the antiplane response of a dipping layer with moving rigid base. When the dip angle of the wedge is of the form Π/2N, N = 3, 5, 7 ..., the exact solution is obtained by no more than geometrical means (Sánchez-Sesma and Velázquez, 1987). Figures 3 and 4 illustrate some numerical results. This model, although much simplified, gives physical insight on the basic mechanisms of local surface wave generation.

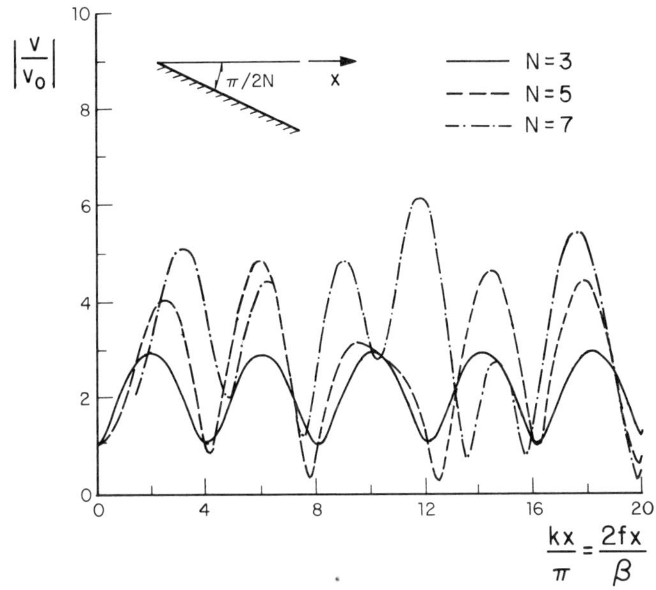

Figure 3. Normalized surface amplitudes versus $2fx/\beta$ for a dipping layer overlaying a rigid moving base with antiplane motion $v_o\ \exp(i\omega t)$. Different angles Π/2N, N = 3, 5, 7. (After Sánchez-Sesma and Velázquez, 1987).

For the more difficult cases of P- or SV-incident waves the orthogonal wave functions developed in classical physics are not separable for the half-space surface due to the coupling of boundary conditions. Lee (1984) overcame this difficulty for a semi-spherical valley by expanding the spherical wave functions further into a power series that matched all the boundary conditions successfully. This approach is, however, limited to small frequencies because the resulting matrix equations, which are infinite, can only be solved approximately.

4. NUMERICAL METHODS

A powerful technique has been developed by Aki and Larner (1970) to treat scattering of SH-waves by irregular interfaces. In the Aki-Larner method, incidence is assumed of a plane single-frequency wave that produces diffracted displacements fields. These fields are represented by super-

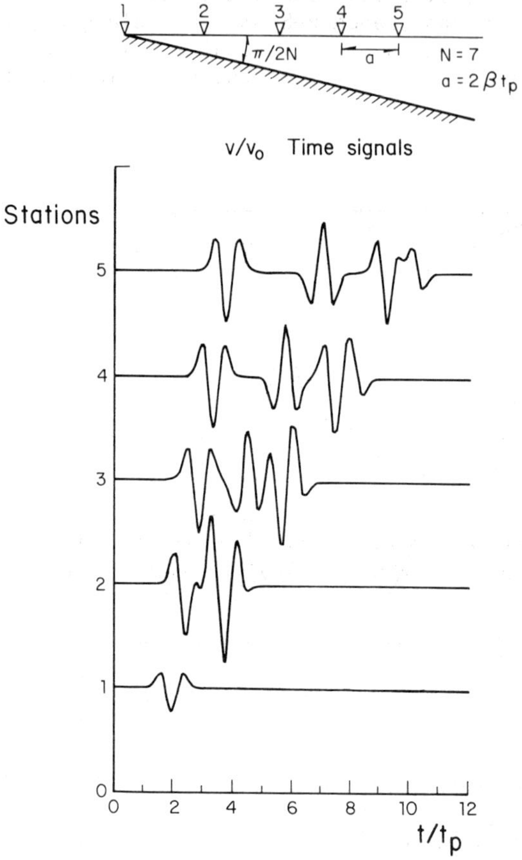

Figure 4. Time signals at five stations on the surface of a dipping layer overlaying a moving rigid base. Equal distance between stations is given by $a = 2\beta t_p$, where β = shear-wave velocity and t_p = reference time. Note that the base antiplane motion is that of station 1. (After Sánchez-Sesma and Velázquez, 1987).

position of plane waves of unknown complex amplitudes propagating in many directions. Inhomogeneous plane waves are allowed. The method is restricted to small-slope irregularities for numerical reasons because it does not include explicitly upgoing waves. Although the representation in terms of plane waves is complete, convergence to the true solution can be very slow. The total motion is obtained from integration over horizontal wave numbers. Under the assumption of horizontal periodicity of the irregularity, the integral is replaced with an infinite sum. Truncation of this sum and application of the interface conditions of continuity of stress and displacement in the wavenumber domain lead to a system of linear equations for the complex scattering coefficients. This discrete-wave-number method has been applied by Bouchon and Aki (1977a,b) to study near seismic source fields in a layered medium with irregular interfaces. Figure 5 shows some results. The method has been extended to time domain computations to study the seismic response of alluvial valleys (Bard and Bouchon, 1980a,b) under incidence of SH, P, and SV waves. Bard and Bouchon showed the generation of local surface waves at the alluvial valley edges. Results are shown in Figure 6. The Aki-Larner method has allowed performing extensive parametric studies (*e.g.* Bard and Bouchon, 1980a,b; Bard and Gariel, 1986) and comparison with observations (Bard and Tucker, 1985). An extension of this powerfull technique is due to Bouchon (1985) in which upgoing waves are explicitly included in the analysis, thus eliminating the restriction of small-slopes. Calculations for irregular layered media show a very good performance of the extended method (Campillo and Bouchon, 1985). The method has been used to model the fields generated by real faults

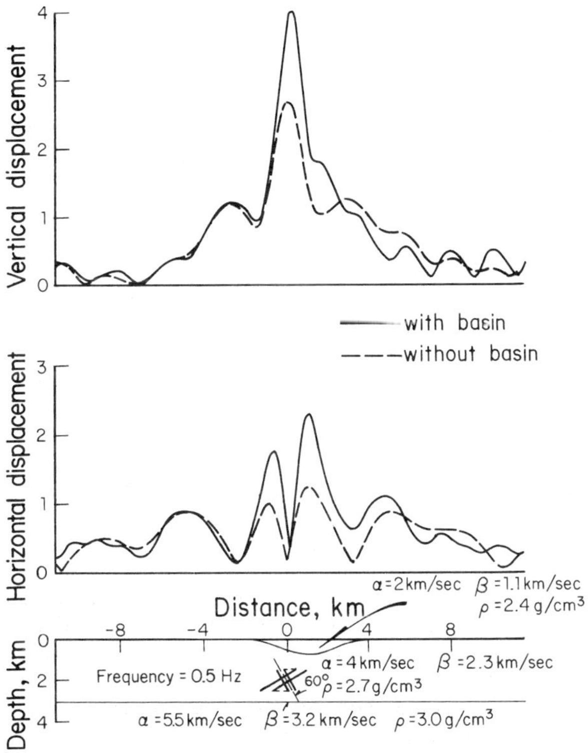

Figure 5. Normalized surface displacements for a 2D dislocation source radiating at 0.5Hz beneath a sedimentary basin. Problem configuration. (After Bouchon and Aki, 1977b).

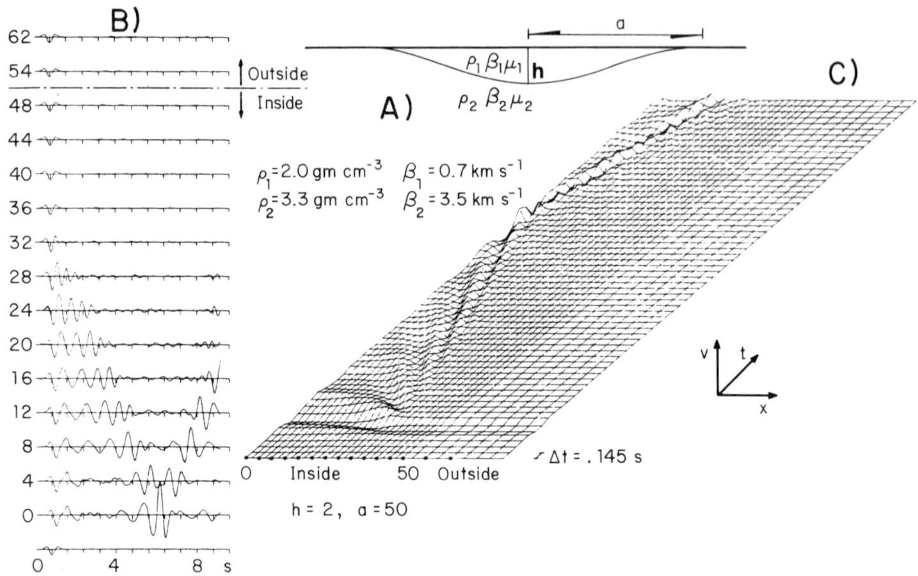

Figure 6. Response of a sine-shaped alluvial valley with maximum depth h = 200 m, half-widht a = 5 km, to a vertically incident SH Ricker wavelet of characteristic period t_p = 0.732 s. (A) Problem configuration and mechanical properties. (B) Traces represent the displacement at surface receivers, spaced from 0 to 6.2 km from the valley center. The *bottom* trace would be the surface displacement signal without the valley. (C) Diagram showing the spatial (x) and temporal (t) evolution of the surface displacement in the valley and in its immediate proximity. The dots indicate the location of the sites where the seismograms of Figure 6B are computed. In both figures, the length units is 100 m and the time unit is 1 s. Only one side of the valley is represented because of the symmetry of the problem. (After Bard and Bouchon, 1980a).

(Bouchon, 1979; Campillo, 1983). It can be used to model ground motion considering together the effects of source, path, and local conditions.

The finite difference method is also a powerful tool in elastic wave propagation studies (Alterman and Karal, 1968; Boore, 1972a; Virieux, 1984). It has been applied to model two-dimensional irregular interfaces (Boore *et al.*, 1971; Boore, 1972b) in the SH case and also for incident P and SV waves upon a sedimentary basin (Harmsen and Harding, 1981) and a step-like topography (Boore *et al.*, 1981). Interesting results have been found concerning the significant generation of Rayleigh surface waves by lateral irregularities. Recent work by Ohtsuki and coworkers (Ohtsuki and Harumi, 1983; Ohtsuki *et al.*, 1984a; 1984b) confirms this phenomenon. They have used a combination of finite differences and finite elements. Figure 7 shows some results that highlight this important effect. A finite difference analysis of axisym-

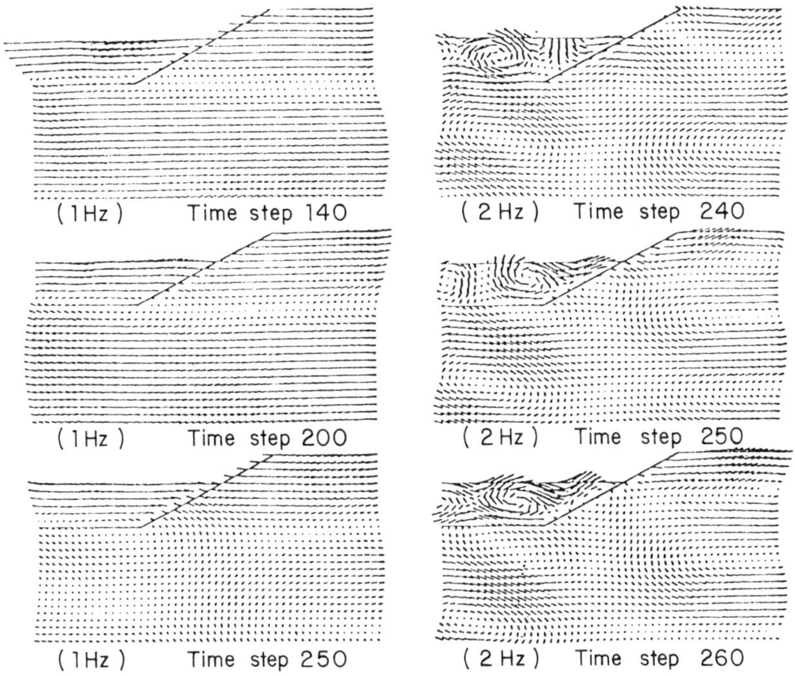

Figure 7. Snapshots of particle displacements in a cliff with a soft layer for vertically incident SV waves of 1Hz(λ/h=4) on the left and for 2Hz(λ/h=2) on the right side. (After Ohtsuki and Harumi, 1983).

metric topographical irregularities has been presented to study the effects of vertically incident shear waves (Liao *et al.*, 1980). Spectral ratios were obtained; comparison with observations gives reasonable agreement. The finite difference method is theoretically unlimited to model details and even nonlinear behavior of materials, but the problem size can easily exceed the capacity of major computing facilities.

The finite element method also allows a detailed description of site topography and layering. It allows calculating the response of two-dimensional soil configurations with truly nonlinear stress-strain relations (Streeter *et al.*, 19874; Joyner and Chen, 1975; Joyner, 1975). The major disadvantages of the method are its low-frequency limit and its high cost. Usually, real time analysis must be shortened to avoid the reflections from the artificial boundaries. Use of different transmitting techniques can reduce the spurious waves to some extent (*e.g.* Smith, 1974, 1975; Ayala and Aranda, 1977; Clayton and Engquist, 1977; Castellani *et al.*, 1981; Liao and Wong, 1981). Successful criteria have been developed for damping out the unwanted reflections by means of non-uniform element size (Day, 1977) or by combining finite elements with a boundary integral representation of the edge conditions in the domain studied (Franssens and Lagasse, 1984). Finite elements have been used to treat problems of irregular layering (Lysmer and Drake, 1972; Drake, 1972; Aranda and Ayala, 1978) and two-dimensional topographical irregularities (Castellani *et al.*, 1982) under idealized conditions. Nevertheless, a realistic wave analysis can be very costly.

Ray methods have also been used to study the ground motion in sedimentfilled basins with irregular interfaces (*e.g.* Jackson, 1971; Hong and Helmberger, 1977; Lee and Langston, 1983; Rial, 1984) or dipping layers (Ziegler and Pao, 1984; Sánchez-Sesma and Velázquez, 1987). An extension of ray theory based on a paraxial approximation of wave equation has been recently applied to study several problems of wave propagation in inhomogeneous media and irregular boundaries (*e.g.* Nowack and Aki, 1984; Madariaga, 1984). These solutions are called Gaussian beams because of the Gaussian shape of the wave amplitude around a central ray. The high frequency character of Gaussian beams place them as a very promissing tool to study site effects on strong ground motion. A simple approximate solution has been recently developed to describe the ground motion on a class of alluvial valleys under incident SH waves (Sánchez-Sesma *et al.* 1987). This solution is based on the existence of a complete family of rays supported by the basin boundaries which are restricted to be formed by two opposite dipping layers of certain angles. Due to its low cost and virtually no upper frequency limit, the method is suited for calculating the time domain seismic response of alluvial deposits if their shape resembles the class of shapes studied. Figure 8 shows the synthetic response of an extended triangular alluvial valley for an incident SH Ricker-type wavelet.

In recent years boundary methods have gained increasing popularity. This is mainly due to the availability of high speed computers. Boundary methods are well suited to deal with wave propagation problems because they avoid the introduction of fictitious boundaries and reduce by one the dimensionality of the problem. These facts yield numerical advantages. Moreover, boundary methods can be used together with the finite element method (Zienkiewics *et al.*, 1977). This allows reducing the size of the region represented with finite elements (*e.g.* Ayala and Gómez, 1979; Shah *et al.*, 1982).

There are two main approaches to the formulation of boundary methods; one is based on the use of boundary integral equations (Cruse and Rizzo, 1968a,b; Brebbia, 1978; Cole *et al.*, 1978; Alarcón *et al.*, 1979); the other, on the use of complete systems of solutions (Herrera and Sabina, 1978; Herrera, 1980). The scattering of incident SH waves from two-dimensional irregular topographies has been formulated with integral equations by Wong and Jennings (1975) for arbitrarily shaped canyon-like profiles and by Sills (1978) for ridges and mixed shapes. This method has been applied with success for calculating the effects of a dipping layer of alluvium of an SH-wave source at the surface (Wong *et al.*, 1977). Results compare favorably with observations during a full-scale low-amplitude propagation test. A powerful approach that combines the boundary integral equation method with finite differences in time has

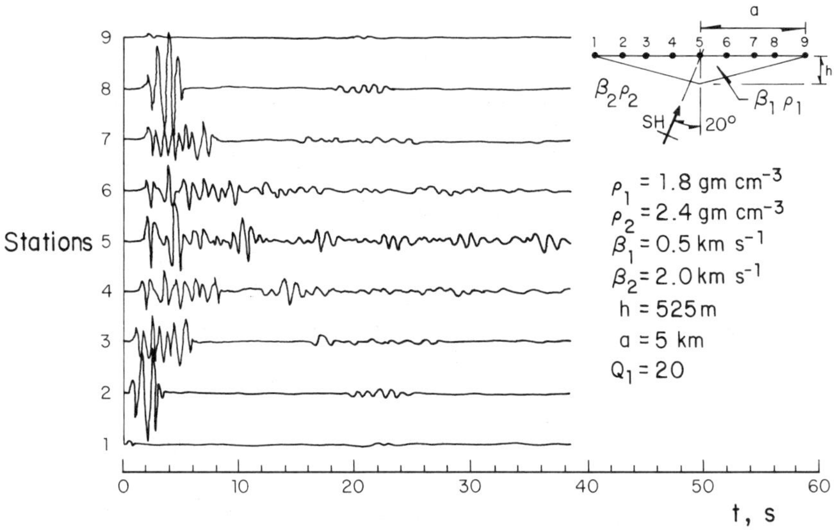

Figure 8. Response of a triangular alluvial valley with depth h = 525 m, half-width a = 5 km, to an oblique (20°) incident SH Ricker wavelet of characteristic period t_p = 0.75 s. Traces represent the displacement of surface receivers marked with dots on the right side. For the alluvial material a quality factor Q = 20 was used.

been presented (Cole *et al.*, 1978) for solving elastodynamic problems. The method's performance was found to be good in a simple numerical problem. A boundary method has been recently developed and applied to solve two-dimensional scattering of harmonic elastic waves by canyons (Sánchez-Sesma, 1978, 1981; Sabina *et al.*, 1979; Sánchez-Sesma and Rosenblueth, 1979; Wong, 1979, 1982; England *et al.*, 1980; Sánchez-Sesma *et al.*, 1982a; 1985a), alluvial deposits (Sánchez-Sesma and Esquivel, 1979; Ize *et al.*, 1981; Dravinski, 1982a,b, 1983; Bravo *et al.*, 1987) and ridges (Sánchez-Sesma and Esquivel, 1980; Sánchez-Sesma *et al.*, 1982b) for different types of waves and shapes of the scatterers. The method consists in constructing the scattered fields with linear combinations of members of a complete family of wave functions (Herrera and Sabina, 1978; Herrera, 1984). These families of functions, which are solutions of the governing equations of the problem, can be constructed in a very general way, with single or multipolar sources having their singularities outside the region of interest. Coefficients of the linear forms thus constructed are obtained from a least-squares matching of boundary conditions. As pointed out by Wong (1982), the method can be considered as a generalized inverse one. In doing this, Wong suggested a procedure that improves the solution numerically. A general framework for the method is given by a recently developed algebraic theory of boundary value problems (Herrera, 1979; 1980a,b; 1984). This approach has been extended by Bravo *et al.* (1987) to deal with stratified alluvial deposits under incidence of SH waves. The field in the layered region is

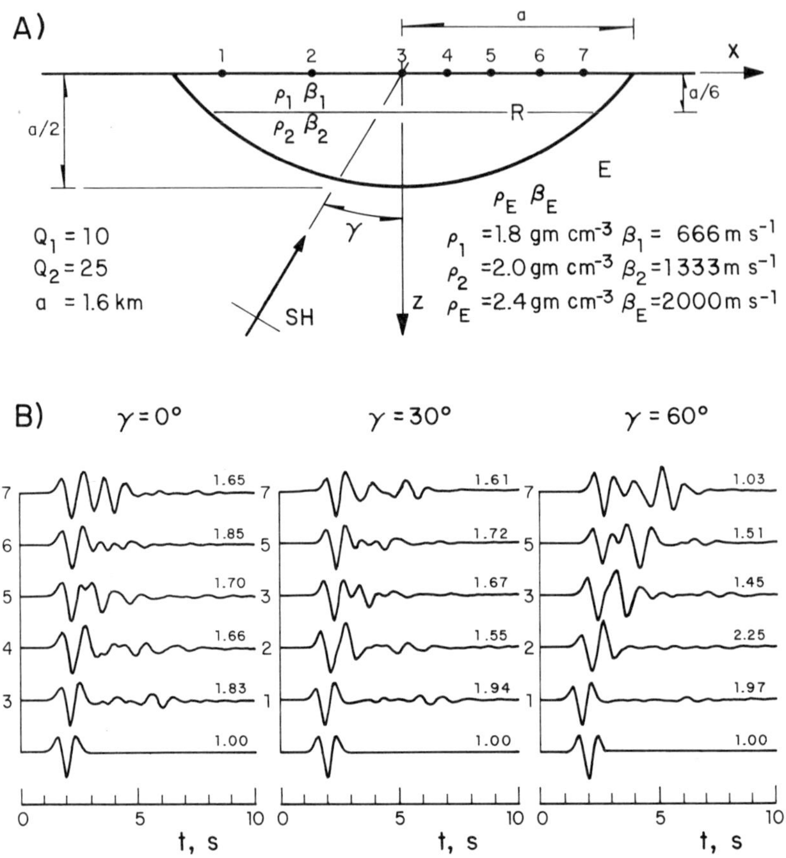

Figure 9. Incidence of SH waves upon a parabolic stratified basin with half-width a = 1.6 km, to incidence of SH Ricker wavelet of characteristic period t_p = 1.0 s. (A) Problem configuration and material properties. (B) Traces represent the normalized displacement of surface receivers marked with dots in figure 9A. They were normalized so that maximum amplitudes in the plot are equal. Amplification factors are given for each trace. Results are given for three incidences. Bottom traces give the incident signal. (After Bravo *et al.*, 1987).

constructed using a discrete wave-number representation whereas the field in the half-space is represented by a finite number of sources. In Figure 9 some results are presented for a simple alluvial valley composed of two strata. The method has been applied to three-dimensional problems (Sánchez-Sesma, 1983; Sánchez-Sesma *et al.*, 1984; Sánchez-Sesma *et al.*, 1987b; Pérez-Rocha and Sánchez-Sesma, 1987a,b). The cases of

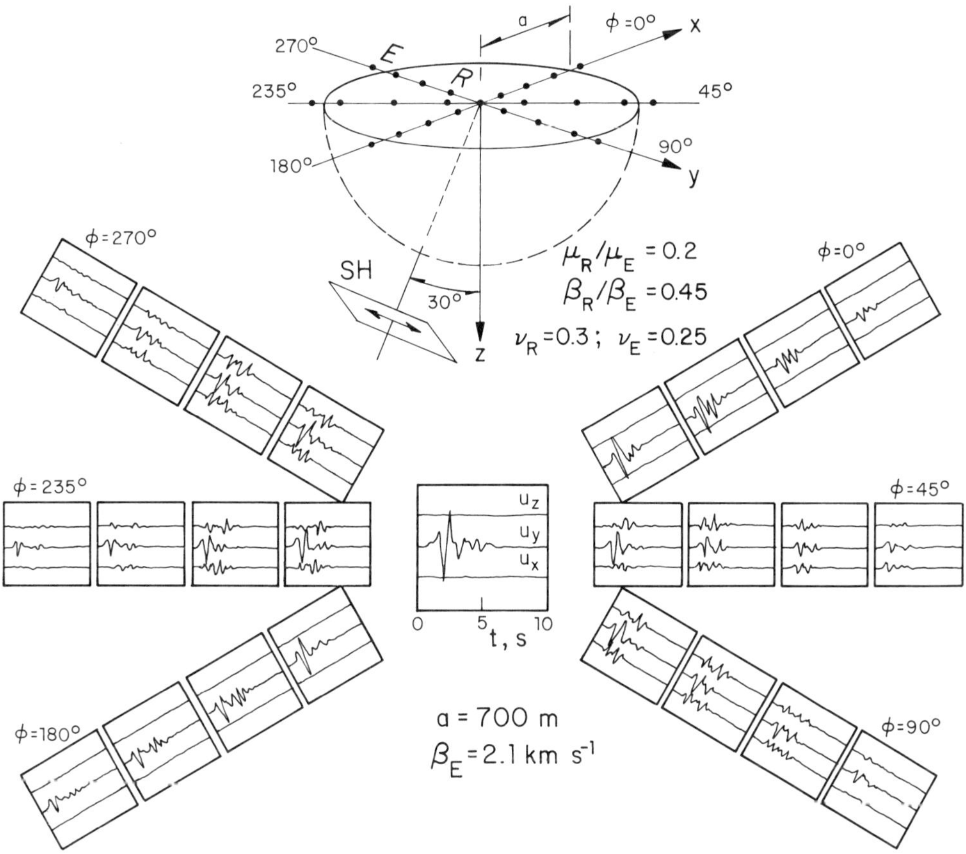

Figure 10. Response of a semi-spherical alluvial valley with radius a = 700 m, to an oblique (30°) incident SH Ricker wavelet of characteristic period t_p = 0.7 s. Traces represent the 3D displacements of surface receivers marked with dots in the upper part. (After Sánchez-Sesma *et al.*, 1987b).

incidence of P, SV, SH, and Rayleigh waves upon axisymmetric irregularities on the surface of an elastic half-space have been formulated by means of an azimuthal decomposition. In it the diffracted and refracted fields have been constructed using multipolar solutions of the reduced Navier equations in spherical coordinates in terms of spherical Bessel and Legendre functions (Takeuchi and Saito, 1972; Aki and Richards, 1980). In Figure 10 some results are displayed of the full three-dimensional response of a semispherical alluvial deposit under incidence of SH waves. Some interesting features can be observed in the synthetics regarding the generation of surface waves and focusing effects. Studies of this type are very costly because of the considerable ammount of computer resources required. However, they throw light on the problem of site effects, give physical insight, and provide results that can be useful to calibrate more efficient procedures.

5. CONCLUDING REMARKS

The seismic response of alluvial valleys and other geological irregularities has been briefly discussed and some of the available methods to deal with the problem reviewed. The different methods have been useful to give physical understanding of site effects. With the advent of supercomputers their capabilities are being extended both in the complexity of the problems and in the possibility of dealing with high frequency signals. Ray methods together with Gaussian beams seem to be powerful tools to deal with high frequencies. On the other hand, experimental techniques can be most useful in the study of site effects. For instance, King and Brune (1981) have used foam rubber models to study the response of sedimentary basins. They have found excellent agreement with analytical solutions and have predicted interesting results for more complicated problems. Advances in measuring techniques, such as laser interferometry, may be useful to deal with controled experiments and improve their predictive capabilities. Regarding full scale experiments, the spread of modern digital recording of strong motion is already producing good quality data from real earthquakes. Well planned experimental settings as that of the Parkfield experiment (Tucker, 1986 personal communication) will serve to test our capacity of predict site effects in strong motion. This will require consideration of source mechanism and wave propagation in the Earth. The research on the quantitative prediction of strong ground motion will prove useful in understanding physical phenomena and improving the practical assessment of seismic hazards.

6. ACKNOWLEDGEMENTS

Thanks are given to E. Rosenblueth for the critical reading of the manuscript and for helpful discussions. The help of M. A. Bravo, L. E. Pérez-Rocha, and M. Suárez in various stages of the work is greatly appreciated. This work was partially supported by the Consejo Nacional de Ciencia y Tecnología, Mexico, under Grants PCCBBNA-021963, PCECCNA-040803 and PCECCNA-040141.

7. REFERENCES

Achenbach, J. D. (1973). Wave propagation in elastic solids, North-Holland Publishing Co., Amsterdam.

Aki, K. (1982). Strong motion prediction using mathematical modeling techniques, Bull. Seism. Soc. Am. 72, S29-S41.

Aki, K. and K. L. Larner (1970). Surface motion of a layered medium having an irregular interface due to incident plane SH waves, J. Geophys. Res. 70, 933-954.

Aki, K. and P. G. Richards (1980). Quantitative seismology, theory and methods, W. H. Freeman and Co., San Francisco.

Alarcón, E., A. Martin and F. Paris (1979). Boundary elements in potential and elasticity theory, J. Computers and Structures 10, 351-362.

Alterman, Z. S. and F. C. Karal, Jr. (1968). Propagation of elastic waves in layered media by finite difference methods, Bull. Seism. Soc. Am. 58, 367-398.

Aranda, G. R. and G. A. Ayala (1978). Modelo numérico eficiente de aplicación en estudios de amplificación dinámica, Proc. Conferencia Centroamericana de Ingeniería Sísmica, II, San Salvador, 127-137.

Ayala, G. A. and G. R. Aranda (1977). Boundary conditions in soil amplification studies, Proc. World Conf. Earthquake Eng., 6th, New Delhi.

Ayala, G. A. and R. Gómez (1979). A general procedure for solving three-dimensional elasticity problems in geomechanics, in Numerical methods in geomechanics, Aachen, 1979. W. Wittke (editor), A. A. Balkema, Rotterdam.

Bard, P. Y. and M. Bouchon (1980a). The seismic response of sediment-filled valleys. Part 1. The case of incident SH waves, Bull. Seism. Soc. Am. 70, 1263-1286.

Bard, P. Y. and M. Bouchon (1980b). The seismic response of sediment-filled valleys. Part 2. The case of incident P and SV waves, Bull. Seism. Soc. Am 70, 1921-1941.

Bard, P. Y. and J. C. Gariel (1986). The seismic response of two-dimensional sedimentary deposits with large vertical velocity gradients. Bull. Seism. Soc. Am. 76, 346-366.

Bard, P. Y. and B. E. Tucker (1985). Underground and ridge site effects: a comparison of observation and theory, Bull, Seism. Soc. Am. 75, 905-922.

Boore, D. M. (1972a). Finite difference methods for seismic wave propagation in heterogeneous materials, in Methods in Computational Physics 11, B. A. Bolt (editor), Academic Press, New York.

Boore, D. M. (1972b). A note on the effect of simple topography on seismic SH waves, Bull. Seism. Soc. Am. 62, 275-284.

Boore, D. M., K. L. Larner, and K. Aki (1971). Comparison of two independent methods for the solution of wave scattering problems: response of a sedimentary basin to incident SH waves, J. Geophys. Res. 76, 558-569.

Boore, D. M., S. C. Harmsen, and S. T. Harding (1981). Wave scattering from a steep change in surface topography, Bull. Seism. Soc. Am. 71, 117-125.

Bouchon, M. (1979). Predictability of ground displacement and velocity near an earthquake fault. An example: the Parkfield earthquake of 1966, J. Geophys. Res. 84, 6149-6156.

Bouchon, M. (1985). A simple, complete numerical solution to the problem of diffraction of SH waves by an irregular surface, J. Acoust. Soc. Am. 77, 1-5.

Bouchon, M. and K. Aki (1977a). Discrete wave number representation of seismic source wave fields, Bull. Seism. Soc. Am. 67, 259-277.
Bouchon, M. and K. Aki (1977b). Near-field of seismic source in a layered medium with irregular interfaces, Geophys. J. R. Astr. Soc. 50, 669-684.
Bravo, M. A., F. J. Sánchez-Sesma, and F. J. Chávez-García (1987). Ground motion on stratified alluvial deposits for incident SH waves, Submitted to Bull. Seism. Soc. Am.
Brebbia, C. A. (1978). The boundary element method for engineers, Pentech Press., London.
Campillo, M. (1983). Numerical evaluation of the near field high-frequency radiation from quasi-dynamic circular faults, Bull. Seism. Soc. Am. 73, 723-734.
Campillo, M. and M. Bouchon (1985). Synthetic SH-seismograms in a laterally varying medium by the discrete wavenumber method, Geophys. J. R. Astr. Soc. 82.
Campbell, K. W. (1985). Strong motion attenuation relations: a ten year perspective, Earthquake Spectra 1, 759-804.
Castellani, A., C. Chesi, and E. Mitsopoulou (1981). An earthquake engineering wave propagation model, Meccanica, Journal of the Italian Association of Theoretical and Applied Mechanics, March, 33-41.
Castellani, A., A. Peano, and L. Sardella (1982). On analytical and numerical techniques for seismic analysis of topographic irregularities, Proc. European Conf. Earthquake Eng. 7th, Athens, Greece, 2, 415-423.
Clayton, R. and B. Engquist (1977). Absorbing boundary conditions for acoustic and elastic wave equations, Bull. Seism. Soc. Am. 67, 1529-1540.
Cole, D. M., D. D. Kosloff, and J. Bernard Minster (1978). A numerical boundary integral equation method for elastodynamics. I, Bull. Seism. Soc. Am. 68, 1331-1357.
Cruse, T. A. and F. J. Rizzo (1968a). A direct formulation and numerical solution of the general transient elastodynamic problem. I, J. Math. Anal. Appl. 22, 244-259.
Cruse, T. A. and F. J. Rizzo (1968b). A direct formulation and numerical solution of the general transient elastodynamic problem, II, J. Math. Anal. Appl. 22, 341-355.
Day, S.M. (1977). Finite element analysis of seismic scattering problems, PhD. Thesis, University of California, San Diego, California, 149 pp.
Drake, L. A. (1972). Love and Rayleigh waves in non-horizontal layered media, Bull. Seism. Soc. Am. 62, 1241-1258.
Dravinski, M. (1982a). Scattering of SH waves by subsurface topography, J. Eng. Mech. Div., Proc. ASCE 108, 1-17.
Dravinski, M. (1982b). Influence of interface depth upon strong ground motion, Bull. Seism. Soc. Am. 72, 597-614.
Dravinski, M. (1983). Amplification of P, SV and Rayleigh waves by two alluvial valleys, Soil Dynamics and Earthquake Engrg. 2, 66-77.
England, R., F. J. Sabina and I. Herrera (1980). Scattering of SH waves by surface cavities of arbitrary shape using boundary methods, Phys. Earth Planet. Ints. 21, 148-157.

Esteva, L. (1977). Microzoning: models and reality, Proc. World Conf. Earthquake Eng., 6th, New Delhi.

Franssens, G. R. and P. E. Lagasse (1984). Scattering of elastic waves by a cylindrical obstacle embedded in a multilayered medium, J. Acoust. Soc. Am. 76, 1535-1542.

Harmsen, S. C. and S. T. Harding (1981). Surface motion over a sedimentary valley for incident plane P and SV waves, Bull. Seism. Soc. Am. 71, 655-670.

Haskell, N. A. (1962). Crustal reflection of plane P and SV waves, J. Geophys. Res. 67, 4751-4767.

Herrera, I. (1979). Theory of connectivity: a systematic formulation of boundary element methods, Applied Math. Modelling 3, 151-156.

Herrera, I. (1980a). Variational principles for problems with linear constraints, prescribed jumps and continuation type restrictions, J. Inst. Maths. and Applics. 25, 67-96.

Herrera, I. (1980b). Boundary methods. A criterion for completeness, Proc. Nat'l. Acad. Sci., U.S.A. 77, 4395-4398.

Herrera, I. (1984). Boundary methods: an algebraic theory, Pitman Adv. Publishing Program, Boston.

Herrera, I. and F. J. Sabina (1978). Connectivity as an alternative to boundary integral equations. Construction of bases, Proc. Nat'l. Acad. Sci., U.S.A., 75, 2059-2063.

Hong, T. L. and D. V. Helmberger (1977). Glorified optics and wave propagation in non planar structures, Bull. Seism. Soc. Am. 68, 1313-1330.

Ize, J. R. England, and F. J. Sabina (1981). Theoretical and numerical study of diffraction of waves by inhomogeneous obstacle, Comunicaciones internas, 291, IIMAS-UNAM, Mexico.

Jackson, P. S. (1971). The focusing of earthquakes, Bull. Seism. Soc. Am. 61, 685-695.

Jennings, P. C. (editor) (1971). San Fernando Earthquake of February 9, 1971, Earthquake Eng. Res. Lab., EERL71-02, Calif. Inst. of Tech., Pasadena, California.

Joyner, W. B. (1975). A method for calculating nonlinear seismic response in two dimensions, Bull. Seism. Soc. Am. 65, 1337-1357.

Joyner, W. B. and A. T. F. Chen (1975). Calculation of nonlinear ground response in earthquakes, Bull. Seism. Soc. Am. 65, 1315-1336.

King, J. L. and J. N. Brune (1981). Modeling the seismic response of sedimentary basins, Bull. Seism. Soc. Am. 72, 1469-1487.

Kupradze, V. D. (1965). Potential methods in the theory of elasticity, Israel Program for Scientific Translations, Jerusalem.

Lee, J. J. and C. A. Langston (1983). Wave propagation in a three-dimensional circular basin, Bull. Seism. Soc. Am. 73, 1637-1655.

Lee, V. W. (1984). Three-dimensional diffraction of plane P, SV & SH waves by a hemispherical alluvial valley, Soil Dyn. and Earthq. Engrg. 3, 133-144.

Liao, Z. P., Y. Baipo, and Y. Yifan (1980). Effect of three-dimensional topography on earthquake ground motion, Proc. World Conf. Earthquake Eng., 7th, Istanbul, 2, 161-168.

Liao, Z. P. and H. L. Wong (1981). A transmitting boundary for discrete methods, Proc. 4th ASCE-EMD Speciality Conf., Purdue University.

Lysmer, J. and L. A. Drake (1972). A finite element method for seismology, in Methods of Computational Physics, 11, B. A. Bolt (editor), Academic Press, New York.

Madariaga, R. (1984). Gaussian beam synthetic seismograms in a vertically varying medium. Geophys. J. R. Astr. Soc. 79, 589-612.

Mow, C. C. and Y. H. Pao (1971). The diffraction of elastic waves and dynamic stress concentrations, Report R-482-PR, The Rand Corporation, Santa Monica, California.

Nowack, R. and K. Aki (1984). The two-dimensional Gaussian beam synthetic method: testing and applications, J. Geophys. Res. 89, 7797-7819.

Ohtsuki, A. and K. Harumi (1983). Effect of topography and subsurface inhomogeneities on seismic SV waves, Int. J. Earthquake Engrg. Struct. Dyn. 11, 441-462.

Ohtsuki, A., H. Yamahara, and K. Harumi (1984a). Effect of topography and subsurface inhomogeneity on seismic Rayleigh waves, Int. J. Earthquake Engrg. Struct. Dyn. 12, 37-58.

Ohtsuki, A., H. Yamahara, and T. Tazoh (1984b). Effect of lateral inhomogeneity on seismic waves, II. Observations and analysis, Int. J. Earthquake Engrg. Struct. Dyn. 12, 795-816.

Pérez-Rocha, L. E. and F. J. Sánchez-Sesma (1987a). Difracción de ondas sísmicas por depósitos tridimensionales de suelos blandos. Instituto de Ingeniería, UNAM.

Pérez-Rocha, L. E. and F. J. Sánchez-Sesma (1987b). On the seismic response of three-dimensional alluvial valleys and surface irregularities, Submitted to Journal of Geophysical Research.

Poceski, A. (1969). The ground effects of the Skopje July 26, 1963 earthquake, Bull. Seism. Soc. Am. 59, 1-29.

Rial, J. A. (1984). Caustics and focusing produced by sedimentary basins. Application of catastrophe theory to earthquake seismology; Geophys. J. R. Astr. Soc. 79, 923-938.

Ruiz, S. E. (1977). Influencia de las condiciones locales en las características de los sismos. Instituto de Ingeniería, UNAM, 387, pp. 65.

Sabina, F. J. and J. R. Willis (1977). Scattering of Rayleigh waves by a ridge, J. Geophys. 43, 401-419.

Sabina, F. J., R. England, and I. Herrera (1979). Theory of connectivity: Applications to scattering of seismic waves. I. SH wave motion, Proc. 2nd International Conf. on Microzonation, San Francisco, California, 2, 813-824.

Sánchez-Sesma, F. J. (1978). Ground motion amplification due to canyons of arbitrary shape. Proc. Int. Conf. on Microzonation, 2nd., San Francisco, California, 2, 729-738.

Sánchez-Sesma, F. J. (1981). A boundary method applied to elastic scattering problems, Arch. Mech. 3, 167-179.

Sánchez-Sesma, F. J. (1983). Diffraction of elastic waves by three-dimensional surface irregularities, Bull. Seism. Soc. Am. 73, 1621-1636.

Sánchez-Sesma, F. J. and E. Rosenblueth (1979). Ground motion at canyons of arbitrary shape under incident SH waves, Int. J. Earthquake Eng. Struct. Dyn. 7, 441-450.

Sánchez-Sesma, F. J. and J. A. Esquivel (1979). Ground motion on alluvial valleys under incident plane SH waves, *Bull. Seism. Soc. Am.* **69**, 1107-1120.

Sánchez-Sesma, F. J. and J. A. Esquivel (1980). Ground motion on ridges under incident SH waves, *Proc. World Conf. Earthquake Engrg. 7th*, Istanbul, **1**, 33-40.

Sánchez-Sesma, F. J., I. Herrera, and M. A. Bravo (1982a). Difracción de ondas P, SV y de Rayleigh en un semiespacio elástico. Instituto de Ingeniería, UNAM, Mexico.

Sánchez-Sesma, F. J., I. Herrera, and J. Avilés (1982b). A boundary method for elastic wave diffraction. Application to scattering of SH waves by surface irregularities, *Bull. Seism. Soc. Am.* **72**, 473-490.

Sánchez-Sesma, F. J., S. Chávez-Pérez, and J. Avilés (1984). Scattering of elastic waves by three-dimensional topographies. *Proc. World Conf. Earthquake Engrg. 8th*, San Francisco, California, **2**, 639-646.

Sánchez-Sesma, F. J., M. A. Bravo, and I. Herrera (1985a). Surface motion of topographical irregularities for incident P, SV and Rayleigh waves, *Bull. Seism. Soc. Am.* **75**, 297-303.

Sánchez-Sesma, F. J. and S. K. Singh (1986). Grandes temblores y sus efectos en el valle de México: observaciones y teoría. Proc. of the Simposium *Los Sismos de 1985: Casos de mecánica de suelos.* Sociedad Mexicana de Mecánica de Suelos, Mexico, September.

Sánchez-Sesma, F. J. and S. A. Velázquez (1987). On the seismic response of a dipping layer. Submitted to *Wave Motion*.

Sánchez-Sesma, F. J., F. J. Chávez-García, and M. A. Bravo (1987a). Seismic response of a class of alluvial valleys for incident SH waves. Submitted to *Bull. Seism. Soc. Am.*

Sánchez-Sesma, F. J., L. E. Pérez-Rocha, and S. Chávez-Pérez (1987b). Diffraction of elastic waves by three-dimensional surface irregularities. Part II. Submitted to *Bull. Seism. Soc. Am.*

Shah, A. H., K. C. Wong, and S. K. Datta (1982). Diffraction of plane SH waves in a half-space. *Int. J. Earthquake Engrg. Struct. Dyn.* **10**, 519-528.

Sills, L. B. (1978). Scattering of horizontally polarized shear waves by surface irregularities, *Geophys., J. R. Astr. Soc.* **54**, 319-348.

Sing, S. K., E. Mena, and R. Castro (1987). Some aspects of source characteristics of the 19 September, 1985, Michoacán earthquake and ground motion amplification in and near Mexico city from strong motion data. Submitted to *Bull. Seism. Soc. Am.*

Smith, W. D. (1974). A nonreflecting boundary for wave propagation problems, *J. Computational Phys.* **15**, 492-503.

Smith, W. D. (1975). The application of finite element analysis to body wave propagation problems, *Geophys. J. R. Astr. Soc.* **42**, 747-768.

Sommerfeld, A. (1949). *Partial differential equations in physics*, Academic Press, Inc., New York.

Sozen, M. A., P. C. Jennings, R. B. Matthiesen, G. W. Housner, and N. M. Newmark (1968). *Engineering Report on the Caracas Earthquake of July 29, 1967.* National Academy of Sciences, Washington, D.C.

Streeter, V. L., E. B. Wylie, and F. E. Richard, Jr. (1974). Soil motion computations by characteristics method, Proc. Am. Soc. Civil Engrg., J. Geotech. Engrg. Div. 100, 247-263

Takeuchi, H. and M. Saito (1972). Seismic surface waves, in Methods in Computational Physics, 11, B. A. Bolt, ed., Academic Press, New York.

Trifunac, M. D. (1971). Surface motion of a semi-cylindrical alluvial valley for incident plane SH waves, Bull. Seism. Soc. Am. 61, 1755-1770.

Trifunac, M. D. (1973). Scattering of plane SH waves by a semi-cylindrical canyon, Int. J. Earthquake Engrg. Struct. Dyn. 1, 267-281.

Virieux, J. (1984). SH-wave propagation in heterogeneous media: velocity-stress finite difference method, Geophysics 49, 1933-1957.

Wong, H. L. (1979). Diffraction of P, SV and Rayleigh waves by surface topographies, Report CE 79-05, Department of Civil Engineering, University of Southern California, Los Angeles, California.

Wong, H. L. (1982). Effect of surface topography on the diffraction of P, SV and Rayleigh waves, Bull. Seism. Soc. Am. 72, 1167-1183.

Wong, H. L. and M. D. Trifunac (1974a). Scattering of plane SH wave by a semi-elliptical canyon, Int. J. Earthquake Engrg. Struct. Dyn. 3, 157-169.

Wong, H. L. and M. D. Trifunac (1974b). Surface motion of a semi-elliptical alluvial valley for incident plane SH wave, Bull. Seism. Soc. Am. 64, 1389-1408.

Wong, H. L. and P. C. Jennings (1975). Effect of canyon topography on strong ground motion. Bull. Seism. Soc. Am. 65, 1239-1257.

Wong, H. L., Trifunac, M. D., and B. Westermo (1977). Effects of surface and subsurface irregularities on the amplitude of monochromatic waves, Bull. Seism. Soc. Am. 67, 353-368.

Ziegler, F. and Y. H. Pao (1984). Transient elastic waves in a wedge-shaped layer, Acta Mechanica 52, 133-163.

Zienkiewicz, O. C., D. W. Kelly, and P. Bettess (1977). The coupling of the finite element method and boundary solution procedures, Intern. J. Num. Meth. Engrg. 11, 355-377.

STRONG MOTION MODELLING OF THE RUPTURING PROCESS OF THE NOVEMBER 23, 1980 IRPINIA, ITALY, EARTHQUAKE

P. Suhadolc (1), F. Vaccari (1) and G. F. Panza (1,2)

(1) Istituto di Geodesia e Geofisica
Universita' di Trieste
34100 Trieste, Italy
(2) International School for Advanced Studies
34100 Trieste, Italy

ABSTRACT. The rupturing process associated with the Irpinia earthquake of November 23, 1980 is modelled. Six strong-motion records, vertical component, obtained very close to the epicentral area, are compared with synthetic accelerograms, constructed by the Rayleigh-waves modal summation. The cut-off frequency used in the computations - 10 Hz - largely exceeds the limits of the presently possible deterministic approach. In the final model, twelve point sources have been used to reproduce the first 30 s of the recordings, and a good agreement between observed and synthetic accelerograms is obtained both in the signal envelopes and in the peak accelerations. A bilateral fracturing propagation model is proposed, in good accord with the interpretations of other Authors. The shallower point sources, in spite of their small seismic moment, contribute in a relevant way to the surface strong motion.

1. INTRODUCTION

Although many recordings have been obtained both from long-period and short-period seismometers and accelerometers, the fault-plane solution and the rupturing process of the November 23, 1980 Irpinia earthquake are still not completely understood (Table I). We can attribute this uncertainty to the complexity of the event itself. The remarkable duration of the strong motion recordings is probably due to the combined effect of many asperities located in a zone roughly defined by the spatial distribution of aftershocks.

In this work we develop a direct method to unravel the rupturing process of this event, based on the construction of synthetic seismograms (Panza, 1985). We are looking for a detailed information about the rupturing process, and this requires the possibility of treating high frequencies. The available strong motion data contain frequencies exceeding 10 Hz. From a deterministic point of view, frequencies higher than a few Hz are of scarce interest, since the

J. Bonnin et al. (eds.), Seismic Hazard in Mediterranean Regions, 105–128.

TABLE I

AUTHOR or AGENCY	COORDINATES Lat.N	Lon.E	ORIGIN TIME h m s	DEPTH km	Θ	δ	λ
ING	40°49'	15 23'	18 34 52.9	18.0	---	--	---
NEIS	40°53'	15°20'	18 34 53.3	10.0	---	--	---
CSEM	40°48'	15°20'	18 34 53.8	10.0	---	--	---
BOSCHI et Al. (1981)	40°44'	14°58'	18 34 53.3	25.0	152°	57°	---
PANZA and CUSCITO (1982)	---	---	---	18.0	114°	69°	277°
KANAMORI and GIVEN (1982)	40°55'	15°22'	18 34 53.8	16.0	129°	45°	270°
SCARPA and SLEJKO (1982)	40°46'	15°18'	18 34 52.8	18.0	118°	64°	258°
DEL PEZZO et Al. (1983)	40°46'	15°18'	18 34 52.5	16.0	118°	64°	258°
DESCHAMPS and KING (1983)	40°48'	15°18'	18 34 52.8	15.0	140°	60°	317°
BRUESTLE and MUELLER (1983)	40°46'	15°18'	18 34 53.8	18.0	138°	60°	250°
NAKANISHI and KANAMORI (1984)	40°55'	15°22'	18 34 53.8	9.8	137°	63°	276°
WESTAWAY and JACKSON (1984)	40°47'	15°20'	---	15.0	135°	60°	274°

Fault-plane solutions of the November 23, 1980 Irpinia earthquake. The value of λ, if not indicated in the referenced paper, was deduced from the focal spheres.

geological structure of the area is not known with the necessary precision. Nevertheless, to be conservative, we will work with frequencies as high as 10 Hz, but in the comparison between synthetic and experimental accelerograms we will give more relevance to envelope fitting rather than to single peaks matching.

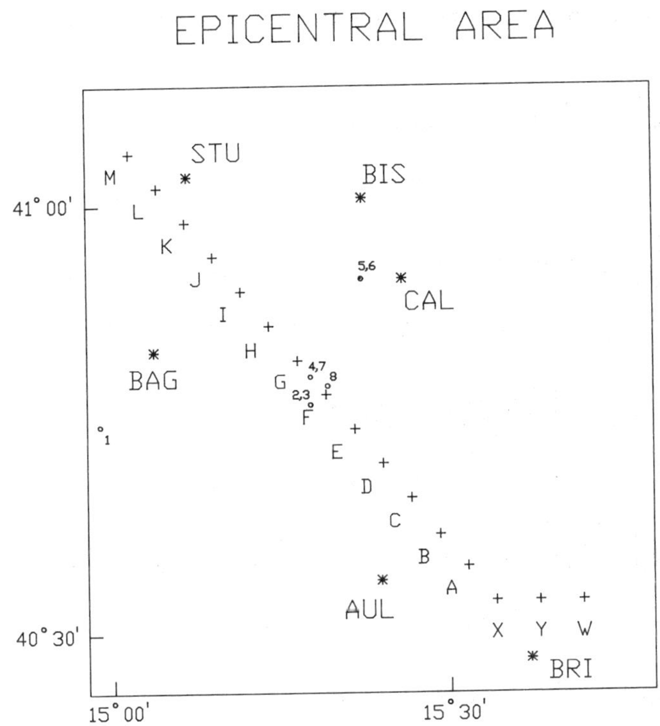

Figure 1. Map showing the epicentral area of the November 23, 1980 Irpinia earthquake, with the six closest strong-motion stations (*). The (+) symbols refer to the location of the point sources used in the investigation of the fracturing process. The epicentral locations of the main event, as proposed in the literature, are represented with numbered circles: (1) Boschi et al. (1981), (2) Bruestle and Mueller (1984), (3) Del Pezzo et al. (1983), (4) Deschamps and King (1983), (5) Kanamori and Given (1982), (6) Nakanishi and Kanamori (1984), (7) Scarpa and Slejko (1982), (8) Westaway and Jackson (1984).

2. THE DATA

The data set used in this paper is represented by the CNEN - ENEL accelerograms (Berardi et al., 1981), recorded very close to the epicentral area. Since we decided to construct complete synthetic seismograms by the Rayleigh-waves modal summation, we can obtain the vertical and the radial components of motion. From the experimental records the radial component is obtained summing the N-S and E-W ones with a proper angle, easily defined only if dealing with a single point source. In case of complex events, like the Irpinia earthquake, this

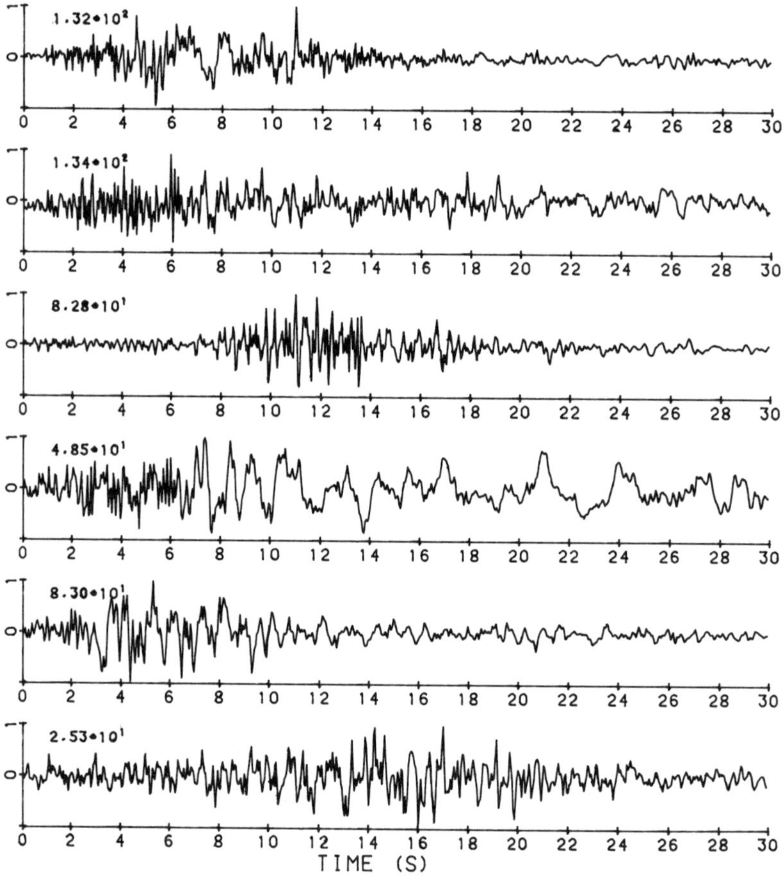

Figure 2a. Experimental strong-motion records, vertical component, obtained at the six stations shown in Fig.1 (from below: Auletta, Bagnoli Irpino, Bisaccia, Brienza, Calitri and Sturno). A Gaussian filter has been applied to these time series, with the same cut-off frequency - 10 Hz - of the synthetic signals. The peak acceleration is in units of cm/s².

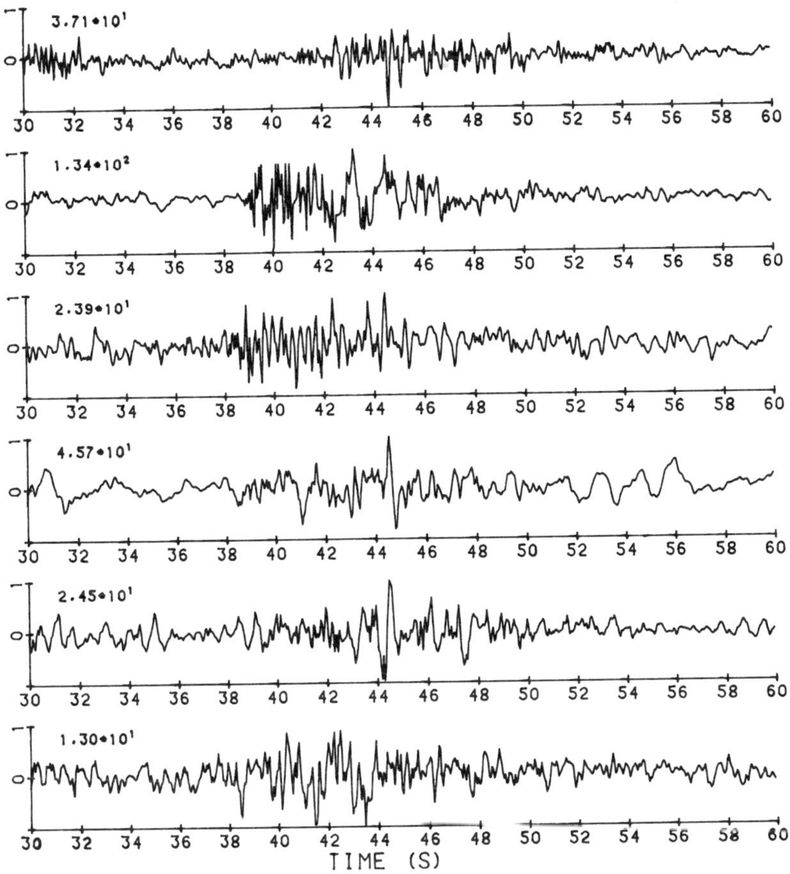

Figure 2b. Same as Fig.2a, but for the next 30 s.

angle is an unknown function of time. Ignoring this function, it is impossible to separate correctly the radial and transversal components recorded on sensors located at source-receiver distances comparable to the fault length (Fig.1). For these reasons we consider only the vertical component accelerograms (Fig.2).

Looking at the recordings, we can easily notice the presence of a strong energy burst, about 40 s after the first one, which in some cases (Calitri, Bisaccia) is comparable in amplitude with the first one. We can also detect several big-amplitude high-frequency peaks, which can be easily interpreted in terms of point sources of very short duration. Furthermore, the accelerations recorded at Auletta are particularly low if compared to the ones obtained at Brienza, a station with almost the same azimuth as Auletta but at a larger distance from the epicentral area. This fact cannot be explàined in terms of different geological setting under the stations or along the source-receiver path (see Tab.II), but, as we will see, can be succesfully used to constrain the orientation of the point sources used to model the strong motion records.

TABLE II

SITE	LITHOTYPE
Auletta	Polygenic conglomerates of deltaic and lacustrian origin, with sandy-clayey cement
Bagnoli	Limestones and dolomitic limestones of the Campania-Lucania platform
Bisaccia	Clays and slates, intercalated with calcarenites and breccias
Brienza	Polygenic conglomerates with sandy groundmass, overlying flyschoid rocks
Calitri	Sandstones and sands, overlying marls and siltitic clays
Sturno	Siltitic clays and marls, interbedded with limestones and quartzitic clays

Characteristic lithotypes at the six accelerometric sites shown in Fig. 1 (from Berardi et al., 1981).

3. WAVEFORM MODELLING

3.1. Single point source

The basic crustal model used to compute synthetic seismograms is the "A" model proposed by Del Pezzo et al. (1983), obtained through the inversion of 913 P-wave travel-times from 60 aftershocks. The model has been modified: at the top, where we have introduced a positive gradient in the P- and S-wave velocities of the first 8 km to obtain a more realistic structural model reproducing the loading effect on sediments, and at the bottom, where more layers have been added to model the lithosphere (Panza et al., 1980). The structural model IRPGRA1 obtained in this way is represented in Fig.3 (see also Suhadolc et al., 1987).

The fault strike, taken to be parallel to the axis of the Apenninic

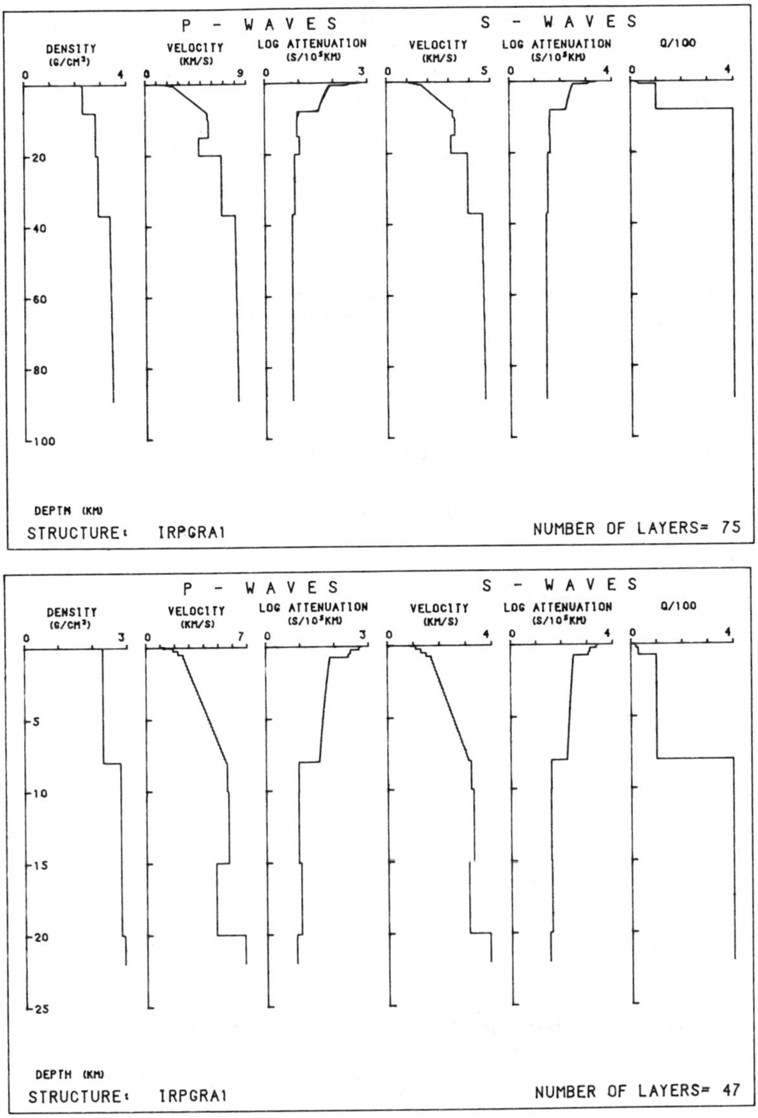

Figure 3. Structural model used to compute synthetic accelerograms. Top: the whole model. Bottom: blow-up of the uppermost layers.

chain (orientation N140°E), and the fault length, taken equal to 70 km, are deduced from the spatial distribution of aftershocks (Deschamps and King, 1984). Initially thirteen evenly spaced locations have been selected along the fault; subsequently three more had to be added near the Brienza station (Fig.1). The spacing of these locations - hereafter called epicenters - is 5.7 km. Point sources of different hypocentral depth are placed in correspondence of each epicenter. For the sake of simplicity, we have chosen to distribute most of the point sources on a vertical plane passing through epicenters A to M and to preserve the fault dip only in the point-source orientations. The distribution of point sources on a single plane with a dip of about 70° may change phases, but does not significantly modify envelopes and peak values of the synthetic signals. Initially, the double-couple orientation has been kept the same for each point source. All the Authors propose a normal fault for the main shock, but there is no agreement on the slip angle: some (e.g. Del Pezzo et al., 1983; Deschamps and King, 1983) propose a relevant left-lateral transcurrent component, others (e.g. Scarpa and Slejko, 1982; Bruestle and Mueller, 1983) a right-lateral one. Finally, Panza and Cuscito (1982), Nakanishi and Kanamori (1984) and Westaway and

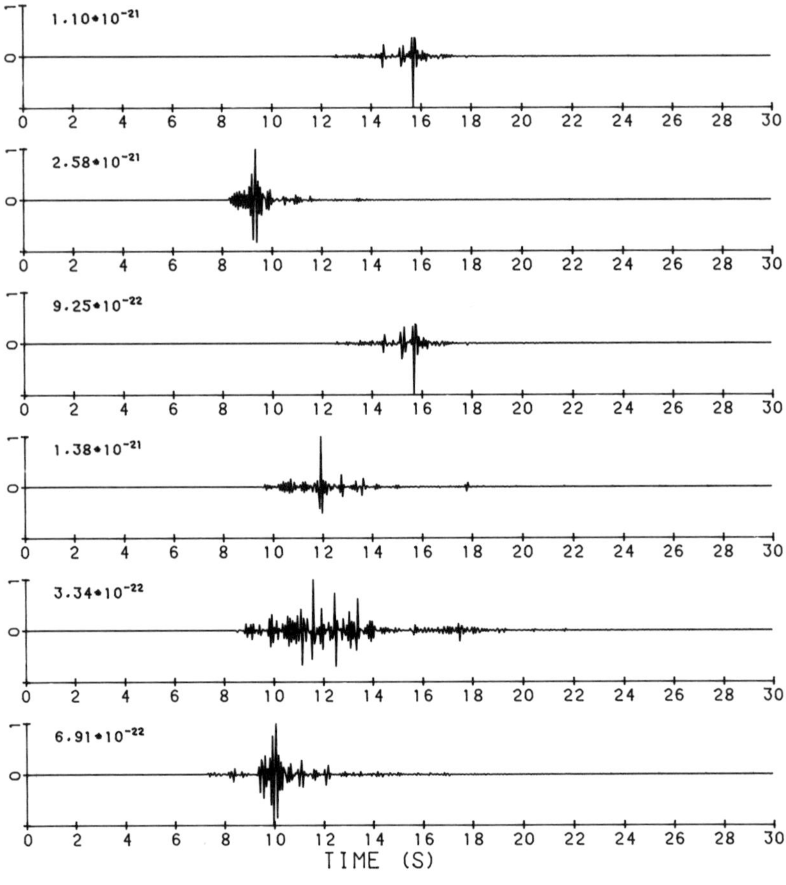

Figure 4. Synthetic accelerograms, vertical component, due to an instantaneous point source located at epicenter E (see Fig.1). Hypocentral depth: 17.5 km, strike N140°E, rake 270°, dip 70°, seismic moment 1 dyn cm. The peak acceleration is in units of cm/s^2.

Jackson (1984) propose mechanisms with almost no transcurrence.
As a first approximation, we decided to start with a rake of 270°, corresponding to a pure normal fault, and a dip of 70°. We placed an instantaneous point source at a depth of 17.5 km in correspondence of epicenter E of Fig.1. The synthetic seismograms, computed at each of the six considered stations, are shown in Fig.4, and they do not resemble the experimental accelerograms of Fig.2 at all. The striking difference in the frequency content of the Fig.2 and Fig.4 signals suggests that sources with finite duration have to be considered. Furthermore, the theoretical accelerations obtained at Auletta are comparable to the ones obtained at Brienza, while they should be three times lower. This cannot be justified by differences in the geologic conditions along the path or near the stations, but rather suggests that changes in the orientation of the point source are necessary. Finally, the modelling of the first 30 s of the experimental records cannot be achieved by means of a single point source, but requires the superposition of several of them. Even considering such a space-time sequence of point sources, the burst of energy, arriving about 40 s after the first, turns out to be well separated in time with respect to the first one. Since we want to concentrate on the first phases of the rupturing process, we will not analyse it here.

As a first step, we have tested the influence of the source duration on the resulting waveforms. Durations of 0.2 s to 1.6 s have been examined. The introduction of the source duration has been made by convolving the time series obtained from an instantaneous source, with the integral of an appropriate Bartlett's triangular window (Kanasewich, 1975). On the basis of the obtained results we can draw the following conclusions: a) a longer source duration does not produce a longer signal, and, as expected, its main effect is to remove the periods shorter than the source duration and to decrease the peak acceleration; b) for source durations greater than 0.4 s the synthetic accelerograms can only be compared with small portions of the experimental records and this is a clear indication of the source complexity. The duration of 0.6 s (see Fig.5) has been initially chosen as the most appropriate, since in the single point-source approximation the spectral contents of the corresponding theoretical signal resembles satisfactorily the observed ones.

The low accelerations at the Auletta station still remain to be explained. The azimuthal dependence of the radiation pattern on the source parameters is well known (e.g. Ben-Menahem and Harkrider, 1964; Panza et al., 1973, 1975a, 1975b), therefore we have to determine which mechanism, compatible with the results of Table I, gives a minimum of radiation in the direction of Auletta. Several point-source geometries have been tested: a minimum of radiation in the Auletta direction has been obtained only for a right-lateral transcurrent component. After some trials, the rake of 240° has been found to give very satisfactory results (Fig.6). This value is constrained only by the radiation corresponding to the 17.5 km deep point source having its epicenter in E and possessing the largest seismic moment. For the sake of simplicity, the rake has been kept constant for all other point sources. The only exception, as will be described in detail in section 3.2,

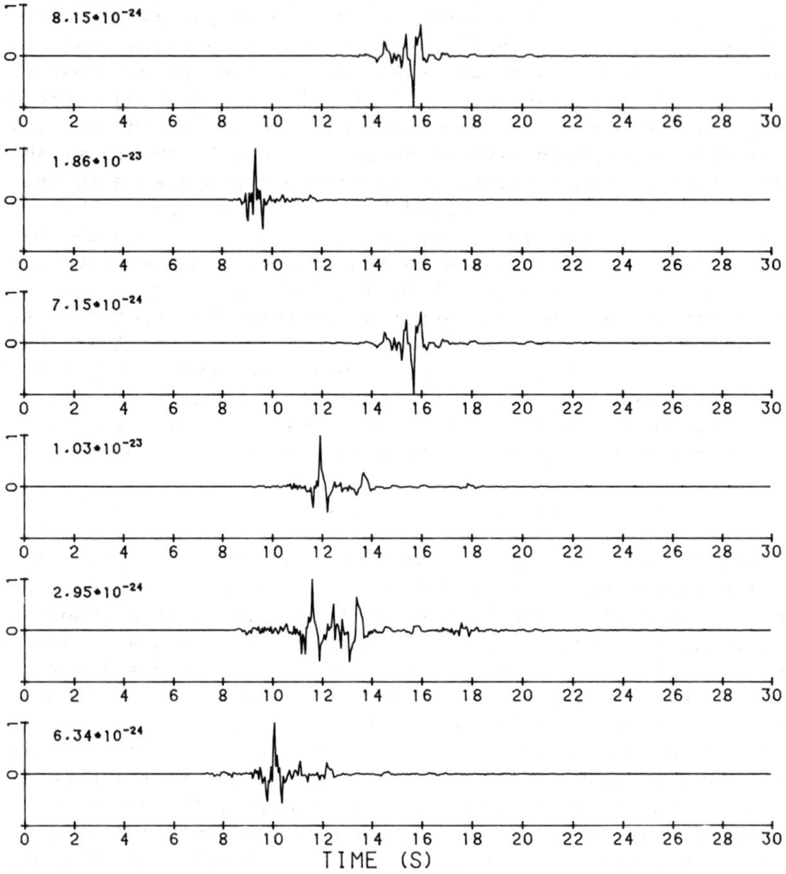

Figure 5. The result of the convolution of synthetic accelerograms of Fig.4 with the integral of a Bartlett's triangular window corresponding to a source duration of 0.6 s.

is represented by a point source, 2 km deep, located in correspondence of epicenter I.

3.2. Superposition of point sources.

In order to increase the duration of the synthetic signal, it is necessary to introduce other point sources properly distributed in correspondence of the epicenters of Fig.1. The temporal sequence of these point sources can be determined only in relative terms, due to the lack of an absolute origin time in the observed accelerograms.

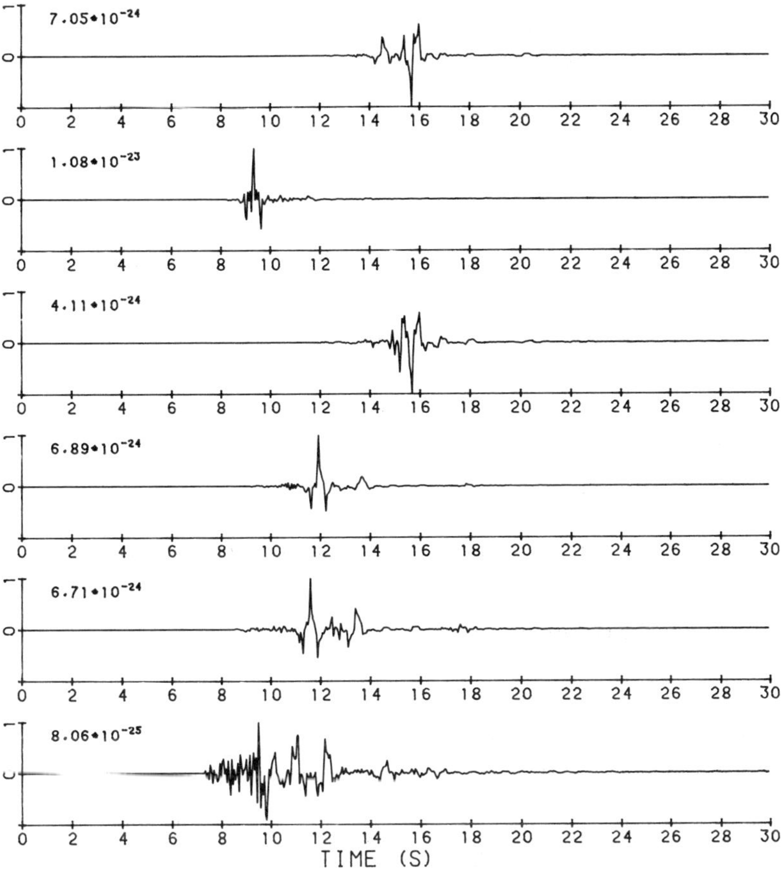

Figure 6. Synthetic accelerograms corresponding to a minimum of energy radiation in the direction of Auletta. Epicenter E, hypocentral depth 17.5 km, strike N140°E, rake 240°, dip 70°, seismic moment 1 dyn cm, source duration 0.6 s.

The definition of the parameters of the point source, with which we simulate the beginning of the rupture, is of fundamental importance, as it influences the choice of the relative weights, w_i, and time shifts, Δt_i, for all the following point sources. One has, therefore, to find in the recorded accelerograms a clear phase or a characteristic waveform that could be reproduced in the theoretical series and used as a reference.

The choice of each weight and of each time shift has been made using an interactive procedure at a graphic work-station, and only the

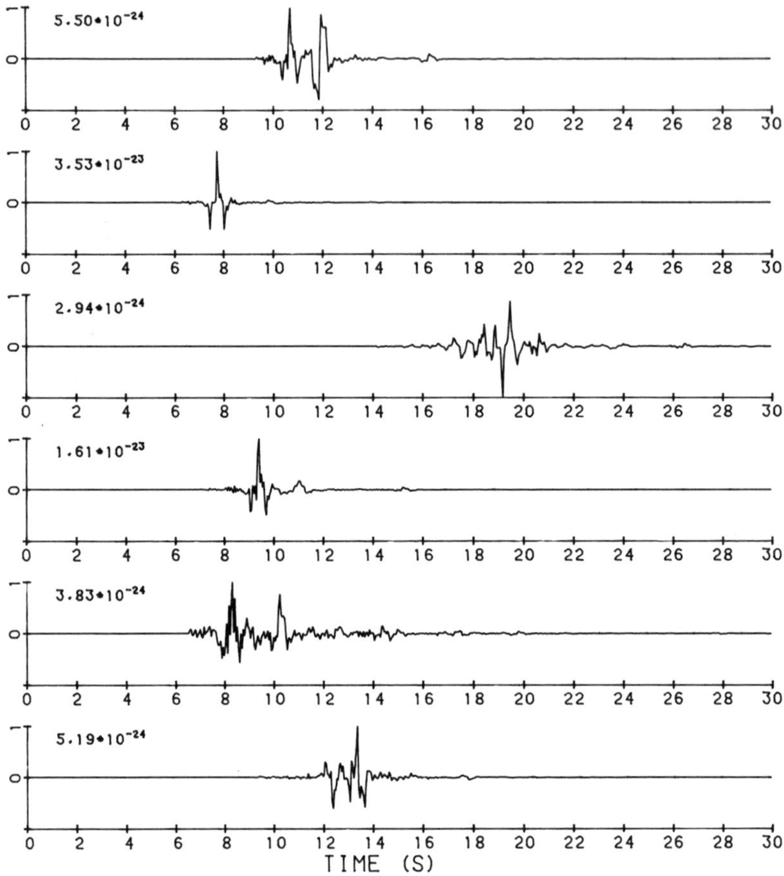

Figure 7. Synthetic accelerograms due to a point source located at epicenter G: hypocentral depth 12 km, strike N140°E, rake 240°, dip 70°, source duration 0.6 s, seismic moment 1 dyn cm, relative weight 1. This point source represents, in our modelling, the nucleation of the fracturing process.

final results are presented in this paper. In any case we have tried, as a general rule, to obtain a good agreement between the relative amplitudes of the theoretical and the experimental accelerograms and, while choosing the Δt_i, to consider rupture velocities smaller than the S-wave velocity in the pertinent layers. This last condition constrains the location of the shock activating the rupturing process in epicenter G. In this way the point sources located, along the fault, both to the NW (like epicenters H and I) and SE (like epicenters F and E) simulate

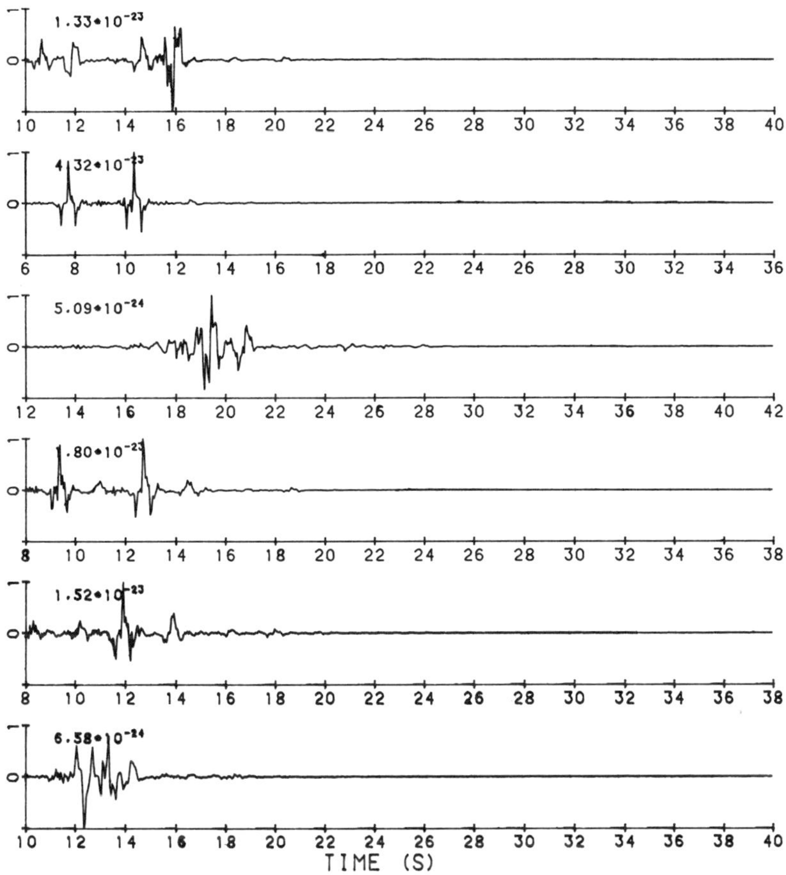

Figure 8. The contribution of a point source located at F has been added to the synthetic accelerations of Fig.7. Strike, rake, dip and source duration same as in Fig.7. Hypocentral depth 15 km, seismic moment 1 dyn cm, relative weight 2, time delay 2.15 s.

the propagation of a bilateral subsonic rupturing process.

Five point sources have been used to model the initial part of the recordings. The time series corresponding to the first of them (Fig.7) have been plotted starting from the origin time and with an unitary weight. On the contrary, all the sums (Fig.8 to 15) have been plotted from a convenient arbitrary time origin. In this way one eventually looses the information relative to the P-wave first arrivals, but this is not a limitation in strong-motion seismology, especially in our case, where very often the accelerographs seem to be triggered by S-waves.

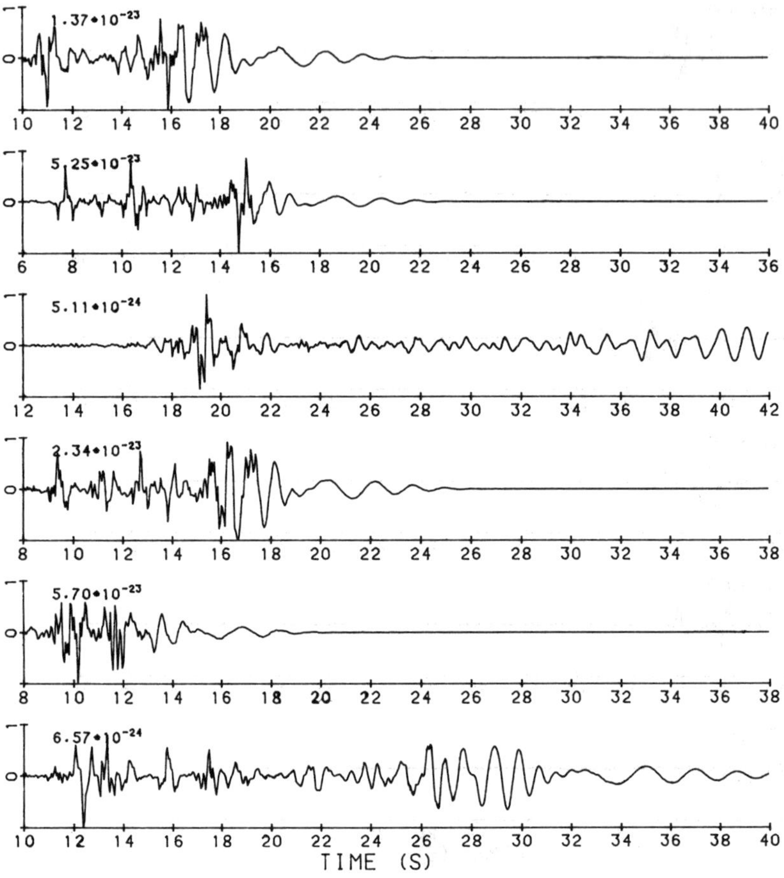

Figure 9. The contribution of a point source located at H has been added to the synthetic accelerations of Fig.8. Strike, rake, dip and source duration same as in Fig.7. Hypocentral depth 2 km, seismic moment 1 dyn cm, relative weight 0.5, time delay 3.8 s.

Without affecting the theoretical signals at the other considered stations, the remarkable peak acceleration observed at Sturno at about 11 s has been modelled with a 2 km deep point source located at epicenter K (Fig.1). To model the long-period oscillation recorded at the same station between 5 and 8 seconds after the starting of the record, it has been necessary to add a signal corresponding to a normal-fault 2 km deep point source, in epicenter I (Fig.1). This is the only point source with a focal mechanism different from all the others: the corresponding fault direction is perpendicular to the Apenninic chain,

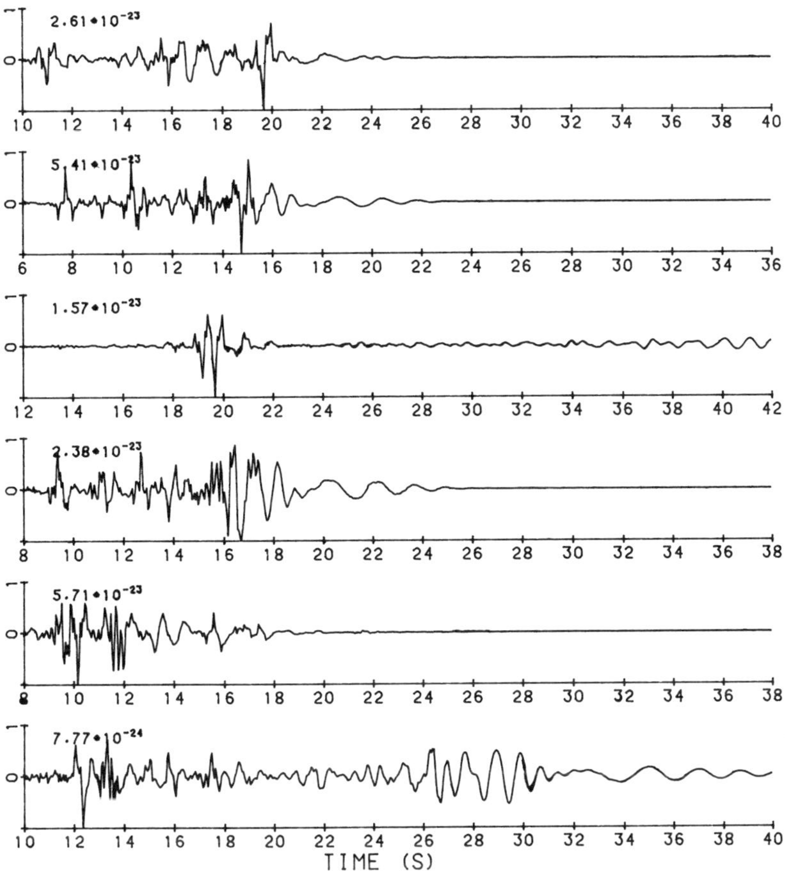

Figure 10. The contribution of a point source located at E has been added to the synthetic accelerations of Fig.9. Strike, rake, dip and source duration same as in Fig.7. Hypocentral depth 17.5 km, seismic moment 1 dyn cm, relative weight 4, time delay 4 s.

and the rake is 270°. This orientation is necessary in order to modify the waveform of the Sturno record without affecting those obtained so far at Bagnoli, Bisaccia and Calitri stations. The existence near epicenter I of a fault orthogonal (anti-Apenninic) to the main one has been suggested by Crosson et al. (1986) to explain the vertical movements observed in the region after the earthquake. The point source located at epicenter I might be interpreted as a forerunner of the rupturing process responsible of the second energy burst (Fig. 2b), recorded with the largest amplitudes just at the stations closest to

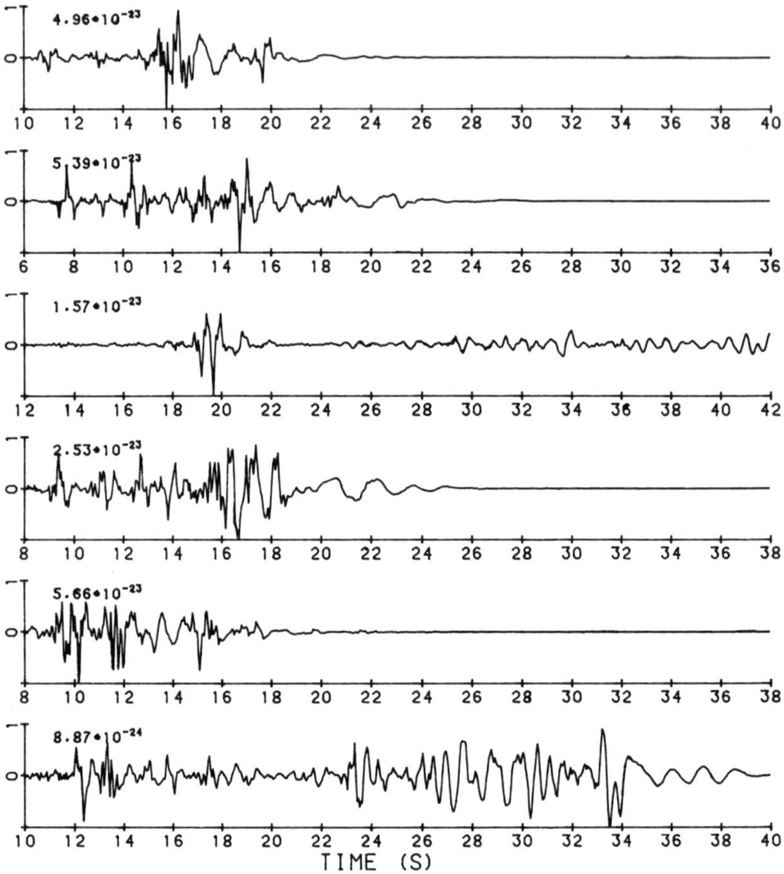

Figure 11. The contribution of a point source located at I has been added to the synthetic accelerations of Fig.10. Strike N130°W, rake 270°, dip and source duration same as in Fig.7. Hypocentral depth 2 km, seismic moment 1 dyn cm, relative weight 0.8, time delay 8.5 s.

this anti-Apenninic fault (Calitri and Bisaccia). The testing of this hypothesis, also taking into account an alternative model recently proposed by Bernard and Zollo (1987), by applying to the second energy burst the same methodology used in this paper, is presently in progress.

The energy maximum, concentrated between 8 and 14 seconds after the starting of the observed Brienza record, cannot be due to point sources located in correspondence of epicenters A to M, as can be deduced from an analysis of the possible P- and S-waves travel times. It must be, therefore, associated with one or more point sources becoming active

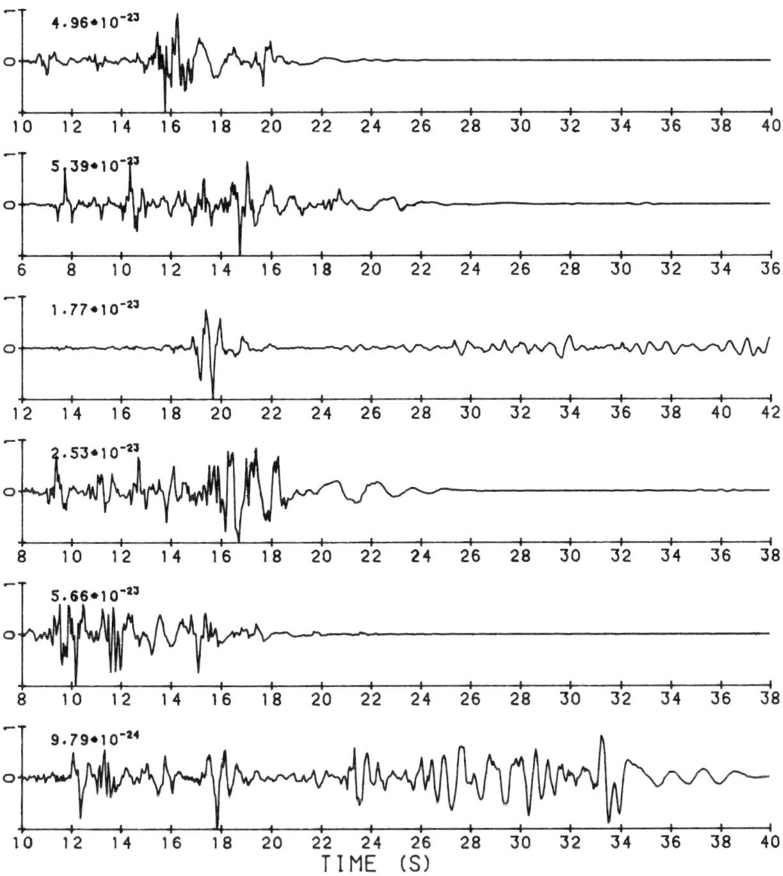

Figure 12. The contribution of a point source located at A has been added to the synthetic accelerations of Fig.11. Strike, rake, dip and source duration same as in Fig.7. Hypocentral depth 6 km, seismic moment 1 dyn cm, relative weight 0.16, time delay 12 s.

many seconds after the first one and close to the station itself. Point sources located in correspondence of epicenters X, Y, W (Fig.1), spaced in time and with different focal depths, allow to obtain a peak acceleration at Brienza three times larger than that at Auletta.

4. THE SOURCE MODEL

In order to obtain the synthetic signals shown in Fig.15 it has been

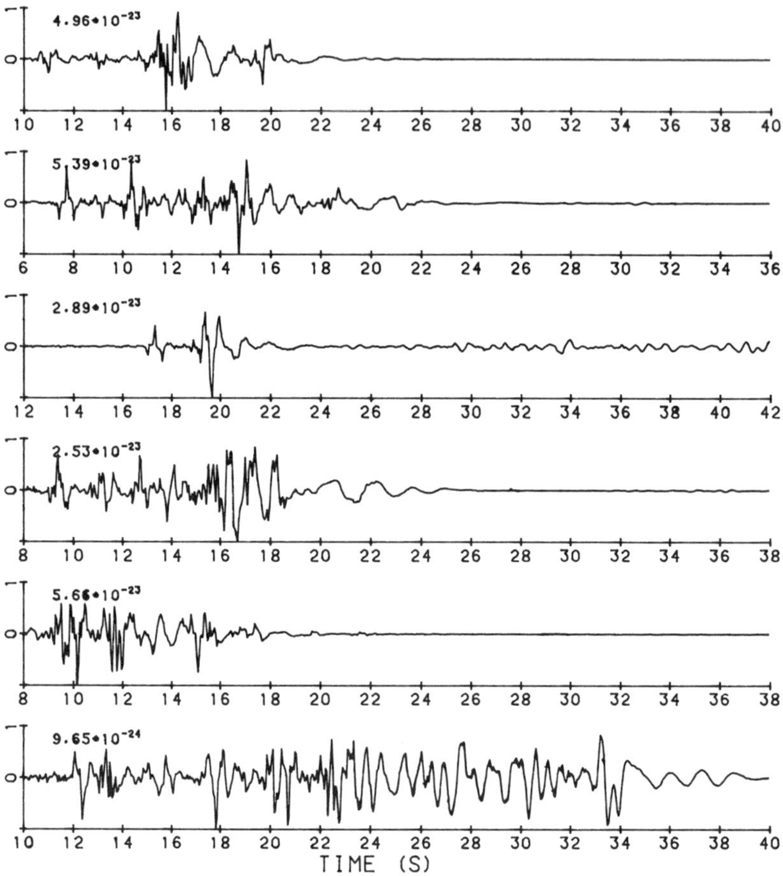

Figure 13. The contribution of a point source located at X has been added to the synthetic accelerations of Fig.12. Strike, rake, dip and source duration same as in Fig.7. Hypocentral depth 2 km, seismic moment 1 dyn cm, relative weight 0.12, time delay 14 s.

necessary to consider twelve point sources, distributed in time and space (Fig.16). Such a sequence of point sources may be considered a schematic representation of asperities (Kanamori and Stewart, 1978; Kanamori, 1981) of different size.

We could well propose that the rupturing process did propagate within the Earth crust with a subsonic velocity in a bilateral fashion, essentially along the path drawn as dotted line in Fig.16. The seismic moment associated with each asperity can be estimated from the following expression:

$$M_{0i} = \frac{a(t)}{\Sigma_j \; p_j \; x_j(t)} \; p_i \qquad (1)$$

where $x_j(t)$ and $a(t)$ are, respectively, the theoretical acceleration due to the j-th point source at time t, and the observed one; p_j is the weight of the j-th point source and the sum runs over the number of point sources used to obtain the final synthetic strong motion records.

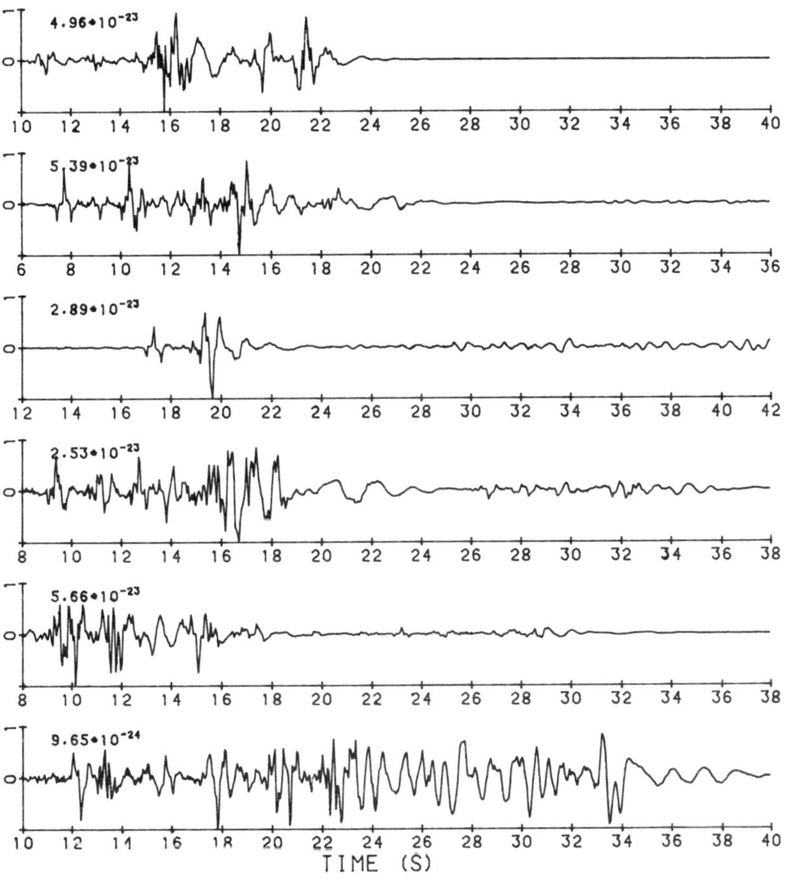

Figure 14. The contribution of a point source located at K has been added to the synthetic accelerations of Fig.13. Strike, rake, dip and source duration same as in Fig.7. Hypocentral depth 2 km, seismic moment 1 dyn cm, relative weight 0.13, time delay 18 s.

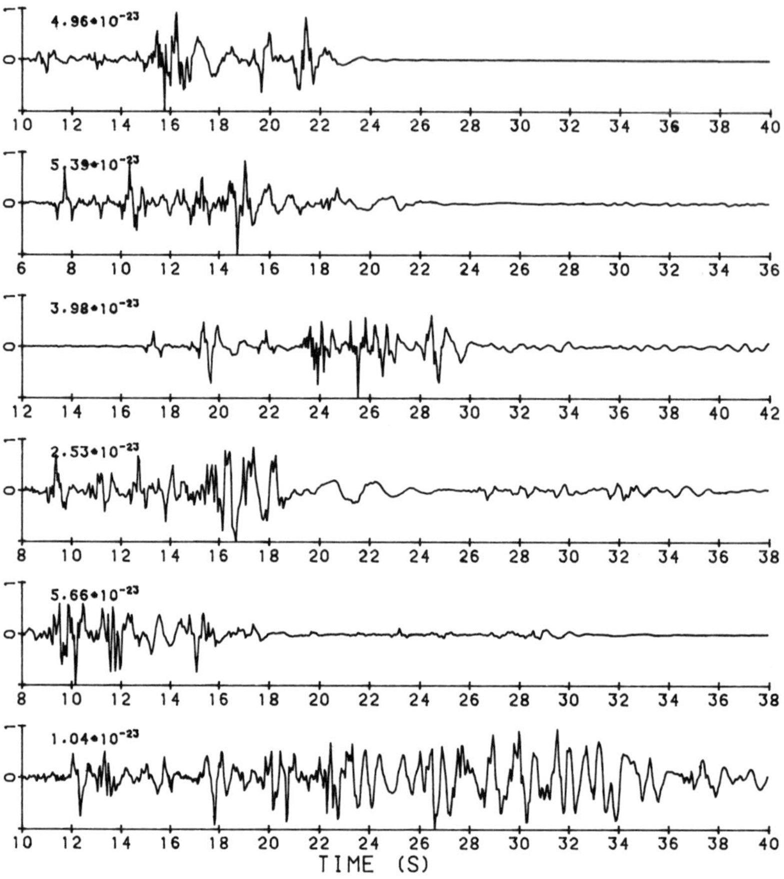

Figure 15. The final synthesis, obtained summing to the accelerations of Fig.14 the contributions of four point sources, located at epicenters Y (4 km deep) and W (5, 4 and 2 km deep). For each point source, the values of strike, rake, dip, source duration and seismic moment are the same as in Fig.7. The relative weights and time delays assigned to them are shown in Fig.16.

Equation (1) represents the scale factor between experimental and theoretical data, when the observed records can be explained in terms of a single point source. The choice of the time t in Eq.(1) is a very critical one and the most reliable results can be obtained comparing well-defined peaks, visible in both theoretical and experimental records. Applying equation (1) to the Sturno case, and considering the sharp peak visible after 11 s of recording, a seismic moment

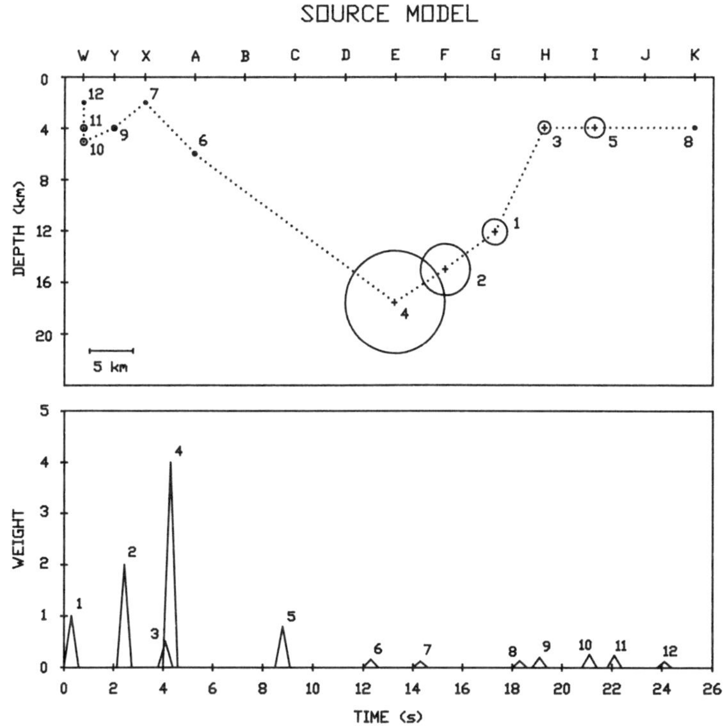

Figure 16. Vertical section (top) showing the spatial distribution of the point sources used to model the rupturing process of the November 23, 1980 Irpinia earthquake; the corresponding temporal sequence is shown at the bottom.

of 10^{25} dyn cm is obtained for the point source located at E, 17.5 km deep. The seismic moments associated with the other point sources can be easily deduced scaling this value by their relative weight.

The fracturing process, deduced from our modelling, nucleates with a point source of relatively small seismic moment, and we cannot exclude earlier nucleations with an even smaller size. As the fracture propagates downwards, the energy radiation increases with increasing rupture length, reaching its maximum at a depth of about 17 km. The point source with the largest seismic moment, simulating the maximum energy release, is located just above the high P- and S-wave velocities which can be, generally, related to the presence of metamorphic rocks (Mueller, 1977; Kern and Schenk, 1985). This maximum in the seismic moment does not coincide with the maximum accelerations obtained at the surface, and this is in agreement with the macroseismic data and with the results of Suhadolc et al. (1987).

The envisaged process seems to us quite illuminating about the physics of the earthquake source and presents some interesting analogies with the models based on crack theory (e.g. Knopoff and Chatterjee, 1982). The fracture propagates also towards the surface. In fact, the later parts of the strong motion records can be modelled by means of some shallow point sources, with varying seismic moment, simulating the stopping of the rupture within the sedimentary cover. The seismic moments of these shallow sources are always smaller than the ones of the point sources simulating the beginning of the rupturing process, but their effect is very relevant in all the records. In particular, they form the dominant part of the Brienza accelerogram.

5. CONCLUSIONS

The construction of complete synthetic accelerograms by the Rayleigh-wave modal summation has been extensively applied to the study of the Irpinia earthquake of November 23, 1980. For this event of high complexity, the gross features of six strong-motion records are reproduced by a sequence of a few point sources properly distributed in space over a time span of about 25 s. The cut-off frequency of our computations - 10 Hz - largely exceeds the limit - a few Hz - above which the experimental signals do not show appreciable coherence (e.g. Bullen and Bolt, 1985) and the present knowledge of the elastic and anelastic properties of the crust does not allow a deterministic modelling of wave propagation.

The fracturing process, deduced on the basis of the space-time distribution of point sources used in the modelling, implies a rupture nucleation around a depth of 10 km, a maximum energy release at around 17 km and a bilateral extension of the rupture towards the surface. The stopping of the rupture near the surface is simulated with a series of shallow sources, that give rise to a relevant signal in the records. A similar effect has been observed, also for a thinner sedimentary cover, by Panza and Suhadolc (1987).

In general, the relative amplitudes of the theoretical accelerograms agree quite well with those of the experimental records. Only the Bagnoli theoretical signal, in spite of the fact that its waveform agrees with the experimental one, is larger than the observed one. A possible explanation could be given supposing that the average elastic and anelastic properties along the path to Bagnoli are significantly different from those that have been found satisfactory for all the other paths. The testing of such a possibility will be the subject of a future investigation.

The waveform fitting can be obviously improved considering a higher number of point sources, but, in our opinion, this would imply a greater use of computer time, without reaching a significant improvement in the understanding of the rupturing process (e.g. Panza and Suhadolc, 1987).

This modelling can be considered satisfactory also from the practical point of view of seismic engineers. In fact, a more detailed fitting of the experimental data would not produce a significant improvement in the modelling and prediction of the form and duration of the shaking, as well as of the peak values of acceleration.

ACKNOWLEDGEMENTS

We are grateful to Mr. M. Gergolet and Mr. S. Zidarich for their help in drawing the figures.

This work has been performed with the financial support of ENEA and MPI (40% and 60%) funds.

REFERENCES

Berardi, R., A. Berenzi and F. Capozza (1981). 'Terremoto campano-lucano del 23 novembre 1980: registrazioni accelerometriche della scossa principale e loro elaborazioni', Technical report CNEN-ENEL, Rome.

Ben-Menahem, A. and D.G. Harkrider (1964). 'Radiation patterns of seismic surface waves from buried dipolar point-sources in a flat stratified Earth', J. Geophys. Res. **69**, 2605-2620.

Bernard, P. and A. Zollo (1987). 'The Irpinia (Italy) 1980 earthquake: detailed analysis of a complex normal faulting', Preprint.

Boschi, E., F. Mulargia, E. Mantovani, M. Bonafede, A.M. Dziewonski and J.H. Woodhouse (1981). 'The Irpinia earthquake of November 23, 1980', EOS **62**, 330 (abstract).

Bruestle, W. and G. Mueller (1983). 'Moment and duration of shallow earthquakes from Love-wave modelling for regional distances', Phys. Earth Planet. Inter. **32**, 312-324.

Bullen, K.E. and B.A. Bolt (1985). 'An introduction to the theory of seismology', Cambridge University Press, Cambridge.

Crosson, R., M. Martini, R. Scarpa and S.C. Key (1986). 'The southern Italy earthquake of 23 November 1980: an unusual pattern of faulting', Bull. Seism. Soc. Am. **76**, 381-384.

Del Pezzo, E., G. Iannaccone, M. Martini, and R. Scarpa (1983). 'The 23 November 1980 southern Italy earthquake', Bull. Seism. Soc. Am. **73**, 187-200.

Deschamps, A. and G.C.P. King (1983). 'The Campania-Lucania (southern Italy) earthquake of 23 November 1980', Earth and Planet. Sci. Lett. **62**, 296-304.

Deschamps, A. and G.C.P. King (1984). 'Aftershocks of the Campania-Lucania (Italy) earthquake of 23 November 1980', Bull. Seism. Soc. Am. **74**, 2483-2517.

Kanamori, H. (1981). 'The nature of seismicity patterns before major earthquakes', in "Earthquake prediction", D.W. Simpson and P.G. Richards editors, Maurice Ewing Ser.4, Am. Geophys. Union, 1-19.

Kanamori, H. and G.S. Stewart (1978). 'Seismological aspects of the Guatemala earthquake of February 4, 1976', J. Geophys. Res. **83**, 3427-3434.

Kanamori, H. and J.W. Given (1982). 'Use of long-period surface waves for rapid determination of earthquake source parameters. 2. Preliminary determination of source mechanism of large earthquakes ($M_s \geq 6.5$) in 1980', Phys. Earth Planet. Inter. **30**, 260-268.

Kanasewich E.R. (1975). 'Time sequence analysis in geophysics', University of Alberta Press, Edmonton.

Kern, H. and V. Schenk (1985). 'Elastic wave velocities in rocks from a lower crustal section in southern Calabria (Italy)', Phys. Earth Planet. Inter. **40**, 147-160.

Knopoff, L. and A.K. Chatterjee (1982). 'Unilateral extension of a two-dimensional shear crack under the influence of cohesive forces', Geophys. J. Roy. astr. Soc. **68**, 7-25.

Mueller, S. (1977). 'A new model of the continental crust', in "The Earth's Crust", J.G. Heacock editor, Am. Geophys. Union, Washington, 289-317.

Nakanishi, I. and H. Kanamori (1984). 'Source mechanism of twenty-six large shallow earthquakes (Ms≥6.5) during 1980 from P-wave first motion and long period Rayleigh wave data', Bull. Seism. Soc. Am. **74**, 805-818.

Panza, G.F. (1985). 'Synthetic seismograms: the Rayleigh waves modal summation', J. Geophys. **58**, 125-145.

Panza, G.F., F.A. Schwab and L. Knopoff (1973). 'Multimode surface waves for selected focal mechanisms - I. Dip-slip sources on a vertical fault plane', Geophys. J. Roy. astr. Soc. **34**, 265-278.

Panza, G.F., F.A. Schwab and L. Knopoff (1975a). 'Multimode surface waves for selected focal mechanisms - II. Dip-slip sources', Geophys. J. Roy. astr. Soc. **42**, 931-943.

Panza, G.F., F.A. Schwab and L. Knopoff (1975b). 'Multimode surface waves for selected focal mechanisms - III. Strike-slip sources', Geophys. J. Roy. astr. Soc. **42**, 945-955.

Panza, G.F., S. Mueller and G. Calcagnile (1980). 'The gross features of the lithosphere-asthenosphere system in Europe from seismic surface waves and body waves', Pure Appl. Geoph. **118**, 1209-1213.

Panza, G.F. and M. Cuscito (1982). 'Influence of Focal Mechanism on Shape of Isoseismals: Irpinia Earthquake of November 23, 1980', Pure Appl. Geoph. **120**, 577-582.

Panza, G.F. and P. Suhadolc (1987). 'Complete strong motion synthetics', in: "Methods of computational physics", Vol. **00**, "Seismic strong motion synthetics", B.A. Bolt editor, In press.

Scarpa, R. and Slejko (1982). 'Some analyses of seismological data. Southern Italy November 23, 1980 Earthquake', Section 2, National Research Council of Italy, Italian Geodynamics Project, Publ. n.**503**, Athens, 15-24.

Suhadolc, P., L. Cernobori, G. Pazzi and G.F. Panza (1987). 'Synthetic isoseismals: applications to Italian earthquakes', this volume.

Westaway, W.M. and J. Jackson (1984). 'Surface faulting in the southern Italian Campania-Basilicata earthquake of 23 November, 1980', Nature **312**, 436-438.

SCALING OF SOURCE AND PEAK GROUND MOTION PARAMETERS FROM MICROEARTHQUAKES AT CAMPI FLEGREI (SOUTHERN ITALY) VOLCANIC AREA

Aldo Zollo and Giuseppe De Natale
Osservatorio Vesuviano
v. Manzoni, 249
80123 Napoli
Italia

ABSTRACT. Microearthquake data recorded at Campi Flegrei, volcanic area, by a three component digital network have been used to study the scaling of the main source and ground motion parameters with seismic moment. Attenuation has been also inferred from coda and direct wave spectra and used to correct source parameter estimates. Due to the small source-to-receiver distances, attenuation doesn't affect seriously the signals radiated by the sources. An inverse method has been applied to compute spectral parameters (low frequency spectral level and corner frequency). Results show a constant stress drop pattern in the range of observed seismic moments ($0.5 \times 10^{18} < M_o < 0.4 \times 10^{21}$). A simple Brune source model (stress drop = 4 bars) with a weak influence of attenuation is able to explain the scaling of source and peak ground motion parameters.

1. INTRODUCTION

The understanding of high frequency radiation and attenuation mechanisms is the main object of strong motion analyses performed by several authors during the last years (Hanks and Mc Guire, 1981 ; Hanks, 1982 ; Boore, 1983 ; Papageorgiou and Aki, 1983 ; Anderson and Hough, 1984 ; Luco, 1985 ; Anderson, 1986 ; Boore, 1986 ; Mc Garr, 1986). Most of these studies were interested in relating the characteristics of the observed spectral shapes to source and propagation effects trying to describe them in the framework of a general theory of seismic radiation. Many evidences exist confirming the predominant contribution of the source properties to the low frequency seismic radiation. Seismic moment and source radius are in fact generally estimated by the asymptotic low frequency spectral level and the low frequency cut-off limit (corner frequency).

Most of the simple sources which are used to interpretate the low frequency radiation (e.g. Brune's model), are not able to explain the dramatic filtering of high frequencies observed on a large amount of acceleration spectra (Joyner and Boore, 1981 ; Anderson and Hough, 1984). This marked decrease of spectral energy is observed to begin from a limit frequency denoted by F_{max} (Hanks, 1982) which generally occurs around 5-15 Hz in the high frequency band. The nature of F_{max} is a controversial issue since it can be explained in terms of both source (Papageorgiou and Aki, 1983; Mc Garr, 1986) or propagation (Hanks, 1982; Boore, 1986) effects.

Recently Anderson and Hough (1984), on the basis of the observation of acceleration spectra of California earthquakes proposed to model the high frequency filtering by an

J. Bonnin et al. (eds.), Seismic Hazard in Mediterranean Regions, 129–148.

exponential function $e^{-\pi k f}$, where k is an attenuation parameter which takes into account variations of the attenuation along the travel path of seismic waves. Anderson and Hough (1984)' s results showed that : 1) k values are weakly dependent on source-receiver distance; 2) k is markedly different from 0 even at very short distances; 3) k values depend on the local geology at the recording site. These observations induced them to conclude that the high frequency seismic radiation is controlled by attenuation mechanisms mainly confined near the recording site (site effect). Furthermore, according to their observations they suggested that F_{max} is an effect of the exponential decay $e^{-\pi k f}$ which produces an apparent F_{max} arround $1/\pi k$ Hz (Boore, 1986) when observed in log-log scale. Data from microearthquakes are suitable to study such a kind of problem due to the involved high frequency range, source rupture expected to be simple and relatively short source to receiver distances.

The present paper concerns the study of the scaling of source and peak ground motion (acceleration and velocity) parameters with seismic moment for microearthquake data ($0.7 < M_L < 3.5$) collected at Campi Flegrei (Southern Italy), volcanic region (Fig. 1) during the recent ground uplift episode (1982-1984).

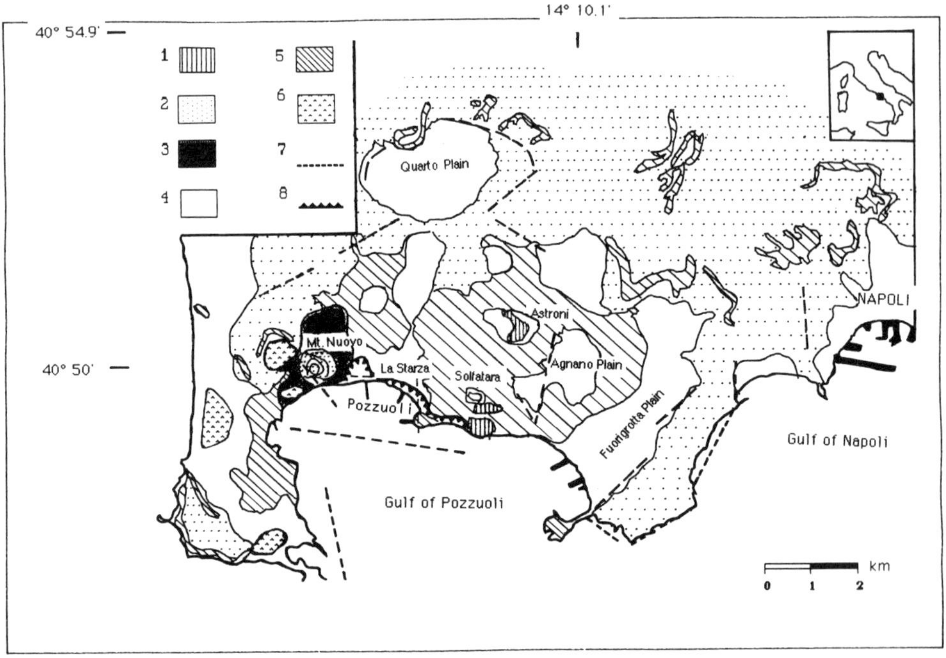

Fig. 1 Volcano-tectonic sketch of the Campi Flegrei area (from Di Vito et al., 1985). 1. Domes and lava flows ; 2. pyroclastic fall deposits (pumices with interbedded ash-levels and unwelded and sometime welded scoriae) ; 3. Pyroclastic-flow deposits (pumices, scoriae and lithic fragments in an ash-matrix, sometime with associated breccia) ; 4. Coastal and internal planes ; 5. Base surge type deposits (pumices, ashes and lithic fragments, with textural facies of sand waves, massive and planar) ; 6. Lake ; 7. Faults and fractures ; 8. limit of the 'La Starza' marine terrace.

In the period January -May 1984 a three-component digital seismic network owned by University of Wisconsin (Madison, USA) was installed in the area in a cooperative scientific project between Osservatorio Vesuviano, Naples and University of Wisconsin. Details about the main instrument characteristics can be found in Del Pezzo et al. (1986). A maximum number of 12 stations were installed during the ground uplift episode which started in the summer of 1982 reaching in two years a maximum of 1.8 m of ground vertical deformation at the center of deformation area. Fig. 2 shows the digital station locations by closed triangles while open ones are the locations of vertical component analog stations of the monitoring network operated by Osservatorio Vesuviano.

Campi Flegrei is an ancient caldera, located few km east of the city of Naples, formed about 30000 years ago after a large ignimbritic eruption. Since that time many eruptions have been occured in the area, the last one in 1538 generating a small volcano, called Mt. Nuovo (Fig. 1). The main purpose of this paper is to review the recent results obtained from spectral analyses of microearthquake data recorded at Campi Flegrei by the digital network.

Such a kind of analysis is mainly addressed to understand the seismic source properties of these microearthquakes and the influence of propagation effects (travel path attenuation and site effects) on the radiated signals.

The results have been used to determine the scaling of the main source parameters (source linear dimension and stress drop), of peak ground acceleration (PGA) and velocity (PGV) with seismic moment.

Furthermore, a stochastic simulation technique of time series (Boore, 1983) has been applied to compare the observed scaling with that one predicted by a proposed model for the spectral shape. The agreement obtained between theory and observations confirms that the adopted model is able to explain the main features of the low and high frequency seismic radiation at Campi Flegrei.

2. LOCATION AND TIME DISTRIBUTION OF THE EARTHQUAKES

More than 10000 earthquakes occured in the period 1982 - 1984 during the ground uplift episode. Fig. 2a shows 450 selected events, located by LQUAKE computer program (Crosson, 1981) using a minimum number of 12 readings (only P from analog stations and both P and S from digital stations). Other selection criteria were :

a) maximum RMS residual = 0.15 sec ;
b) maximum azimuthal gap between the recording stations = 150° ;
c) maximum hypocentral error = 0.5 km .

A half-space velocity model (Vp = 3 km/sec ; Vs = 1.7 km/sec) was chosen for the locations according to several results by travel times inversion and direct measurements made in holes drilled at the borders of the Campi Flegrei zone by AGIP (Italian Petroleoum Agency). All of these measurements confirm the lack of evidence for marked discontinuities in the first few km of the crust beneath the seismic network. The selected sample of seismicity shown in fig. 2a, b is representative of the whole sequence. The earthquakes are concentrated in a small volume (4 x 4 x 4 km3), roughly corresponding to the aera of maximum uplift. No spatial variation of seismicity has been observed during the time of ground uprising (De Natale and Zollo, 1986). As can be seen in fig. 2a,b earthquakes appear aligned along the coast line where the main tectonic feature of the area is present, namely "La Starza" marine terrace. This was formed about 5000 years ago by a large magmatic intrusion and most of the recent volcanic activity occured in the surrounding zone (Rosi et al., 1983).

Fig. 3 shows some focal mechanisms inferred from P-wave polarities and obtained using a minimum number of 15 recording stations (Gaudiosi and Iannaccone, 1984). Most

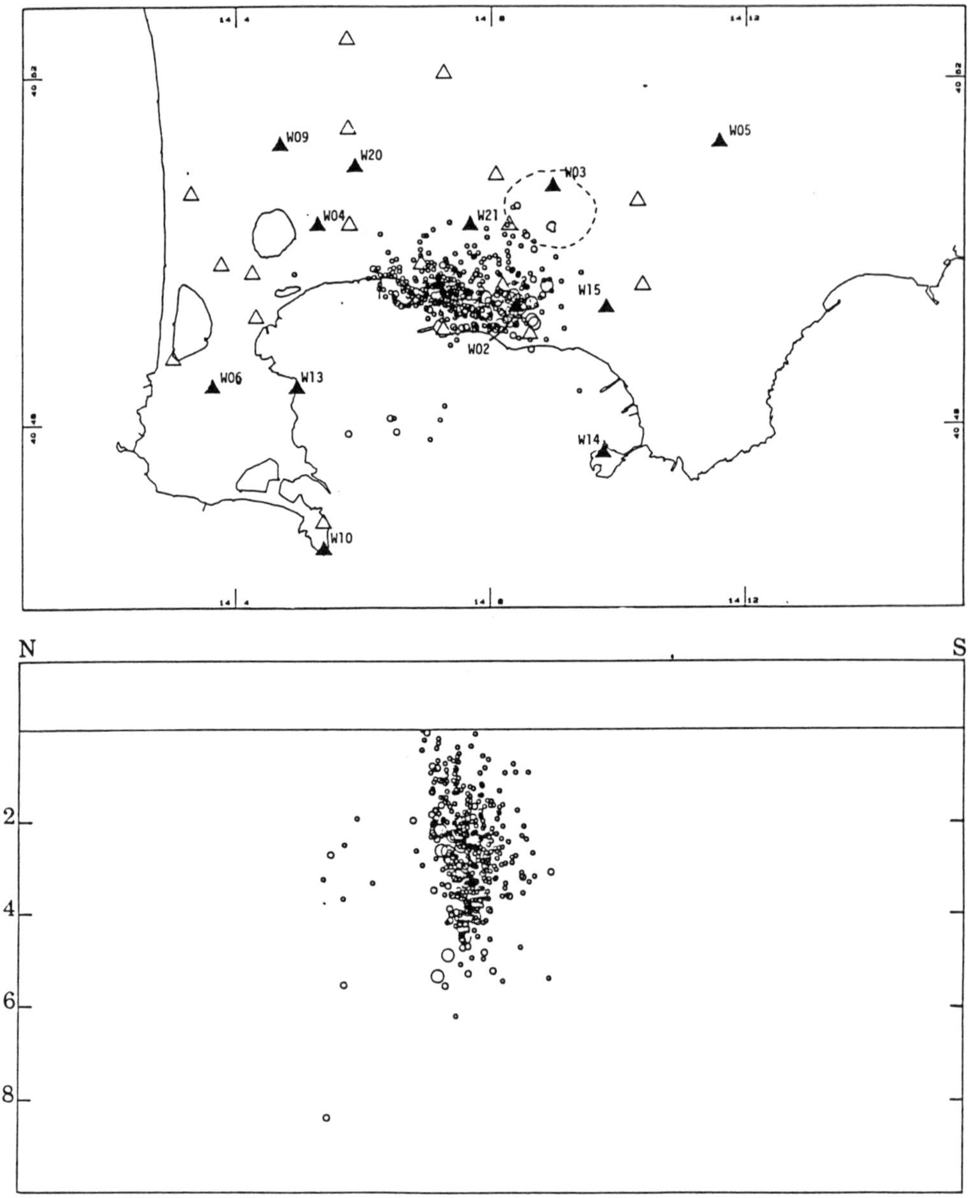

Fig. 2 a) Epicentral map of the 450 selected earthquakes (see the text). Open triangles indicate the monitoring seismic network ; the black ones are the digital three component stations; b) North-South cross section of the events shown in fig. 2a.

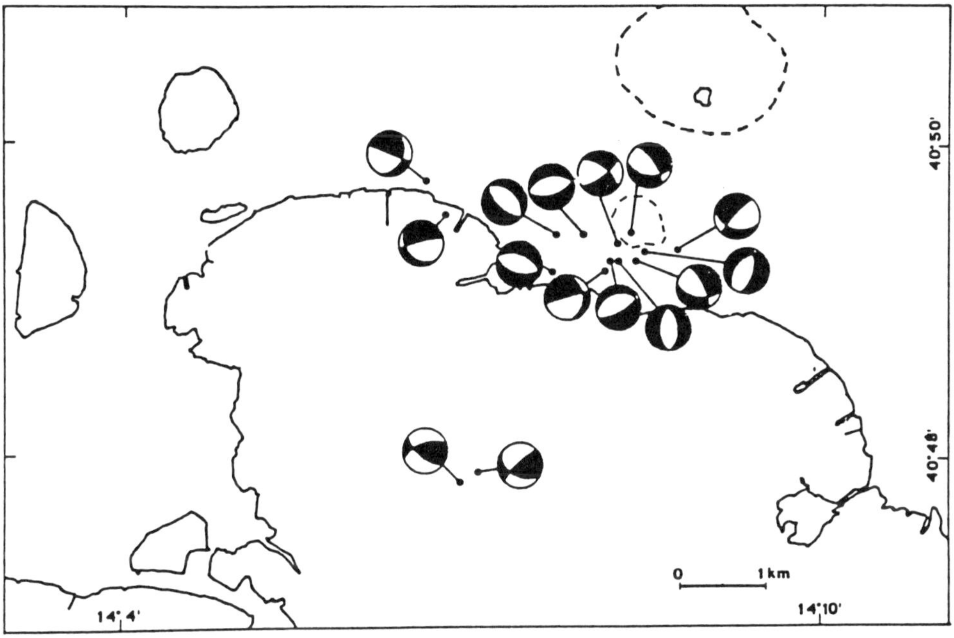

Fig. 3 Fault plane solutions from P-wave polarities for 15 selected events occured in the period 1983-1984 (lower hemisphere, Wulff projection ; dark sectors indicate compression).

of these focal mechanisms close to the Starza structure are normal faults dipping 30° - 60° with minor strike-slip components, while two events in the Gulf of Pozzuoli show compressive mechanisms.

Time distribution of events was analyzed by De Natale and Zollo (1986) by using the theory of stochastic processes. Their results show that the events occured clustered in time with no simple causal relationship among the cluster occurences and energies. These clusters were characterized by a maximum of energy in the central part of the subsequence as expected by a rupture mechanism of consecutive faulting in a highly fractured medium.

3. ATTENUATION ANALYSIS

The seismic signals emitted by earthquakes are strongly affected by attenuation and scattering along the ray path from source to receiver. An adequate knowledge of the propagation attenuation is required for the proper determination of seismic moment and other parameters of interest for the characterization of the seismic sources. Anelastic attenuation along the travel path produces amplitude and phase distortion on waveforms and spectra of seismic signals radiated by the sources. Attenuation is parameterized by the quality factor Q, that represents a dimensionless measure of the internal friction. It is given by $1/Q = - \Delta E/2\pi E$

where $\Delta E/E$ is the fraction of energy lost in a wave cycle due to imperfections in the elasticity of the material. Several methods have been applied to estimate Q in the Campi Flegrei area. Details can be found in Del Pezzo et al. (1985) and De Natale et al. (1986b) .

The first method uses the coda waves of seismograms. They are supposed to be S waves single scattered by hetereogeneities uniformly distributed in a random medium (Aki and Chouet, 1975). Single scattering assumption is appropriate when the mean distance between scatterers is greater than the scatterer-to-receiver distances. This is true for coda wave travel times $t_c < 0.8(n\sigma v)^{-1}$ where n is the number of scatterers pet unit volume, σ is the scatterer cross section and v is the wave velocity (Gao et al., 1983). Castellano et al. (1984) estimated a mean free path (the average travel distance between the scatterers) of the order of 50 km for the Campi Flegrei area. This means that we can strictly apply the single scattering assumption for travel times < 20 sec which is close to the upper limit of the observed coda wave durations (30 sec) (Del Pezzo et al., 1985).

Under this hypothesis the theory gives the following relationships for the amplitude Fourier spectrum A(f/t) evaluated at the time t along the coda of seismograms (Aki and Chouet, 1975) :

$$(1) \quad A(f/t) = A_o(f)\, t^{-1}\, e^{-\pi f\, t/Q_c}$$

$$(2) \quad \ln(A\, t) = \ln A_o - \pi\, f\, t\, /Q_c$$

Theory also requires small hypocentral distances as compared to coda wave travel paths. According to the observed range of travel distances, S wave velocity and coda durations the volume under investigation has the shape of a hemisphere with a radius of about 20 km.

When digital data are availabe equations (1) and (2) can be used to detect automatically the onset and the end of coda on seismograms and to estimate Q_c by the slope of the quantity ln (A t). At a fixed frequency this function shows a general pattern as a function of the time along the seismogram (fig. 4) : an increase is observed up to the point where the maximum (corresponding to the main S arrival) is reached ; then it decays linearly due to attenuation (eq. 2) and increases slightly, when coda amplitudes fall in the noise region. This last feature is due to the noise whose spectrum multiplied by t must increase linearly with time.
128 earthquakes, recorded at stations WO3 and W20 (fig. 2), were selected for this analysis on the basis of the signal-to-noise ratio.

Measured coda Q ranges from about 100 at 1.7 Hz to 300 at 16 Hz. A quite constant pattern of Q at high frequencies is observed for Campi Flegrei data, differently form the tectonic regions where similar estimates have been performed. This can be explained by considering the total anelastic attenuation as due to both scattering and intrinsic (dissipative) mechanisms. Following Dainty (1981) the total Q is related to intrinsic (Q_i) and scattering (Q_s) attenuation by :

$$Q^{-1} = Q_i^{-1} + Q_s^{-1}$$

where $Q_s^{-1} = g\, v/f$, g is the turbidity (i.e. the inverse of mean free path), v is the wave velocity and f is the frequency.

When scattering mechanism is predominant, a frequency dependent Q is expected. This is generally observed in tectonic environments. On the contrary a constant Q vs frequency means that intrinsic dissipation is the main attenuation mechanism.

It is notewhorthy to observe that such a behaviour is common to several volcanic areas (see fig. 5). This would indicate stronger intrinsic attenuation in volcanic zones probably due

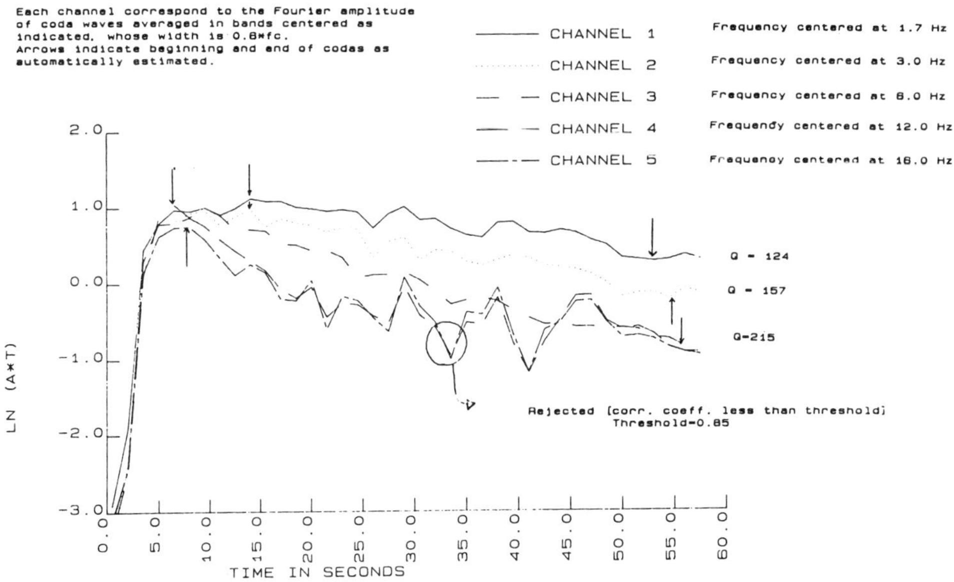

Fig. 4 Moving-window Fourier spectrum multiplied by time (formula (2)) as a function of time.

TABLE 1

f	Q_c	sd	se	Q_c	sd	se	Q_c	sd	se
1.7	97	47	16	90	47	16	105	46	19
3.0	156	67	22	144	65	21	157	73	30
6.0	299	102	24	285	69	16	293	76	18
12.0	288	68	14	317	148	29	280	61	15
16.0	299	106	15	284	55	11	270	44	9

f = frequency ; sd = standard deviation ; se = standard error.

to the presence of magmatic bodies.

Source parameter estimates are generally obtained from direct waves which are less contaminated by propagation effects. Since direct and coda waves sample different volumes of the propagation medium, they could be affected by different attenuation in the case of hetereogeneous Q distribution in the crust.

To check the stability of Q values estimated from coda waves we performed Q estimates from direct SH spectra at low and high frequency.

40 earthquakes recorded by a minumum of 4 digital stations (about 300 SH records) were considered for these analyses (Del Pezzo et al., 1985) De Natale et al., 1986c).

As a first order approximation we assumed Q = const. for direct waves at low

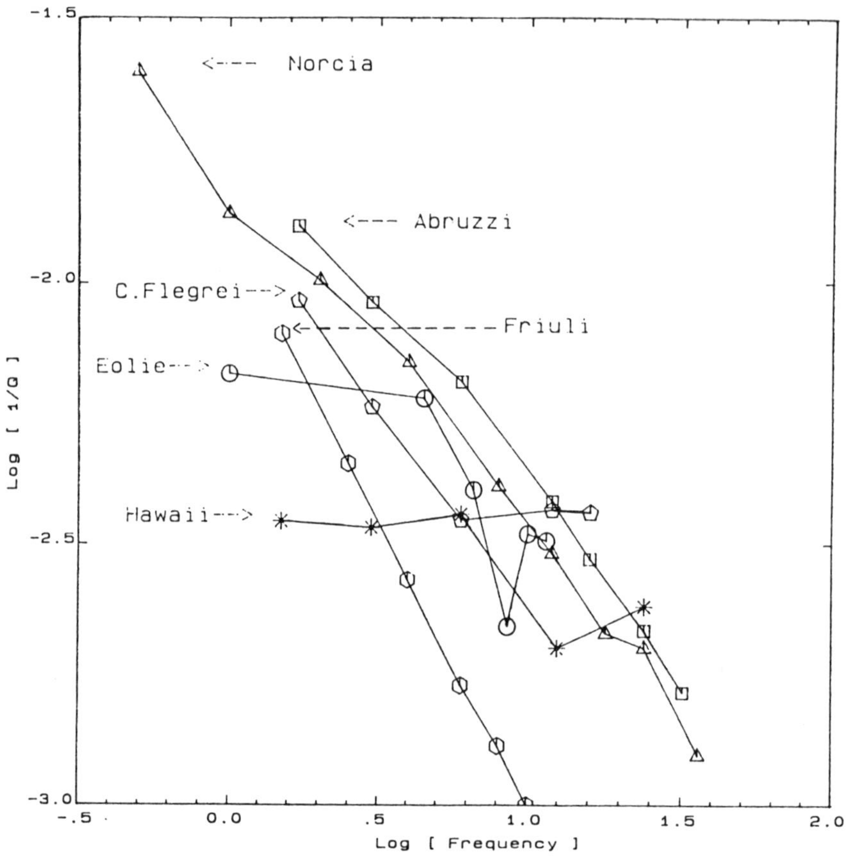

Fig. 5 Quality factor from coda wave analysis (Qc) as a function of frequency for three volcanic areas (Campi Flegrei Del Pezzo et al. (1985), Eolie (Del Pezzo et al. (1983)) and hawaii (Chouet (1976))) and three italian tectonic areas (Abruzzi (Del Pezzo and Scarcella (1986)), Norcia (Del Pezzo and Zollo (1984), Friuli (Rovelli (1982))).

frequencies. In this case the SH displacement spectrum for frequencies smaller than the corner frequency can be written:

(3) $A(f) = S_0\, e^{-\pi f t / Q}$

(4) $\ln A(f) = \ln S_0 - \pi f t / Q$

where S_0 is the low frequency spectral level related to seismic moment (Brune, 1970).

Equations (3) and (4) were applied to estimate Q from the slopes of the observed SH displacement spectra at low frequencies.

Table 2

Q_s estimates for the listed stations

Station	Q_s	Standard error	# events
W03	143	86	11
W20	76	34	25
W21	134	87	6
W04	172	124	20
W15	94	79	29
W05	158	75	6
W10	216	98	4
W14	61	11	6

Total Average $\langle Q_s \rangle = 110\ (\pm 50)$.

Tab. 2 lists the Q estimates for each station and averaged over the earthquakes and stations. We stress the fact that the average Q is, within the errors, in a good agreement with the low frequency coda Q.

The attenuation of direct waves at high frequencies has been determined applying the method proposed by Anderson and Hough (1984) by considering the acceleration SH spectra. On the ground of a large amount of observations Anderson and Hough proposed to model the high frequency decay observed on S wave acceleration spectra using the exponential function $e^{-\pi k f}$. k is an attenuation parameter which can be directly measured by the slope of acceleration spectra plotted in log-linear scale. It takes into account both travel path and site attenuation.

SH acceleration spectra were used to compute k for Campi Flegrei area (De Natale et al., 1986c). The obtained average value was k = 0.015 + - 0.020 which is (errors included) three times smaller than the ones found by Anderson and Hough (1984) from Californian data and by De Natale et al. (1986a) from Italian data. We believe this is due to the very short source-receiver distances involved (less than 8 km) together with absence of marked site effects, as can be deduced by two different observations. First, k values at very short distances (<2 km) are scattered around 0 (fig. 6) meaning that no particular high frequency site attenuation is present ; second, the average k value (0.015) is almost equivalent to the high frequency coda Q (250-300) when a mean distance of 6 km is considered. Finally, from these attenuation studies we conclude that :

1) Q from coda waves provides also a good estimate for low and high frequency direct S wave attenuation in the area due to the lack of strong filtering site effects.

2) Given the very short distances the measured Q values do not produce large distortion of the signals radiated by the source.

4. INVERSION OF SPECTRAL DATA AND SOURCE PARAMETER DETERMINATION

Objective estimation of spectral parameters represents a fundamental tool to compute source parameters (seismic moment, source radius, stress drop).

The method used in the paper is based on the linearization of the theoretical model :

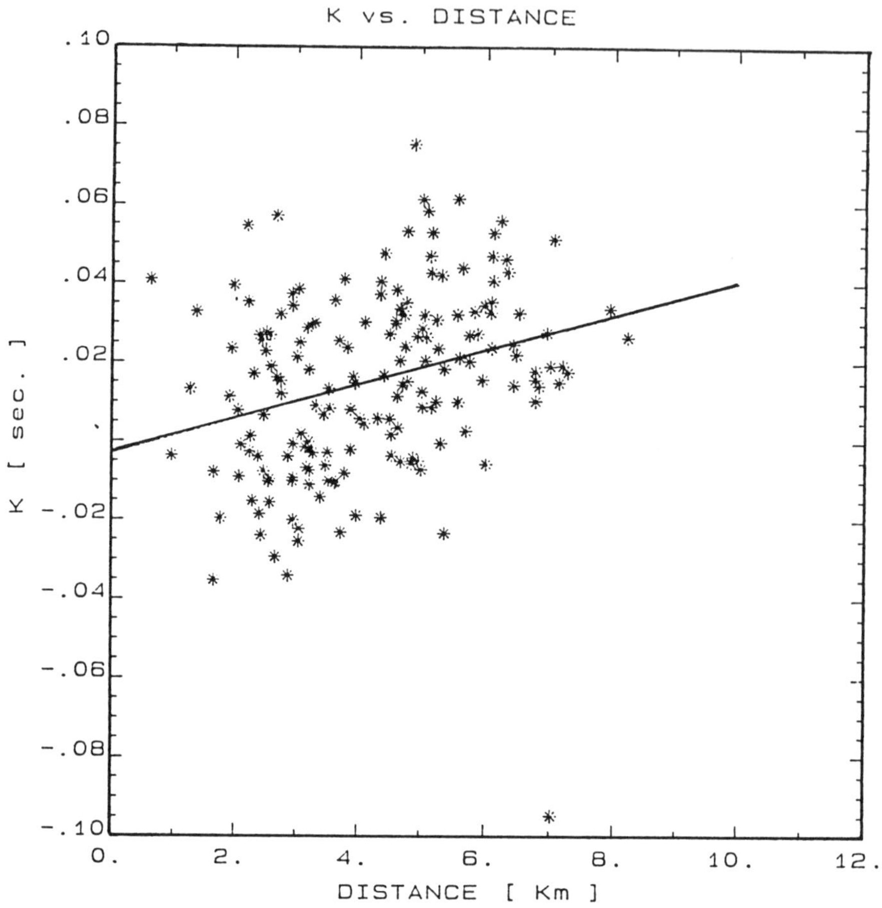

Fig. 6 Distance distribution of the spectral decay parameter k (see the text). The least squares straight line is shown by continous line.

(5) $\text{Log } S(f) = \text{Log } S_O - 0.43\ \pi ft/Q - 0.5\ \text{Log}\ [1 + (f/fc)\ 2\gamma]$

representing the spectral shape for a given set of parameters S_o, f_c, γ, Q, t, where :
S_o is the low frequency spectral level ;
f_c is the corner frequency ;
γ is the high frequency decay of the source spectrum ;
Q is the quality factor ;
t is the travel time of the considered wave.
We fixed t from the location, and Q by the results of the previous analyses. S_o, f_c and g were determined by inversion of spectral data, consisting of the following steps :

1) Assume a trial estimate $x^* = (S_o^*, f_c^*, \gamma^*)$ and compute residuals between observed and computed spectral amplitudes.

2) Expand (3) in Taylor's series about the trial estimate x* and form the equation (4) $A\ \delta x = \delta y$ where : δx are the first order corrections to the starting model ; δy are the residuals ; A is the matrix of partial derivatives of the spectral amplitudes with respect to changes in the model parameters, computed in x_o.

3) Invert (4) by Singular Value Decomposition to obtain the first order corrections :
$\delta x = V_k\ S_k^{-1}\ U_k^T\ \delta y$;
$x = x^* + \delta x$ is the new estimate.

4) Compute errors on the spectral parameters by the covariance matrix (5) $C_X = V_k S_k^{-2} V_k^T$.
Steps 1) through 3) are repeated iteratively up to convergence to a stable solution of minimum residual is reached. Synthetic tests with noised data and applications to real data showed this method to be very stable and efficient.

The objectiveness of this method, and the possibility of taking properly into account the attenuation, makes it better than the generally used fitting on spectra by eye ; furthermore, for routinely applications, it permits to do fast automatic process of large amounts of data. Examples of P and SH displacement spectra fitted by the model (5) are shown in fig. 7.

Using above procedure we computedthe seismic moment (M_o), the fault radius (r) using Madariaga (1976) and Brune (1970) source models and stress drop for a selected sample of microearthquakes ($0.7 < M_l < 3.5$), recorded at least by 4 stations (about 600 P radial ans SH displacement spectra) (del Pezzo et al., 1986).

Source parameters have been computed by weighted logarithmic averages of the quantities :

$$M_o = \frac{4\ v_{p,s}^3\ RSo}{F\ R^{P}_{\vartheta\varphi},SH}$$

$$(6)\quad \Gamma_B = 0.37\ v_s / f_c^s \qquad \text{(Brune's model)}$$

$$\Gamma_M = 0.32\ v_s / f_c^P = 0.21\ v_s / f_c^s \qquad \text{(Madariaga's model)}$$

where :
ρ = 2.2 g/cm
v_p = 1.73 vs = 3 km/sec
R = source-receiver distance
F = free surface operator = 2
$R^P_{\upsilon\varphi}$ = average P radiation pattern for normal faulting = 0.4
(Boore and Boathwright, 1984)
$R^{SH}_{\upsilon\varphi}$ = average SH radiation pattern = 0.25 (Boore and boathwright, 1984). Static stress drop was evaluated from source radius according to the formula :

$$\Delta\sigma = 0.44\ M_o / r^3.$$

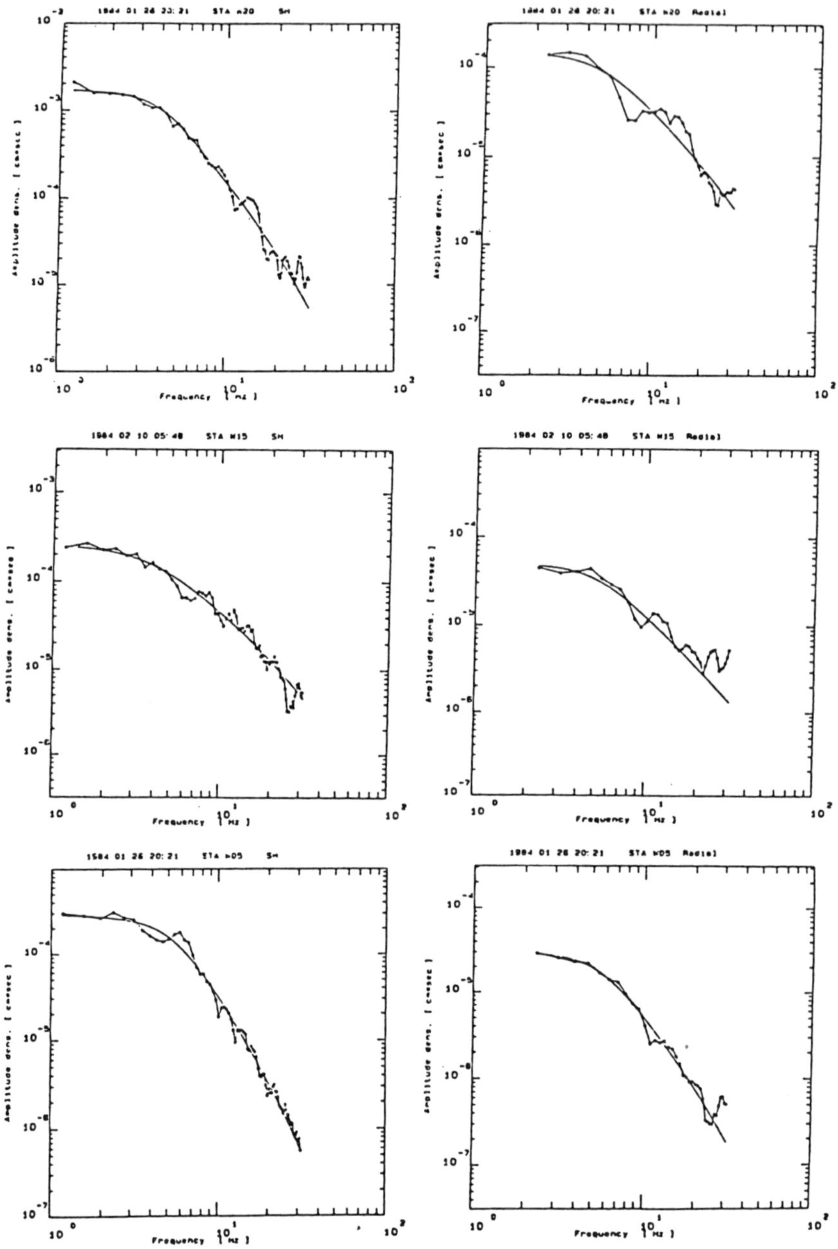

Fig. 7 Examples of P-radial and SH smoothed displacement spectra. Best fit models from the inversion procedure are also shown.

Individual weights for S_o and f_c were defined as $1/\sigma$, where σ is the standard deviation as inferred by the covariance matrix (5). Results are summarized in fig. 8, 9 and table (3). Madariaga's source radii were computed by both P radial and SH data, while Brune's radii by SH data only.

The most interesting result of this analysis is the constancy of the stress drop within the investigated range of seismic moments ($0.5 \times 10^{18} < M_o < 0.5 \times 10^{21}$). In fig. 9 the Brune stress drop is plotted as a function of seismic moment ; we note the absence of any correlation, with an almost constant stress drop of about 4-5 bars. In this range of seismic moments a sharp decrease of stress drop is generally observed in many other areas (Chouet et al., 1978 ; Archuleta et al., 1982 ; Brune et al., 1985).

This behaviour has been interpreted by some authors in terms of source effects, as due to the existence of a characteristic scale of fracturing (barriers) (Aki, 1986). An alternative interpretation explains the decrease of stress drop below a certain threshold of seismic moment as due to the effect of strong anelastic attenuation in the first layers of the crust, that would cut off high frequencies from the observed spectra, biasing their estimates for small earthquakes (Anderson, 1986). The constant pattern of stress drop observed at Campi Flegrei, together with the results obtained from the attenuation studies, indicate the absence, in this area, of particular attenuation effects in addition to the whole path attenuation with values inferred by previous studies. Furthermore, a detailed visual inspection of spectral data showed no evidence for an high frequency cut-off limit (F_{max}) in the analyzed frequency range (1-20 Hz).

5. STOCHASTIC SIMULATION OF PEAK GROUND MOTION

The scalling of peak ground motion values with seismic moment is strictly related to the properties of the radiated source spectra, and the propagation effects. So, it can provide an independent information about source mechanisms and attenuation effects ; furthermore, peak values are quantities directly related to the ground motion and then less affected by errors due to data processing techniques (as for example seismic moment and corner frequency by spectral measurements).

Theoretical values of peak ground motions can be computed, once a model for the amplitude spectrum is assumed, by the random vibrations theory (Hanks and Mc Guire, 1981). Based on the previous results about source parameters and attenuation properties, we assumed, for SH waves, a spectral model of the form :

$$(6) \quad A(\omega) = C\, M_o\, S(\omega,\omega_c)\, e^{-k\,\omega/2} / R$$

where C is a constant given by

$$(7) \qquad C = \langle R^{SH}_{\vartheta\varphi} \rangle\, F / (4\,\pi\,\rho\, v_s^3)$$

The source spectrum is :

$$(8) \quad S(\omega,\omega_c) = \omega^n / (1 + (\omega/\omega_c)^2)$$

where n = 1,2 for velocity and acceleration, respectively, and $\omega_c = 2\,\pi\, f_c$. This last one is related to the stress parameter through the expression :

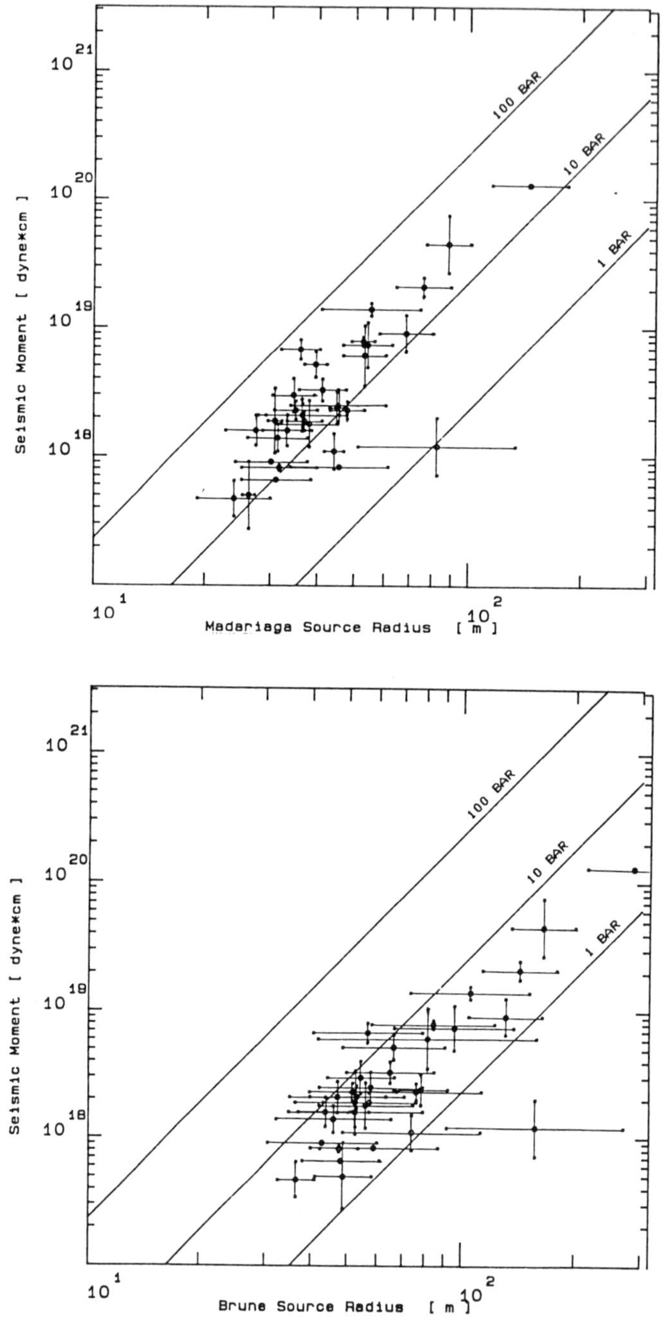

Fig. 8 Seismic moment vs source radius using Madariaga (a) and Brune source (b) models. Bars indicate the errors (see the text).

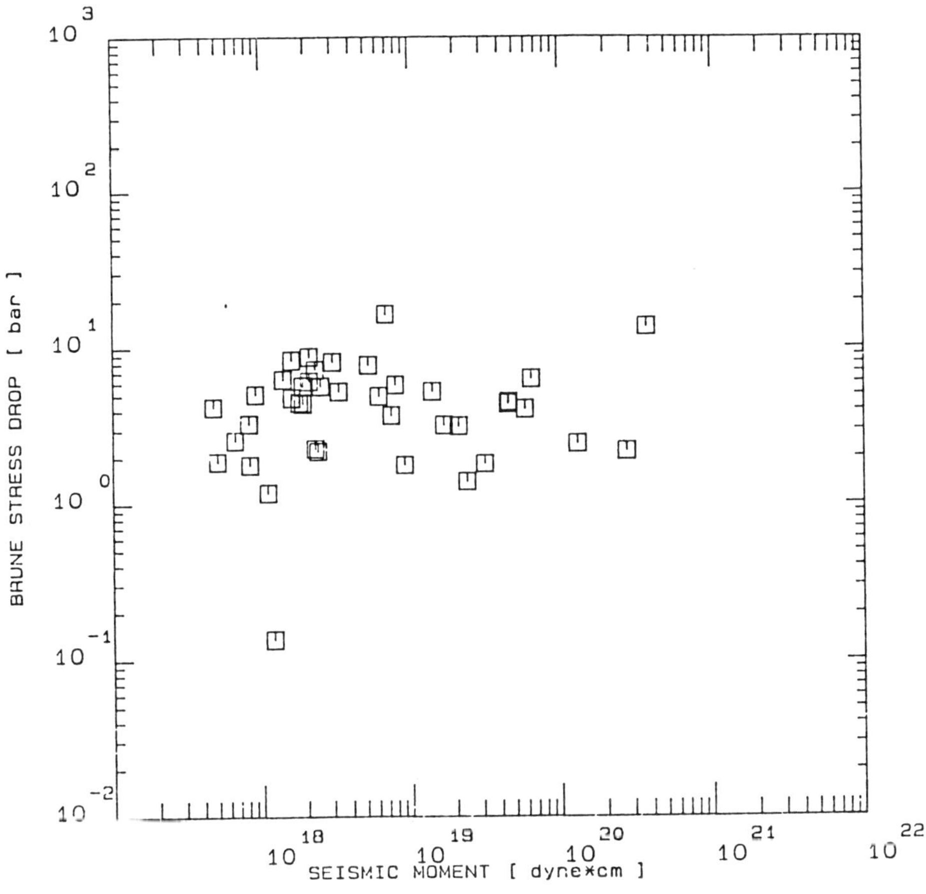

Fig. 9 Brune stress drop as a function of seismic moment.

(9) $f_c = 4.9\ 10^6\ V_s\ (\Delta\sigma/M_o)^{1/3}$

where f_c is in Hz, V_s in km/sec, $\Delta\sigma$ in bars and M_o in dyne-cm.

To compute theoretical peak velocities and accelerations based on the model (7) we used the stochastic simulation procedure as suggested by Boore (1983) in fig. (10). Observed SH seismograms from the 40 earthquakes analyzed in the previous section, were instrument corrected and band pass filtered between 3 and 20 Hz by a two poles Butterworth filter ; the same filtering was applied to theoretical spectra. Observed peak values for each individual earthquake were averaged over all the stations and normalized to a 1 Km distance (De Natale et al., 1986c).

The comparison between simulated and observed peak velocities and accelerations is shown in fig. 10. We can see that the observed peak values show a good coherence and a very good agreement with the theoretical predictions. So, we conclude that the observed seismic radiation, both in velocity and in acceleration is very well predicted by a simple

Brune source model with an exponential decay $e^{-\pi kf}$ (k = 0.015), with a stress drop $\Delta\sigma = 4$ bars.

6. DISCUSSION AND CONCLUSION

One of the main results pointed out by both attenuation and source parameters studies at Campi Flegrei is the weak influence of the attenuation on the radiated seismic signals. We bellieve this is mainly due to the small source-receiver distances (1 - 8 Km) involved,

TABLE 3

N	TIME	M_L	$\langle M_o \rangle$ (dyne*cm)	s.d.	f_c (Hz)	s.d	$\Delta\sigma_B$ (bars)	s.d.
01	01150331	1.1	$.83 \times 10^{18}$	$.10 \times 10^{17}$	11.1	4.3	1.8	2.1
02	01161900	1.6	$.85 \times 10^{19}$	$.71 \times 10^{18}$	11.3	3.5	5.7	5.6
03	01162351	1.0	$.66 \times 10^{18}$	$.14 \times 10^{17}$	13.5	3.2	2.6	1.8
04	01170258	1.6	$.11 \times 10^{19}$	$.35 \times 10^{18}$	8.8	3.7	1.2	1.5
05	01170448	1.3	$.12 \times 10^{19}$	$.62 \times 10^{18}$	7.4	3.0	0.1	0.2
06	01180206	1.3	$.91 \times 10^{18}$	$.27 \times 10^{17}$	15.2	5.1	5.1	5.1
07	01190453	1.9	$.79 \times 10^{19}$	$.61 \times 10^{18}$	7.7	2.9	5.8	6.5
08	01211339	-	$.82 \times 10^{18}$	$.40 \times 10^{17}$	13.7	1.6	3.3	1.2
09	01221345	1.5	$.24 \times 10^{19}$	$.68 \times 10^{18}$	8.3	1.3	2.2	1.2
10	01230543	3.4	$.13 \times 10^{21}$	$.19 \times 10^{19}$	4.2	1.0	2.4	2.1
11	01240248	2.4	$.45 \times 10^{20}$	$.18 \times 10^{20}$	5.2	0.7	4.5	4.6
12	01240424	1.6	$.19 \times 10^{19}$	$.33 \times 10^{18}$	11.4	3.2	4.5	3.9
13	01242250	0.7	$.23 \times 10^{19}$	$.37 \times 10^{18}$	12.6	3.4	7.3	5.9
14	01262021	2.6	$.45 \times 10^{20}$	$.23 \times 10^{20}$	4.6	0.8	4.4	3.4
15	01272343	1.5	$.21 \times 10^{19}$	$.44 \times 10^{18}$	12.3	3.5	6.2	5.5
16	01272353	1.7	$.23 \times 10^{19}$	$.36 \times 10^{18}$	8.5	3.4	2.3	2.7
17	01282126	1.9	$.33 \times 10^{19}$	$.61 \times 10^{18}$	10.0	2.7	5.3	4.3
18	01282351	2.5	$.27 \times 10^{21}$	$.74 \times 10^{20}$	2.9*	0.7	2.2	1.7
19	01290422	1.6	$.16 \times 10^{19}$	$.43 \times 10^{18}$	14.9	2.5	8.4	4.8
20	01300022	1.5	$.47 \times 10^{18}$	$.15 \times 10^{18}$	17.8	2.0	4.2	2.0
21	01300040	1.6	$.14 \times 10^{19}$	$.34 \times 10^{18}$	14.2	5.0	6.3	6.9
22	01300044	1.8	$.21 \times 10^{19}$	$.60 \times 10^{18}$	13.8	4.1	8.8	8.2
23	01300601	2.3	$.14 \times 10^{20}$	$.16 \times 10^{19}$	9.2	2.6	5.2	5.7
24	01310036	2.1	$.21 \times 10^{20}$	$.35 \times 10^{19}$	5.5	1.7	3.2	2.2
25	01310220	1.7	$.16 \times 10^{19}$	$.44 \times 10^{18}$	12.4	5.1	4.8	6.1
26	02100548	1.9	$.91 \times 10^{19}$	$.30 \times 10^{19}$	6.4	2.6	1.8	1.3
27	02101116	1.6	$.30 \times 10^{19}$	$.89 \times 10^{18}$	12.0	2.4	8.2	5.6
28	02110400	0.9	$.50 \times 10^{18}$	$.30 \times 10^{18}$	13.3	2.3	1.9	1.5
29	02112222	2.0	$.24 \times 10^{20}$	$.12 \times 10^{20}$	6.6	1.2	1.4	1.6
30	02150159	1.9	$.68 \times 10^{19}$	$.12 \times 10^{19}$	11.5	3.8	16.4	16.6
31	02150523	2.1	$.64 \times 10^{20}$	$.33 \times 10^{20}$	4.7	0.6	6.3	6.3
32	02162205	1.9	$.52 \times 10^{19}$	$.12 \times 10^{19}$	9.8	3.0	7.8	7.5
33	02172041	1.9	$.74 \times 10^{19}$	$.30 \times 10^{19}$	6.8	2.5	3.7	4.3
34	02180402	1.9	$.61 \times 10^{19}$	$.33 \times 10^{19}$	8.0	5.3	4.9	10.2
35	02190034	1.8	$.19 \times 10^{19}$	$.11 \times 10^{19}$	12.4	4.6	5.7	7.2
36	02191511	2.4	$.17 \times 10^{20}$	$.39 \times 10^{19}$	5.5	1.2	3.2	1.6
37	02200111	1.7	$.18 \times 10^{19}$	$.75 \times 10^{18}$	11.6	3.3	4.5	4.3
38	02202329	2.8	$.58 \times 10^{20}$	$.22 \times 10^{20}$	3.8	0.4	4.1	2.3
39	02220130	3.7	$.37 \times 10^{21}$	$.23 \times 10^{20}$	2.8	0.2	13.6	3.1
40	02221720	2.2	$.31 \times 10^{20}$	$.11 \times 10^{20}$	4.6	0.9	1.8	1.6

* f_c estimated only from P-radial components

s.d. = standard deviation

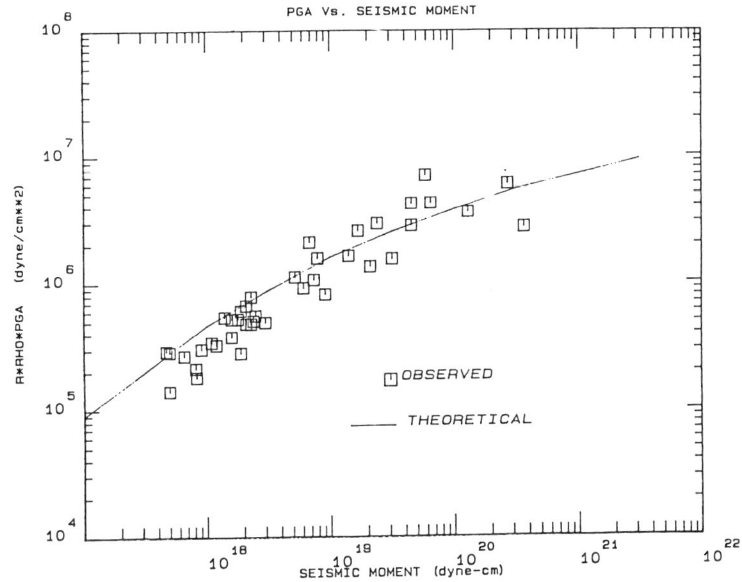

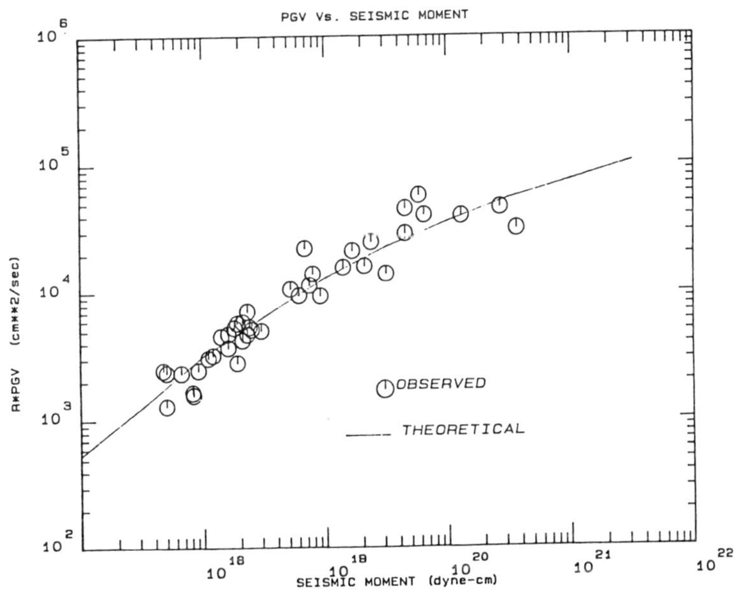

Fig. 10 Scalling of peak ground motion acceleration (a) and velocity (b) with seismic moment. Continous lines show the pattern predicted by the adopted model (see the text).

and to the absence of strong attenuation effects near the recording site. This last feature is evidenced by the small average value of the k decay parameter (Anderson and Hough, 1984) and its distribution around 0 for the smallest source-receiver distances ($R \rightarrow 0$).

Source parameter estimates have been carried out by an inverse method, based on the least squares fitting of displacement spectra. This method presents several advantages with respect to the traditional fitting by eye of amplitude spectra : first, it gives an objective estimation of spectral parameters ; second, it takes into account in a natural way the attenuation that, if not considered, can seriously bias the corner frequency estimates ; furthermore, the inverse method provides error estimates on the spectral parameters which can be used as weighting factors when source parameters are obtained by averaging over many stations.

Using this method, seismic moments, source radii and stress drops for 40 earthquakes were determined. The results show a quite constant Brune's stress drop, around 4 bars, a value considerably lower than the values inferred for large earthquakes (around 100 bars) (Hanks and Mc Guire, 1981). The constant stress drop in this range of seismic moments ($0.5 \times 10^{18} < M_0 < 0.5 \times 10^{21}$) is quite uncommon. This means that the cube of source dimension scales linearly with seismic moment as expected for self-similar fracturing processes (Aki, 1969). This evidence contradicts the most generally observed high frequency cut-off for microearthquakes ($M_o < 10^{22}$ dyne cm) which produces a sharp decrease of stress drop as a function of seismic moment (Chouet et al., 1978 ; Archuleta et al., 1982 ; Brune et al., 1985) This is attributed to both source (Papageorgiou and Aki, 1983 ; Mc Garr, 1986) or local site attenuation (Anderson and Hough, 1984 ; Boore, 1986) effects.

For Campi Flegrei data, the constant stress drop pattern over about 2.5 orders of seismic moment, the lack of a characteristic F_{max} up to 20-22 Hz and the weak high frequency decay due to attenuation indicate that neither source complexities nor local site or travel path attenuation seriously affect the seismic radiation. Scalling of strong motion parameters (PGA and PGV) vs seismic moment shows a quite similar pattern with a pronounced decay of peak values for small magnitude events.

On the ground of previous analyses a simple spectral model has been used to simulate the observed scaling of PGA and PGV. This model predicts a simple Brune source radiation at a constant stress drop of 4 bars and a low high frequency exponential attenuation with $k = 0.015$.

The agreement between the observed and predicted scaling of PGA and PGV confirms the validity of the adopted model to explain the general features of the seismic radiation in the area.

REFERENCES

Aki K., 1967. 'Scalling law of seismic spectrum'. *J. Geophys. Res.*, 72, 1217-1231.

Aki K., 1986. 'Physical theory of earthquakes', *this volume.*

Aki K., Chouet B., 1975. 'Origin of coda waves : source, attenuation', and scattering effects. *J. Geophys. Res.*, 80, 3332-3342.

Anderson, J.G., 1986 'Implications of attenuation for studies of the earthquake source', in *Earthquake Source Mechanics*, Geophys. Monograph 37 (A.G.U., Washington 1986) 311-318.

Anderson, J.G., and Hough, S., 1984. 'A model for the shape of the Fourier amplitude spectrum of acceleration at high frequencies', *Bull. Seism. Soc. Am., 74*, 1969-1994.

Archuleta R.J., Cranswick E. Mueller C., Spudich P., 1982. 'Source parameters of the 1980 Mammoth Lakes, California, earthquakes sequence'. *J. Geophys. Res., 87*, 4595-4607.

Boore, D.M., 1983. 'Stochastic simulation of high frequency ground motions based on seismological models of the radiated spectra', *Bull. Seism. Soc. Am., 73*, 1865-1894.

Boore, D.M., 1986. 'The effect of finite bandwidth on seismic scaling relationships', in *Earthquake Source Mechanics*, Geophys. Monograph 37 (A.G.U., Washington 1986), pp 275-284.

Boore, D.M. and Boatwright, J., 1984. 'Average body wave radiation coefficients', *Bull. Seism. Soc. Am., 74*, 1615-1621.

Brune, J.N., 1970. 'Tectonic stress and the spectra of seismic shear waves', *J. Geophys. Res.,75*, 4997-5009 (1971. Correction, *J. Geophys. Res., 76*, 5002).

Castellano, M., Del Pezzo, E., De Natale, G. and Zollo A., 1984. 'Seismic coda Q and turbidity coefficient at the Phlegraean Fields volcanic area : preliminary results', *Bull. Volcan..,* 47, 219-224.

Crosson, R.S., 1981. 'LQUAKE : a computer program for hypocenter locations'. *Techn. Rep.* Univ. Wash. (Seattle).

Chouet, B., 1976. 'Source, scattering and attenuation effects on high frequency seismic waves'. Mass. Inst. Technology, *Ph. D. Thesis*.

Chouet, B., Aki K., Tsujura M., 1978. 'Regional variation of the scalling law of earthquake source spectra', *Bull. Seism. Soc. Am., 68*, 49-80.

Dainty, A.M., 1981. 'Scattering model to explain seismic Q observations in the lithosphere between 1 and 30 Hz', *Geophys. Res.* lett., 8, 1126-1128.

Del Pezzo, E., Ferulano, F., Giarrusso, A., and Martini, M., 1983. 'Seismic coda Q and scalling law of the source spectra in the Aeolian Islands, Southern Italy', *Bull. Seism. Soc. Am., 73*, 97-108.

Del Pezzo, E., Scarcella, G., 1986. 'Three component coda Q in the Abruzzi-Molise region, Central Appenines', *Annales Geophysicae, 5*, 589-592.

Del Pezzo, E. and Zollo A., 1984. 'Attenuation of coda waves and turbidity coefficient in Central Italy', *Bull. Seism. Soc. Am., 74*, 2665-2659.

Del Pezzo, E., De Natale, G., Scarcella, G. and Zollo A., 1985. 'Q_C of three component seismograms of volcanic microearthquakes at Campi Flegrei volcanic area, Southern Italy', *Pageoph, 123*, 683-696.

Del Pezzo, E., De Natale, G., Martini, M. and Zollo, A., 1986. 'Source parameters of microearthquakes at Phlegraen Fields (Southern Italy) volcanic area', *Phys. Earth Planet Int.*, in press.

De Natale, G., Madariaga, R., Scarpa, R. and Zollo, A., 1986a. 'Source parameter analysis from strong motion records of the Friuli (Italy) earthquake sequence (1976-1977)', submitted to *Bull. Seism. Soc. Am.* for publication.

De Natale, G., Iannaccone G., Martini, M. and Zollo A., 1986b. 'Seismic sources and attenuation properties at Campi Flegrei volcanic area', Submitted to *Pageoph* for publication.

De Natale, G., Faccioli, E. and Zollo A., 1986c. 'Scaling of peak ground motions from digital recordings of small earthquakes at Campi Flegrei, Southern Italy', Submitted to *Pageoph* for publication.

De Natale, G. and Zollo, A., 1986. 'Statistical analysis and clustering features of the Phlegraean Fields earthquake sequence (May 1983 - May 1984)', *Bull. Seism. Soc. Am., 76* , 801-814.

Di Vito, M., Lirer, L., Mastrolorenzo, G. Rolandi, G. and Scandone, R., 1985. 'Volcanological map of Campi Flegrei'. University of Naples and Italian Ministry of Civil Protection.

Gao, L.S., Lee, L.C., Biswas, N.N. and Aki, K., 1983. 'Comparison of the effects between single and multiple scattering on coda waves for local earthquakes', *Bull.*

Seism. Soc. Am., *73*, 337-389.

Gaudiosi, G. and Iannaccone G., 1984. 'Preliminary study of stress pattern at Phlegraean Fields as inferred from local mechanisms', *Bull. Volcan..*, *47*, 225-231.

Hanks, T., 1982.'f_{max}', *Bull. Seism. Soc. Am.* , *72*, 1867-1879.

Hanks, T. and Mc Guire, R., 1981. 'The character of high frequency strong motion', *Bull. Seism. Soc. Am.*, *71*, 2071-2095.

Luco, J., 1985. 'On strong ground motion estimates based on the models of the radiated spectrum", *Bull. Seism. Soc. Am.*, *75*, 641-650.

Madariaga, R., 1976. 'Dynamics of an expanding circular fault', *Bull. Seism. Soc. Am.*, *66*, 639-666.

Mc Garr, A., 1986. 'Some observations indicating complications in the nature of earthquake scaling', in *Earthquake Source Mechanics*, Geophys. Monograph 37 (A.G.U., Washington 1986), pp. 217-225.

Papageorgiou A.S., Aki K., 1983. 'A specific barrier model for the quantitative description of inhomogeneous faulting and the prediction of strong ground motion. 1. Description of the model', *Bull. Seism. Soc. Am.*, *73*, 693-722.

Rosi, M., Sbrana, A. and Principe, C., 1983. 'The Phlegraean Fields : structural evolution, volcanic history and eruptive mechanisms', *Journ. Volc. Geoth. Res.*, *17*, 237-288.

Rovelli, A., 1982. 'On the frequency dependence of Q in Friuli from short period digital records', *Bull. Seism. Soc. Am.*, *72*, 2369-2372.

CHAPTER III

ENGINEERING SEISMOLOGY AND SEISMIC RISK

NUMERICAL MODELLING OF THE DYNAMICS OF SATURATED ANELASTIC SOILS

D. Aubry (*), H. Modaressi (**)
(*) Ecole Centrale de Paris
92295 Châtenay Malabry
France
(**) Bureau de Recherches Géologiques et Minières
45100 Orléans
France

ABSTRACT. The behaviour of anelastic saturated soils during earthquakes is very complex. In this paper we present a coherent method combining an elastoplastic multimechanism constitutive law with a modified two phase model based on the Biot theory. The variational formulation of the whole problem is presented and the boundary and interface conditions are discussed. This technique is then applied to the analysis of local site effects. The infinite domain corresponding to the soil surrounding the domain of study is modelled using the paraxial approximation. The variational formulation of the coupling between finite and infinite domains is also presented. Finally a sediment-filled valley is chosen as example and its seismic response for different soil behaviours is studied.

INTRODUCTION. In classical soil mechanics text books the fundamental role of the pore water is always emphasized but the modelling is often limited to the quasistatic Terzaghi consolidation theory. The literature on soil dynamics is however very large but mainly about laboratory properties or rather empirical methods and the area of the numerical modelling of saturated soil behaviour when submitted to seismic loadings has emerged only very recently.

Continuous progress in the modelling of cyclic soil properties based on the concept of effective stresses including volume dilatancy or compaction together with the implementation of a Biot-like model for the two phases porous medium allows now to develope a sound approach to such a material. The following themes will be discussed in this paper :

1. A two-phase model for saturated soils
2. Remarks on soil behaviour
3. The numerical model
4. Seismic loading and absorbing boundaries
5. Applications to local site effects in earthquake engineering
6. Conclusions
7. References

J. Bonnin et al. (eds.), Seismic Hazard in Mediterranean Regions, 151–171.

1. A TWO-PHASE MODEL FOR SATURATED SOILS

Several papers have been dedicated to the mechanical description of porous media either following the Terzaghi or Biot theory. The Terzaghi theory is one dimensional and limited to static loadings but the essential concept of effective stresses is introduced. The non linear soil behaviour is also taken into account in this approach. The Biot theory does not use the concept of effective stresses and is usually limited to linear elasticity of the porous matrix. However it is fully three dimensional and is valid in the dynamic case.

1.1. Balance equations

The model that we present now is somewhat intermediary between the preceeding theories :

- the concept of effective stresses is used together with the non linear behaviour of soils,
- it is three dimensional and applies in the dynamic case.

Let $\mathbf{u}_s$ be the solid skeleton displacement that is the average displacement of the grains in a so-called representative volume element of soil and let $\mathbf{u}_f$ be the average displacement of the pore fluid. Then the relative displacement $\mathbf{u}_{rf}$ of the fluid phase with respect to the solid phase is :

$$\mathbf{u}_{rf} = n\,(\,\mathbf{u}_f - \mathbf{u}_s\,)$$

where n is the porosity of the soil. The momentum balance equation for the bulk material is given by :

$$\mathbf{Div}\,\boldsymbol{\sigma}' - \mathbf{grad}\,p + \rho\,\mathbf{g} = \rho\,\partial^2_{tt}\,\mathbf{u}_s + \rho_f\,\partial^2_{tt}\,\mathbf{u}_{rf}$$

where $\boldsymbol{\sigma}'$, p , $\mathbf{g}$ and ρ_f stand respectively for the effective stresses, the pore pressure, the gravity vector and the pore fluid density. The bulk density ρ is depending on the porosity n:

$$\rho = (1\text{-}n)\,\rho_s + n\,\rho_f$$

Numerical computations have shown that for a frequency range similar to the characteristic frequencies of earthquakes the relative acceleration of the fluid could be neglected in the above equation so that the following equation can be used :

$$\text{(1)} \qquad \mathbf{Div}\,\boldsymbol{\sigma}' - \mathbf{grad}\,p + \rho\,\mathbf{g} = \rho\,\partial^2_{tt}\,\mathbf{u}_s$$

A more refined model has been implemented by Modaressi (1987)in his thesis but this one will be sufficient for our purpose here. The momentum balance equation of one of the two phases is required. The pore fluid momentum is considered next which gives a generalization of the Darcy equation :

$$(2) \quad \partial_t \mathbf{u}_{rf} = \boldsymbol{K} \; (- \mathbf{grad}\, p + \rho_f \mathbf{g} \; - \; \rho_f \; \partial^2_{tt} \mathbf{u}_s)$$

still neglecting a term related to the relative movement of the fluid and where $\boldsymbol{K}$ stands for the permeability tensor of the porous medium.

The coupling between the two phases is already revealed by the introduction of the pore pressure in the equation (1). The second type of coupling is shown when $\mathbf{u}_{rf}$ is eliminated by using the mass balance equation of the bulk material :

$$(3) \quad \mathrm{div}\, (\partial_t \mathbf{u}_s \; + \; \partial_t \mathbf{u}_{rf}) \; = \; - \partial_t p \; / \; Q$$

where Q is related to the fluid and solid compressibilities by the following relation :

$$(4) \quad 1 / Q = \; n / K_f + (1\text{-}n) \; / K_s$$

Eliminating $\partial_t \mathbf{u}_{rf}$ between (2) and (3) and using the divergence operator the following equation is obtained :

$$(5) \quad \mathrm{div}\, (\partial_t \mathbf{u}_s \; - \rho_f \boldsymbol{K} \, \partial^2_{tt} \mathbf{u}_s) + \mathrm{div}\, (\boldsymbol{K} \, (- \mathbf{grad}\; p + \rho_f \mathbf{g})) = - \, \partial_t \, p \;\; / \, Q$$

which is the final equation for the hydrodynamical part.

1.2. Boundary conditions

The choice of the boundary conditions adapted to the model is crucial either from a practical point of view or for the virtual work equations which are a necessary tool for the finite element model.

1.2.1. Mechanical boundary conditions. In a standard manner the boundary of the domain under study is split into two parts. On the first part Γ_u the displacements are imposed as function of the time variable. On the complementary part Γ_σ surface forces are given which must be equal to the total traction vector in the material. This boundary condition is rather well adapted to the model and furthermore is consistent with the pasting of a saturated and a dry zone which is often necessary in situ.

1.2.2. Hydrodynamical boundary conditions. Another partition of the boundary is needed : $\partial\Omega = \Gamma_p \cup \Gamma_\phi$. On Γ_p the pore pressures are given with respect to the time variable. On Γ_ϕ flux boundary conditions are imposed, which is equivalent to an imposed relative velocity of the fluid. For instance a fully impermeable wall will impose the vanishing of this relative velocity on it.

1.3. Initial conditions

The initial values of the displacements and velocities of the skeleton must be given together with the initial pore pressures. More delicate is the computation of the initial effective stresses and also of the intial state of the material. Very often a first computation of these quantities is necessary which is not the easiest part of the analysis.

2. REMARKS ON THE ANELASTIC SOIL BEHAVIOUR

Mechanical soil behaviour is highly non linear even for low stress level. Experimental results show that when the cyclic shear strain amplitude is less than 10^{-6} the soil response is rather reversible and non linear elasticity due to the effects of the confinement may describe rather well the stress-strain curve. Between 10^{-6} and 10^{-4} irreversible deformations begin to appear but the cycles are rather stable. Then an approach like the linear equivalent modulus with hysteretic damping may be of some value. For larger shear strain amplitudes the loops get modified at each cycle due either to the densification of the material under drained conditions or to pore pressure increase under undrained conditions.

The prediction of the pore pressure increase and of permanent deformations can only be done by using an incremental constitutive equation in the framework of elastoplasticity theory with due account of internal variables like the density of the material. In such an approach stress increments are related to strain increments the relationship being driven by a yield function such as the following :

$$f^m(\boldsymbol{\sigma}', r_n, e, \mathbf{n}) = \|\boldsymbol{\tau}_n\| + \sigma'_{nn}\; r_n\; F(p, e)\; \mathrm{tg}\,\Phi$$

where $\boldsymbol{\sigma}'$, r_n, e, Φ stand respectively for the effective stresses, the degree of mobilization of the shear friction mechanism, e the void ratio and Φ the friction angle of the material and where the following definitions have been used for a plane with unit normal $\mathbf{n}$:

$$\sigma'_{nn} = \mathbf{n} \,.\, \boldsymbol{\sigma}' \,.\, \mathbf{n}$$

$$\tau_n = \boldsymbol{\sigma}' \,.\, \mathbf{n} - \sigma'_{nn}\; \mathbf{n}$$

The function $F(p, e)$ is introduced to take into account the so-called critical state of the material. Essentially F tends to unity when the current void ratio goes to its critical void ratio which is an intrinsinc property of the solid skeleton at a given confining pressure. The evolution of the plastic strain rates is given by :

$$\partial_t \gamma^p = \sum_{\mathbf{n}} \lambda^n \{ \mathbf{b}_n \otimes_s \mathbf{n} \}$$

with :

$$(\mathrm{w} \otimes_s \mathbf{n})_{ij} = 1/2\, (\mathrm{w}_i\, \mathrm{n}_j + \mathrm{w}_j\, \mathrm{n}_i)$$

The plastic sliding vector $\mathbf{b}_n$ is given by :

$$\mathbf{b}_n = \boldsymbol{\tau}_n / \| \boldsymbol{\tau}_n \| + (\operatorname{tg} \Psi + \| \boldsymbol{\tau}_n \| / \sigma'_{nn}) \ \mathbf{n}$$

where Ψ is the dilatancy angle of the material. The plastic multiplier λ^n can be computed by using the stress-strain rates relationship :

$$\partial_t \boldsymbol{\varepsilon} = \boldsymbol{A} \ \ \partial_t \boldsymbol{\sigma}' \ + \partial_t \gamma^p$$

together with the condition of continuous plastic loading :

$$f^m (\boldsymbol{\sigma}', r_n, e , \mathbf{n}) = 0 \ , \ \text{and} \ : \ \partial_t f^m (\boldsymbol{\sigma}', r_n, e , \mathbf{n}) \ = 0$$

The generalization to cyclic loading can be performed using cyclic loading functions with a memory of the last loading reversal.

Several soils have been modelled with such a constitutive equation including sands and clays and a good agreement with observed experimental investigations is generally obtained. From the point of view of the coupling between the solid skeleton and the pore fluid the two following fundamental aspects of the model should be noted :

- effective stresses are implied in the above equations which is unavoidable if characteristic soil behaviour is to be described,
- a dilatancy rule is incorporated in the plastic strains which is also fundamental to predict correctly the volumetric strains.

3. NUMERICAL MODELLING

The successive following steps are required for the numerical modelling of the above sets of equations :

- weak form or virtual work form equations of the coupled problem,
- discretization with respect to space variables by the finite element method,
- time discretization by a step by step scheme,
- numerical solution of the non linear system at each time step.

In what follows the following notations will be used throughout for vector fields $(\mathbf{v}_1, \mathbf{v}_2)$, or tensor fields $(\boldsymbol{\tau}_1, \boldsymbol{\tau}_2)$ defined over the domain Ω or its boundary Γ :

$$(\boldsymbol{\tau}_1, \boldsymbol{\tau}_2)_\Omega = \int_\Omega \boldsymbol{\tau}_1 : \boldsymbol{\tau}_2 \, d\Omega = \Sigma_{ij} \int_\Omega \boldsymbol{\tau}_{1\,ij} \cdot \boldsymbol{\tau}_{2\,ij} \, d\Omega$$

$$(\mathbf{v}_1 , \mathbf{v}_2)_\Omega = \int_\Omega \mathbf{v}_1 \cdot \mathbf{v}_2 \, d\Omega = \Sigma_i \int_\Omega \mathbf{v}_{1i} \cdot \mathbf{v}_{2i} \, d\Omega$$

$$< \mathbf{v}_1 , \mathbf{v}_2 >_\Gamma = \int_\Gamma \mathbf{v}_1 \cdot \mathbf{v}_2 \, d\Gamma = \Sigma_i \int_\Gamma \mathbf{v}_{1i} \cdot \mathbf{v}_{2i} \, d\Gamma$$

3.1. Weak formulation of the differential equations

Let V_S be the virtual displacement field defined by :

$$V_S = \{ \mathbf{w}_S \ / \mathbf{w}_S \ \text{smooth on } \Omega \ / \ \mathbf{w}_S = \mathbf{0} \ \text{ on } \Gamma_{us} \}$$

Let Q be similarly a virtual pore pressure field defined by :

$$Q = \{ q \ / q \ \text{smooth on } \Omega \ / q = 0 \ \text{ on } \Gamma_p \}$$

Then the variational formulation of the coupled problem reads :

$$(6) \quad (\rho\, \partial_{tt}\mathbf{u}_S \ , \mathbf{w}_S)_\Omega + (\sigma' , \varepsilon(\mathbf{w}_S))_\Omega \ - (p , \text{div}\, \mathbf{w}_S)_\Omega = (\rho\, \mathbf{g}, \mathbf{w}_S)_\Omega + \langle \mathbf{T}, \mathbf{w}_S \rangle_{\Gamma\sigma}$$

$$(7) \quad - (\rho_f \mathbf{K}\, \text{div}\, \partial_{tt}\mathbf{u}_S, q)_\Omega \ - \ (\text{div}\, \partial_t \mathbf{u}_S, q)_\Omega - (\partial_t p \,/\, Q, q)_\Omega - \ (\mathbf{K}\mathbf{grad}\ p, \mathbf{grad}\ q)_\Omega$$
$$= \langle \varphi, q \rangle_{\Gamma\varphi} - \ (\mathbf{K}\mathbf{grad}\ (\rho_f\, \mathbf{g}\, \mathbf{X}), \mathbf{grad} q)_\Omega$$

and the $\mathbf{V_P}$ problem is the following :

$\mathbb{V}_{\mathbb{P}}$: Find $\mathbf{u}_S(t) \in V_S$, $p(t) \in Q$ and ($\sigma'(t)$, $\alpha(t)$) solutions of (6),(7) and satisfying the above constitutive equation.

Remark : It is interesting to notice that this formulation is well adapted to the choice of the boundary conditions either mechanical or hydrodynamical.

3.2. Finite element approximation of the problem

A classical finite element formulation is used for the approximation of the above mentionned problem. The displacement field $\mathbf{u}_S$ (t), and the pore pressure are approximated by the corresponding field $\mathbf{u}_{sh}(t)$, and $p_h(t)$. More precisely to each node of the finite element mesh there corresponds a shape function w_J equal to one at node J and to zero at the other nodes of the mesh. Then the following expansion is used:

$$\mathbf{u}_{sh} = \Sigma_{Ii}\, \mathbf{u}_{sIi} \cdot w_{sI}\, \mathbf{e}_i \qquad , \ p_h = \Sigma_I\, p_I \cdot w_{fI}$$
$$\mathbf{w}_{sh} = w_{sJ} \cdot \mathbf{e}_j \qquad , \ q_h = w_{fJ}$$

When these expressions are plugged into the variational formulation of the problem the following equations are obtained :

$$(8)\quad \Sigma_{Ii}\, \partial_{tt}\mathbf{u}_{sIi}\ \ \delta_{ij}\ \ (\rho\, w_I, w_J)_\Omega + (\boldsymbol{\sigma}', \boldsymbol{\varepsilon}(w_J \cdot \mathbf{e}_j))_\Omega$$
$$- \Sigma_I\, p_I\, (w_I, \mathrm{div}\, w_J\, \mathbf{e}_j)_\Omega = (\rho\, \mathbf{g}, w_J \cdot \mathbf{e}_j)_\Omega + <\mathbf{T}, w_J \cdot \mathbf{e}_j>_{\Gamma\sigma}$$

$$(9)\quad - \rho_f \Sigma_{Ii}\, \partial_{tt}\mathbf{u}_{sIi}\, (\mathbf{K}\, \mathrm{div}\, w_I\, \mathbf{e}_i, w_J)_\Omega - \Sigma_{Ii}\, \partial_t \mathbf{u}_{sIi}\, (\mathrm{div}\, w_I\, \mathbf{e}_i, w_J)_\Omega$$
$$- \Sigma_I\, \partial_t p_I\, (1/Q\, w_I, w_J)_\Omega - \Sigma_I\, p_I\, (\mathbf{K}\, \mathbf{grad}\, w_I, \mathbf{grad}\, w_J)_\Omega$$
$$= <\varphi, w_J>_{\Gamma\varphi} - (\mathbf{Kgrad}\, (\rho_f\, \mathbf{g}\, \mathbf{X}), \mathbf{grad}\, w_J)_\Omega$$

3.3. Time discretization

The finite element method provides for the spatial approximation of the dependant variables. The full time discretization is obtained by using a step by step Newmark scheme with prediction and correction with the classical parameters β and γ. The predictor for the displacements and velocities is :

$$\mathbf{u}_{n+1}^* = \mathbf{u}_n + \Delta t\, \mathbf{v}_n + \Delta t^2 (1 - \beta/2)\, \mathbf{a}_n$$
$$\mathbf{v}_{n+1}^* = \mathbf{v}_n + \Delta t\, (1 - \gamma)\, \mathbf{a}_n$$
$$\sigma'_{sn+1}{}^* = \sigma'_{sn} + \Delta t\, D^{ep} : \varepsilon\, (\mathbf{u}_{n+1}^* - \mathbf{u}_n)$$

The corrector itself is given by :

$$\mathbf{u}_{n+1} = \mathbf{u}_{n+1}^* + \Delta t^2\, \beta\, \mathbf{a}_{n+1}$$
$$\mathbf{v}_{n+1} = \mathbf{v}_{n+1}^* + \Delta t\, \gamma\, \mathbf{a}_{n+1}$$
$$\sigma'_{sn+1} = \sigma'_{sn} + \Delta t\, D^{ep} : \varepsilon\, (\mathbf{u}_{n+1} - \mathbf{u}_n)$$

3.4. Numerical solution of the non linear system of equations

The algebraic system obtained at each time step must be solved iteratively. A quasi-Newton type iterative scheme has been implemented. Special care must be dedicated to the criteria for stopping the iterations.

4. THE INCIDENT SEISMIC FIELD AND THE ABSORBING BOUNDARY CONDITIONS

To study soil response to seismic loading the spatial variations of the seismic free field must be known in advance and this is only possible when the source itself is included in the model (Bouchon (1979)). Such an approach is certainly possible theoretically but should be very costly especially when the source is very far from the site under study. It is the reason why the soil engineers assume that the free field is known only at certain points called control points usually chosen at the free surface. This specification is often too crude to describe the seismic incident field necessary to completely identify the seismic driving forces.

On the other hand very often in geotechnical engineering the soil domain is very large compared to the characteristic sizes of the obstacle like an embedded foundation and thus has to be modelled as an unbounded domain. This is the source of many numerical difficulties especially with the finite element method where the mesh is necessarily bounded. Absorbing boundaries must be implemented to get rid of the subsequent spurious wave reflexions against the mesh boundaries. These aspects are discussed in the following sections.

4.1. The seismic incident field

In this section the implementation of an absorbing boundary will be briefly presented. Various other numerical techniques have been reported in the literature but none of them seems to be able to deal with non linear material properties. The unbounded domain under study Ω is first of all divided into two zones : the bounded interior domain Ω_s and the unbounded exterior Ω'_s as shown on the following figure. Inside the inner zone the heterogeneities are complex. The following assumptions are made :

(i) A further partition of the inner domain is implied to deal with the saturated soil : inside Ω_{sb} sediments are modelled as a two phase anelastic material whose description has been the topic of the preceeding chapters. This zone is assumed to be situated in the upper part of Ω_s that is close to the free surface. The domain Ω_{sm} is made of an anelastic one phase material. The domain Ω_{st} is made of an elastic bulk sediment (one phase) and its purpose is to put a nice transition zone between the anelastic soil close to the free surface and the deep elastic layers. In some circumstances one of the two domains Ω_{sm} or Ω_{sb} may be absent.

(ii) The unbounded exterior zone is assumed to have a very simple structure like horizontal layering with viscoelastic properties. Wave propagations in this domain may be analyzed by a Thomson-Haskell approach (Tsakalidis (1985)).

These simplifying assumptions may certainly be modified and generalized but clearly set the stage for a consistant analysis of site amplification under seismic loading which consists in the study of the focusing or defocusing of waves due to local lateral heterogeneities.

The full set of equations of the problem is the following :

4.1.1. <u>Inside Ω'_s</u>. The displacement field $\mathbf{u'}_s$ is a solution of the elastodynamics equations:

$$(1) \qquad (\lambda+\mu)\,\mathbf{grad}\ \mathrm{div}\ \mathbf{u}_{s'} + \mu\,\Delta\mathbf{u}_{s'} = \rho\,\partial_{tt}\mathbf{u}_{s'}$$

together with free surface boundary conditions. At infinity the field $\mathbf{u'}_s$ must match with the

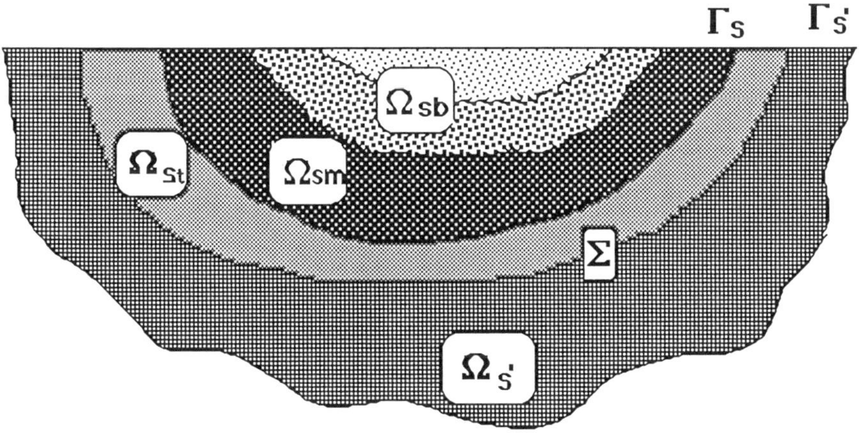

Figure 1 : Geometry of the different domains

free field $\mathbf{u}_i$ assumed to be given :

$$\text{(2)} \qquad \lim \mathbf{u}_{S'} = \mathbf{u}_i \qquad \text{when } |\mathbf{x}| \to \infty$$

The stress tensor inside Ω'_S is the sum of the intial static stresses $\sigma^0{}_{S'}$ and of the dynamic stresses :

$$\text{(3)} \qquad \sigma_{S'} = \sigma_{S'}{}^0 + \sigma_{S'}(\mathbf{u}_{S'})$$

4.1.2. <u>The interface Σ between Ω_S and Ω'_S</u>. At the interface between the inner and exterior zones the displacement and stress vector continuity must be imposed :

$$\text{(4)} \qquad \begin{cases} \mathbf{u}_S = \mathbf{u}_{S'} \\ \mathbf{t}_S + \mathbf{t}_{S'}(\mathbf{u}_{S'}) = 0 \end{cases}$$

Now if it is recalled that the sediments mechanical behaviour is assumed to be linear elastic in the neighbourhood of Σ then it is possible to write :

$$\mathbf{t}_{S'}(\mathbf{u}_{S'}) = \sigma_{S'}(\mathbf{u}_{S'}) \,.\, \mathbf{n}_{S'} = \lambda \; (\text{div}\, \mathbf{u}_{S'}) \;\; \mathbf{n}_{S'} + 2\,\mu\; \varepsilon\,(\mathbf{u}_{S'}) \,.\, \mathbf{n}_{S'}$$

4.1.3. Inside Ω_{sm}. The mechanical behaviour of the soil is assumed to be elastoplastic and the following set of equations is implied :

$$(5) \qquad \mathbf{Div}\, \sigma + \rho \mathbf{g} = \rho\, \partial_{tt} \mathbf{u}_s$$

where $\mathbf{g}$ is the gravity vector. Furthermore if $\varepsilon(\partial_t \mathbf{u}_s)$, stands for the rate of strains tensor the constitutive equation may be summarized by :

$$(6) \quad \begin{cases} f\,(\sigma,\alpha) \leq 0 \\ \partial_t \sigma = \mathbf{D}^{ep}\,(\sigma, \alpha\ , \varepsilon(\,\partial_t \mathbf{u}_s)\,)\ : \partial_t \varepsilon(\mathbf{u}_s) \\ \partial_t \alpha = \mathbf{L}\ (\sigma, \alpha\ , \varepsilon(\,\partial_t \mathbf{u}_s)\,) \end{cases}$$

where f stands for the yield function, $\mathbf{D}^{ep}$ the elastoplastic matrix and $\mathbf{L}$ the flow rule of the internal variable α.

4.1.4. Inside Ω_{sb}. The mechanical behaviour of the sediments is described by the model which has been discussed in the preceeding section. The corresponding equations will not be recalled here. It is however important to underline the continuity requirement at the interface between the saturated two phase zone Ω_{sb} and the one-phase zone Ω_{sm} : the two displacement fields must be equal, and there is an imposed pore pressure boundary condition (assuming the Ω_{sm} domain to be very permeable) and a stress vector continuity equation :

$$\begin{cases} \mathbf{u}_{sb} = \mathbf{u}_{sm}, \quad p = 0 \\ \mathbf{t}_{sb} + \mathbf{t}_{sm} = 0 \end{cases}$$

4.1.5. Initial conditions in each domain.

- In $\Omega_{s'}$

$$(7) \quad \begin{cases} \mathbf{u}_{s'}\,(0) = 0 \\ \sigma_{s'}\,(0) = \sigma'_0 \quad \text{such that :} \quad \mathbf{Div}\, \sigma_{s'}{}^{0} + \rho \mathbf{g} = 0 \\ \partial_t \mathbf{u}'(0) = 0 \end{cases}$$

- <u>In Ω_{sm}</u>:

$$(8) \quad \begin{cases} \mathbf{u}_s(0) = 0 \\ \sigma_s(0) = \sigma_0, \ \alpha(0) = \alpha'_0 \quad \text{with :} \quad \mathbf{Div}\,\sigma_s^0 + \rho\mathbf{g} = 0 \\ \partial_t \mathbf{u}(0) = 0 \end{cases}$$

- <u>In Ω_{sb}</u>: The initial conditions have already been mentionned in the preceeding sections.

4.2. Paraxial approximation of wave propagations in the outer domain

Many techniques have been published in the literature to deal with absorbing boundaries and with unbounded domains. In the frequency domain the so-called consistant boundaries work very nicely but are frequency dependant and seem to be a little bit costly in the time domain.

In the transitory case Lysmer and Kuhlmeyer (1969) or White et al. (1977) have proposed the use of absorbing dashpots at the transition boundary Σ. Smith (1974) too has proposed an ingenious way to deal with that problem. In our opinion the paraxial approximation seems to be the more interesting because of its ability to deal with more inclined incident waves on the boundary by simply refining the approximation (Claerbout(1976) , Engquist and Majda (1977) , Clayton and Engquist (1977), Cohen and Jennings (1983)). The principles to build this kind of boundary is briefly recalled now :

4.2.1. <u>Spectral impedance of the boundary Σ</u>. The initial step consists in computing in the frequency and wave number domain the local impedance of the exterior domain Ω'_s that is the relationship between the stress and displacement vector. A local Fourier transform of the elastodynamics equations with respect to the time and the space variables in the tangent plane to the boundary is performed after which by using the Sommerfeld radiation conditions and the boundary conditons the required relationship is found :

$$(9) \qquad \mathbf{t}_0 = \boldsymbol{A} \quad (|\xi'|, \omega) \ \hat{\mathbf{u}}'_0(\xi', \omega)$$

where $\boldsymbol{A}$ stands for the spectral impedance, $\mathbf{t}_0$ the spectral stress vector, ξ' the wave vector, ω the angular frequency, and $\hat{\mathbf{u}}'_0$ the transformed displacement on the boundary. By the inverse Fourier transform the following impedance in the physical time-space domain is computed :

$$(10) \quad \mathbf{t}_0(\mathbf{x}',t) = (2\pi)^{-3/2} \int \boldsymbol{A} \quad (|\xi'|, \omega)\, \hat{\mathbf{u}}'_0(\xi', \omega) \quad \exp[\,i\,(\xi'\mathbf{x}' + \omega t)]\, d\xi' d\omega$$

4.2.2. <u>Paraxial approximation of the impedance</u> The preceeding equation gives the surface forces acted by the outer domain on the inner one when waves are radiating in the outer domain. In the general case this impedance is not local in the sense that this surface force not only depends on the displacement at a given point but also on the displacement field on the whole boundary Σ. The basic idea of the paraxial equation is to approximate the impedance for waves propagating close to the normal to the interface Σ which means essentially that the wave vector amplitude $\|\xi'\|$ should be rather small.

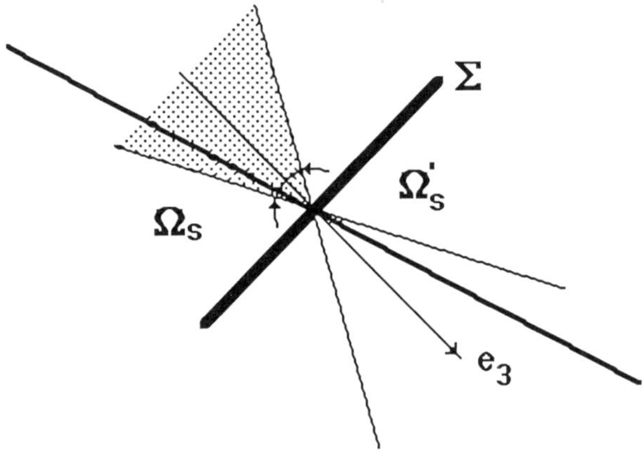

Figure 2 : Wave propagation close to the normal vector to the boundary Σ

An expansion according to this principle gives at order zero :

(11) $$\mathbf{t}_{S'} = \rho\, c_p\, \partial_t u_3\, \mathbf{e}_3 + \rho\, c_s\, \partial_t \mathbf{u}_{S'}$$

This relationship may be written as an operator equation :

(12) $$\mathbf{t}_{S'} = A_0(\partial_t \mathbf{u}_S)$$

At the order one the impedance is more complex and involves a relation between stress rates and accelerations :

(13) $$\partial_t\, \mathbf{t}_{S'} = A_1(\partial_{tt}\mathbf{u}_S, \partial_t \mathbf{u}_S, \mathbf{u}_S)$$

Now that the approximate space and time impedance has been computed the final matching equations between the outer and inner domains may be implemented.

4.3. Variational formulation of the coupling between the two domains

It is first recalled that the neighbourhood of the interface Σ as well as the outer domain are assumed to be linear elastic it is correct to introduce the radiated field here as the difference between the total displacement and the incident field. The radiated field satisfies the following boundary condition :

(14) $\mathbf{u'}_r = \mathbf{u}_{s'} - \mathbf{u}_i$ satisfies the Sommerfeld radiation condition

Then on the interface Σ the paraxial equation with order zero gives for the radiated field :

(15) $$\mathbf{t}_{s'}(\mathbf{u'}_r) = A_0(\partial_t \mathbf{u'}_r)$$

and with order one :

(16) $$\mathbf{t}_{s'}(\mathbf{u'}_r) = A_1(\partial_{tt}\mathbf{u'}_r, \partial_t \mathbf{u'}_r, \mathbf{u'}_r)$$

The stress vector reciprocity relation on Σ gives then e.g. for the order zero approximation:

(17) $$\mathbf{t}_s(\partial_t \mathbf{u}_s) = -\mathbf{t}_{s'}(\mathbf{u'}_i) - A_0(\partial_t \mathbf{u}_s) + A_0(\partial_t \mathbf{u'}_i)$$

In the domain Ω_{sb} where the saturated porous material lays the variational formulation is the one proposed in the preceeding chapters. The matching condition between the saturated porous material and the one phase sediment has be shown to be satisfied so that it is enough to elaborate the required equation at the interface Σ.

Let then $\mathbf{w}$ be a virtual displacement in Ω_{sm} the principle of virtual work in this domain reads :

(18) $$(\rho \;\; \partial_{tt}\mathbf{u}_s, \mathbf{w})_{\Omega sm} + (\boldsymbol{\sigma}_s, \boldsymbol{\varepsilon}(\mathbf{w}))_{\Omega sm} - \langle \mathbf{t}_s, \mathbf{w}\rangle_\Sigma = (\rho\, \mathbf{g}, \mathbf{w})_{\Omega sm}$$

By plugging equation (17) into this formulation the following final equation is obtained in the case of the order zero approximation :

(19) $$(\rho \;\; \partial_{tt}\mathbf{u}_s, \mathbf{w})_{\Omega sm} + (\boldsymbol{\sigma}_s, \boldsymbol{\varepsilon}(\mathbf{w}))_{\Omega sm} + \langle A_0(\partial_t \mathbf{u}_s), \mathbf{w}\rangle_\Sigma$$
$$= (\rho\, \mathbf{g}, \mathbf{w})_{\Omega sm} + \langle -\mathbf{t}_{s'}(\mathbf{u'}_i) + A_0(\partial_t \mathbf{u'}_i), \mathbf{w}\rangle_\Sigma$$

In the case of the order one paraxial approximation equation (18) is transformed into :

(20) $$(\rho \;\; \partial_{tt}\mathbf{u}_s, \mathbf{w})_{\Omega sm} + (\boldsymbol{\sigma}_s, \boldsymbol{\varepsilon}(\mathbf{w}))_{\Omega sm} - \langle \mathbf{t}_{s'}, \mathbf{w}\rangle_\Sigma = (\rho\, \mathbf{g}, \mathbf{w})_{\Omega sm}$$

with the following differential equation for the stress vector at the interface :

(21) $$\partial_t \mathbf{t}_{s'} = \partial_t \mathbf{t}_{s'}(\mathbf{u'}_i) - A_1(\partial_{tt}\mathbf{u'}_i, \partial_t \mathbf{u'}_i, \mathbf{u'}_i) - A_1(\partial_{tt}\mathbf{u}_s, \partial_t \mathbf{u}_s, \mathbf{u}_s)$$

The seismic loading appears explicitely in the equation (19) although it is implicit in equation (21). In either case these equations will be discretized by a finite element similar to the one described above and the time integration will be performed along the lines of the Newmark integrator.

5. APPLICATIONS TO THE ANALYSIS OF LOCAL SITE EFFECTS

As mentionned several times in this paper local site amplification is concerned with the influence of heterogeneities either vertical or horizontal with size of the order of a few tens of meters to let us say a few kilometers. This phenomenon has been studied very extensively in the past recent years. Very often in earthquake engineering the seismic incident field is modelled as P or SH plane wave propagating vertically. This assumption is good only when the source is very deep and the mechanical properties of the deposits decrease continously upwards. This simplifying assumption implies no out of phase displacement at a given horizontal level when the heterogeneities are only vertical. It is often necessary to take into account inclined waves and we refer to Luco (1980) for an interesting discussion on reasonable assumption not as crude as the preceeding ones.

As a numerical technique of analysis the finite element method has been scarcely used to study the influence of the topography on wave diffractions and site amplifications. The soil mechanical behaviour is very often assumed to be linear in this kind of study although Joyner (1975) and Joyner and Chen (1975) addressed such a topic but with a very simple soil model. It is the purpose of this section to show some recent applications of the more realistic anelastic saturated soil model presented here. The presentation is limited to the more typical results more details being to be found in H. Modaressi doctoral thesis.

The geometry of the problem is shown in the underneath figure for a sedimentary valley already studied by P.Y. Bard et al. (1980) in the elastic case. The signal is a Ricker wavelet very often proposed in the literature. In all the cases the incident field is a vertically propagating plane SV wave.

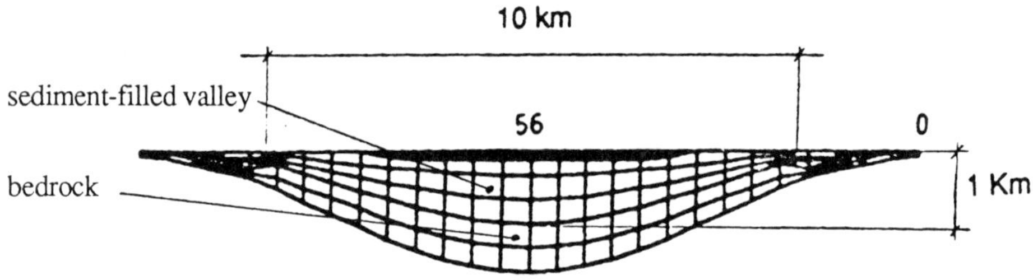

Figure 3: Geometry of the mesh used for the sedimentary valley

Results to check the adequacy of the paraxial approximation in the underlying halfspace are first presented :

5.1. Linear elastic sedimentary valley

The time history of the vertical and horizontal displacements at the free surface of the valley either at the center or at the lateral boundary with the underlying halfspace are shown in the following figure. It is observed that the SV wave when reaching the surface triggers surface waves which cause displacement amplification towards the center. As already observed by Bard and Bouchon the seismic signal is also considerably lengthened. Acceleration response spectra show an increase by a factor of three.

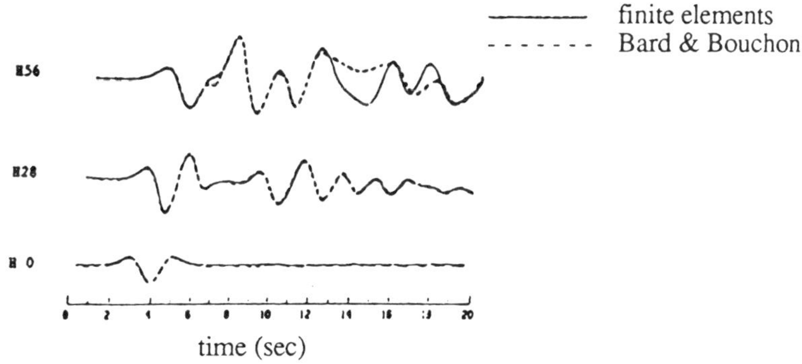

Figure 4: Horizontal displacements profile and history. Elastic case.

5.2. Anelastic sedimentary valley

Inside the sedimentary valley it is now assumed that the first 120 meters are made of an anelastic material with a constitutive model of the type presented in the section 2. The time history of displacements are shown on the following figures. A marked amplification of these quantities is observed together with irreversible displacements. In particular there is a net important settlement due to the densification of this kind of material when submitted to cyclic loading. The amount of vertical permanent displacement is directly correlated to the initial density of the soil. Acceleration response spectra show a further amplification of the signal which is characteristic of such soft sediments.

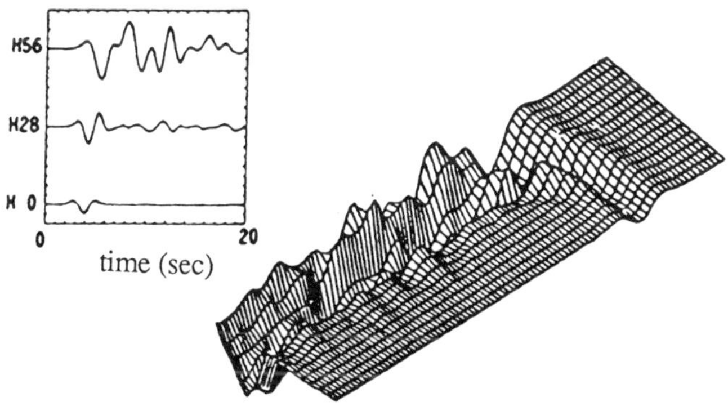

Figure 5a : Horizontal displacements profile and history.
Anelastic case.

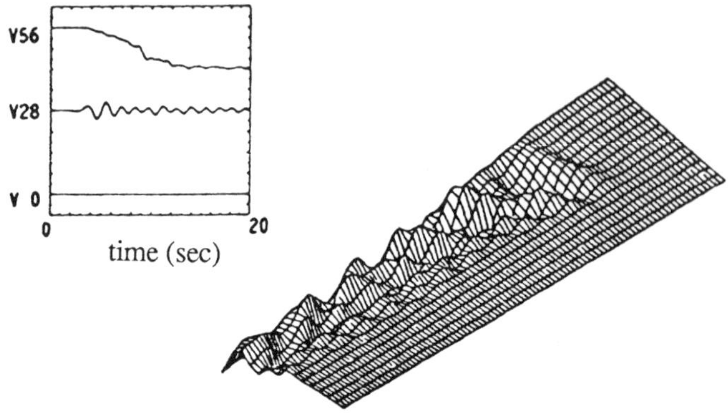

Figure 5b : Vertical displacements profile and history. Anelastic case.

5.3. Elastic saturated sedimentary valley

To summarize qualitatively the saturated soil model presented in this paper it is recalled that the main ingredients are the introduction of the pore pressure which decreases the effective stresses in the material on one hand and which has a dissipation effect on the other hand. In the chosen site here the the underlying bedrock is assumed to be very pervious. Other boundary conditions can obviously be analyzed with the same ease.

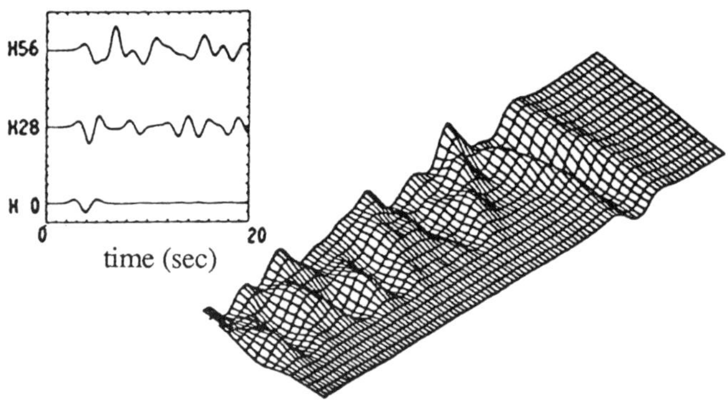

Figure 6a : Horizontal displacements profile and history.
Saturated elastic case.

The time history of the displacements still at the surface shows the decrease of the horizontal movements certainly due to the dissipative nature of the porous medium. On the contrary the vertical displacements are amplified which can be due to the fact that the water is much less compressible than the porous soil. The response spectra show that the frequency content of the horizontal accelerations is rather decreased.

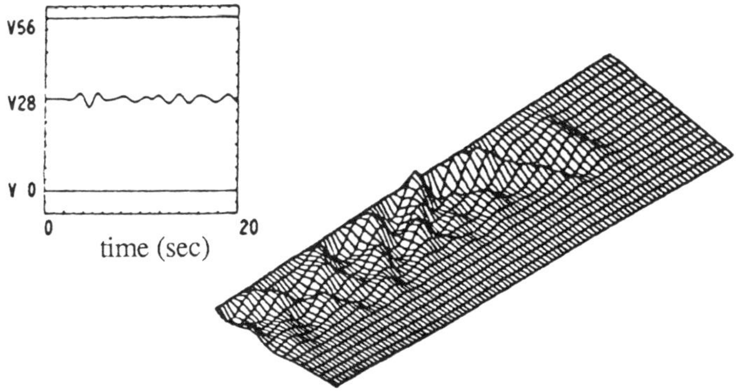

Figure 6b : Vertical displacements profile and history.
Saturated elastic case.

5.4. Anelastic saturated sedimentary valley

In this case the amplification is not so large as in the one phase case. The rapid loading sollicitation does not permit any filtration of the water and the vertical settlement is not as marked as in the corresponding one-phase case. However the new important phenomenon is the induced local liquefaction of the sediments which has been observed so many times during earthquakes or in laboratory experiments.

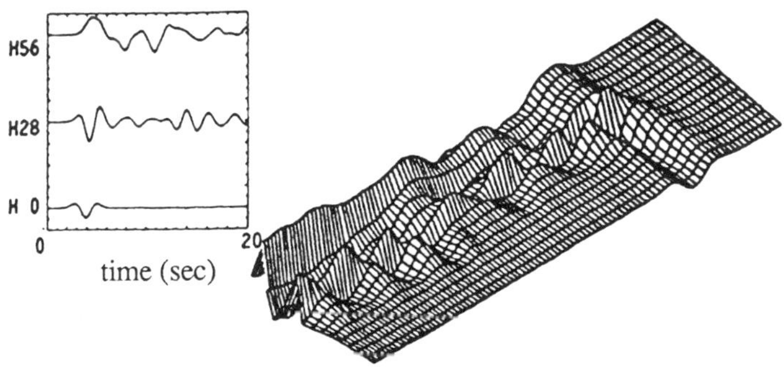

Figure 7a : Horizontal displacements profile and history.
Saturated anelastic case.

The liquefaction happens at a depth of about 30 meters and is explained by the pore pressure increase and consequent effective stress decrease. As the soil strength is described in terms of the latter this phenomenon can induce the rupture of the soil where it occurs.

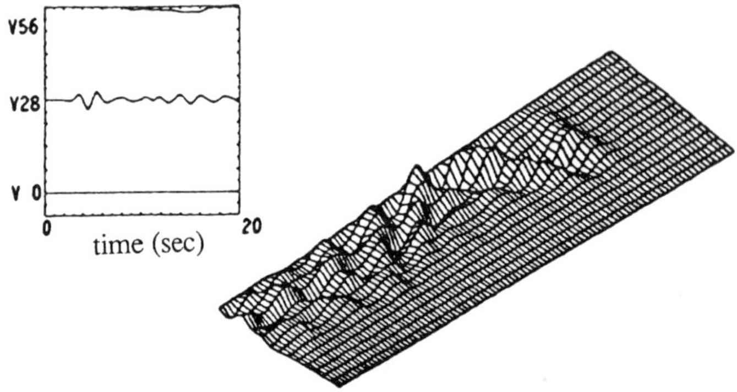

Figure 7b : Vertical displacements profile and history.
Saturated anelastic case.

It is more clearly seen in the stress strain diagrams where the soil strength is completely lost after a few cycles at that depth. The cycles in the effective stress space also show the very low final effective stress. Mohammadioun et al. (1984) have observed that the maximum amplitude of the horizontal accelerations for major events in soft soils is scarcely higher than 0.5g but the corresponding value of the vertical acceleration may increase to 1.7g. It seems that only the saturated case can reproduce this phenomenon as the one phase case creates much larger amplifications.

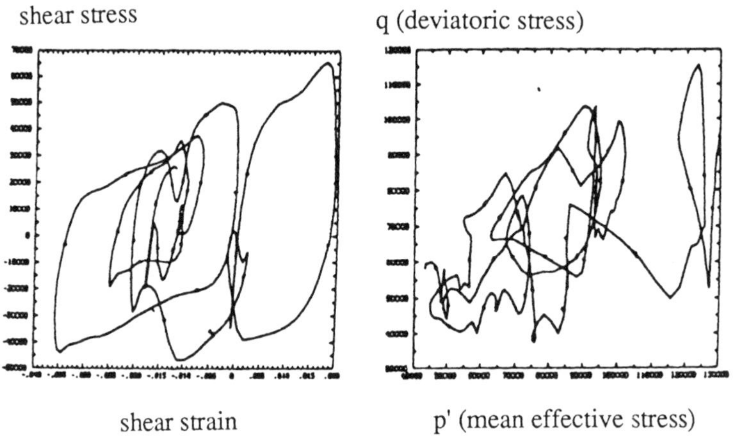

Figure 8 : Effective stress-strain and stress path diagrams.
Induced liquefaction.Saturated anelastic case.

6. CONCLUSIONS

New developments in the realistic modelling of saturated anelastic soils give now the possibility of very refined analyses where most of the environmental soil conditions can be reproduced. The cost of the analysis is still high but the advent of recent supercomputers shows that this will not be a large handycap in the near future compared to the informations that can be obtained from this kind of analysis.

7. REFERENCES

Aubry D., Hujeux J.C., Lassoudière F. & Meimon Y. (1982) 'A double memory model with multiple mechanisms for cyclic soil behaviour', *Int. Symp. Num. Mod. Geomech.*, Balkema.3-13.

Aubry D., Chouvet D., Modaressi H. & Mouroux P. (1985) 'Local amplification of a seismic incident field through an elastoplastic sedimentary valley', *Numerical Methods in Geomechanics* (Ed. Kawamoto, Ichikawa), Balkema, 421-428.

Aubry D., Chouvet D., Modaressi A. & Modaressi H. (1985) 'GEFDYN : Logiciel d'analyse du comportement mécanique des sols par éléments finis avec prise en compte du couplage sol-eau-air'. *Rapport EdF/Région d'équipement Chambéry.* Ecole Centrale de Paris.

Bard P.Y. & Bouchon M. (1980) 'The seismic response of sediment-filled valleys.Part 1:The case of incident SH waves'. *Bull. Seism. Soc. Am.*, vol **70**, n° 4, 1263-1286.

Bard P.Y. & Bouchon M. (1980) 'The seismic response of sediment-filled valleys.Part 2:The case of incident P and SV waves'. *Bull. Seism. Soc. Am.*, vol **70**, n° 5, 1921-1941.

Bard P.Y. (1982) 'Diffracted waves and displacement field over two-dimensional elevated topographies'. *Geophys. J. R. Soc.*, 731-760.

Bard P.Y. (1983) 'Les effets de site d'origine structurale en sismologie; Modélisation et interprétation,Application au risque sismique'. Thèse d'état. Université Scientifique et Médicale et l'Institut national Polytechnique de Grenoble.

Bard P.Y. (1985) 'Les effets de site d'origine structurale: Principaux résultats expérimentaux et théoriques'. *Génie parasismique* (V. Davidovici ed.), Presses ENPC, 224-238.

Bouchon M. (1979) 'Effect of topography on surface motion '. *Bull. Seism. Soc. Am.*, vol **63**, n° 3, 615-632.

Bear J. & Pinder G.F. (1978) 'Porous medium deformation in multiphase flow'. *Jou. of the Eng. Mech. Div.*, ASCE, vol. **104**, n° EM4, 881-894.

Bear J. & Pinder G.F. (1983) 'Porous medium deformation in multiphase flow'. *Jou. Geotech. Eng.*, vol **109**, N° 5, 736-737.

Biot M. A. (1941) 'General theory of three-dimensional consolidation'. *Journal of Applied Physics*, vol **12**, 155-164.

Biot M. A. (1962) 'Mechanics of deformation and acoustic propagation in porous media'. *Journal of Applied Physics*, vol **33**, n° 4,1482-1498.

Biot M. A. (1962) 'Theory of propagation of elastic waves in a fluid-saturated porous solid I- Low frequency range'. *Journal of Acoustical Society of America*, vol **28**, n° 2,168-178.

Biot M. A. (1962) 'Theory of propagation of elastic waves in a fluid-saturated porous solid II- Higher frequency range'. *Journal of Acoustical Society of America*, vol **28**, n° 2,179-191.

Boelle J.L. (1983) 'Mesure en régime dynamique des propriétés mécaniques des sols aux faibles déformations'. Doctoral thesis, Ecole Centrale de Paris.

Boore D.M., Harmsen S.C. & Harding S.T. (1981) 'Wave scattering from a step change in surface topography'. *Bull. Seism. Soc. Am.*, vol **71**, n° 1, 117-125.

Boore D.M. & Joyner W.B. (1982) 'The emprical prediction of ground motion'. *Bull. Seism. Soc. Am.*, vol **72,** n° 6, S43-S60.

Claerbout J.F. (1976) 'Fundamentals of Geophysical Data Processing'. McGraw-Hill Inc.

Clayton R. & Engquist B. (1977) 'Absorbing boundary conditions for acoustic ans elastic wave equations'. *Bull. Seism. Soc. Am.*, vol **67**, n° 6, 1529-1540.

Chouvet D. (1983) 'Calcul elastoplastique d'interaction dynamique sols-Structure ; Application aux ouvrages de soutenement'. Doctoral thesis. Ecole Centrale de Paris.

Cohen M. & Jennings P.C. (1983) 'Silent boundary methods for transient analysis'. *Computational Methods for Transient Analysis. (Ed. Belyschko,Hughes),* Elsevier SciencePublishers.

Dravinski M. (1982) 'Influence of interface depth upon strong ground motion'. *Bull. Seism. Soc. Am.*, vol **72**, n° 2, 597-614.

Dravinski M. (1983) 'Amplification of P,SV ans Rayleigh waves by two alluvial valleys'. *Soil Dynamics & Earthquake Engineering*, vol **2**, n° 2, 66-77.

Elgamal A.W.M., Abdel-Ghaffar A.M & Prevost J.H. (1985) 'Elasto-plastic earthquake-shear response of one-dimensional earth dam models'. *Earthquake Eng. & Struc. Dyn.*, vol **13**, 617-633.

Engquist B. & Majda A. (1977) 'Absorbing boundary conditions for the numerical simulation of waves'. *Mathematics of Computaion,* vol **31**, n°139, 629-651.

Engquist B. & Majda A. (1979) 'Radiation boundary conditions for acoustic and elastic wave calculations'. *Communications on pure and applied Mathematics*, vol **XXXII**, n°3, 313-357.

Harmsen S. & Harding S. (1981) 'Surface motion over a sedimentary valley for incident plane P and SV waves'. *Bull. Seism. Soc. Am.*, vol **71**, n° 3, 655-670.

Hujeux J.C. (1985) 'Une loi de comportement pour le chargement cyclique des sols. Génie Para- sismique' (V. Davidovici ed.), *Presses ENPC*, 287-302.

Ishihara K. (1982) 'Evaluation of soil properties for use in earthquake response analysis'. *Int. Symp. on Num. Mod. in Geomech.*, Zurich, 237-259.

Johnson L.R. & Silva W. (1981) 'The effects of unconsolidated sediments upon the ground motion local earthquakes'. *Bull. Seism. Soc. Am.*, vol **71**, n°1, 127-142.

Joyner W.B. & Chen A.T.F. (1975) 'Calculation of nonlinear ground response in earthquakes'. *Bull. Seism. Soc. Am.*, vol **65**, n° 5, 1315-1336.

Joyner W.B. (1975) 'A method for calculating nonlinear seismic response in two dimensions'. *Bull. Seism. Soc. Am.*, vol **65**, n° 5, 1337-1357.

Joyner W.B., Warrick R.E. & Fumal T.E. (1981) 'The effects of quaternary alluvium on strong ground motion in the Coyote lake, California earthquake of 1979'. *Bull. Seism. Soc. Am.*, vol **71**, n° 4, 1333-1349.

King J.L. & Brune J.N. (1981) 'Modeling the seismic response of sedimentary basins'. *Bull. Seism. Soc. Am.*, vol **71**, n° 5, 1469-1487.

Lee K.L. & Albaisa A. (1974) 'Earthquake induced settlements in saturated sands'. *Jou. of the Geotech. Eng. Div. ASCE*,GT4, 387-406.

Lee V.W. (1984) 'Three dimensional diffraction of plane P,SV and SH waves by a hemispherical alluvial valley'. *Soil Dynamics & Earthquake Engineering*, vol **3**, n° 3.

Luco J.E. (1980) 'Linear soil-structure interaction'. Report UCRL-15272, Lawrence Livermore National Laboratory.

Lysmer J. & Kuhlemeyer R.L. (1969) 'Finite dynamic model for infinite media'. *Jou. of the Mech. Div. ASCE*, 859-877.

Modaressi H., Aubry D. & Mouroux P. (1986) 'Wave propagation in a saturated porous media'. *8th European conference on Earthquake engineering*, Lisbon.

Modaressi H. (1987) 'Modélisation numérique de la propagation des ondes dans les milieux poreux anélastiques', *Doctoral Dissertation* , Ecole Centrale de Paris.

Mohammadioun B. & Pecker A. (1984) 'Low-frequency transfer of seismic energy by superficial soil deposits and soft rocks'. *Earthquake Eng. & Struc. Dyn.*, vol **12**, 537-564.

Murakami H., Shioya S., Yamada R. & Luco J.E. (1981) 'Transmitting boundaries for time-harmonic elastodynamics on infinite domains'. *Int. Jou. for Num. Meth. in Eng.*, vol **17**, 1697-1716.

Ohtsuki A. & Harumi K. (1983) 'Effects of topography and subsurface inhomogeneities on seismic SV waves'. *Earthquake Eng. & Struc. Dyn.*, vol **11**, 441-462.

Ohtsuki A., Yamahara H. & Harumi K.(1984) 'Effects of topography and subsurface inhomogeneities on seismic Rayleigh waves'. *Earthquake Eng. & Struc. Dyn.*, vol **11**, 37-58.

Ohtsuki A., Yamahara H. & Tazoh T.(1984) 'Effects of lateral inhomogeneities on seismic waves, II: Observation and analysis'. *Earthquake Eng. & Struc. Dyn.*, vol **12**, 795-816.

Pecker A. (1985) 'Réponse d'un profil de sol en champ libre'. *Génie parasismique* (V. Davidovici ed.), Presses ENPC, 313-322.

Prévost J.H. (1982) 'Nonlinear transient phenomena in saturated porous media'. *Comp. Meth. in Appl. Mech. and Eng.*, vol **30**, 3-18.

Sanchez-Sesma F.J. (1983) 'Diffraction of elastic waves by three-dimensional surface irregularities'. *Bull. Seism. Soc. Am.*, vol **73**, n° 6, 1621-1636.

Sandhu, R.S., Wilson, E.L. (1969) 'Finite element analysis of flow in saturated porous elastic media', *J. Eng. Mech. Div.*, ASCE, **95**, 641-652.

Smith W.D. (1974) 'A nonreflecting plane boundary for wave propagation problems'. *Jou. of Computational Physics*, vol **15**, 492-503.

Toki K. & Sato T. (1977) 'Seismic response analysis of surface layer with irregular boundaries'. *Proc. 6th World Conf. on Earthquake Eng.*, vol **2**, 81-86, Newdelhi, India.

Toki K. & Sato T. (1983) 'Seismic response analysis of ground with irregular profiles by the boundary element method'. *Natural Disaster Science*, vol **5**, n° 1, 31-52.

Tsakalidis C. (1985) 'Diffraction d'ondes sismiques sur les structures sur pieux et fonctions de Green du sol multicouche'. Doctoral thesis, Ecole Centrale de Paris.

White W., Villiappan S. & Lee I.K. (1977) 'Unified boundary for finite dynamic models'. *Jou. of Eng. Mech. Div. ASCE*, 949-964.

Zienkiewicz O.C. & Shiomi T. (1984) 'Dynamic behaviour of saturated porous media ; The generalized Biot formulation and its numerical solution'. *Int. Jour. Num. and Anal. Meth. Geomech.* vol **8**,71-96.

STATE OF THE ART IN THE CALCULATION OF A REFERENCE MOTION FOR THE ANTI-SEISMIC DESIGN OF CRITICAL STRUCTURES

Bagher Mohammadioun
Commissariat à l'Energie Atomique
Institut de Protection et de Sûreté Nucléaire
C.E.N. Fontenay-aux-Roses, B.P. n° 6,
92265 Fontenay-aux-Roses Cédex, France.

ABSTRACT. The methodology put into practice in the analysis of seismic hazard on the site of a nuclear facility relies upon a deterministic approach and endeavors to account for the particularities of every site considered insofar as available data and techniques allow. The calculation of a seismic reference motion for use in the facilities' design calls upon two basic sets of data. Regional seismicity over the past millennium, from historical sources, revised while preparing the seismotectonic map of France, is fundamental to this analysis. It is completed by instrumental data from the last quarter century. A collection of strong-motion accelerograph data from seismic areas worldwide reflects a variety of source characteristics and site conditions. A critical overview of current practice in France and elsewhere highlights shortcomings and areas of particular need both in experimental data and in methodology, and namely the scarcity of near-field data, the predominance of California records, and inaccurate approaches to integrating soil effects into ground-motion calculations.

1. BACKGROUND

One of the aims in engineering seismology is to predict a strong reference motion for the site of a given critical structure (NPP, dam, etc.) to be used in the design of said installation. This need has been increasingly felt by engineers over the past decades and has motivated the instrumentation of seismic high-risk areas throughout the world, but such a coverage is a recent development, pioneered in the western United States in the 1930's, but dating back only a few years in the majority of other high-priority zones. Although the number of recordings retrieved can now be considered, globally, as statistically valid, major earthquake recordings in the near field, where severe structural damage was effectively incurred, continue to be rare. When they do exist, each often tends to be a case study in itself, displaying the preponderant effects of individual contexts and raising new queries, without necessarily having yielded conclusive answers to

J. Bonnin et al. (eds.), Seismic Hazard in Mediterranean Regions, 173–193.

the previous ones. Indeed, the question may well be raised as to whether a simplified, purely statistical approach to the problem is either desirable or even conceivable. Figure 1 shows the magnitude and focal distance distributions for those records that are currently readily available. The large number of records for magnitude 6.5 relate for the most part to the 1971 San Fernando earthquake. Although the distance distribution shows some records for distances under 10 km, many of these correspond to smaller-magnitude events.

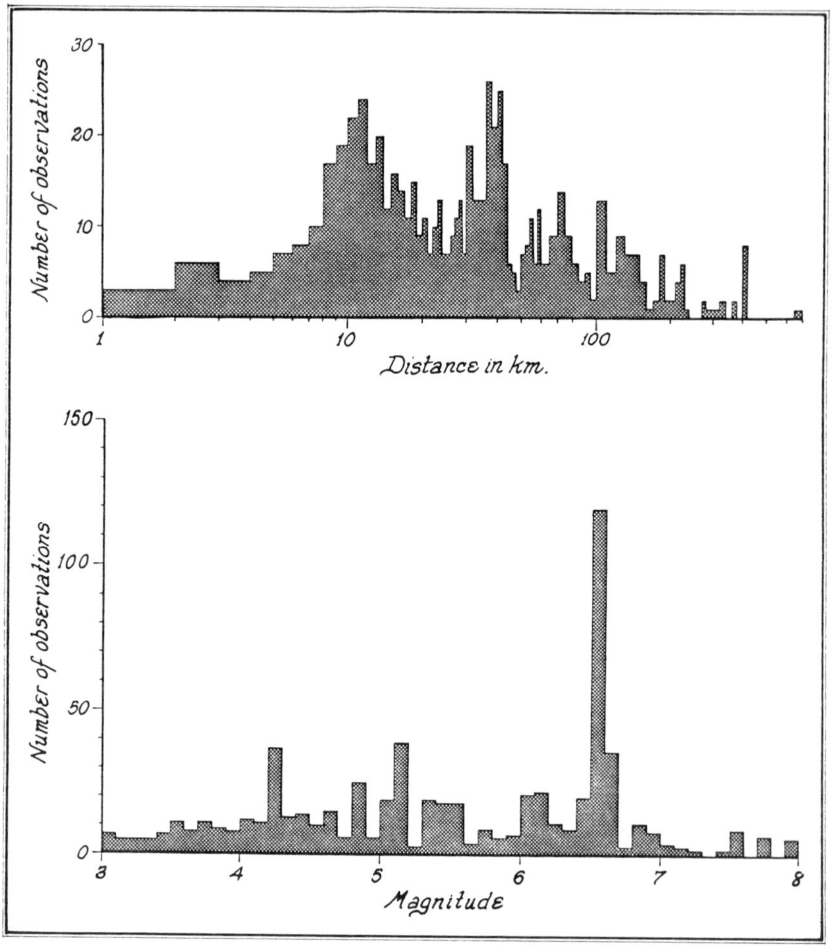

Figure 1. Magnitude and focal distance distributions for records contained in a strong ground motion data base.

From such collections of data, engineers have devised correlations linking certain parameters of motion to the earthquake's magnitude and the distance between the earthquake's focus and the recording site ; these correlations stem invariably from simplifications of a theoretical formulation in linear elasticity. If $S_0(t)$ is taken to represent the function defining the seismic waves generated at the source, $Z_R(t)$, the recording obtained at focal distance R, and A(t), and T(t), respectively, the transfer function of the media traversed between the source and the upper surface of the bedrock beneath the recording site and that of the most superficial layers underlying the station itself, the following equation may be written, supposing linear elasticity :

$$Z_R(t) = S_0(t) * A(t) * T(t) \qquad (1)$$

The Fourier transform of equation (1) is accordingly seen to be :

$$Z_R(f) = S_0(f) \cdot A(f) \cdot T(f) \qquad (2)$$

A simplification of equation (2) commonly called upon for engineering purposes is :

$$Z_R(f) = C \cdot 10^{\alpha M} \cdot R^{-n} \qquad (3)$$

where $10^{\alpha M}$ stands for the source function (M being the magnitude) and R^{-n}, the attenuation function. C, α, et n, correlation coefficients, become frequency-dependent when used to compute whole spectra rather than merely peak values :

$$PSV = C\ 10^{\alpha M}\ R^{-n} \qquad (4)$$

$$\log PSV = K + \alpha M - n \log R \qquad (5)$$

where PSV represents the maximum value of pseudo relative velocity for any given frequency and percentage of damping.

1.1. Calculation of Ground Motion Using an Empirical Approach

A common practice in engineering seismology, when modal analysis is performed in view of the anti-seismic design of structures, is to use, as the input motion, a set of spectra with progressive values of damping. These spectra are initially defined in terms of shape (either standard or site-adapted), then anchored to a predetermined value of peak ground acceleration (PGA). PGA values are derived from type-(3) correlations based on actual strong motion data, and can be read from parametered curves such as those depicted on Figure 2.

Now, the application of predictions of this kind to the case of strong earthquakes near the causative fault is clearly a debatable procedure. This is true not only because the number of near-source records obtained to date is small, as stated previously, but also

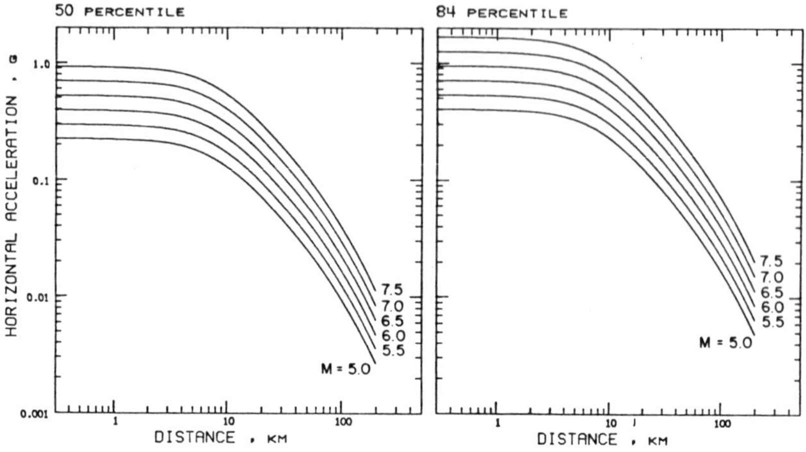

Figure 2. Predicted values of peak horizontal acceleration for 50 and 84 percentiles as fonctions of distance and moment magnitudes (from Joyner and Boore, 1981).

because the attenuation laws originally conceived for moderate earthquakes are being extrapolated to account for the effects of strong ones. Figures 3 and 4 compare PGA values actually recorded for two recent major earthquakes (Imperial Valley 1979 : $M_L \simeq 6.5$; Chile : $m_b = 6.7$, $M_S \simeq 8$), with PGA prediction curves. For both events, the discrepancies observed are considerable both level-wise and in the very nature of the variation.

The high degree of variability encountered in PGA's is easily understood in the light of the multiple and complex factors capable of having significant influence upon accelerations. Indeed, the high-frequency content of strong motion, responsible for large values of acceleration is thought to be ascribable to an irregular propagation of the rupture front along the fault plane (source effect), but it is also strongly affected by the mechanical properties of the soil and rocks underlying the recording site. The Fourier acceleration spectrum is quite often flat between the corner frequency (f_c) - the lower frequency limit - and the maximum frequency (f_{max}), beyond which the spectrum begins to fall off (see the diagram on Figure 5). The former is believed to be related to the size of the source, whereas the latter would be affected both by certain source characteristics and by the soil behavior.

1.2. Determining Ground Motion by Theoretical Simulation

Parallel to these empirical predictions, seismologists have in turn begun to focus their attention on strong motion studies in recent years as a consequence of the growing demand for a more rigorous approach to

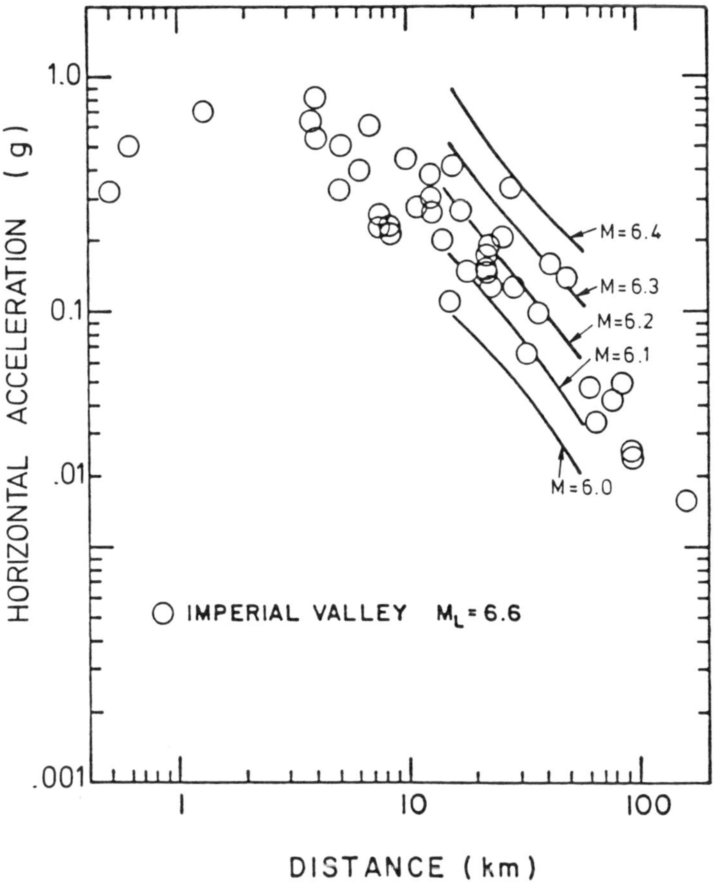

Figure 3. Peak horizontal accelerations for the 1979 Imperial Valley earthquake against closest distance to the fault compared with predicted values of PGA (from Boore and Porcella, 1980).

safety problems where the integrity of critical structures is concerned. Over the past ten years, significant progress has been made in the theoretical simulation of seismic sources and wave propagation. An important factor contributing to this progress has been the acquisition of strong-motion recordings from the immediate vicinity of causative faults. However, the recording stations involved were generally situated at the ground surface, and most often on alluvial deposits. The recorded motion, therefore, reflects not only the source and wave path characteristics, but also possible non-linear effects of the uppermost soil layers at the recording site. The inverse problem thus set forth consists in deducing the source function from the end-product recording. This process presupposes both large (tens of kilometers) and

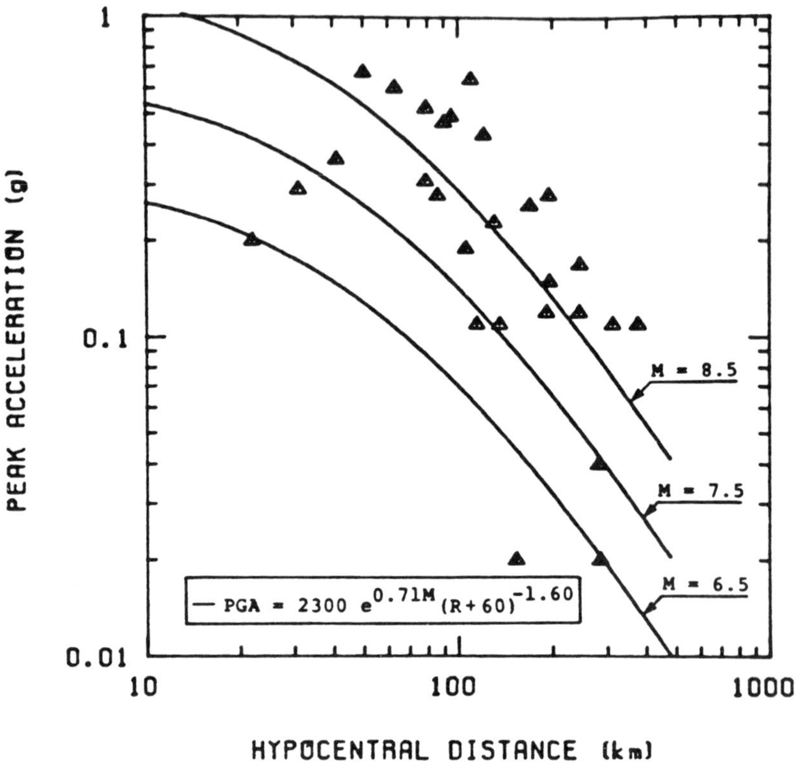

Figure 4. The largest peak horizontal accelerations at each strong motion station recorded for the 1985 Chilean earthquake compared with predicted values of PGA (from Wyllie et al., 1986).

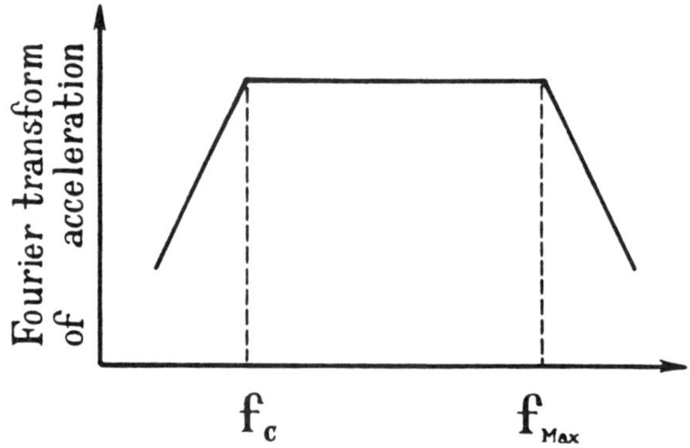

Figure 5. Diagram of a Fourier acceleration spectrum.

small-scale (tens of meters) corrections, the former in the realm of basic seismology, the latter in that of soil dynamics.

In this field, computational routines have been devised that operate either analytically or by numerical simulation, but many difficulties have yet to be overcome, particularly insofar as high frequencies are concerned - a sensitive area in engineering seismology. Furthermore, owing to the high degree of complexity of these theoretical studies, the solutions proposed rely exclusively on the principle of linear elasticity, whatever the stress level, and the introduction of non-linearities anywhere along the line would require computational techniques not yet mastered or too costly.

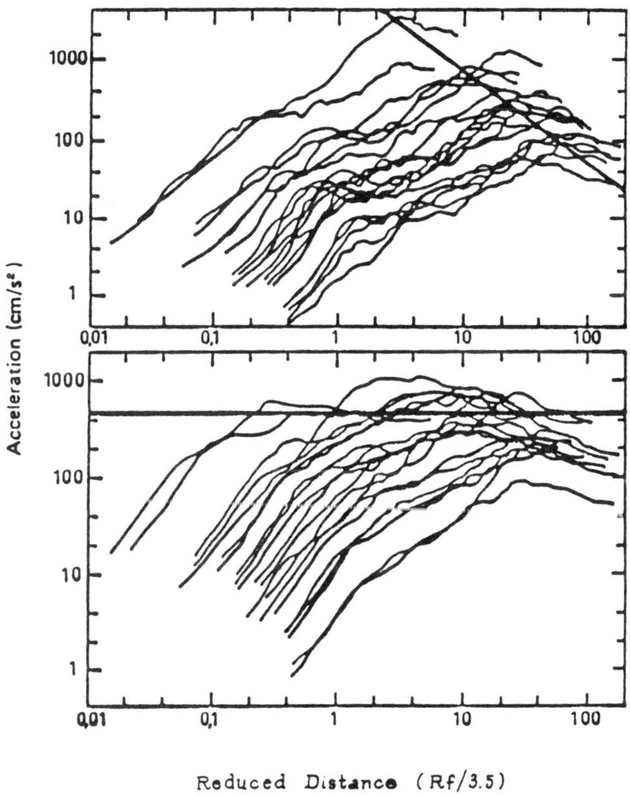

Figure 6. Acceleration spectra (20% damping) from recordings of the Imperial Valley earthquake of October 15, 1979 versus reduced distance (Rf/ 3.5) : vertical components (above) and horizontal components (below) (from Mohammadioun and Pecker, 1984).

1.3. Modification of Bedrock Motion by Superficial, Young Deposits

If, in general, linear solutions may be accepted wherever propagation in rock is involved, they should, on the contrary, be rejected for loose soil subjected to strong ground motions. However, in order to take into account this aspect of soil behavior, variable with the level of motion and individual soil properties, it would be necessary to establish the threshold beyond which linear or equivalent linear calculations become unsuitable for alluvia or other poorly consolidated materials.

The Imperial Valley series (1979) affords an unprecedented opportunity to investigate ground motion in the near field, as well as

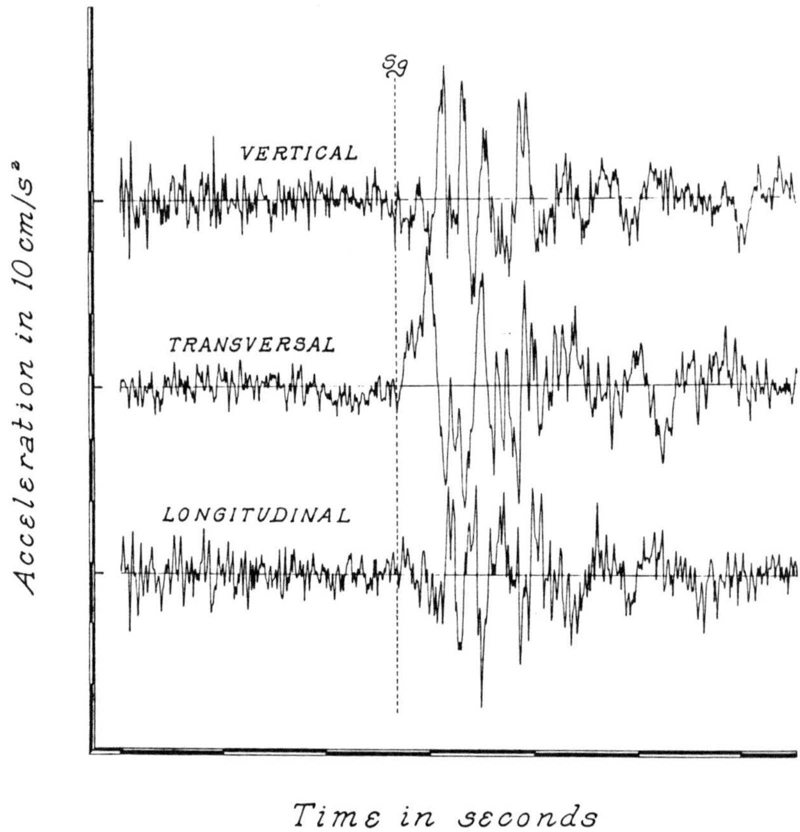

Figure 7. Three components of ground acceleration recorded in the bedrock, 166 meters below the surface, at the McGee Creek site : Bishop, California earthquake of November 23, 1984.

the problem of level-dependent soil behavior, thanks to the geometry of its strong-motion array (at right angles to the fault), to the large number of recordings retrieved for the different events, and to the deep layer of alluvia overlying the site. Acceleration spectra computed for the main event, when examined versus distance on Figure 6, demonstrate the fact that, in alluvial contexts, horizontal acceleration may be effectively restrained to a limiting value, consequently displaying little or no variation with distance, although ground displacement may be significantly enhanced (it will be noted that vertical motion, on the other hand, undergoes no modification). Although seismologists, invoking "source effect", can come up with a good explanation for each individual case, another alternative consists in assuming that linear calculations used to reproduce the behavior of

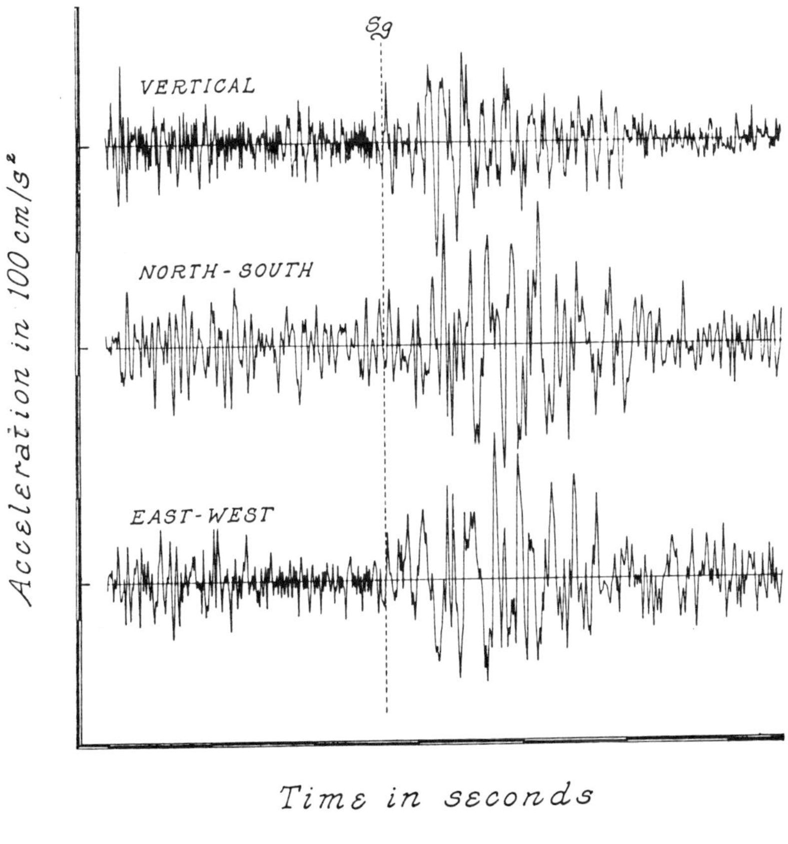

Figure 8. Three components of ground acceleration recorded at the surface atop a layer of glacial till, at the McGee Creek site : Bishop, California earthquake November 23, 1984.

superficial layers with weak mechanical resistance are not necessarily valid in the very near field.

Precisely in an attempt to ascertain the limits of applicability for certain linear or equivalent linear codes, it seems essential to verify different hypotheses by confronting them with experimental results. One possibility calls for a seismic downhole array, where motion at different levels, and in different geological formations, can be measured directly. Such data, by contributing useful information on the arrival of various waves, can aid in elaborating the most representative model of propagation. In 1984, an experiment of this type was undertaken in the highly-seismic, west-central California area of Mammoth Lakes (site, notably, of three major earthquakes –

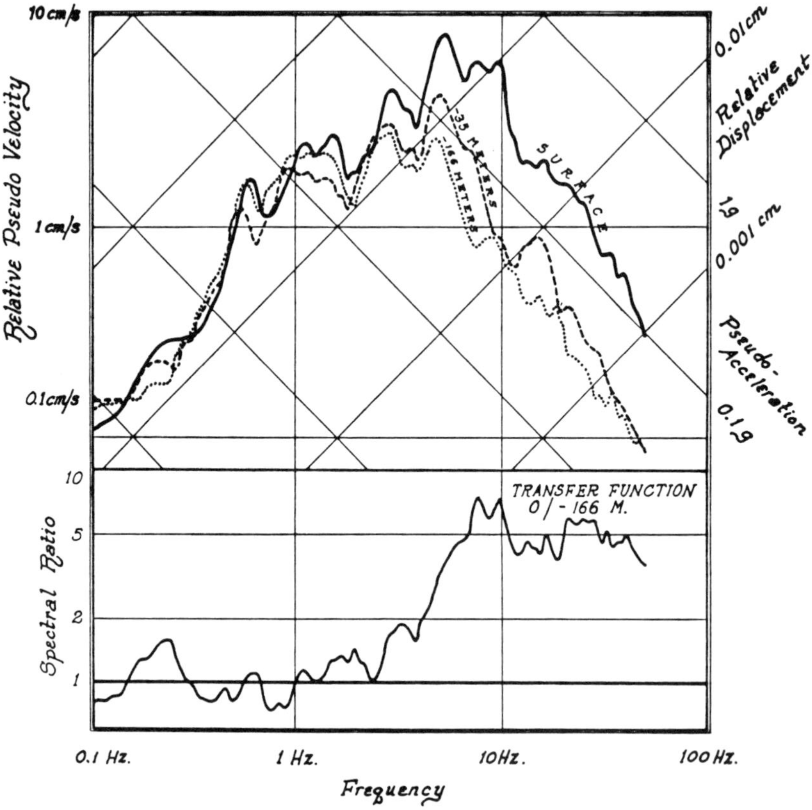

Figure 9. Response spectra (5% damping) of the vertical component of motion recorded at three levels, on the McGee Creek site, for the Bishop, California earthquake of November 23, 1984 ; transfer function of the overlying layer of sediment.

- magnitudes m_b 6.1, 5.7, and 5.6 - in 1980 and one m_b 5.7 in 1981, followed by prolonged and intense aftershock activity) by the United States Nuclear Regulatory Commission (USNRC). Within the framework of an agreement concluded between the NRC and the French Commissariat à l'Energie Atomique (CEA), the Institut de Protection et de Sûreté Nucléaire (IPSN), with its consultants, is collaborating with the NRC in the analysis and interpretation of data from this experiment. Instrument packages containing both accelerometers and velocity transducers were installed at the surface and in two downholes with respective depths of 35 and 166 meters, both of which holes reached into the metamorphic bedrock, some 31 meters below the surface. The overlying sedimentary layer consists of strongly heterogeneous glacial till.

Data has already been processed for a widely-felt earthquake that occured in late 1984 (m_b = 5.6) and its largest aftershock (m_b = 4.8), together with smaller events from early 1985 ; examples of the recordings obtained in the bedrock (at -166 meters) and at the surface are shown in Figures 7 and 8. As might have been expected in view of the nature of the sediments involved and the moderate level of acceleration measured in the bedrock (significantly lower than 0.1 g, due to a relatively large focal distance of at least 20 km.), no non-linear behavior was observed in the sediments, which, it must be stressed, can in no way be taken to represent alluvia (a similar study is to be conducted on a typical alluvial site : work on this project is slated to begin in early 1987). Figure 9 shows response spectra at three levels for the main event, together with the surface/166-meter transfer function. It is interesting to note that the feedback effect at -35 meters from the layer above was by no means negligible. In late July, 1986, a new swarm of events took place : the largest of these had m_b values of 6.0 , 5.6, and 5.5, and a good sampling of lesser magnitudes is also to be found. These new data, which should become available over the coming months, by supplementing the work already accomplished, can certainly be expected to prove of great benefit in calibrating appropriate models and techniques for investigating various aspects of the problem of predicting strong ground motion in the near field.

2. SEISMIC HAZARD ASSESSMENT ON NUCLEAR FACILITY SITES

The Institut de Protection et de Sûreté Nucléaire (IPSN) of the Commissariat à l'Energie Atomique (CEA) has developed and put into practice a method analyzing seismic hazard on the sites of nuclear facilites. Based upon a deterministic approach, this method endeavors to account for the particularities of every site considered insofar as available data and techniques allow. These data include two main categories of information : historical seismicity and recorded strong ground motion.

2.1. Historical Seismicity

Data on the historical seismicity of France over approximately the past 1000 years, updated annually and presented in computer file form, initially prepared as part of the Seismotectonic Map of France project, are the cornerstone of the methodology used for seismic hazard analysis. The possibility of tracing backwards in historical time and of compiling a millenarian data file can throw considerable light on geodynamic evolution, a very slow process by definition, and lead to a recognition of potential source zones and their respective sizes and thence, eventually, to assigning characteristic earthquakes to specific regions. These data are supplemented by a second file containing instrumental data over approximately the past quarter of a century, procured by a national monitoring network. This second file, although covering a limited lapse of time, provides elements for evaluating certain highly useful characteristics of earthquakes, such as focal mechanisms, which cannot be deduced from the first file.

2.2. Recordings of Strong Ground Motion

A collection of accelerograms of significant earthquakes obtained worldwide constitutes the second facet of the data base. In assembling this collection, the IPSN has not limited its activity to the procurement of data from other sources, but rather has itself contributed by recording the aftershocks of important earthquakes in Europe and the Mediterranean area at large : Friuli in Italy, El Asnam in Algeria, and the Swabian Alps in Germany, among others. In order to be able to make best use of the records, they must first undergo various types of computer processing and, in particular, a correction for the filtering effect introduced by the recording instrument. Considerable effort is being devoted within the framework of the European Association of Earthquake Engineering (EAEE) so as to agree upon uniform processing routines in the perspective of creating a European data bank.

In its present form, the IPSN base contains several categories of data : uncorrected acceleration, corrected acceleration (also, in certain instances, corrected velocity and displacement), response spectra, and Fourier spectra. A detailed list of all the information available is presented in a catalogue. Table I summarizes, region by region, the number of each type of data currently held.

2.3. Determining Seismic Reference Motions for Design Purposes

The procedure consists in a seismotectonic approach singularly well suited to the French seismic environment, characterized by well-documented historical seismicity and involving scattered seismic events in an intraplate context, where earthquake/fault relationships are difficult to establish (faults often being buried beneath thick layers of sediments).

The process of determining reference earthquakes for individual sites calls for the following steps :

1) A critical analysis of historical and instrumental seismicity for the whole area surrounding the site, with, as the case may be, consultation of the original historical documents and the plotting of isoseismals.

2) The delimitation of seismotectonic provinces and identification of the active faults present, whenever possible.

3) The displacement of earthquake sources according to the following criteria :

   a) Translation directly beneath the site of the largest earthquake assigned to the province to which the site belongs ;

GEOGRAPHICAL CLASSIFICATION		NUMBER OF SEISMIC EVENTS	UNCORRECTED RECORDINGS	CORRECTED ACCELERATIONS VELOCITIES DISPLACEMENTS	RESPONSE SPECTRA	FOURIER SPECTRA
THE AMERICAS	UNITED STATES	79	433	1155	1155	1155
	MEXICO	18	210	81	81	81
ITALIAN EARTHQUAKES	ANCONA	9	51	50	49	46
	FRIULI	33	387	170		
	NORCIA	4	30	28		
	CAMPANO LUCANO	5	57			
	UMBRIA	2	24			
	LAZIO ABRUZZO	11	144			
OTHER EUROPEAN EARTHQUAKES		13	60	53	52	53
EL ASNAM EARTHQUAKES		99	371	90	90	
JAPANESE EARTHQUAKES		53	177	177		
IRAN GUADELOUPE		2	14	3	3	3

Table I. A summary of information contained in the strong-motion data base.

b) Translation to that point along the border of all adjacent provinces that is nearest the site, of the largest earthquake assigned to each ;

c) Whenever the earthquakes are shown to be associated with a recognized active fault, translation along that fault of the largest earthquake ascribed to it, to the point nearest the site.

Following this analysis, one or more hypothetical events termed Séisme Maximal Historique Vraisemblable (SMHV), or "maximum plausible historical earthquake", capable of producing the strongest effects on the site, are defined. This first level of earthquake, however, is only the initial stage in evaluating seismic hazard and is not used for design purposes : it is a second, severer level of event, known as the Séisme Majoré de Sécurité (SMS), or "safety design earthquake" (as for SMHV, the site in question may be assigned more than one), the intensity of which has been increased by one degree on the MSK scale with respect to the SMHV event from which it derives, that is adopted as a basis for the design of facilities to be constructed on the site.

As mentioned above, the coverage of seismically active zones with strong-motion networks is quite a recent development, dating back a few decades in the western United States and often only a few years in many other high-priority areas. Although a statistically valid number of recordings has now been retrieved, these tend to be restricted to a relatively limited number of regions : the United States and Japan, together with sundry events from Latin America, Taiwan, and around the Mediterranean (Italy, in particular, operates a very dense network and has obtained many recordings in recent years), among others. For many other areas, including countries like France, knowledge of significant earthquakes mainly takes the form of macroseismic intensities (often deduced from descriptions in historical sources). Importantly, very few evaluations of magnitude are available inasmuch as this scale was devised only in 1935.

Under these circumstances, a way has been sought to bridge the gap between the mass of qualitative data from the past and the present demand for precise quantitative parameters to meet scientific and engineering needs. The strong-motion data base just described has providentially afforded the opportunity to accomplish just that by correlating the three most essential parameters that are macroseismic intensity I, (local) magnitude M, and focal distance R :

$$M = BI + B' \log R + B'' \pm \sigma \quad , \qquad (6)$$

coefficients B, B', et B" being estimated by means of a regression analysis. Using 576 observations in the United States, with intensities ranging from III to X on the Mercalli Modified scale, B, B', B", and σ were evaluated at 0.55, 2.2, -1.14, and ± 0.43 respectively. (Figure 10). Although this computation was carried out almost exclusively with data from California, the results obtained have since been shown to apply in many instances to data from almost anywhere else (provided shallow-focus earthquakes are considered). Accordingly, for historical earthquakes, wherever both intensity and distance can be determined, an hypothesis of magnitude can be made. Nevertheless, it is advisable that these coefficients be recalculated each time a statistically valid sampling of seismic observations is to be had for a given region.

In regions like France, the only vestiges of truly strong earthquake motion from centuries past take the form of macroseismic intensities. For this reason, we believe in the absolute neccesity of relating the spectrum not to a single parameter of notion (PGA for instance), but rather to the intensity, with magnitude and focal distance as variables.

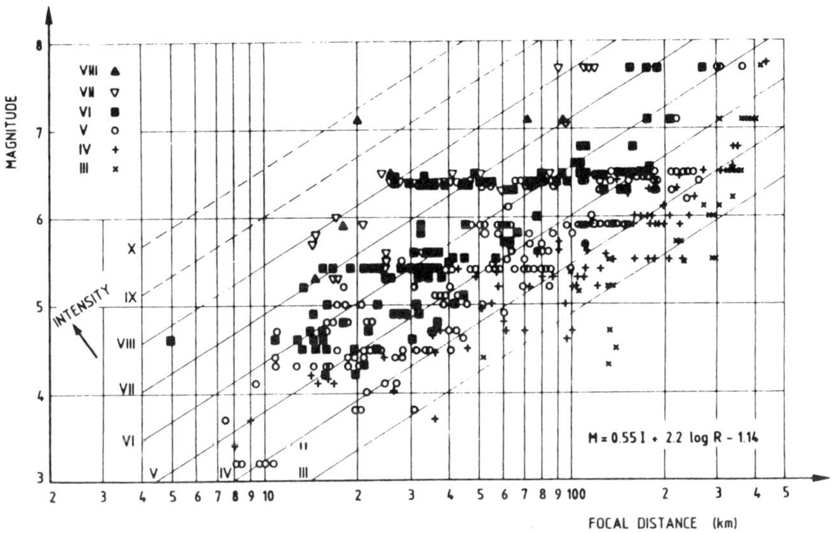

Figure 10. Correlation of intensities with magnitude and focal distance computed using 576 observations of strong ground motion from the continental United States.

The response spectra contained in the data base were sorted into several classes depending upon the intensity that was assigned to the recording site. For each of these classes, coefficients K, α , and n from equation (5), as well as the standard deviation σ , were computed for 46 frequencies between 0.07 and 25 Hz. and for five percentages of damping : 0%, 2%, 5%, 10%, and 20%. (To avoid biasing the results, only a moderate number of recordings from the San Fernando series were retained ; also, all the records used were from the free field).

Figure 11 compares the horizontal components recorded for an actual earthquake (Imperial Valley, 1940) with a corresponding synthetic spectrum, and the latter plus one standard deviation.

In view of the fact that existing strong-motion data are continually increasing in numbers, new additions to the data base may be expected to permit the computation of improved sets of coefficients, the creation of heretofore unevaluated intensity classes, or further refinements of the method, including the isolation of other significant factors such as soil conditions.

The application of the above method to obtain reference spectra for individual sites includes :

1) SMHV spectra :

a) SMHV site intensity and focal distance are determined through a seismotectonic analysis ; when no magnitude value is available, one is computed using equation (6).

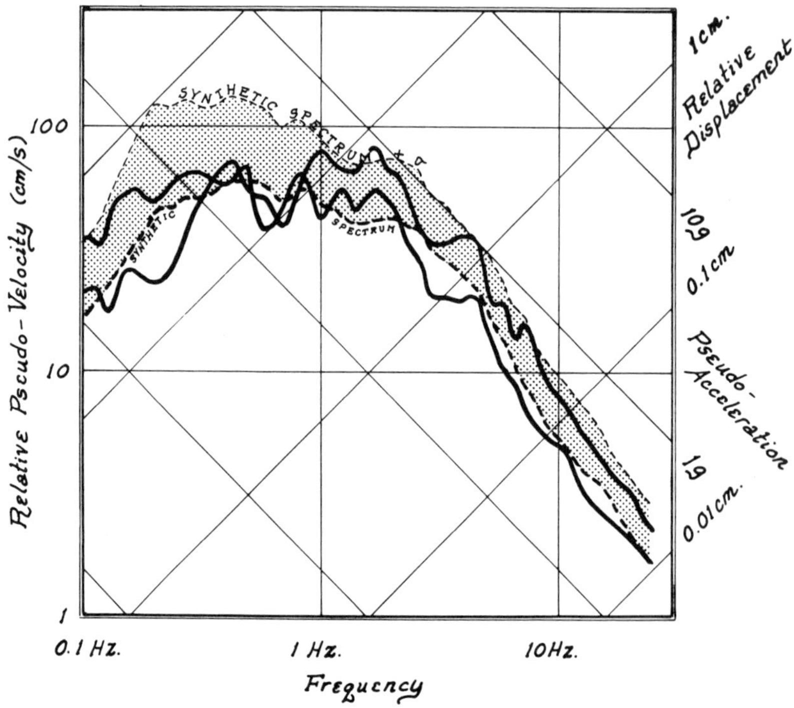

Figure 11. A comparison of synthetic spectra with those obtained for an actual earthquake (intensity VIII).

b) Now response spectra for different levels of damping are calculated with equation (5), and the coefficients for the same intensity class as the site's SMHV.

2) SMS Spectra. These spectra are derived from the preceding (SMHV) by applying increase factors derived from a statistical study of the ratios of response spectra corresponding to intensities differing by one degree. These factors, which are frequency-dependent, show values of about 2 near 1 Hz., of 1.5 at high frequencies, and of 3 in the low-frequency band (Figure 12). In cases where such coefficients are not available, a flat conversion factor of 2 is resorted to.

3. LIMITATIONS TO THE METHODOLOGY CURRENTLY PUT INTO PRACTICE AND FUTURE RELATED RESEARCH AND DEVELOPMENT WORK

The methodology described in the preceding paragraphs, founded upon a statistical basis, brings into play two sets of data, one relative to the seismotectonic setting and the other consisting of strong ground motion from elsewhere. Any shortcomings encountered are due either to the nature of the present input data or to the actual formulation of the method itself.

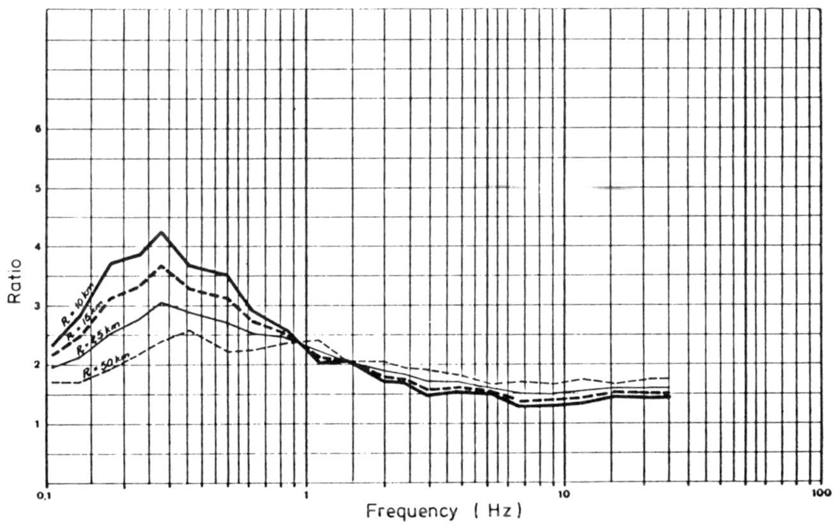

Figure 12. Increase factors used to compute SMS spectra from SMHV spectra.

The Data. Historical data are frequently quite spotty for certain regions of the globe. For most others, tectonic and neotectonic data are very incomplete, hence the difficulty in identifying active faults and in assessing their potential, and also in defining seismotectonic provinces.

The majority of well-documented, strong-motion data comes from the United States ; for a satisfactory balance with respect to seismotectonic style, ample quantities of data from other regions will accordingly need to be acquired. Furthermore, near-field recordings ($R \leqslant 10$ km) continue to be rare (almost all short-distance records shown on Figure 1 correspond to small-magnitude events). In order to predict ground motion very near the causitive fault, where severe structural damage is likely to be incurred, arrays concentrated in the immediate vicinity of frequently activated sources (such as the Imperial Valley array) should become more widespread.

The Methodology. The generalized method currently applied is found wanting in certain specific areas. The first of these we might mention, alluded to above, is the problem of predicting ground motion in the near field : here, magnitude is no longer a valid representation of the size of the source function and the attenuation cannot be accounted for in the same manner as at greater distance. It is therefore essential to devise innovative approaches (having recourse to theoretical simulation, for example). Secondly, progress needs to be accomplished in the realm of expressing the effects of local geology at the site. Existing codes for calculating soil dynamics must be validated, or perfected as the case may be, on the strength of pertinent experimental data.

In order to contribute to solving the aforementioned problems, as well as others where current knowledge and practice are deemed insufficient, the IPSN, frequently in collaboration with research teams of recognized competence, has undertaken punctual studies in some key areas :

1) Seismotectonic analysis. An effort is being made to define source zones through the comprehensive use of data from all related branches of earth sciences. Particular attention is being paid to the determination of stress regime.

2) Strong-motion data. A joint effort, undertaken by members of the EAEE, seeks to favor data exchange in Europe and promote the preparation of a uniform data base, available to all interested parties.

3) Ground motion in the near field. Use is being made of techniques developed by seismologists in the area of theoretical simulation.

4) Influence of alluvial layers on ground motion. A downhole experiment in California is in progress, with instrument packages at the surface, at the upper boundary of the bedrock, and deeper within that same layer, conducted jointly with the United States Nuclear

Regulatory Commission, for the purpose of studying the influence of the overlying soil layers on the ground motion observed atop the bedrock. The recording of both strong and weak levels of excitation is expected to contribute to circumscribing the limits on linear behavior.

5) Probabilistic analysis. The application of a probabilistic approach to seismic hazard analysis in France has at present progressed to the stage of feasibility studies ; two aspects are being investigated. A study conducted for southeastern France addresses the problem of low levels of probability associated with the upper levels of intensity and their sensitivity to different hypotheses considered. A second, statistically-based study seeks to evaluate the residual risk due to aftershocks. From a large mass of data from all over the world, curves have been computed for the probability of occurence of aftershocks of different magnitudes versus the time elapsed since the main event.

4. CONCLUSION

Despite important progress achieved over the last decade in the realm of theoretical seismology, current state-of-the-art for the engineering profession in computing ground motion has remained largely empirical. French practice in the assessment of reference ground motion is pragmatic in that it endeavors to found the analysis on reliable and abundant data and to reflect the specificities of each site. Historical data, carried back as far as may be, is well documented in France, and plays a preeminent role in French methodology. As to determining the actual spectra for a site, the process, still in common use, of selecting a standard spectrum, which is then scaled to individually-determined values of acceleration, has not been retained. Rather, the principle has been respected whereby the spectral shape itself must vary from one case to another : a big step has been taken towards the detemination of true site-adapted spectra by relating the response spectrum to macroseismic intensity, magnitude, and focal distance.

BIBLIOGRAPHY

Bernard, P. and R. Madariaga (1984). 'A New Asymptotic Method for the Modeling of Near-Field Accelerograms', Bull. Seism. Soc. Am. **74**, pp. 539-557.

Boore, D. and Porcella (1980). 'Peak Acceleration from Strong- Motion Records : A Postscript', Bull. Seism. Soc. Am. **70**, pp. 2295-2297.

Bouchon, M. and K. Aki (1977). 'Discrete Wave Number Representation of Seismic Source Fields', Bull. Seism. Soc. Am. **67**, pp. 259-277.

Godefroy, P. and A. Levret (1985). 'Nature et définition de l'aléa sismique', Génie Parasismique, V. Davidovici, editor, Presses de l'Ecole Nationale des Ponts et Chaussées, Paris, pp. 159-177.

Goguel, J., J. Vogt, and C. Weber (1985). 'La Carte Sismotectonique de la France', Génie Parasismique, pp. 187-192.

Goula, X. (1980). 'A Feasibility Study Concerning the Probabilistic Approach in Seismic Assessment', Proceedings CSNI Specialist Meeting, Lisbon, Portugal.

Goula, X., D. Hamaide, and B. Mohammadioun (1986). 'Analyse comparative des méthodes de traitement d'accélérogrammes', Premier Colloque National de Génie Parasismique, Saint-Rémy-lès-Chevreuse, France, pp. 2/1 to 2/10.

Hamaide, D., B. Mohammadioun, and G. Mohammadioun (1986). 'Données de mouvements forts actuellement disponibles au Département d'Analyse de Sûreté de l'Institut de Protection et de Sûreté Nucléaire', Premier Colloque National de Génie Parasismique, pp. 2/61 to 2/75.

Hanks, T.C. (1982). 'f_{max}' Bull. Seism. Soc. Am. **72**, pp. 1867-1879.

Joyner, W. and D. Boore (1981). 'Peak Horizontal Acceleration and Velocity from Strong-Motion Records Including Records from the 1979 Imperial Valley, California, Earthquake,' Bull. Seism. Soc. Am. **71**, pp. 2011-2038.

Levret, A. and B. Mohammadioun (1984). 'Determination of Seismic Reference Motion for Nuclear Sites in France', Engineering Geology **20**, pp. 25-38.

Massinon, B. and J. Vogt (1985) 'Sismicité historique et sismicité instrumentale', Génie Parasismique, pp. 179-186.

Ministère de l'Industrie, Service Central de Sûreté des Installations Nucléaires (1981). Règle fondamentale de sûreté 1.2.c. : Détermination des mouvements sismiques à prendre en compte pour la sûreté des installations.

Mohammadioun, B. and A. Pecker (1984). 'Low-Frequency Transfer of Seismic Energy by Superficial Soil Deposits and Soft Rocks', Earthquake Eng. Struct. Dyn., **12**, pp. 737-764.

Mohammadioun, B. (1985a). 'Calcul du mouvement du sol : Etat de l'art', Génie Parasismique, pp. 104-116.

Mohammadioun, B. (1985b). 'Pratique utilisée en France en matière d'analyse du risque sismique et travaux de recherche associés', Proceedings of an International Symposium on Safety Codes and Guides (NUSS) in the Light of Current Safety Issues, organized by the

International Atomic Energy Agency, Vienne, October 29 - November 2, 1984, IAEA-SM-275/56, pp. 639-651.

Mohammadioun, B. (1986). 'A Comparative Study of Imperial Valley (1979) Main Shock and First Aftershock Recordings as Evidence of Ground Motion Level Dependent Non-Linearities', Earthquake Eng. Struct. Dyn. **14**, pp. 317-320.

Mohammadioun, G. and J.-P. Faye (1985). 'Etude des séquences de répliques consécutives à un grand tremblement de terre en fonction du temps et de la magnitude', Génie Parasismique, pp. 131-139.

Mohammadioun, G., X. Goula, and H. Ferrieux (1985). 'Signal Processing and Spectral Analysis of the Aftershocks of the 1980 El Asnam Earthquake - Engineering Application', Physics of the Earth and Planetary Interiors **38**, pp. 174-188.

Wyllie, L., B. Bolt, M. Durkin, J. Gates, D. McCormick, P. Smith, N. Abrahamson, G. Castro, L. Escalante, R. Luft, R. Olson, and J. Valenas (1986). 'The Chile Earthquake of March 3, 1985', Earthquake Spectra **2**, pp. 249-513.

RECORDING SMALL REGIONAL EARTHQUAKES ON A LARGE STRUCTURE: THE CATHEDRAL OF STRASBOURG

M. Cara, C. Labat, M. Frogneux, G. Wittlinger, J.M. Holl
Institut de Physique du Globe de Strasbourg
5 rue R. Descartes
F-67084 Strasbourg Cedex
France

ABSTRACT. Three magnitude 2.8 - 3.8 earthquakes located at 99-186 km from Strasbourg were recorded on different points of the Cathedral during the 1985 summer. Three 3-component seismometers and a telemetry system were used for this experiment. One seismometer was installed beneath the Crypt, the second one on different keystones of the main Nave and the third one in the Spire, 103 m above the ground floor. Significant vibrations of this large structure are excited by the earthquakes. In particular, large resonance effects have been put in evidence for the main tower at 1.2 Hz with a magnification of the motion of two orders of magnitude when passing from the Crypt to the recording point located in the Spire. Fourier analysis of the observed vibrations are presented and are qualitatively discussed.

1 INTRODUCTION

The Cathedral of Strasbourg is a sandstone large structure, 142 m high. Its construction started in 1190 and ended in 1439 near the fall of the Gothic period. At this time and up to the XIX th century, it was the highest man made structure. One of it particularities is to be completely built in sandstone, including the Spire with its famous conical upper part designed by Jean de Hültz from Köln (figure 1). Since it was built, the Cathedral has actually more suffered from storms, fire and human actions than from earthquakes. During the large Basel earthquake in 1356, the main tower was not yet built and it seems that only ornamental parts of the building suffered. The estimated M.S.K.intensity reached VII in Strasbourg (Vogt, 1979). In 1728, a local event near Lahr caused a rather serious shaking with cracks appearing in the vaults of the main Nave and again serious damage to the ornamental parts (J. Vogt, personal communication).

During the 1985 summer several measurements of vibrations of the Cathedral were made. Most of the observed vibrations are induced by the action of the wind and by human activities - including the ringing of the large bells, one of them weighting 9 tons. In addition to those vibrations, we were lucky enough to record three regional earthquakes on different points of this large structure. If several measurements of vibrations were made on the Cathedral in the past, it is likely that we made the first records of vibrations induced by earthquakes. The purpose of this short note is to give a description of the special instrumental setting we employed and to present the preliminary observations.

J. Bonnin et al. (eds.), Seismic Hazard in Mediterranean Regions, 195–203.

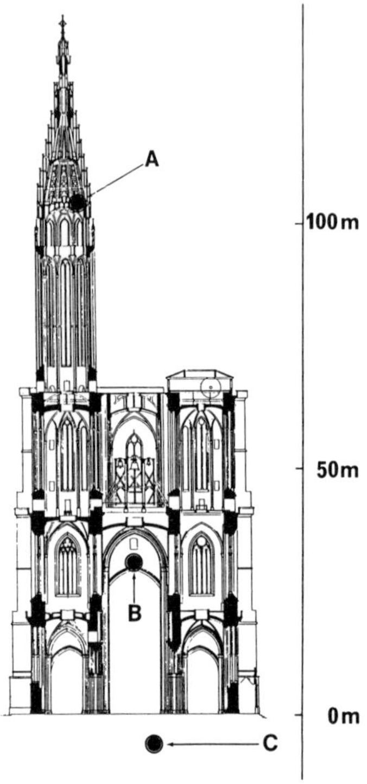

Figure 1. Cross-section view of the Cathedral of Strasbourg. Points A, B and C indicate the location of the seismometers.

2 THE INSTRUMENTAL SETTING

During the passive seismic recording experiment we did from july to november 1985 use was made of three triaxial seismometers of 1 second free period operating with a 0.6 damping factor. We failed in identifying any teleseismic signal during this experiment but, surprisingly, Lg waves coming from small regional earthquakes were fairly well recorded.

The main difficulty in such an experiment was to use a triggering device appropriate to the important level of the vibrational noise on such a large structure. Any local triggering system, as the one we experimented in a first stage, leads to fill up the recording tape very quickly due with a large number of false event detections. For this reason we have eventually decided to trigger the recording device from an external source of information using the local permament network of the Institute of Physique du Globe with stations installed at a few tenth of kilometers from Strasbourg in the Vosges mountains. To facilitate this operation radio links were established in FM mode between the seismometers and a local central point in the Cathedral. A second radio link was established between the Cathedral and the main building of the Institut of Physique du Globe where the recording on trigger mode was done (figure 2). The conversion from analog to digital PCM signal was made at the intermediate point in the Cathedral. The radio link between this point and the IPG building was made in digital mode. All the records were made in PCM mode using the recorders designed by Holl and Speisser (1981).

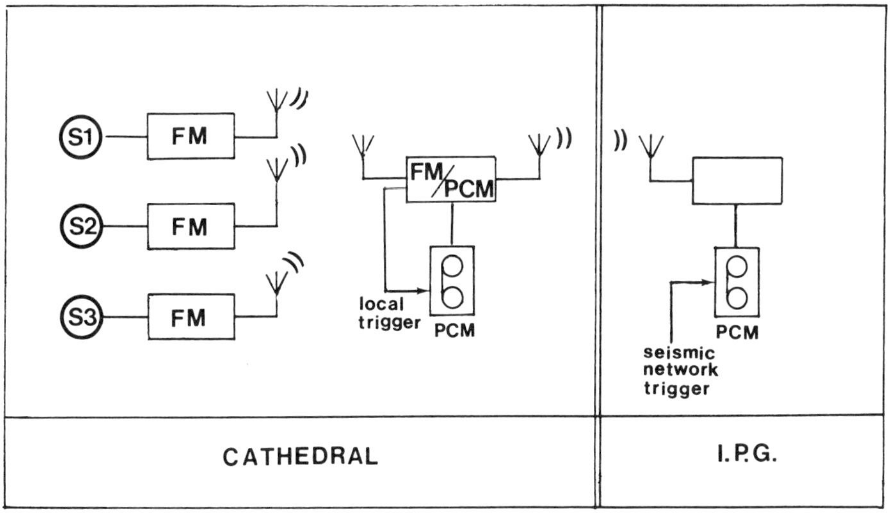

Figure 2. Instrumental setting in the Cathedral and the main building ofI.P.G.. S1, S2 and S3 represent the 1 Hz 3-component seismometersinstalled at different points of the monument. The intermediateFM/PCM conversion device was installed beneath the roof of the main Nave. Two audio tape recorders were used, one trigered by the Cathedral seismometers and one trigered by the IPG seismic network.

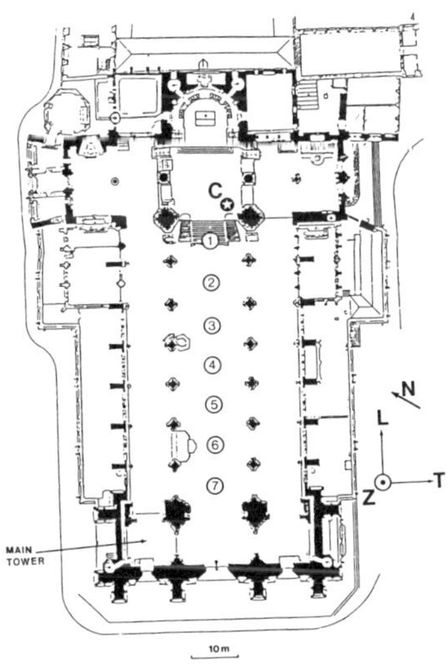

Figure 3. Map view of the Cathedral and orientation of the seismometer components Z, L, T. The location of the seismometer in the Cryptis indicated by the letter C.

During the experiment, the seismometers were set in different locations. One 3-component seismometer was permanently installed in a well, 3 m below the crypt. It gives us the reference motion of the ground beneath the Cathedral. The two other 3-component seismometers were installed on the monument itself. Because our recording system was limited to 8 channels, one of these seismometers was operated with two horizontal components only. In an attempt to observe large amplification effects, the two mobile seismometers were installed on different keystones of the Nave of the Cathedral at 30 m and in the main Spire at 103 m above the ground floor. Figure 3 shows the location of the seismometers in a map view together with the orientation of the different components Z, L and T. The seismometers used in these experiments are 1 Hz Mark Product L43D instruments.

3 THE REGIONAL EARTHQUAKES RECORDS

The three regional earthquakes recorded from July 3 to November 13 1985 are located at distances ranging from 99 to 186 km from the Cathedral of Strasbourg (Table I, figure 4).

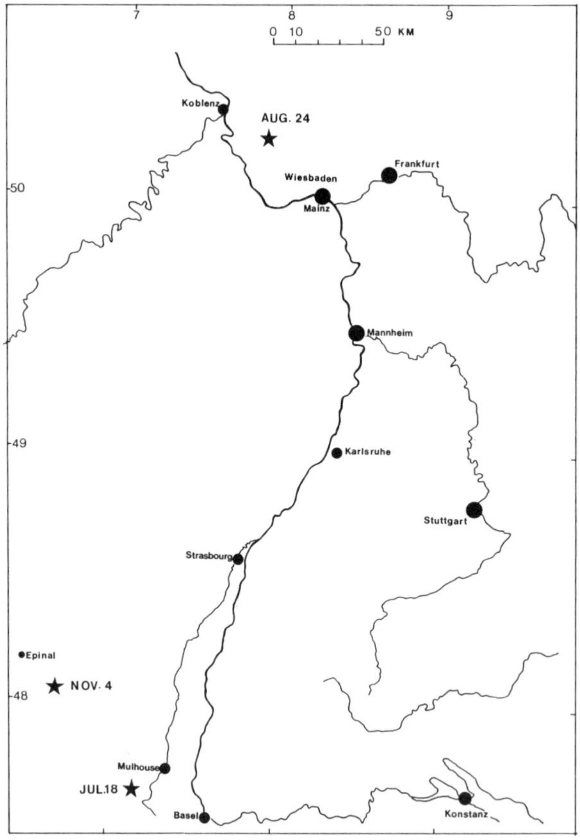

Figure 4. Map of the surroundings of Strasbourg showing the location ofthe three 1985 earthquakes presented in this study.

The three events all occured early in the morning from 7:04 to 8:34, local time. The nature of the signal and the time of occurence makes no doubt that these events are of seismic origin.

date	H	MIN	S	Latitude	Longitude	Magnitude
July 18, 1985	6	34	5.9	47.66	7.16	2.8
August 24, 1985	6	8	40.4	50.25	7.89	3.
November 4, 1985	6	4	14.0	48.06	6.68	2.9

Table 1: epicenters of the regional earthquakes analysed in this study

The largest motion was recorded for the august 24, Magnitude= 3.8 event. The

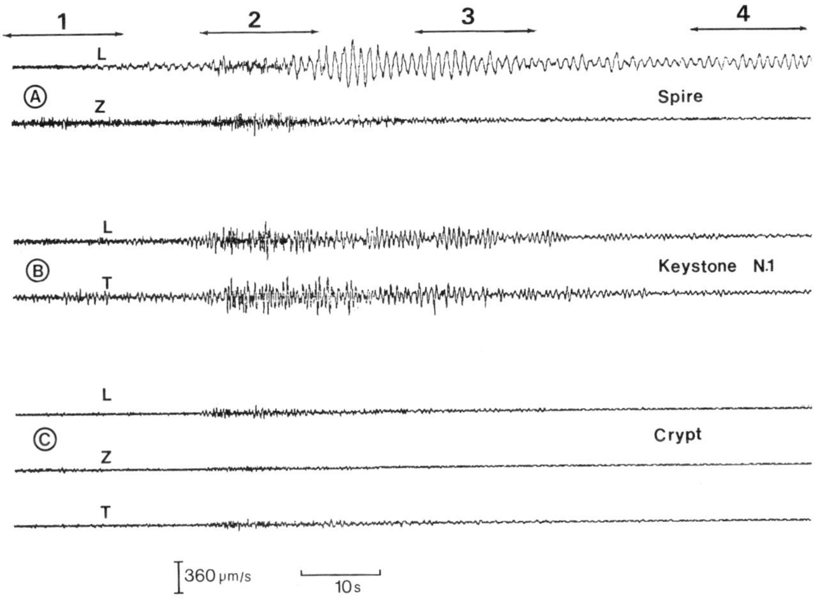

Figure 5. Aug. 24, 1985, magnitude 3.8 earthquake. The groups of traces A, B and C correspond to the different recording points. See caption for figure 3 for the orientation of Z, L and T. The numbers on the upper trace indicate the different time windows used for time-frequency analysis.

corresponding records are presented in figure 5. This example is also the only one for which we have a record of the motion of the Spire excited by an earthquake, unfortunately on two components only (L and Z). The largest motion is observed at point A along the L component with a clear free oscillation of 0.8 sec.

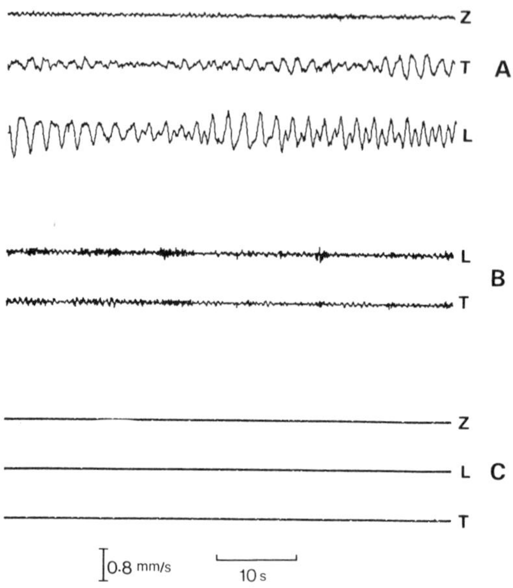

Figure 6. Vibrations due to the wind in points A, B and C. See caption for figure 5.

In order to make a comparison with other types of vibrations we show in figure 6 the resulting action of the wind at points A, B, and C. These records were made on november 5, 1985 during a storm while the wind was blowing mainly from SW (i.e parallel to the L direction). Note that negligible motion is observed on the vertical component on point A and that the vibration is not visible at point C beneath the Crypt.

4 SPECTRUM ANALYSIS AND COMMENTS

The Fourier spectrum of the records displayed in figure 5 and 6 for the L component in the main tower are presented in figure 7. Note, on the one hand, that the large peaks excited by the wind action at 0.7 and 1.5 Hz are only weakly visible on the seismic spectrum. On the other hand, the large 1.2 Hz peak excited by the seismic Lg wave is not visible on the wind spectrum. The spectrum observed for the wind is confirmed by other records with two marked peaks at 0.7 and 1.5 Hz. We have no other seismic spectrum for the L component in the main tower to confirm this 1.2 Hz peak.

Indirect confirmation of this peak can be made, however, by looking at the vibration of the keystones. In case of the August 24th earthquake, figure 8 displays a time-frequency analysis of the motion at point B - the Nave keystone n°1 on the map displayed in figure 3. Before the motion induced by earthquake starts (fig. 8 - window 1), the low amplitude noise spectrum is observed. At the begining of the earthquake motion (figure 8, window 2) the spectrum presents an envelope which is similar to that of the whole motion observed in the Crypt for this earthquake (fig. 7-C). As time goes on, the 1.2 Hz peak grows up and eventually dominates the spectrum (figure 8 - window 4). It appears quite likely that the high

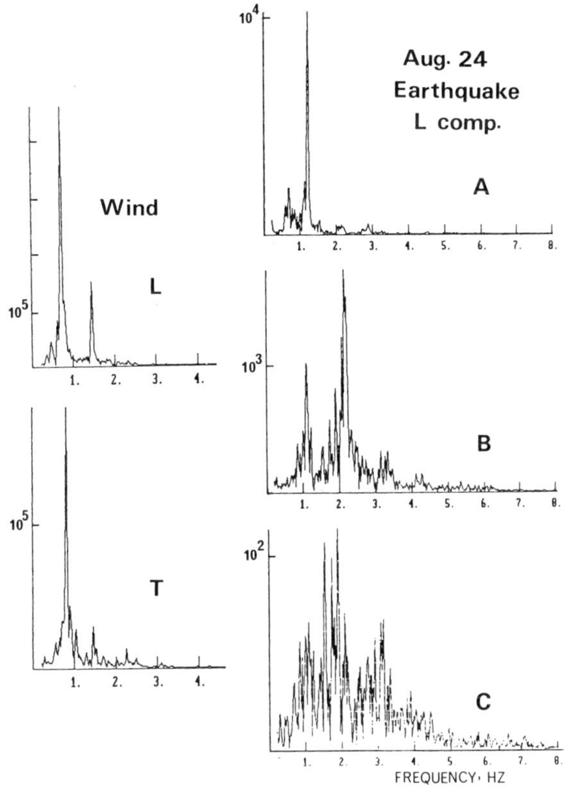

Figure 7. Fourier amplitude spectra of the ground motion for the records displayed in figure 6 (common arbitrary units for all spectra). Horizontal vibration of the Spire due to a strong wind (components L and T) and L-component spectra for recording points A (Spire), B (keystone N° 1) and B (Crypt) for the August 24, 1984 event.

Q large free motion of the Spire acts as a forcing force on the rest of the building so that the 1.2 Hz motion dominates the motion of the Nave after 1 minute of vibrations. The same 1.2 Hz peak is also clearly observed for the July 18th earthquake on the component L for the keystones n°1 and 7. If these 1.2 Hz peaks are interpreted as forced vibrations of the Nave caused by the oscillation of the Spire, one has an indirect confirmation of the observation made for the August 24 earthquake. No such observation, however, is made for the November 4, 1985 earthquake indicating thus that different excitations of free vibrations of the structure have to be expected for earthquakes of similar magnitudes at close epicentral distances.

Other inferences can be made by comparing the amplitude of the motion at point A or B to that of point C beneath the Crypt. It is in particular possible to estimate an order of magnitude for the transfer functions relating the ground motion to the motion of the different points of the structure. If we consider for example the resonance frequency of 1.2 Hz for point A in the Spire, the magnification reaches about 100 on the L component. The vertical component spectrum exhibits even larger magnification near 1.3 Hz were a large

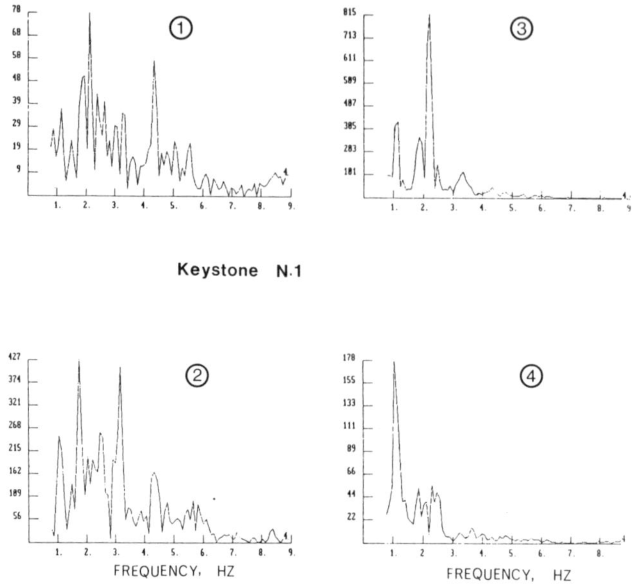

Figure 8. Time-frequency analysis of the vibration along L component on keystone n°1 for the earthquake of August 24, 1985. The circled numbers are related to the different time windows marked on the top of figure 5.

resonnance effect is also observed at point A. The resonance peaks are less pronounced for the recording points B on the top of two keystones of the main Nave. Magnifications of about 10 are observed between the Crypt and the keystone n°1 near 2 Hz for the L and T components. On the Z component the magnification reaches a factor of about 3 in the range 2-4 Hz.

Detailed examination and modeling of the observed vibration spectrum remains well beyond the scope of the present paper. In particular, it appears clear that understanding the way by which the different modes are excited in the main tower, depending on the source of vibration - earthquake or wind- would necessitate extensive finite element computations in the way followed for example by Fanelli (1986) for antic monuments in Italy. In addition to such a numerical modelization approach, let us just stress out here that a lot can be learned on the actual dynamical behavior of large structures under realistic earthquake situation from direct measurement of small earthquakes as we have done.

For the sake of illustration let us conclude by making some oversimplified estimates for the 1356 Basel earthquake, a reference earthquake in the upper Rhine graben area. At the time of this earthquake, the main Nave was built but the Cathedral was not yet completed, the first 66m only being built. An estimated magnitude for this earthquake is 6 1/4 (Ahorner and Rosenhauer, 1978). According to the macroseismic epicenter, South of Basel (Mayer Rosa and Cadiot, 1979), the epicentral distance from Strasbourg is approximately 120 km. Correcting for the difference in distances by using the 1.66 log(D) term of the IASPEI magnitude formula (Vanek et al., 1962) and comparing with the motion due to the 3.8 earthquake of August 24 1985, one can get a rough order of magnitudes for the motion of the structure at points where the August 24 earthquake was recorded. At point A in the

Spire, a linear prediction of the motion leads for example to an amplitude of 30 mm peak to peak. In order to get a better estimate one should of course take into account much more parameters in this estimate. In addition to the likely non-linear effects for this rather large amplitude, one should take into account the differences between the frequency content of the source spectra of the two earthquakes and also the likely differences of attenuation along the two paths. Due to the difficulties for estimating the non-linear effect on a large stone structure, due to our poor knowledge of the focal mechanism of the Basel earthquake, due also to large uncertainties in the scaling laws near 1 Hz (Gusev, 1983), and finally due to our poor knowledge of lateral variations of Lg wave attenuation it is doubtful, however, that much more precise estimates could be made from available data for this earthquake.

Acknowledgements

This work was conducted in close cooperation with the Laboratoire Régional de l'Equipement de Strasbourg, thanks to a financial support from the "Service des Monuments Historiques de France". Enthusiastic help provided by the staff and the students of the Institut de Physique du Globe de Strasbourg during the "field work" was decisive for the success of this experiment.

REFERENCES

Aki, K., 1987. Physical Theory of earthquakes, this volume.

Cara,M., Cote, A., Durand, J., Granet, M., Hoang Trong, P. and Holl, J.M., 1986. Le réseau national de surveillance sismique (ReNaSS), *Bull. liaison Labo P. et Ch., 143*, 30-39.

Fanelli, M., 1986. Indagi dinamische sui monumenti antichi: alcune riflessioni e qualche risultato, *Construzioni, anno 35, n. 372*.

Gusev, A. A., 1983. A descriptive statistical model of earthquake source radiation and its application to an estimation of short-period strong motion, *Geophys. J. R. astron. Soc., 74*, 787-808.

Holl, J.M. and Speisser, 1981. Stations sismologiques numériques autonomes, *report, I.P.G., Strasbourg*.

Mayer-Rosa, D. and Cadiot, B., 1979. A review of the 1356 Basel earthquake: basic data, *Tectonophysics, 53*, 325-333.

Vanneck,J., Zatopeck, A. et al., 1962. Standardization of magnitude scales. *Izv. Akad. Nauk. S.S.R., Ser. Geofiz., 2*, 153-158.

Vogt, J., 1979. Les tremblements de terre en France, *Mémoire BRGM n° 96*, Orléans.

SYNTHETIC ISOSEISMALS: APPLICATIONS TO ITALIAN EARTHQUAKES

P. Suhadolc (1), L. Cernobori (1),
G. Pazzi (1), and G.F. Panza (1,2)

(1) Istituto di Geodesia e Geofisica
Universita' di Trieste,
34100 Trieste, Italy
(2) International School for Advanced Studies
34100 Trieste, Italy

ABSTRACT. Synthetic isoseismals have been computed for two large Italian earthquakes - November 23, 1980 Irpinia and May 6, 1976 Friuli - by vector composition, in the horizontal plane, of the SH-wave and P-SV-wave displacement fields. For pratical purposes - significant saving of computer time - the computations have been performed with a cut-off frequency of 0.1 Hz. This is not a limitation since, when the cut-off frequency is increased to 1 Hz, the shape of the isolines does not change significantly. In the presence of a relevant sedimentary cover, the amplitude ratio between the SH-waves and the radial component of P-SV-waves displacement field depends on the hypocentral depth and on the elastic parameters around the source. Such dependence can be empirically described by a single adimensional parameter. For both events the shape of the observed isoseismals is essentially controlled by the SH-waves. The macroseismic field of the November 23, 1980 Irpinia earthquake can be explained in terms of a single point source with a focal depth ranging between 8 and 12 km, and a focal mechanism averaging the very complex rupture process characterizing this event (Suhadolc et al., 1987). In the case of the Friuli earthquake, a remarkable fit between experimental and theoretical data has been obtained with a focal mechanism in good agreement with that proposed by Cipar (1981) for a 7 km deep point source.

1. INTRODUCTION

The macroseismic information obtained through the quantification of the effects and damages produced by an earthquake and represented by isoseismal lines is perhaps the oldest scientific tool to study earthquakes (e.g. Mallet, 1862). An enormous amount of isoseismal maps exists especially in countries with a long seismological tradition, as for example Italy (CNR, 1985). These data have been processed, up to now, using empirical formulas (e.g. Blake, 1941; Shebalin, 1959;

J. Bonnin et al. (eds.), Seismic Hazard in Mediterranean Regions, 205–228.

Sponheuer, 1960) in order to deduce a rough estimation of the depth of the source and attenuation laws (e.g. Anderson, 1978; Chandra, 1979). Shebalin (1972) was among the first who suggested that macroseismic data contain also information on the source geometry and rupturing processes. In the last years the development of algorithms capable of synthetizing, over a very broad band of frequencies, wave propagation in vertically heterogeneous media (e.g. Panza et al., 1986) has allowed a quantitative interpretation of ground motion during an earthquake. A first application of synthetic seismograms to the study of observed isoseismals has been presented by Panza and Cuscito (1982), who considered only SH-waves, up to a maximum frequency of 0.1 Hz. In this paper we extend their method to the combined effect of the SH-wave and P-SV-wave components of the displacement field. The experimental data considered are due to the Italian earthquakes of November 23, 1980 in Irpinia and of May 6, 1976 in Friuli.

2. METHODOLOGY

The complete synthetic seismograms used in this study are obtained with the normal mode summation method (Liao et al., 1978; Panza, 1985). For shallow sources, surface waves dominate the ground motion at moderate distances - tens of kilometers - and at periods greater than 1 or 2 s (Hanks, 1975; Herrmann, 1977; Herrmann and Nuttli, 1975 a,b; Kawasaki, 1978) and therefore a very efficient way of obtaining synthetic signals of relevance for our problem is to consider only a few modes. We have

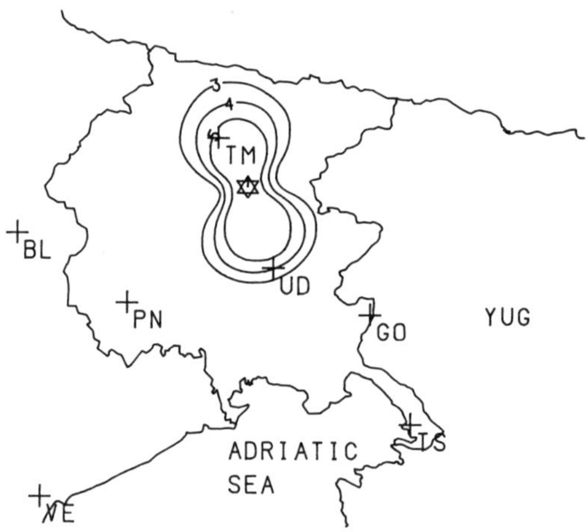

Figure 1a. Displacement field (cm) for the radial component of P-SV-waves corresponding to a point source: strike 225°, dip 10°, rake 60°, and depth 6.5 km. Seismic moment $M_0=10^{25}$ dyn cm. Cut-off frequency 1 Hz. Structural model FRIUL5b (see Fig.9). BL = Belluno, GO = Gorizia, PN = Pordenone, TM = Tolmezzo, TS = Trieste, UD = Udine, VE = Venezia.

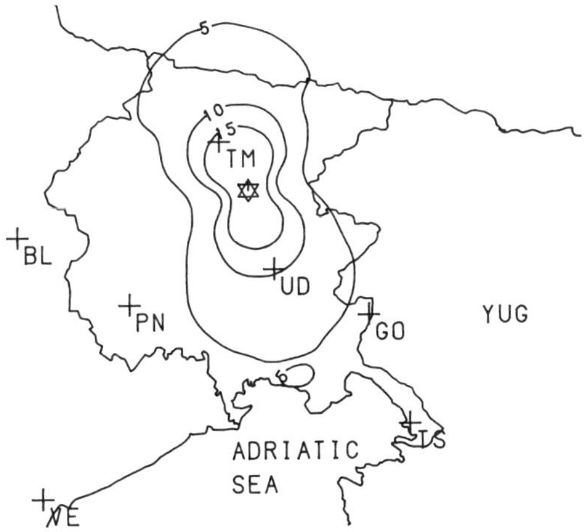

Figure 1b. Same as Fig. 1a, but for the velocity field (cm/s).

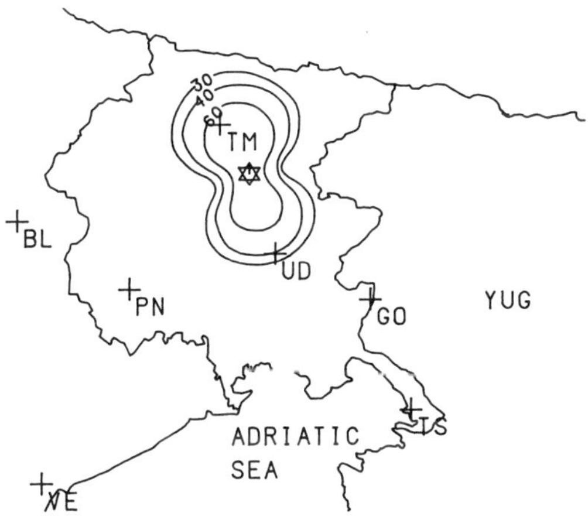

Figure 1c. Same as Fig. 1a, but for the acceleration field (cm/s^2).

decided to construct theoretical isoseismals using the equal value contours of the displacement field due to a point source as in Panza and Cuscito (1982). This choice, even if one usually empirically relates intensity to acceleration (e.g. Trifunac and Brady, 1975; Murphy and O'Brien, 1977), is an appropriate one. In fact, for frequencies as high as 1 Hz, the shape of isolines for displacement, velocity and acceleration is approximately the same and does not vary with frequency (Fig.1a to 1c). Therefore, to construct theoretical isoseismals, we consider the vector sum of the radial and transverse components of displacement and use its maximum zero-to-peak value to represent the distribution in space of ground shaking. As a preliminary step, it is

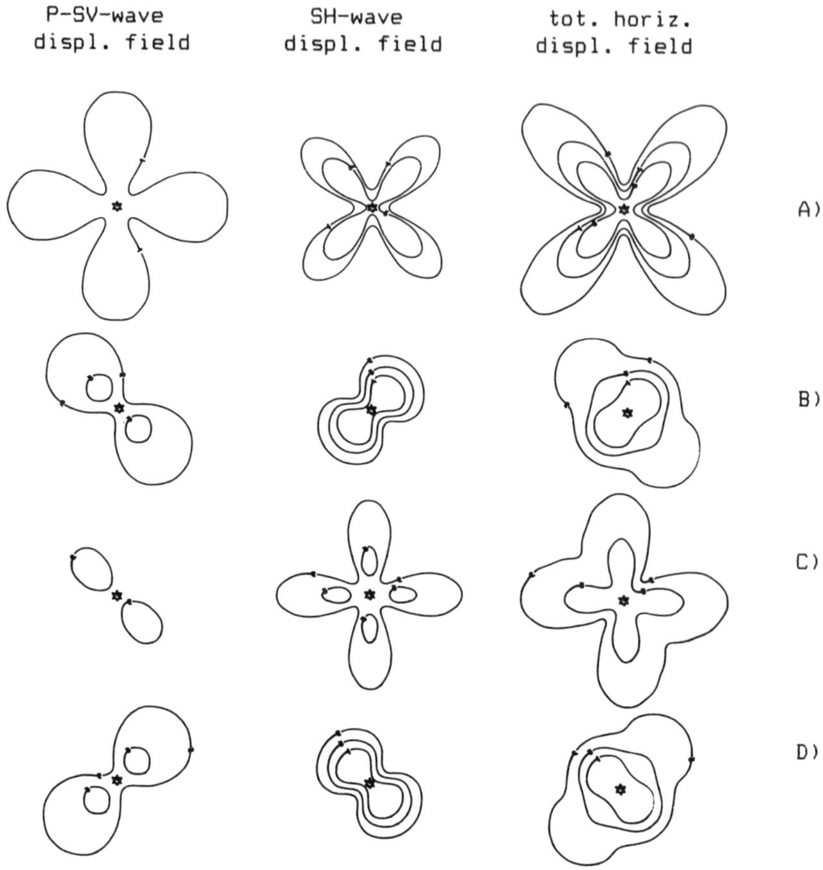

Figure 2. Displacement fields (cm) corresponding to the four elementary tensors (Aki and Richards, 1980) for a point source 17 km deep with a strike of 225°. Starting from the left, the first column represents the P-SV-wave displacement field, the second the SH-one, while the third represents the total horizontal displacement field. Each row corresponds to a different mechanism, from the top to the bottom A) dip 90°,rake 0°; B) dip 90°,rake 90°; C) dip 45°, rake 90°; D) dip 0°,rake 0°. Seismic moment = 10^{26} dyn cm. Cut-off frequency 0.1 Hz. Structural model FRIUL5B (see Fig.9).

very useful to analyse the isoseismals corresponding to the four fundamental moment tensors (Aki and Richards, 1980). An example, for a hypocentral depth of 17 km and for a fault strike oriented 225°, is shown in Fig.2. The SH-wave radiation prevails in the case of a vertical fault with a zero rake and in the case of an inverse fault with a dip of 45°; the total horizontal displacement field has four lobes in both cases. The efficiency in the radiation of P-SV and SH-waves is about the same when the fault is vertical (rake 90°) or horizontal (rake 0°). In these last two cases the total horizontal displacement field has two lobes and the elongation changes in going from the inner to the outer isoseismals.

2. THE NOVEMBER 23, 1980 IRPINIA EARTHQUAKE

On November 23, 1980 at 18h 34m GMT, an Ms = 6.9 (USGS) earthquake occurred in Irpinia (southern Italy). In spite of the fact that this event (M_o=2 10^{26} dyn cm, Deschamps and King, 1983) has been one of the best monitored seismic events in the last years, there is still disagreement about its focal parameters (see Table I in Suhadolc et al., 1987, this volume). Only the orientation of the principal plane, which is related to a normal fault with an Apenninic trend is well constrained. On the contrary, there is a large spread of proposed values for the direction of slip. The hypocentral depth is not very well constrained either. This is probably due in part to the different methods and frequency ranges used to study this event, in part to the complexity of the source (e.g. Berardi et al., 1981; Suhadolc et al., 1987).

Structural effects due to the very complicated geological situation of the Irpinia region have to be considered as well. The epicentral area is placed in the Campania-Lucania Apennines, a nappes structure generally trending NW-SE, formed and emplaced in the Cretaceous-Lower Pliocene interval. Underneath the Apennines the maximum crustal thickness is of about 40 km, while the lithospheric one is decreasing from the Adriatic to the Tyrrhenian Sea from about 110 km to 30 km (Calcagnile and Panza, 1981). The Bradanic foretrough located on the Adriatic side of the Apennines is filled with Pliocene and Quaternary sediments reaching the maximum thickness of about 3 km. This sedimentary cover has undergone vertical displacements only, therefore is less disturbed than the Apenninic nappes. The sediments of the Bradanic foretrough are piled up on the top of the Apulian foreland, which is lowered by faults. From the inversion of phase velocities of Rayleigh waves, Calcagnile and Panza (1980) obtain, for the Apulian plate, an S-wave velocity of about 4.6 km/s for the high-velocity subcrustal layers, "the lid", and a very well-marked asthenospheric low-velocity channel with S-wave velocity of about 4.2 km/s. For a detailed geological description of these areas see, for example, D'Argenio et al. (1975), Pescatore (1981), Ciaranfi et al. (1983).

The observational data are shown in Fig.3 (Postpischl, 1981). With the exception of the inner isoseismal, the elongation along the Apenninic chain is quite evident. Panza and Cuscito (1982) have explained this trend simply in terms of source geometry. They also showed that the theoretical isoseismals are slightly perturbed if one takes into account the main lateral heterogeneity in the lithosphere-asthenosphere system of the area - the boundary between the Apenninic chain and the Bradanic foretrough.

The orientation of the point source approximating the main event is quite well determined, except for the rake (see Suhadolc et al., 1987,

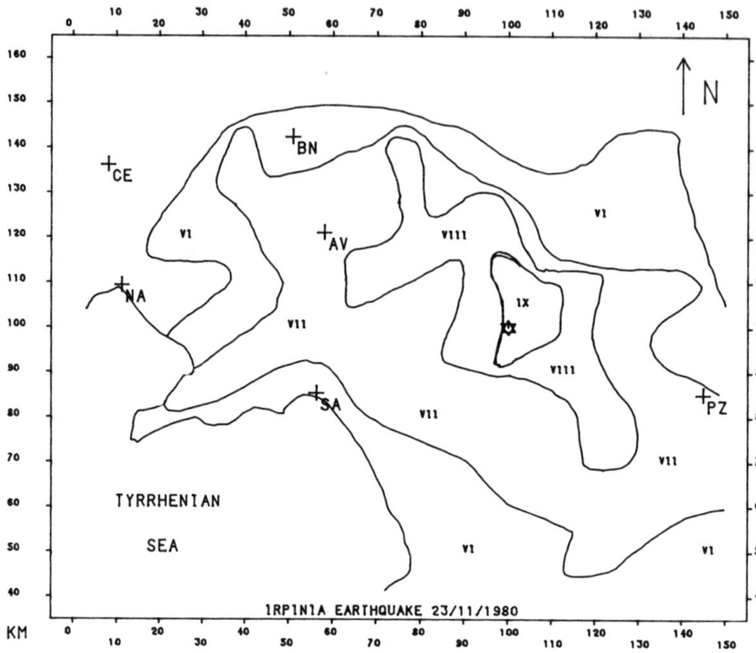

Figure 3. Observed isoseismals of the November 23, 1980 Irpinia earthquake, MSK scale (after Postpischl, 1981). AV = Avellino, BN = Benevento, CE = Caserta, NA = Napoli, SA = Salerno, PZ = Potenza.

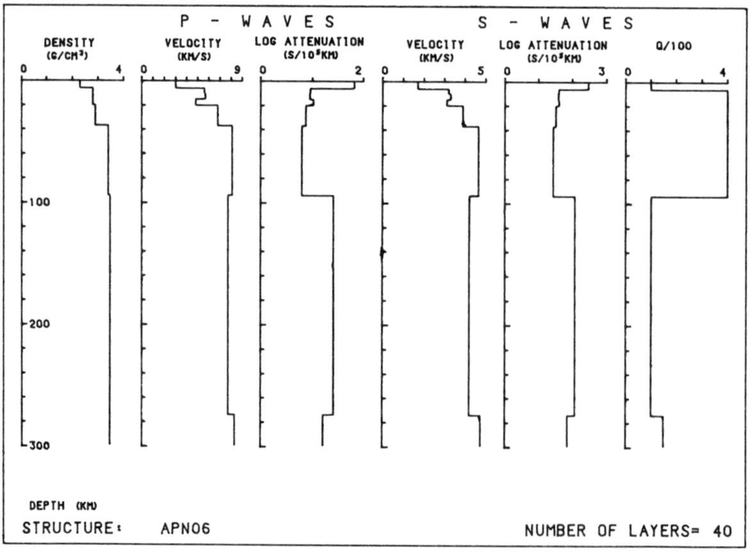

Figure 4. Structural model APN06.

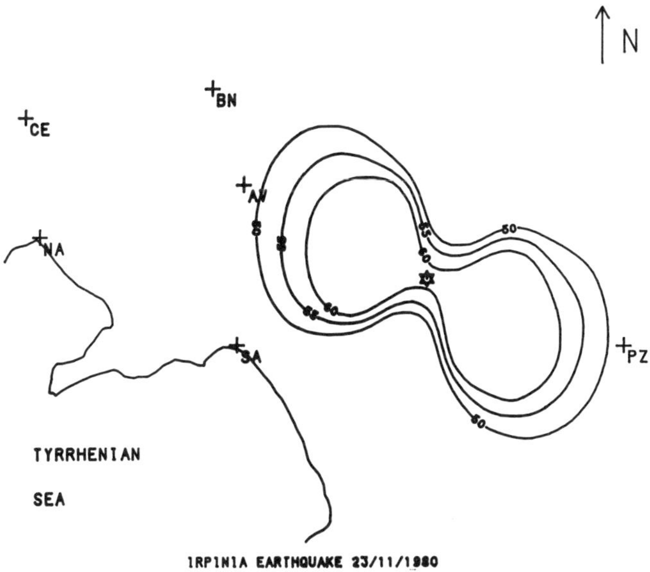

Figure 5. Displacement field (cm) for SH-waves corresponding to a point source: strike 115°, dip 70°, rake 275°, and depth 17.5 km. $M_o = 10^{27}$ dyn cm. Structural model APN06. Cut-off frequency 0.1 Hz.

this volume). Preliminary tests, made for various values of rake, have shown that an acceptable fit between the theoretical and observed isoseismals is reached only when considering almost pure normal faulting. The value of the rake has been therefore kept fixed at 275° in this study. This value can be taken as representative of the point source averaging the whole complex rupture process of the Irpinia event (e.g. Suhadolc et al., 1987; Bernard and Zollo, 1987).

In the present work we focus on the effects of variations in hypocentral depth, elastic and anelastic parameters of the uppermost layers, and their thickness. Starting from the structural model "A" of Del Pezzo et al. (1983), which has been extended to the upper mantle according to Panza et al. (1980), the model shown in Fig.4 has been obtained. This model, APN06, is meant to represent the Apenninic zone. The Bradanic foretrough, model BRA02, is modelled assuming for the low-velocity and highly-dissipative superficial layer a thickness of 2 km, while for the Apenninic zone the thickness of such a layer has been tentatively set equal to 6 km. The adopted models are obviously representative of average situations, and this, for the moment, is one of the main limitations of our approach. Nevertheless, in the frequency range considered, this approximation is acceptable, as we will show, at least to reproduce the main features of the observed isoseismals. The source model considered initially is almost the same as the one used by Panza and Cuscito (1982): focal depth 17.5 km, dip 70°, strike 115°, rake 275°. The SH-wave displacement field for the structure APN06

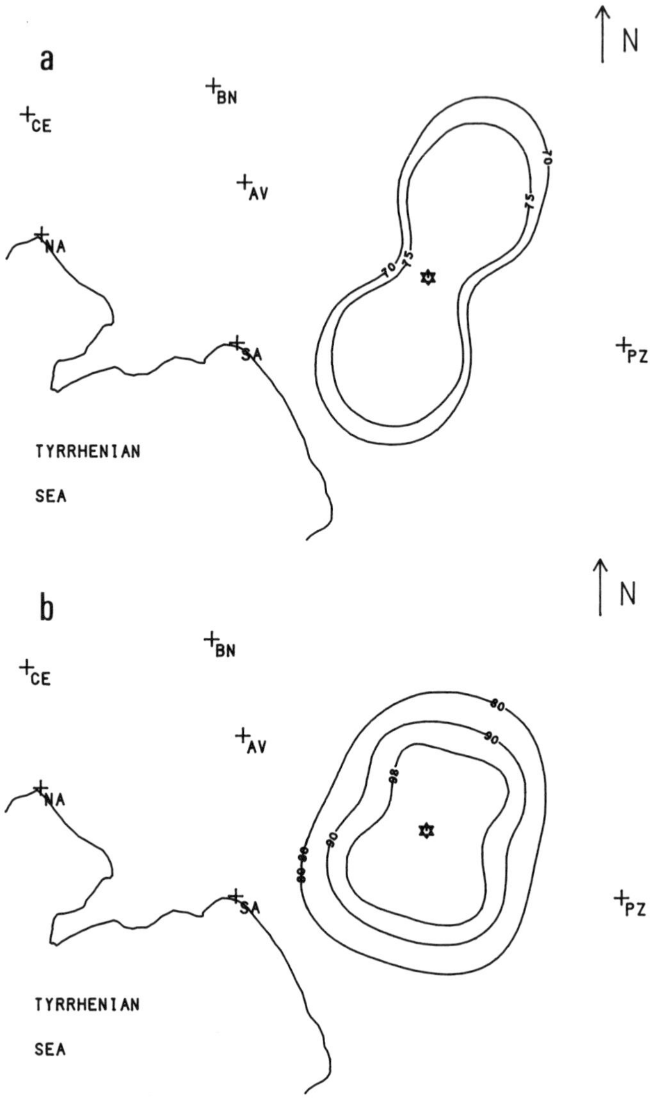

IRPINIA EARTHQUAKE 23/11/1980

Figure 6. (a) same as Fig. 5a, but for P-SV-waves. (b) same as for Fig.5 but for the total horizontal-displacement field amplitude.

(Fig.5) shows an elongation in the direction of the Apenninic chain. When the radial component of the P-SV-wave displacement field (Fig.6a) is added, the situation changes radically and the total horizontal displacement field looses the Apenninic elongation (Fig.6b), since the P-SV-wave displacement field, oriented perpendicularly to the mountain

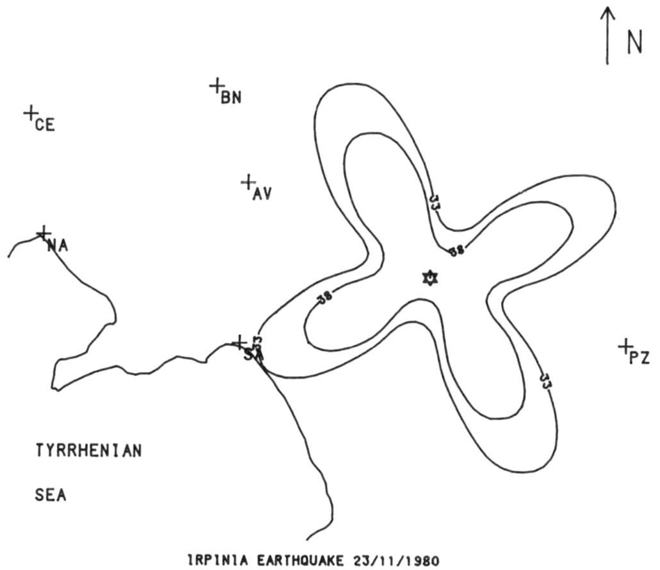

Figure 7a. Displacement field (cm) for SH-waves corresponding to a point source with fault mechanism as in Fig. 5, focal depth 6 km, structural model BRA02.

chain, dominates. The same is true if the structure BRA02 is used.

Within the space of allowable solutions, the orientation and shape of the P-SV-wave displacement field are not very sensitive to changes in the fault plane parameters. We can conclude that, for the considered event, the elongation of the experimental isoseismals is mainly determined by the SH-wave displacement field.

An accord with the experimental data can be therefore achieved varying the source parameters in such a way that the amplitude ratio between the two fields changes by enhancing the displacements due to the SH-waves. On the base of an analysis of eigenfunctions for P-SV- and SH-components of motion and taking into account the expression for the radiation pattern for the two displacement fields (Harkrider, 1970), the simplest way to achieve this effect, keeping the other source parameters fixed, is the change in the hypocentral depth.

At a hypocentral depth of 6 km, for the structure APN06, the SH-wave displacement field has two lobes with a remarkable elongation parallel to the mountain chain, quite similar to the one of Fig.5, while for the structure BRA02 a clear elongation cannot be seen, since the displacement pattern has four lobes (Fig.7a). The radial component of the P-SV-wave displacement field computed for the Apenninic structure is almost six times smaller than the corresponding one computed for the Bradanic structure and it is negligible in comparison with the SH-wave displacement field. As a result, the pattern of the total horizontal field for the model APN06 (Fig.7b), is almost coincident with the

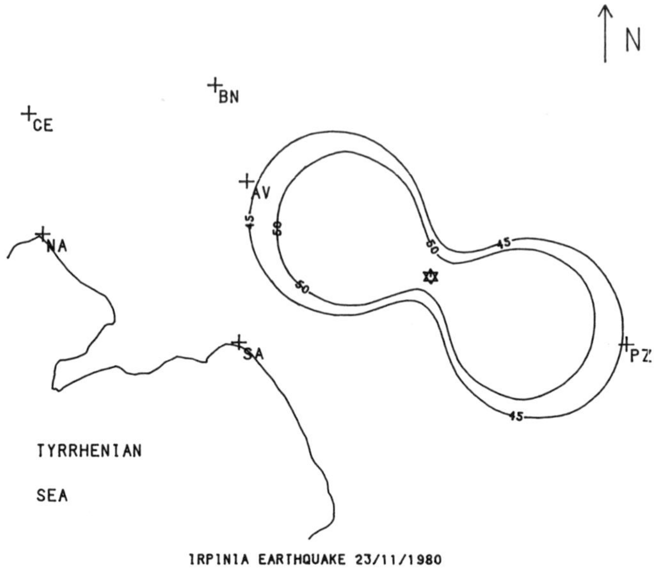

Figure 7b. Total horizontal displacement field (cm) corresponding to a point source with fault mechanism as in Fig. 5, depth 6 km. Structural model APN06.

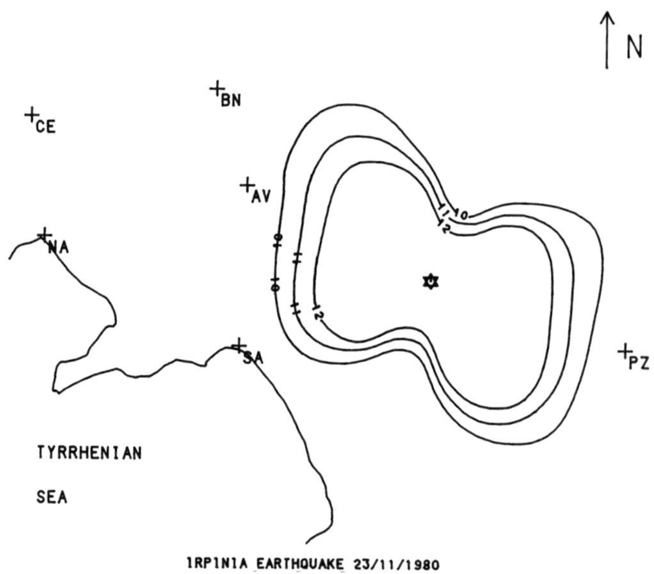

Figure 7c. Total horizontal displacement field (cm) corresponding to a point source with fault mechanism as in Fig. 5, depth 8 km. $M_o = 10^{26}$ dyn cm. Structural model APN06.

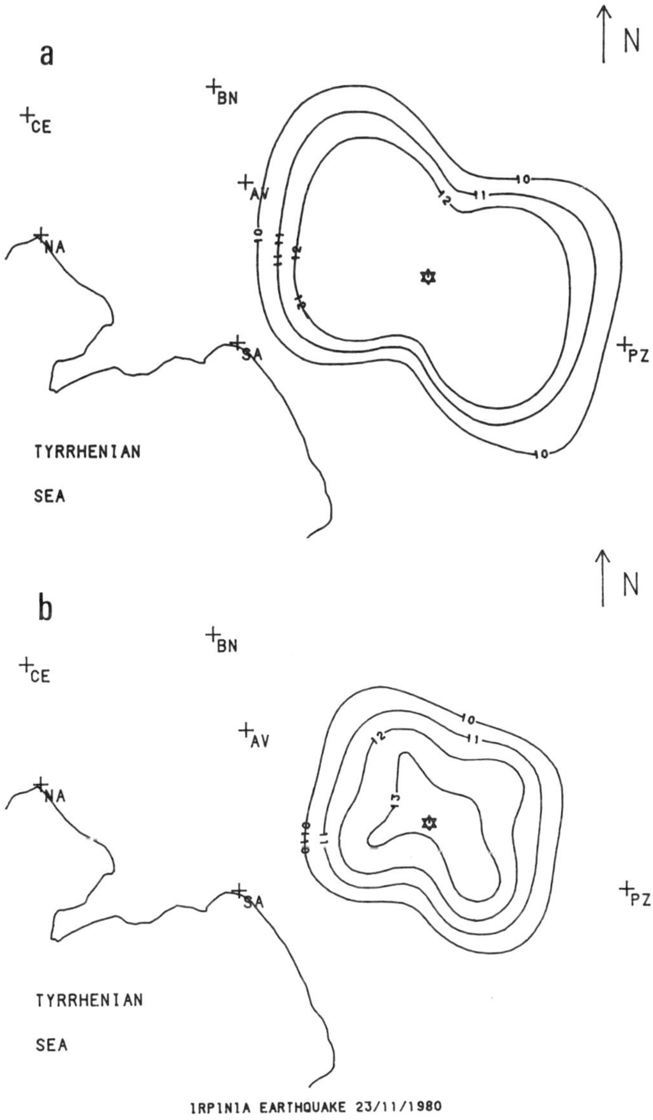

Figure 8. (a) same as Fig. 7c but for depth equal to 10 km. (b) same as (a) but for focal depth equal to 12 km. Structural model APN08.

SH-wave one. An even better accord with experimental data is reached if the displacement field is computed for a source depth of 8 km (see Fig.7c). On the contrary, if the model BRA02 and hypocentral depths of 6 km and 8 km are used, the displacement field due to SH-waves does not dominate. The fit does not improve if the seismic source is placed even deeper (10 km). We can conclude that, keeping the fault mechanism fixed, the isoseismal elongation in the fault direction is obtained by placing the seismic source near the bottom of the low-velocity superficial layers.

Most of the damaged area is located in the Apenninic zone, for which the total thickness of the nappes is reported to be of the order of 15 km (D'Argenio et al.,1975). This makes reasonable to consider structural models with a thicker sedimentary cover than that of model APN06. As a first step we have considered a low-velocity superficial layer 8 km thick (model APN08). Fig.8a and Fig.8b show the total horizontal displacement field computed, with the model APN08, for hypocentral depths of 10 km and 12 km, respectively. These focal-depth values are in excellent agreement with the results of Nakanishi and Kanamori (1982), based on P-wave first-motion polarities and long-period Rayleigh-wave data, and with the values given by CSEM and NEIS.

Up to now we have been considering a single low-velocity layer overlying faster crustal material. In such a way a very strong velocity contrast - 1.5 km/s for the S-wave velocity - is present at the interface between the sedimentary layer and the basement. A realistic modelling of a thicker low-velocity cover requires to take into account the pressure effect. For this purpose we have introduced a linear gradient in the elastic and anelastic properties of the uppermost 15 km. At the surface, the P- and S-wave velocities are 1.55 km/s and 0.9 km/s and increase linearly to 5.6 km/s and 3.25 km/s respectively. The computation, for the hypocentral depth of 17.5 km, gives a total horizontal field pattern which does not resemble at all the experimental one. The contribution of the SH-wave displacement field is, in fact, negligible with respect to the radial component of the P-SV one. A similar situation is found also when the source is placed 15 km deep. When assuming a hypocentral depth of 12 km, the SH-wave displacement field starts to dominate the P-SV one, and the total horizontal field has a slight elongation in the direction of the Apenninic chain. This elongation becomes stronger, if the hypocentral depth is decreased. As a last variant we have considered the model IRPGRA1, proposed by Suhadolc et al. (1987), in which the linear gradient in elastic and anelastic properties extends only over the first 8 km of structure. An Apenninic elongation is obtained for hypocentral depths not exceeding 9 km.

A general conclusion which can be drawn from our analysis is that, in the presence of a relevant cover of low-velocity material, the parameters influencing the elongation of the total horizontal displacement field are, beside the point-source orientation, the hypocentral depth, the thickness of the superficial low-velocity materials, as well as the elastic properties of the crustal structural model. In such cases, for pratical purposes, it is extremely useful to define a single parameter capable of describing the influence, on theoretical isoseismals, of the following quantities: hypocentral depth, average thickness of the low-velocity superficial layer(s), and the variation in the distribution versus depth of the crustal elastic parameters. A suitable quantity turns out to be the ratio $R = V_B/V_A$, where V_B and V_A are, respectively, the average S-wave velocity, weigthed with the thickness, of the crustal layers below and above the source. The total horizontal displacement field has a striking elongation in the fault direction as long as R is greater than 1.7. This result is confirmed also by computations made for the Bradanic structure (BRA02), where very shallow sources - hypocentral depth less than 3 km - are necessary to get an Apenninic elongation.

4. THE MAY 6, 1976 FRIULI EARTHQUAKE

On May 6, 1976 at 20h 00m GMT, an Ms = 6.5 earthquake occurred in the Friuli region, in NE Italy. The location of the epicenter of the main shock is not well determined, but there is a general accord around the values: 46.36° N, 13.28° E, focal depth 9±2 km (NEIS; ISC; CSEM; Cipar, 1980; Rogers and Cluff, 1979). The seismic moment and the time function duration were found to be 2.9 10^{25} dyn cm and 4.5 s, respectively (Cipar, 1981). Concerning the fault parameters, most of the Authors agree on a thrust-fault mechanism with a compressive axis oriented in a north-south direction (Console, 1976; Ebblin, 1976; Mueller 1977; Cipar, 1980; Lyon -Caen, 1980; Stoll, 1980; Zonno and Kind, 1984; Barbano et al., 1985). A summary of the available fault plane solutions is given in Table I.

The Friuli region can be divided into several topographic and geological elements (Amato et al., 1976; Cavallin and Martinis, 1976; Cavallin et al., 1984). In the north there are the Carnic Alps constituted of Paleozoic rocks, more to the south there are the Mesozoic Tolmezzo Alps in the west and Julian Alps in the east. Then the mountains become lower in elevation (Prealps) and are underlain by Cenozoic Flysch and molasse deposits. In the south of the region, down to the Adriatic Sea, there is the Friuli Plain of Quaternary sediments.

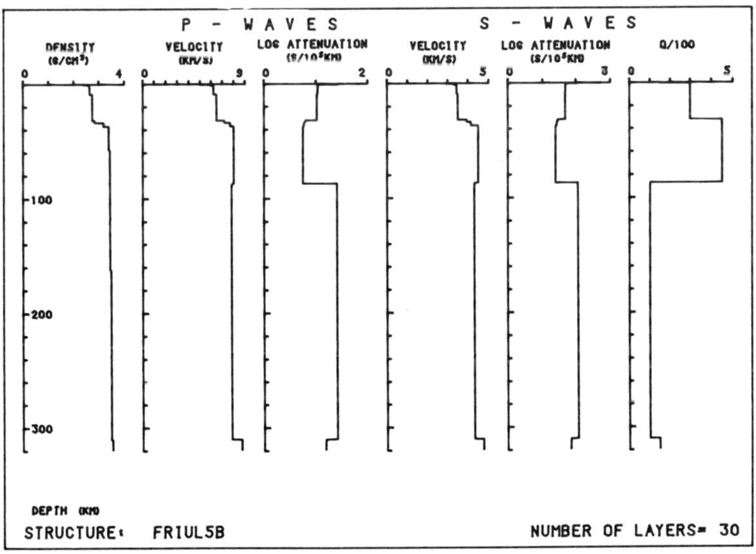

Figure 9. Structural model FRIUL5B.

TABLE I

AUTHORS		STRIKE	DIP	RAKE
CONSOLE	(1976)	260°	14°	92°
EBBLIN	(1976)	205°	18°	39°
MUELLER	(1977)	225°	10°	60°
CIPAR	(1980)	260°	15°	100°
LYON-CAEN	(1980)	222°	16°	52°
SLEJKO & RENNER	(1984)	230°	22°	63°
THIS PAPER	(1987)	280°	30°	115°

Fault-plane solutions of the May 6, 1976 Friuli event.
The values of rake and dip, if not indicated by the Authors, were deduced from the focal spheres.

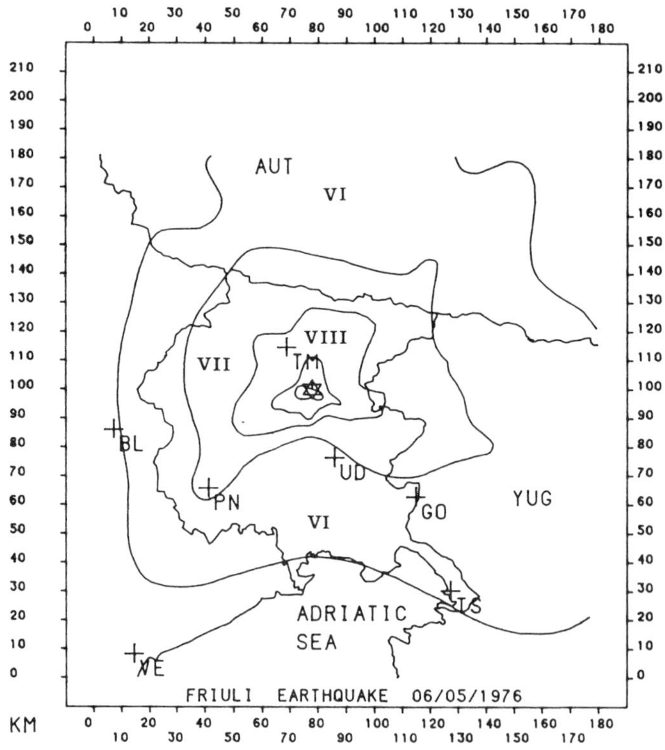

Figure 10. Observed isoseismals of the May 6, 1976 Friuli earthquake, MSK scale (after Giorgetti, 1976).

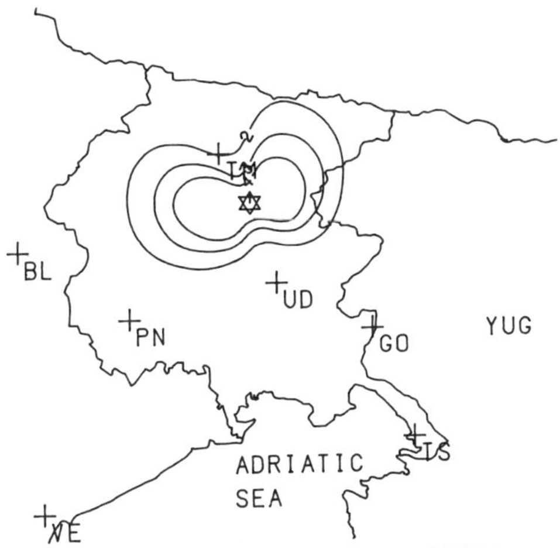

Figure 11a. Displacement field (cm) for SH-waves corresponding to a point source with strike 225°, dip 10°, rake 60°, and depth 7 km. Structural model FRIUL5B. $M_o = 10^{26}$ dyn cm. Cut-off frequency 0.1 Hz.

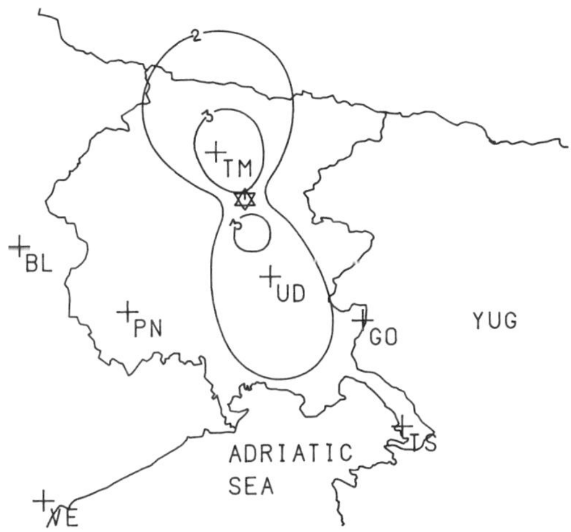

Figure 11b. Same as Fig 11a but for P-SV-waves.

Near the Italy-Yugoslavia border the structural alignments change from the northwest-southeast Dinaride trends to the more east-west trending structures of the Southern Alps, constituted by the Carnic, Tolmezzo and Julian Alps together with the Prealps in the west.

The chosen structural model (FRIUL5B) shown in fig.9 is mainly representative of the central part of the region (Angeheister et al.,

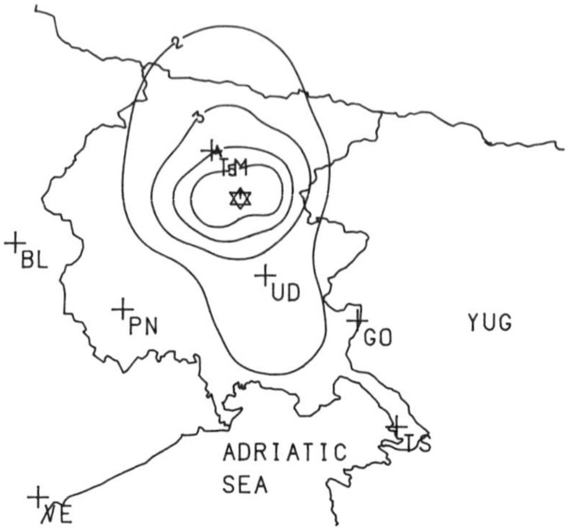

Figure 11c. Same as in Fig 11a but for the total horizontal displacement field (cm).

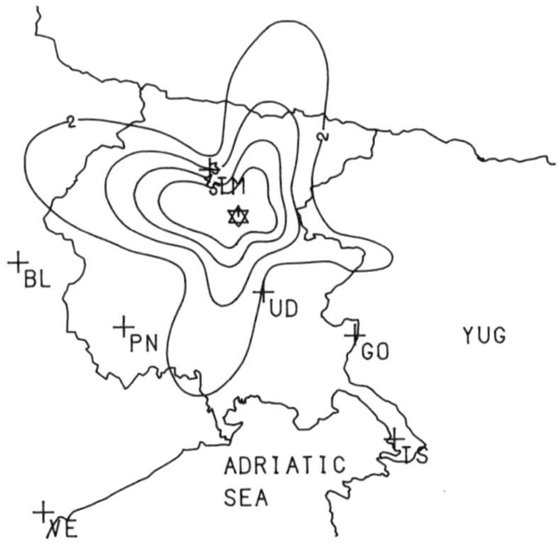

Figure 12. Displacement field (cm) for SH-waves corresponding to a point source with strike 225°, dip 30°, rake 60°, and depth 9 km.

1972; Brambati et al., 1980; Calcagnile and Panza, 1981; I.E.S.G., 1981; Panizza et al., 1981; Verri, 1982). It contains some hundreds of meters of sediments. The almost total absence of low-velocity and highly-dissipative sediments makes unnecessary, as we will show, the use of the parameter R defined in the previous section.

The experimental isoseismals (MSK scale) shown in Fig.10 are drawn after Giorgetti (1976).

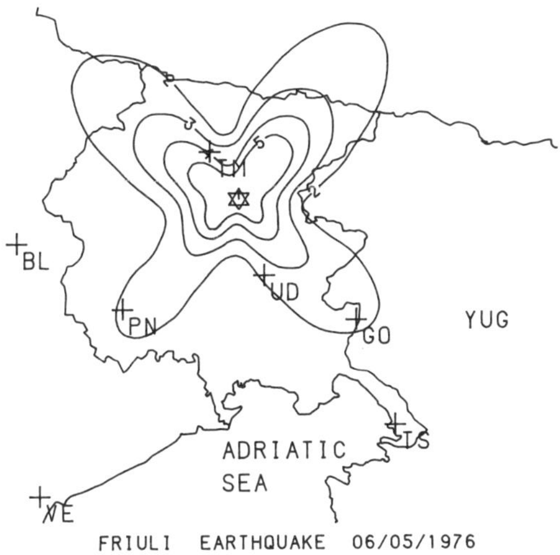

Figure 13a. Displacement field (cm) for SH-waves corresponding to a point source with strike 280°, dip 30°, rake 115°, and depth 7 km.

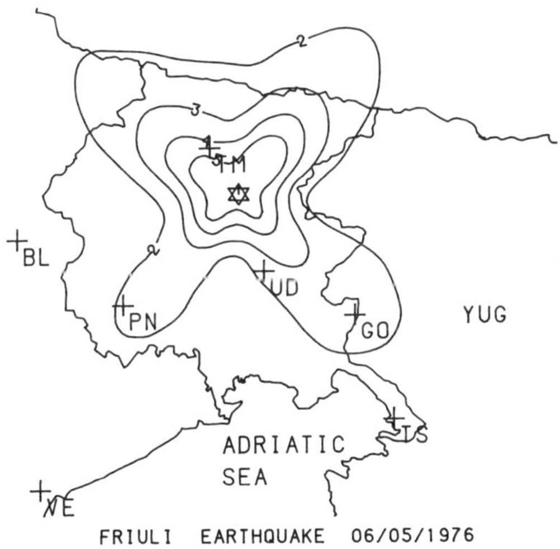

Figure 13b. Same as in Fig 13a but for the total horizontal displacement field (cm).

The initial source model used in our computations is the one proposed by Mueller (1977) with a hypocentral depth of 7 km. The results of this first trial are shown in Fig.11a to Fig. 11c. There is absolutely no similarity between these and the experimental isoseismals. Other synthetic isoseismals have been therefore computed, varying the fault parameters in the range allowed by the fault-plane solutions (Table I). The strike has been varied between 225° and 280°, with the condition of keeping fixed the NNW-SSE orientation of the axis of maximum compression. The dip has been incremented from 10° to 30°. The variations of strike and rake, keeping the dip at 10°, do not produce satisfactory results, while increasing the dip to 30° allows to generate isoseismals, Fig.12, which are somehow resembling the observed ones (see the lobes oriented SE and SW). At this stage an increase of the strike leads to a better agreement between the theoretical and the experimental isoseismals (Fig.13a and Fig.13b). The lobes in the SW and SE directions, the strong intensity attenuation toward the south and the absence of nodes in the northern part have been successfully synthetically reproduced. The fault plane parameters used to construct Fig.13 are in excellent agreement with the ones proposed by Cipar (1981). The fit is not excellent in the areas east and west of the epicenter, but in these sparsely populated regions few data exist, and a reconstruction of the experimental isoseismals is difficult.

Other variants of the structural model for the Friuli region, including some with a low-velocity channel in the crust, were used to check the stability of the obtained result. In the cases considered, the shape of the synthetic isoseismals has a very weak dependence on the input structure and is therefore mainly determined by the source orientation.
This conclusion does not contradict the results obtained for the Irpinia earthquake, since the structural models for the Friuli area contain an extremely thin sedimentary cover.

4. CONCLUSIONS

The main result of this study is that the shape of the experimental isoseismals for the earthquakes of November 23, 1980 (Irpinia) and May 6, 1976 (Friuli) is mainly due to the SH-wave radiation pattern. In the presence of a relevant sedimentary cover, the amplitude ratio between the SH-wave and the radial component of P-SV-wave displacement fields depends on the hypocentral depth and the distribution of elastic parameters around the source. If the source is close to the bottom of the sedimentary cover, the SH-wave displacement field largely dominates the radial component of the P-SV one, while if it is deeper the opposite is true. Therefore, in the point-source approximation, it is possible to infer from the experimental isoseismals, once the fault geometry is constrained, in which depth range to locate the source radiating the maximum of the energy reaching the surface. The problem arising from the trade-off between the source depth and the distribution of elastic parameters with depth can be easily solved, for pratical purposes, computing the ratio $R = V_B/V_A$, where V_B and V_A are, respectively, the average S-wave velocities, weighted with the thickness, of the crustal layers below and above the source. The SH-wave radiation pattern is dominant as soon as R is greater than 1.7.

The obtained results seem to imply that most of the energy felt at the surface during the November 23, 1980 Irpinia earthquake, comes from a depth not exceeding about 8 - 12 km. This value is consistent with some previous studies based on teleseismic data (see Table I in Suhadolc

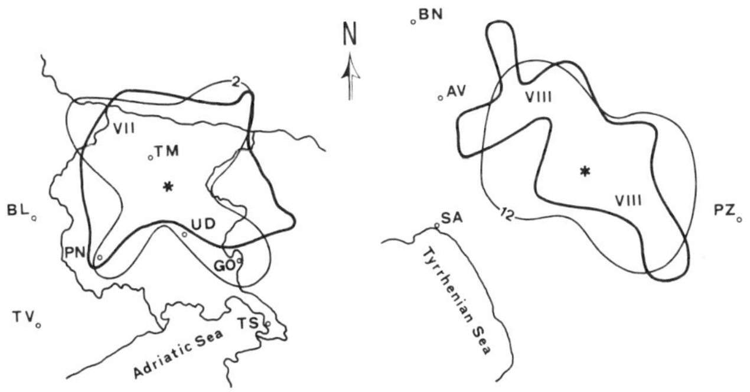

Figure 14. Comparison of synthetic (thin line) with observed (heavy line) isoseismals. Left: Friuli event - the 2 cm theoretical isodisplacement of Fig.13b is compared with the observed isoseismal enclosing macroseismic intensities VII or greater. Right: Irpinia event - the 12 cm theoretical isodisplacement of Fig.8a is compared with the observed isoseismal enclosing macroseismic intensities VIII or greater.

et al., 1987, this volume).

The observed isosesmals of the May 6, 1976 Friuli earthquake have been succesfully reproduced using a structural model with an extremely thin sedimentary cover and a point source 7 km deep. The fault plane parameters used in this case are consistent with those proposed by Cipar (1981).

For both events, the agreement between experimental and theoretical isoseismals is quite remarkable, as shown in Fig.14. As already shown by Panza and Cuscito (1982), the main difference between the theoretical and experimental patterns can be reasonably associated with local effects and not with large scale lateral variations. Anyway the local effects play a secondary role in comparison with that due to the source geometry.

If the far field value of the seismic moment, M_o, is used to obtain the zero-to-peak maximum displacement corresponding to a given intensity, highly misleading values may be obtained in the case of complex events such as the Irpinia one. For example, in the case of Fig.14, an unrealistic value of about 25 cm would correspond to the intensity VIII isoline. If, however, the near-field seismic moment determination, made from observed accelerograms (Suhadolc et al., 1987, this volume) is used, a maximum zero-to-peak displacement between 1 and 2 cm is obtained in corrispondence of the isoline of intensity VIII. On the other hand, for the relatively simpler and smaller Friuli event the use of the far-field seismic moment value seems to be still appropriate. In agreement with the displacement value obtained from the integrated accelerograms observed at Tolmezzo (TM), the resulting maximum displacement, which can be associated also in this case to the intensity VIII, is between 1 and 2 cm.

The reliability of the method described in this paper needs to be further tested against other instrumentally studied events. Should these tests be satisfactory, it will be worthwhile to apply the method to the study of large historical earthquakes. Italy has a huge number of historical documents referring to earthquakes. From their analysis it has been possible to reconstruct the spatial distribution of the macro-seismic intensities for many large events of the past (CNR, 1985). The computation of theoretical isoseismals can thus provide a very powerful tool to retreive the fault-plane solutions of old earthquakes. The informations obtained in such a way could be extremely helpful in the study and mitigation of seismic risk and in the understanding of the tectonic setting.

Moreover, once this method is proved to be reliable, it could serve as a powerful tool, after the occurrence of a disastrous earthquake, in the relief action. Very often civil Authorities do not possess enough informations about the spatial distribution of damages and have a hard time to decide promptly where to send aids and rescue teams. In these situations, on the basis of a first rough estimate of the event's focal parameters, which will be increasingly easy to obtain with the advent of satellite telemetered broad-band seismic networks, and starting from a pre-stored multi-mode spectra (Panza, 1985) of a pertinent structural model, synthetic isoseismals could be quickly computed. Their features might give the civil Authorities a quick first information in order to decide how to distribute in space rescue resources, with the obvious result of reducing losses of human lives.

ACKNOWLEDGEMENTS

We are grateful to Mr. M. Gergolet, Mr. S. Zidarich and Mr. G. Cavicchi for their help in drawing the figures.

This work has been performed with the financial support of MPI 40%, 60%, and ENEA funds.

REFERENCES

Aki, K. and P.G. Richards (1980). 'Quantitative seismology', W.H. Freeman and company, San Francisco.

Anderson, J.C. (1978). 'On the attenuation of modified Mercalli Intensity with distance in the United States' Bull. Seism. Soc. Am. **68**, 1147-1179.

Amato, A., P.F. Barnaba, I. Finetti., G. Groppi, B. Martinis and A. Muzzin (1976). 'Geodynamic outline and seismicity of Friuli Venezia Giulia Region', Boll. Geof. Teor. Appl. **19**, 217-256.

Angenheister, G., H. Bogel, P. Gebrande, P. Giese, P. Schmidt-Tomè and W. Zeil (1972). 'Recent investigations of surficial and deeper crustal structures of the Eastern and Southern Alps', Geol. Runds. **61**, 349-395.

Barbano, M.S., R. Kind, G. Zonno (1985). 'Focal parameters of some Friuli earthquakes (1976-1979) using complete theoretical seismograms', J. Geophys. **58**, 349-395.

Berardi, R., A. Berenzi and F. Capozza (1981). 'Terremoto campano-lucano del 23 novembre 1980: registrazioni accelerometriche della scossa principale e loro elaborazioni', Technical report CNEN-ENEL, Rome.

Blake, A. (1941). 'On the estimation of focal depth from macroseismic data', Bull. Seism. Am. **31**, 225-231.

Bernard, P. and A. Zollo (1987). 'The Irpinia (Italy) 1980 earthquake: detailed analysis of a complex normal faulting, Preprint.

Brambati, A., E, Faccioli, G.B. Carulli, F. Cucchi, R. Onofri, S. Stefanini and F. Ulcigrai (1980). 'Studio di microzonazione sismica dell'area di Tarcento (Friuli)', Regione Autonoma Friuli-Venezia Giulia and Universita' degli Studi di Trieste.

Calcagnile, G. and G.F. Panza (1980). 'Upper-mantle structure of the Apulian plate from Rayleigh waves', Pure Appl. Geophys. **118**, 823-830.

Calcagnile, G. and G.F. Panza (1981). 'The main characteristics of the lithosphere-asthenosphere system in Italy and surrounding regions', Pure Appl. Geoph. **119**, 865-879.

Cavallin, A. and B. Martinis (1976). 'Geological setting' in: "The Friuli Earthquake, May 6, 1976: Geology", B. Martinis editor, Boll. Geof. Teor. Appl. **19**, 758-761.

Cavallin, A., F. Giorgetti and B. Martinis (1984). 'Geodynamic outlines of North-Eastern Italy and seismogenetic implications', Boll. Geof. Teor. Appl. **26**, 69-92.

Chandra, U. (1979). 'Attenuation of intensities in the United States', Bull. Seism. Soc. Am. **69**, 2003-2024.

Ciaranfi, N., M. Guida, G. Iaccarino, T. Pescatore, P. Scandone, P. Pieri, L. Rapisardi, G. Ricchetti, I. Sgrosso, M. Torre, L. Tortorici, E. Turco, R. Scarpa, M. Cuscito, I. Guerra, G. Iannacone, G.F. Panza and P. Scandone (1983). Elementi sismotettonici dell'Appennino meridionale', Boll. Soc. Geol. It. **102**, 201-222.

Cipar, J. (1980). 'Teleseismic observation of the 1976 Friuli, Italy earthquake sequence', Bull. Seism. Soc. Am. **70**, 963-983.

Cipar, J. (1981). 'Broadband time domain modeling of earthquakes from Friuli, Italy', Bull. Seism. Soc. Am. **71**, 1215-1231.

CNR (1985). 'Atlas of isoseismal maps of Italian earthquakes', D. Postpischl editor, C.N.R, Progetto finalizzato Geodinamica, Sottoprogetto Rischio sismico e ingegneria sismica.

Console, R. (1976). 'Focal mechanisms of some Friuli earthquakes (1976). Boll. Geof. Teor. Appl. **19**, 549-558.

D'Argenio, B., T. Pescatore and P. Scandone (1975). 'Structural pattern of the Campania-Lucania Apennines', Quaderni de "La Ricerca Scientifica" **90**, Structural Model of Italy, CNR, Roma, Italy.

Del Pezzo, E., G. Iannaccone, M. Martini, and R. Scarpa (1983). 'The 23 November 1980 southern Italy earthquake', Bull. Seism. Soc. Am. **73**, 187-200.

Deschamps, A. and G.C.P. King (1983). 'The Campania-Lucania (southern Italy) earthquake of 23 November 1980', Earth and Planet. Sci. Lett. **62**, 296-304.

Ebblin, C. (1976). 'Orientations of stresses and strains in the piedmont area of eastern Friuli, NE Italy', Boll. Geof. Teor. Appl. **19**, 559-579.

Giorgetti, F. (1976). 'Isoseismal map of the May 6, 1976 Friuli earthquake', Boll. Geof. Teor. Appl. **18**, 707-714.

Hanks, T.C. (1975). 'Strong Ground Motion of the San Fernando, California, Earthquake: Ground Displacement', Bull. Seism. Soc. **65**, 193-225.

Harkrider, D.G. (1970). 'Surface waves in multilayered media. Part II. Higher mode spectra and spectral ratios from sources in plane layered Earth models', Bull. Seism. Soc Am. **60**, 1973-1987.

Herrmann, R.B. (1977). 'Research Study of Earthquake Generated SH Waves in the Near Field and Near-Regional Field', Final report, contract DACW 39-76-c-0058, Waterways Experiment Station, Vicksburg, Miss.

Herrmann, R.B. and O.W. Nuttli (1975a). 'Ground-motion Modelling at Regional Distances for Earthquakes in a Continental Interior. I. Theory and Observations', Earth. Eng. Struct. Dyn. **4**, 49-58.

Herrmann, R.B. and O.W. Nuttli (1975b). 'Ground-motion Modelling at Regional Distances for Earthquakes in a Continental Interior. II. Theory and Observations', Earth. Eng. Struct. Dyn. **4**, 59-72.

Italian Explosion Seismology Group and Institute of Geophysics, ETH Zuerich (1981). 'Crust and upper-mantle structures in the Southern Alps from deep seismic sounding profiles (1977, 1978) and surface wave dispersion analysis', Boll. Geof. Teor. Appl. **23**, 297-330.

Kawasaki, I. (1978) 'The near field Love wave by exact Ray Method', J. Phys. Earth **26**, 211-237.

Liao, A., F. Schwab and E. Mantovani (1978). 'Computation of complete theoretical seismograms for torsional waves', Bull. Seism. Soc. Am. **68**, 317-324.

Lyon-Caen, H. (1980). 'Seismes du Frioul (1976): modelles de source a l'aide de sismogrammes synthétiques d'ondes de volume', Ph.D. Thesis Université Paris VII.

Mallet, R. (1862). 'The great Neapolitan earthquake of 1857', London.

Mueller, G. (1977). 'Fault-plane solution of earthquake in Northern Italy, 6 May 1976, and implication for the tectonics of the Eastern Alps', J. Geophys. **42**, 343-349.

Murphy, J.R, L.J. O'Brien (1977). 'The correlation of peak ground acceleration amplitude with seismic intensity and other physical parameters', Bull. Seism. Soc. Am. **67**, 877-915.

Nakanishi, I. and H. Kanamori (1984). 'Source mechanism of twenty-six large shallow earthquakes (Ms≥6.5) during 1980 from P-wave first motion and long period Rayleigh wave data', Bull. Seism. Soc. Am. **74**, 805-818.

Panizza, M., D. Slejko, G. Bartolomei, A. Carton, D. Castaldini, M. Demartin, R. Nicolich, U. Sauro, E. Semenza and L. Sorbini (1981) 'Modello sismotettonico dell'area fra il lago di Garda e il monte Grappa', Atti del Convegno su "Sismicità dell'Italia: stato delle conoscenze e qualità della normativa", Udine 12-14 maggio 1981, Prog. Fin. Geodinamica CNR.

Panza, G.F. (1985). 'Synthetic seismograms: the Rayleigh waves modal summation', J. Geophys. **58**, 125-145.

Panza, G.F., S.T. Mueller and G. Calcagnile (1980). 'The gross features of the lithosphere-asthenosphere system in Europe from seismic surface waves and body waves', Pure Appl. Geoph. **118**, 1209-1213.

Panza, G.F. and M. Cuscito (1982). 'Influence of Focal Mechanism on Shape of Isoseismals: Irpinia Earthquake of November 23, 1980', Pure Appl. Geoph. **120**, 577-582.

Panza, G.F., P. Suhadolc and C. Chiaruttini (1986). 'Exploitation of broadband networks through broadband synthetic seismograms' Ann. Geophys. **4B**, 315-328.

Pescatore, T. (1981). 'Lineamenti strutturali dell'Appennino campano-lucano', Rend. Soc. Geol. It. **4**, 49-54.

Postpischl, D.(1981) 'Il terremoto del 23.11.1980, rilievo macrosismico: stato di avanzamento al 27.1.1981', Convegno straordinario del Progetto Finalizzato Geodinamica, CNR Roma, 27-28 gennaio 1981.

Rogers T.H. and L.S. Cluff (1979). 'The May 1976 Friuli Earthquake (Northeastern Italy) and interpretations of past and future seismicity', Tectonophysics **52**, 521-532.

Shebalin, N.V. (1959). 'Determination of focal depth from macroseismic data with consideration of the low velocity layer', Problems in engineering seismology 2: Akad. Nauk SSSR Inst. Fiziki Zemli Trudy **5**,172.

Shebalin, N.V. (1972). 'Macroseismic information on source parameters of large earthquakes', Phys. Earth. Plan. Inter. **6**, 316-323.

Siro, L. and D. Slejko (1984). 'Modello sismotettonico dell'area friulana: considerazioni e proposte per la scelta delle aree sismogenetiche in funzione di diversi livelli di mitigazione del rischio', Atti dell'incontro sul tema "Finalità ed esperienze della rete sismometrica del Friuli-Venezia Giulia", Trieste, 7 dicembre, 1984. Published by Regione Autonoma Friuli-Venezia Giulia, 245-263.

Slejko, D. and G. Renner (1984). 'La sequenza sismica iniziata col terremoto del 6 maggio 1976 in Friuli', Atti dell'incontro sul tema "Finalità ed esperienze della rete sismometrica del Friuli-Venezia Giulia", Trieste, 7 dicembre 1984. Published by Regione Autonoma Friuli-Venezia Giulia, 75-91.

Sponheur, W. (1960). 'Methoden zur Herdtiefenbestimmung in der Makroseismik', Freiberger Forschungshefte **C88**, Akademie Verlag, Berlin.

Stoll, D. (1980). 'Determination of focal parameters of five 1976 Friuli earthquakes from Rayleigh wave spectra', Boll. Geof. Teor. Appl. **22**, 3-12.

Suhadolc, P. (1981) 'Spatial distribution of the aftershock sequence relative to the September 16, 1977 Friuli earthquake', Boll. Geof. Teor. Appl. **23**, 331-348.

Suhadolc, P. (1982) 'Time evolution of the aftershock sequence relative to the September 16, 1977 Friuli earthquake', Boll. Geof. Teor. Appl. **24**, 57-65.

Suhadolc, P. (1986). 'Synthetic accelerograms: case examples from Irpinia and Friuli, Italy', Proceedings of the international symposium on "Engineering geology problems in seismic areas", Bari, 13-19 aprile, 1986, In press.

Suhadolc, P., F. Vaccari and G.F. Panza (1987). 'Strong motion modelling of the rupturing process of the November 23, 1980 Irpinia, Italy, earthquake', this volume.

Trifunac, M.D. and A.G. Brady (1975). 'On the correlation of seismic intensity scales with the peaks of recorded strong motion', Bull. Seism. Soc. Am. **65**, 139- 162.

Verri, G. (1982). 'Modelli per la mappatura delle variabili del territorio', XXIX Congr. Naz. Ordine Ing., Venezia, 1982.

Zonno, G. and R. Kind (1984). 'Depth determination of North Italian earthquakes using Grafenberg data', Bull. Seism. Soc. Am. **74**, 1645-1660.

PROBABILISTIC TOOLS IN EARTHQUAKE HAZARD FORECAST : A CASE STUDY FOR TURKEY

Polat Gülkan*
Department of Nuclear Engineering
Hacettepe University
06532 Beytepe, Ankara, Turkey

and

Mustafa Erdik
Earthquake Engineering Research Center
Middle East Technical University
06531 Ankara, Turkey

ABSTRACT. This article is concerned with the determination of seismic hazard primarily for engineering purposes. The nature of ground motions caused by earthquakes is quite complex but for purposes of code requirements, preliminary site evaluation for critical facilities or for quantifying the earthquake threat to communities, it is necessary to be able to make a statistical estimate of the magnitude of future seismic events. Characterization of the earthquakes may be in terms of intensity, peak ground acceleration or some other index. We present at first the methodology for making a probabilistic forecast of the seismic index of interest, discussing the problems associated with the steps that this forecast involves. In the next part we illustrate the application of the methodology to Turkey which is situated at the northeastern corner of the Mediterranean. Estimates of intensity and acceleration are presented in the form of contour maps for the country.

1. BACKGROUND

The present state of knowledge regarding the design of engineering structures to withstand the effects of earthquakes is still less than perfect because the underlying natural phenomenon does not lend itself to straight-forward and mathematically precise formulations. This statement is intended more to justify current research in all fields of earthquake engineering than as a denial of the output of the research conducted within the last four decades. We cite seismic hazard assessment as an example field in which perhaps more than in any other, engineers and earth scientists need to exchange information in an intimate way in order to build coherent models formulating the effects of as-yet-to-occur earthquakes on existing or future structures. Such "crystal balls" are not easy to produce.

* Currrent address : Basler & Hofmann, 8029 Zürich, Switzerland

J. Bonnin et al. (eds.), Seismic Hazard in Mediterranean Regions, 229–253.

A number of fundamental questions including the following must be answered before hazard at a given point can be quantified: (1) Where are earthquakes likely to occur ? (2) What might be the largest of these ? (3) How frequent would be these earthquakes ? (4) How would the vibration effects be transported to other locations with the earthquake waves ?

Geologic tools alone are not sufficient to answer all of these questions. A careful blending of past data addressing some of these points with statistical tools and incorporating uncertainty (which is mathematically equivalent to taking into consideration what we do not know for certain) will provide a better view of the future. This vision, or assessment of hazard, is as free of blurs as the ingredients which constitute its basic fabric.

In this paper, we review at first the probabilistic formulation for seismic hazard assessment. While we avoid digressing into lengthy discussions of some facets of this step, we do underline a few basic topics of contention. This part is followed by an implementation of the methodology to the northeastern part of the Mediterranean where palpable prognostications are presented for Turkey. In this way we intend to establish a link with the remaining articles in this volume for a full description of the seismic hazard in the entire Mediterranean. The organization of the text follows these two areas of emphasis : in Section 2 we discuss the concept and quantification of seismic hazard for a point and in Section 3 we illustrate the application of the theory for Turkey, a country with a land area of about 0.8 M km^2. The forecast of the seismic threat is expressed in terms of intensity or acceleration contours.

2. SEISMIC HAZARD AT A POINT

The "classical" papers on seismic hazard (designated as risk in these early works) dealt with increasingly more refined mathematical techniques which yielded a statement concerning the likelihood of some descriptive earthquake parameter occurring within some future time span [Cornell (1968), Cornell and Vanmarcke (1969), Der Kiureghian and Ang (1977)]. Not wishing to emphasize problems that arise in the application of these techniques at the expense of related geologic studies that form the core of these predictions, we refer to Donovan and Bornstein (1978) and Yücemen (1981) for some basic assessments.

A convenient way of describing how seismic hazard analysis is performed for a region containing a collection of points of interest deployed over some matrix, and an array of "sources" which contribute to the hazard at each of these points, is to formulate it for a single point influenced by a single source. The extension of this formulation to the multi-point and multi-source reality is then straightforward.

2.1. Assumptions

The fundamental assumptions which we will use are the following :

(a) The relation between the logarithm of the cumulative number of occurrences of earthquakes larger than a given intensity (or magnitude) is linear, at least for the intensity (or magnitude) values of interest. There is evidence that for larger magnitudes a higher order approximation may be more appropriate, but available data usually does not justify such refinements.

(b) An upper limit exists for the largest magnitude of the earthquake any source can engender. This follows even from the heuristic notion that a limit must exist for the amount of energy that can be stored within the earth's crust before it ruptures.

(c) The obeyance of the ground motion over distance can be expressed continuously in the form of analytical or numerical relationships. This attenuation property of the earthquake waves is a fertile ground for controversy because an all-seeing law can not be formulated on the basis of current state-of-the-art. Regardless of the purported accuracy of the information that is employed in the derivation of such laws, they represent best-fit curves to a collection of data. We will consider this spread in the following.

(d) Earthquake occurrence follows a Poissonian law with regard to time.

2.2. Computation of the probability of occurrence of a seismic event of interest

The formal computation of the steps involved in a seismic hazard analysis, expressed here in terms of intensity for illustration purposes, may be summarized as follows :

(a) Computation of the probability FI(i) that at the source (e.g. the fault zone) the intensity of the earthquake that occurs will not exceed some value i. The probability of the complementary event is

(1) $$GI(i) = 1 - FI(i)$$

(b) Computation of the probability that given the event i at the source the intensity of the same event felt at the site (or point) at some distance from it will be exceeded. We will call this number $GI(i_s)$.

(c) Computation of the probability of experiencing GI (i_s) within a given time period, such as one year. The inverse of this number is the average recurrence period.

In the following we will give the mathematical expressions that correspond to these steps. Necessarily, specific forms will be assumed for some of the relationships that are involved, but it is hoped that the algebra will not obscure the underlying physical reasoning. For similar derivations we refer to Merz and Cornell (1973) and Gülkan and Yücemen (1977).

Step 1 : Computation of FI (i) and GI (i)

The intensity relationship for earthquakes in a given source between the lower limit $i = i_o$ and the upper limit $i = i_1$ may be taken to be of the form

(2) $$\log n = a + b\,(i - i_o)$$

Equation (2) implies that the number of earthquakes equal to or less than i_o is $n_o = \exp(\ln 10\,(a))$, and that there is one earthquake whose intensity transcends or is equal to i_1.

The number of earthquakes with intensity falling between $i = i_o$ and $i = i$ is

(3) $$n_G = n_o - n$$

and the total number of earthquakes in the sample is

(4) $$n_T = n_o - n_{i_1}$$

We can now show that the required probability FI (i) is

(5) $$FI(i) = \frac{n_G}{n_T}$$

or, substituting equations (2), (3) and (4) into (5)

(6) $$FI(i) = \frac{1 - \exp(b \ln 10 (i - i_o))}{1 - \exp(b \ln 10 (i_1 - i_o))}$$

We therefore have

(1') $$GI(i) = 1 - \frac{1 - \exp(b \ln 10 (i - i_o))}{1 - \exp(b \ln 10 (i_1 - i_o))}$$

or denoting

(7) $$k_{i1} = \frac{1}{1 - \exp(b \ln 10 (i_1 - i_o))}$$

(8) $$GI(i) = 1 - k_{i_1} + k_{i_1} \exp(b \, 1n10 (i - i_o))$$

Step 2 : Computation of GI(i_S)

We require a correlation between the intensity at the point of origin, or source, of the earthquake and that at the point of interest, which we call the site. Such a correlation should ideally be based on data collected within the region containing the site so that the transport characteristics of the geologic structure are reflected properly, but in many situations deviations from this ideal are necessary. It is also possible to work with the variation of an instrumentally measured index of the ground motion such as the peak ground acceleration or velocity. Representation of such data in the form of curves grouped according to magnitude or site conditions is a prevalent practice of hazard studies.

For illustration let us continue to work with a functional form of the attenuation of intensity. We stress that the following relationship has no implied superiority over other possible forms :

(9) $$i_s = i - w + \varepsilon$$

Here

- i = epicentral intensity
- i_S = site intensity
- w = a function dependent on distance from the epicenter (see Table 3 for some specific forms)
- ε = an error term reflecting the scatter in the data of equation (9).

As we have already pointed out that functional relations such as the equation above are only best curve fit correlations with some scatter of the observed data about these equations. Esteva (1976) was among the first to call attention to the fact that the residuals about the least squares fit represented by this equation are lognormally distributed with mean zero and a standard deviation of σ. In the analysis it is necessary to consider this error term explicitly.

The probability that the site intensity of i_S will be exceeded is found by substituting equation (9) into equation (8) after solving the former for i :

(10) $$GI(i_s) = (1 - k_{i1}) + k_{i1} \exp\{b \ln 10 (i + w - \varepsilon) - i_o\}$$

Denoting $\beta = b \ln 10$

(10') $$GI(i_s) = (1 - k_{i1}) + k_{i1} \exp\{\beta(i + w) - \beta\varepsilon - i_o\}$$

ε is normally distributed so that multiplication of equation (10') with the complementary cumulative distribution function of the standard Gaussian distribution may be achieved in three steps :

(a) Multiplication with the probability distribution function
(b) Integration over ε to yield the cumulative distribution
(c) Consideration of the integration limits to obtain the complementary function

(11) $$GI(i_s) = \int_x^\infty [(1 - k_{i1}) + k_{i1} \exp\{\beta(i + w) - \beta\varepsilon - i_o\}] \frac{1}{\sigma\sqrt{2\pi}} \exp(-\frac{\varepsilon^2}{2\sigma^2})\, d\varepsilon$$

where

(12) $$x = i_s + w - \varepsilon < i_1$$

So the limiting value of ε, which we will denote as z becomes

(13) $$z = i_s + w - i_1$$

If we denote

(14) $$\Phi^*(x) = \int_x^\infty \frac{1}{\sigma\sqrt{2\pi}} \exp\{-\frac{\varepsilon^2}{2\sigma^2}\}\, d\varepsilon \qquad = p(X \geq x)$$

Equation (11) can then be written in a considerably simplified form if we assume a specific form of w. One particular form which has found favor is (see Section 3) :

$$w = c_1 + c_2 \ln d$$

where d denotes the distance parameter. The final form of Equation (11) then becomes

(15) $$GI(i_s) = (1 - k_{i1})\, \Phi^*(\frac{z}{\sigma}) + k_{i1}\, \Phi^*(\frac{z}{\sigma} - \beta\sigma) \exp(\beta^2\sigma^2/2).$$
$$.\exp(\beta c_1) \exp(\beta i_o)\, d^{-\beta c_2} \exp(-\beta i_s)$$

Step 3 : Accounting for random occurrence

From our assumption of the Poisson character of earthquake occurrences, the number of future earthquakes during a given interval of time Δt causing site intensities larger than i_s may be calculated as

$$(16) \qquad p(N) = \frac{1}{N!} \{GI(i_s)\, \nu\, \Delta t\}^N \exp(-GI(i_s)\, \nu\, \Delta t)$$

Here

N = number of earthquake occurrences with intensity i_S
ν = total annual number of earthquakes
Δt = time interval

The annual probability that no earthquake causing a local intensity of i will occur is found by substituting N = 0 and Δt = 1 (year) into equation (16). The complementary number

$$(17) \qquad p(i_s) = 1 - \exp(-GI(i_s)\nu)$$

then becomes the probability that i_S will be exceeded at least once during any year-long period. For small values of GI(i) equation (17) may be written as

$$(18) \qquad p(i_s) = GI(i_s)\nu$$

When there exist more than one source of seismic threat, possibly with a different rate ν, a summation on equation (18) is indicated. For an array of points, the procedure outlined here must be repeated for each point. The inverse of equation (18) is the average period of recurrence for intensity i_S.

2.3. Discussion

From the previous section it may appear that the determination of the seismic threat posed by a single source is a rather straightforward exercise once the underlying statistical concepts can be fulfilled. In reality the entire process involves the simultaneous consideration of many different types of data each of which is afflicted with some defect. It is not our purpose to delve into the matter of uncertainties in seismic hazard analysis : for this purpose we refer to Veneziano (1982). In the following we give a brief discussion of some of the major points of consideration.

2.3.1. Definition of the source zones. Location of a seismic source is determined from the hypocentral position of past earthquakes and from geologic seismological information. The types of source characterizations that have won acceptance are the point, line and area representations at some particular depth. The rate of seismicity is then assumed to be constant within each source. Well defined fault zones are best described as line sources because historical hypocenters are constrained to a narrow band clustered around them. In the case of long fault zones displaying different rates of activity it would be permissible to break them up into several shorter line sources. Even then a map of hypocenters overlaid with a tectonic map is subject to different interpretations each of which will have some subtle influence on the final hazard assessment. Most probabilistic hazard studies assume that future earthquakes will occur in well-defined and homogenous source zones. This assumption implies that the modeled seismic rates change abruptly at source boundaries with the result that the predicted intensity levels may differ by unreasonable amounts at sites a few kilometers apart (Bender, 1986). Such homogenous characterizations of source zones are an approximation, and their full extent is usually not well understood. Different ways of expressing the source of the seismic hazard is therefore indicated.

2.3.2. Maximum earthquakes. In the previous development we made reference to a maximum source intensity i_S (for events of the past). Though obscured by the statistics this number will have an important effect on the derived site intensity. The question that needs to be answered is : what is the maximum intensity (or magnitude) of the future event that will influence the value of b in equation (2) ? Professional experience in assigning maximum values to source zones seems to indicate the prevalence of two practices :

(a) Maximum historical earthquake procedure.

For active faults with a long recorded seismic history and rather short recurrence intervals between large events, this may be expected to give good estimates. In practice the intensity of the largest historical earthquake is increased by an appropriate margin to yield the quantity of interest.

(b) Fault rupture length procedure.

This involves the use of regression analysis of the surface rupture length versus intensity or magnitude. Worldwide data suggests that for faults with total length less than about 200 km, the maximum rupture length may be 1/6 of the total length, and for faults longer than 1000 km 1/3 or 1/2 may rupture.

The former procedure is perceived as being more appropriate for areal source characterizations while the latter is more appropriate for line sources. Inasmuch as the implicit purpose of the hazard analysis itself is to find the maximum effect engendered by the maximum cause, it is apparent that there is an element of the vicious circle here.

2.3.3 Earthquake occurrence model. The parameters of the earthquake occurrence model required for each source are the magnitude or intensity distribution and the mean rate of occurrence. The scope of this distribution involves the maximum intensity "cut-off" as well.

Data on earthquake size is commonly discretized at every half intensity value. This representation allows the use of a discrete distribution model, and is advantageous in that it avoids the log-linear fitting in equation (2) which may be unacceptable in some cases. It is known, for instance, that such a linear curve fitting implies larger events than could reasonably be associated with the sources, so quadratic forms of the fit have been advanced as a possible remedy against over-estimating the cut-off (Shlien and Toksöz T, 1970).

Once the seismic sources have been located, it is commonly assumed that from each source earthquakes occur according to a Poisson process. This means that the following conditions need to be satisfied :

(a) Earthquake occurrences are independent in space and time
(b) There is a vanishingly small probability that two seismic events will take place at the same time and at the same place.

The first suggests that seismic events do not influence one another, and the second condition is known to comply with the physics of the problem if aftershock occurrences are eliminated from the statistics. Other statistical models of occurrence have been advanced, but the Poisson assumption is adequate for a major part of engineering applications.

2.3.4. Attenuation relationship. As we pointed out earlier, the general decrease in the severity of the ground shaking with increasing distance from the epicentral region is termed attenuation. It has been observed that attenuation of ground motion varies in different parts of the world. For instance, in the western United States the decrease of intensity (here we use the word "intensity" in its generic sense in that it expresses more than a rating according to some scale) is more rapid than in the rest of the country. For California, where instrumental strong motion data is available for a relatively large number of earthquakes Donovan and Bornstein (1987) have presented the following relationship for acceleration attenuation :

(19) $$\text{peak acceleration (cm/s}^2) = b_1 \exp (b_2 M) (X)^{-b_3}$$

In this equation

M = earthquake magnitude (Richter or local)

(20)
$b_1 = 2\ 154\ 000\ (R)^{-2.10}$
$b_2 = 0.046 + 0.445 \log R$
$b_3 = 2.515 - 0.486 \log R$
R = distance to the energy center on the "causative" fault in km
X = a function of R

An attenuation relationship such as equation (19) is a best fit to the existing data, and exhibits considerable scatter. This dispersion of the actual data should be included in the hazard analysis, as we have illustrated earlier. In equation (19), the logarithmic standard deviation in the peak ground acceleration ranges from 0.3 to 0.5.

As an example of an intensity attenuation we give the relationship derived by Cornell and Merz (1975) for northeastern United States :

(21) $$I_{site} (MM) = I_{epicenter} + 3.22 - 2.99 \log R$$

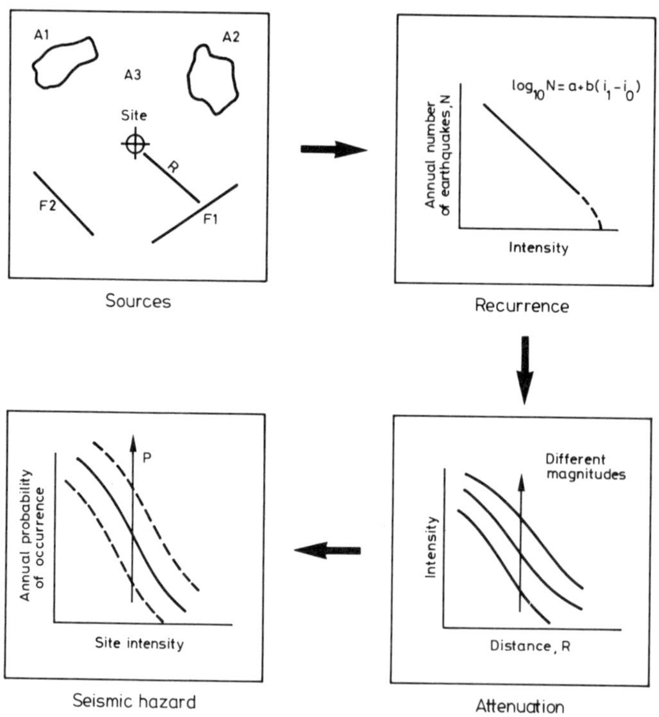

Figure 1. Steps in Hazard Forecast Studies

Where sufficient instrumental data is lacking, the site intensity can be converted to the peak ground acceleration which is an instrumental value. The relationship below has been made available by Murphy and O'Brian (1977) who also provide a collection of worldwide data :

(22) $\log a_p = 0.25\ I_s + 0.25$

Because of the paucity of strong motion data for many parts of the world, engineers are frequently forced to rely on empirical relations of this kind which all too often retain the specific regional coloring of the area from which they are obtained. Our objective in this section has been to stress the condiderable amount of uncertainty contained in our ability to predict the size of earthquake waves at some distance from the point at which they are released. For further details we cite only two compendia out of what is actually a large collection of research output : OECD/NEA (1983), EERI (1983).

2.3.5. A final note. The essential steps of a particular type of seismic hazard forecast are represented in Figure 1. There exists consensus among professionals as to what these steps should be. It is the treatment of the data comprising the ingredients where professional opinion exhibits divergence, and it is not uncommon to arrive at hazard quantifications that differ by an order of magnitude by different, and equally consistent, interpretations of the same body of raw data. Reliance on expert opinions appears to be the logical recourse to take (TERA Corp., 1980). The purpose of this discussion has been to emphasize that the outcome of a seismic hazard study, computed with mathematical exactness, should be regarded more as representative of a range of values than as specific quantities. Just as it would not be consistent with the nature of the problem at hand to invoke significant accuracy concerning the likelihood of future seismic events, it would not be justifiable to view it merely as a mathematical exercise. Probabilistic hazard analysis provides logic and greater consistency over procedures based solely on deterministic approaches in that geologic, geophysical, and seismological data are used in a more consistent manner. It also yields the information required for well-balanced engineering designs that compare costly designs for greater resistance against economical designs with higher risks of economical loss.

3. SEISMIC HAZARD IN TURKEY

In this section we bring together the data base required for making a hazard assessment for a wide area, such as part or whole of a country. The difference here, as we pointed out earlier, is that a summation is done to reflect the contribution from different sources, and multiple, rather that a single, points are considered. Seismic hazard of Turkey has been a topic of interest over the last ten years (Gülkan and Gürpinar, 1977 ; Gurpinar *et al.*, 1979 ; Erdik *et al.*, 1985). Our format in the following conforms to Figure 1.

3.1. Earthquake database

For Turkey, it can be stated that a reliable seismological data base exists only for the last two decades. The earthquake data utilized for the determination of the seismic source regions in Turkey and for the assessment of the frequency-magnitude characteristics of the earthquakes in these seismic sources are based mainly on the following catalogues :

For the period 1913-1970 : Alsan *et al.* (1975)

For the period 1970-1982 : NOAA Earthquake Data File (U.S. National Oceanic and Atmospheric Administration (NOAA) Data File and Bulletins of Preliminary Determination of Epicenters, U.S.G.S., N.E.I.S.).

The first catalogue includes relocated earthquakes between the coordinates of 35.5 -

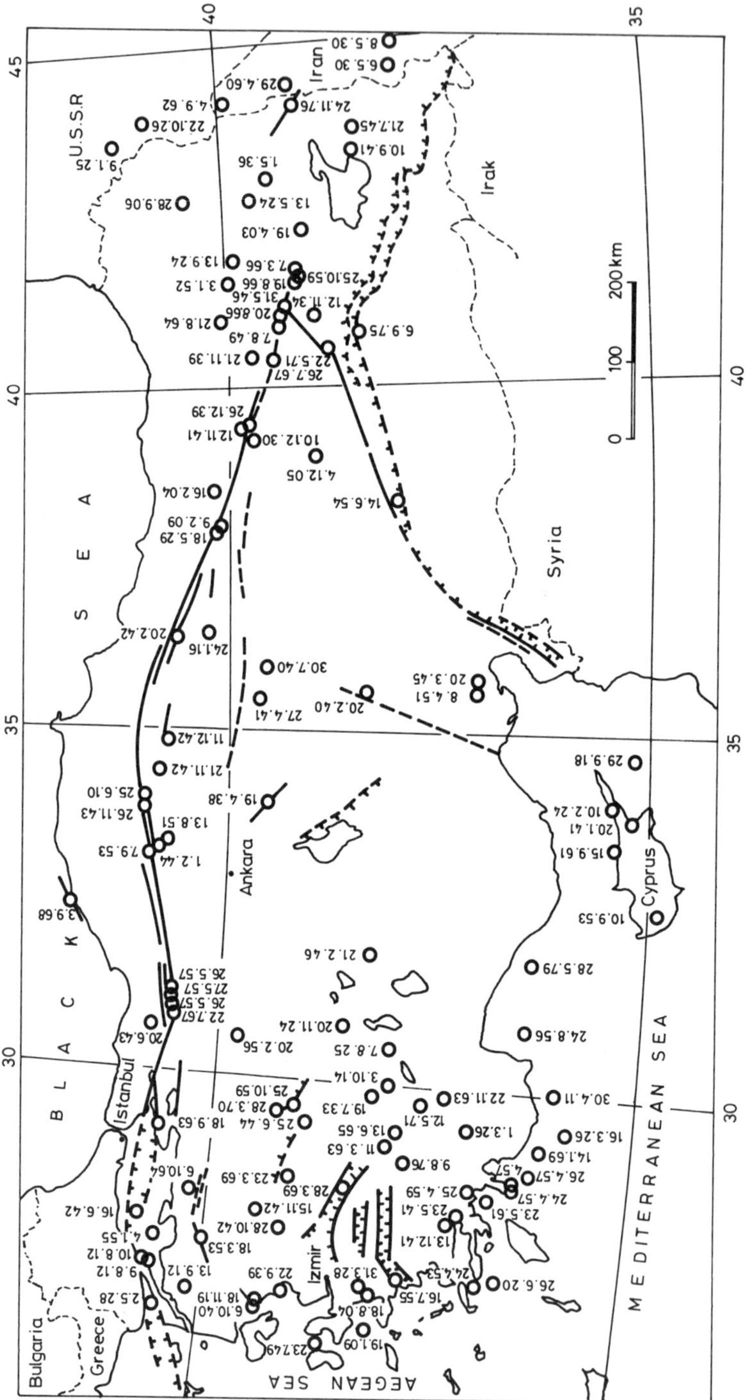

Figure 2. Recent Seismicity of Turkey

42.5 N and 25.5 - 45.0 E. The data base consists of entries from the Bulletins of International Seismological Summary (ISS) and Bureau Central International de Séismologie (BCIS).

Both catalogues provide the instrumental locations of the earthquakes which are known to include uncertainties for small events and/or for earlier reporting periods.

3.2. Seismic sources

Seismic sources are defined on the basis of the uniformity of geotectonic features, homogeneity of earthquake occurrences and the consistency of focal mechanisms. Solution of this problem require the full utilization of the geologic, tectonic, paleoseismologic data, the historical macro-and instrumental seismic records and micro-earthquake survey results. Yet, the state-of-the-art does not describe a standard general practice and the methodologies employed in definition of the seismic sources and assessment of maximum earthquake potentials can be subjective.

Portrayal of the seismicity and the tectonics of a region provides the first and essential clues towards the seismic source regionalization. Figure 2 indicates the recent seismicity of Turkey. The inherent inhomogeneity and dispersion of the total seismologic data contained in catalogues hinders the detailed correlation of the epicentral locations with all neo-tectonic features.

For the correlation of the seismic activity and the neo-tectonic elements of Turkey the catalogue of the large events $(M \geq 5.9)$ and damaging events (epicentral intensity, $I \geq VII$) was used. In Figure 2 the macroseismic locations of these large and damaging events are plotted together with the neo-tectonic features which correlate well with the distribution of the macroseismic epicenters.

Turkey lies with the Mediterranean sector of the Alpine-Himalayan orogenic system. The Alpine orogeny is produced as a result of the compressional motion between Europe and Africa, whereas the Himalayan orogeny has resulted from the India-Asia collision. The distribution of seismicity within the Alpine-Himalayan system is not homogeneous, the seismic activities being mostly concentrated along the plate margins. According to Morgan (1968), Le Pichon (1968) and McKenzie (1970) only three major plates are involved in the tectonics of the Mediterranean region. These are the African, Eurasian and Arabian plates. The situation in the eastern Mediterranean is, however, somewhat more complicated. The cause of the local increase in seismic activity of this region is attributed to the existence of smaller but rapidly moving plates. Various plate tectonics models for the eastern Mediterranean region have been proposed. Those proposed by McKenzie (1972), Alptekin (1973), Dewey *et al.* (1973), Dewey and Sengör (1979) and Papazachos (1974) are provided in Figure 3. These models will be discussed here briefly.

According to the McKenzie model (McKenzie, 1970 and 1972), the Aegean part of Greece, Crete and the western part of Turkey constitute the Aegean plate. The Turkish or the Anatolian plate contains most of Turkey and Cyprus. The northern boundary of the Anatolian plate is the North Anatolian fault (NAF) which is an active east-west right-lateral strike-slip fault. The southern boundary of the Anatolian plate appears to join the southern boundary of the Aegean plate south of southwest Turkey and to run south of Cyprus into the Gulf of Iskenderun to join the NAF east of Erzincan (about 10 km east of Karliova). Between the Gulf of Iskenderun and Karliova this boundary actually follows the East Anatolian fault (EAF) which is an active left-lateral strike-slip fault. In the west, the boundary between the Aegean and the Anatolian plates is rather poorly defined. McKenzie (1972) places this boundary slightly east of the east-west trending graben complexes of western Anatolia. However, Alptekin (1973) points out the shortcomings of this boundary by questioning the mechanisms with which such a boundary causes the graben system in western Anatolia and he combines the two plates to call it the "Aegean-Turkish Plate".

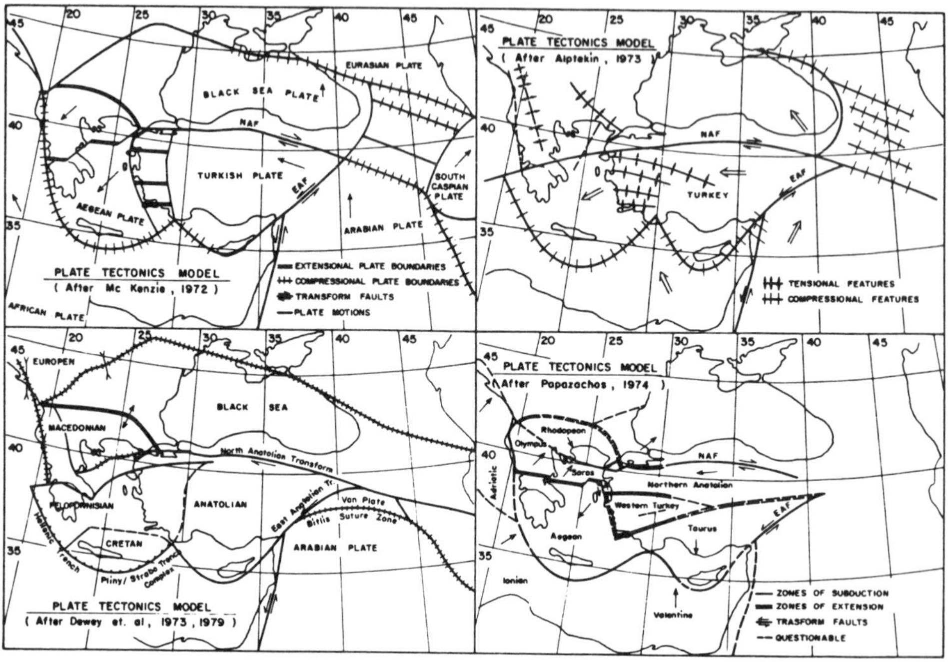

Figure 3. Seismotectonic Models for Turkey and Neighboring Areas

The seismotectonic model proposed by Papazachos (1974) provides a somewhat different picture. Based on the geographic and vertical distribution of the foci of the earthquakes between 1901 and 1971 as well as on the fault plane solutions of the earthquakes which occurred between 1948 and 1969, Papazachos (1974) defined several other smaller blocks (plates) within the Aegean region. From north to south these blocks include the Rhodopean, Olympus, Saros, Northern Anatolian, Western Turkey, Taurus and Aegean.

Dewey and Sengör (1979) suggest an alternative solution for the Aegean tectonics. They do not agree with McKenzie's (1972) Aegean plate but rather propose the Peloponnesian plate in the west of the Anatolian plate whose boundary is rather vaguely defined but definitely does not correspond to that of McKenzie's (1972) boundary. In the north of the Peloponnesian plate they also define the Macedonian plate which is torn off from the Black Sea plate as a result of the obstructive locking geometry caused by the bending of the Anatolian fault in the northern Aegean.

The Aegean plate, on the other hand, is bounded in the north by the western extension of the North Anatolian fault, the so-called "Anatolian Trough". The southern boundary of the Aegean plate passes through the Hellenic Trench and the Pliny/Strabo Trench complex, south of Crete and Rhodes, respectively (McKenzie, 1979 ; Dewey and Sengör, 1979). The fault plane solutions of McKenzie (1972) and Alptekin (1973) indicate that the southern margin of the Aegean plate is characterized by thrust faults suggesting the overriding of this plate onto the African plate. The relative motion rate between the African and the Aegean plates is in the order of 3.7 cm/year (McKenzie, 1972) ; 2,5 cm/year (Papazachos and Comninakis,

1971) or 2.6 cm/year (Le Pichon, 1968). The direction of the motion of the Aegean plate is towards southwest relative to the European plate (McKenzie, 1970). The relative plate motion in the eastern Meditterranean, however, is determined not by the African-Europe motions but by the north compressive motion that is occurring between the Arabian plate on the one hand and the Eurasian plate on the other (McKenzie, 1970 ; Dewey *et al.*, 1973).

The Arabian plate, although thrusting under the Van plate along the Bitlis zone, pushes the Van plate northward (Dewey *et al.*, 1973), thus producing a north-south compression in eastern Anatolia. The NAF takes up the westward motion of the Anatolian plate relative to the Black Sea plate and the EAF takes up the motion of the Anatolian plate relative to the Van plate. The westward motion of the Anatolian plate relative to Africa is taken up by the subduction along the Hellenic trench.

There are differences between the various plate tectonics models for the eastern Mediterranean including Turkey, but the neotectonic features of Turkey are rather well established. The boundaries between the Black Sea, Anatolian, Van and Arabian plates are responsible for the major seismicity of Turkey. In western Anatolia, the east-west trending grabens account for most of the seismic events of this region. The proposed models and the associated seismicity suggest that certain seismic sources may be delineated along the micro-plate boundaries. Figure 4 is a synthesis of the common features of these models.

Brief descriptions of the seismic sources adopted for Turkey are given below. The corresponding members are in Figure 5.

Anatolian Through (Zone No. 1)

The Anatolian Trough is a well established plate boundary in the northern Aegean including Saros and Sporades Troughs and terminating at the Gulf of Corinth (Morelli *et al.*, 1975; McKenzie, 1972). The Anatolian Trough corresponds to the western continuation of the northern strand of the NAF. It also includes the Sea of Marmara, the Izmit-Sapanca graben, and the Gemlik-Iznik graben north of Bursa.

North Anatolian Fault (Zone No. 2)

The NAF is a morphologically distinct and scismically active right-lateral strike-slip fault. It has a well developed surface expression for most of its length of about 1000 km. The width of the shear zone ranges between a few hundreds meters and a few kilometers. The actual limits of the fault to the east and to the west are not well defined. To the east the fault zone seems to terminate 10 km east of Karliova where it meets the EAF. To the west the termination of the zone is not clear. Allen (1969) places this terminal point west of Bolu, Ketin (1969) extends the zone to the Aegean Sea, north of Edremit.

Caldiran (Zone No. 2')

The Caldiran seismic zone accounts for several earthquakes east of Karliova (e.g. Varto and Caldiran) which produced right-lateral surface rupturcs but lacks the continuity and uniformity of the NAF.

Abul Samsar (Zone No. 3)

The Abul Samsar fracture zone extends in the NE-SW direction traversing the geological structures of the Caucasus. According to Nowroozi (1971), the Abul Samsar fracture zone accounts for most of the seismic activity represented by a transverse trend of epicenters within the Lesser Caucasus. Although there is no sound geological evidence for the existence of the Abul Samsar fracture zone in the northeastern part of Turkey, the zone is identified on

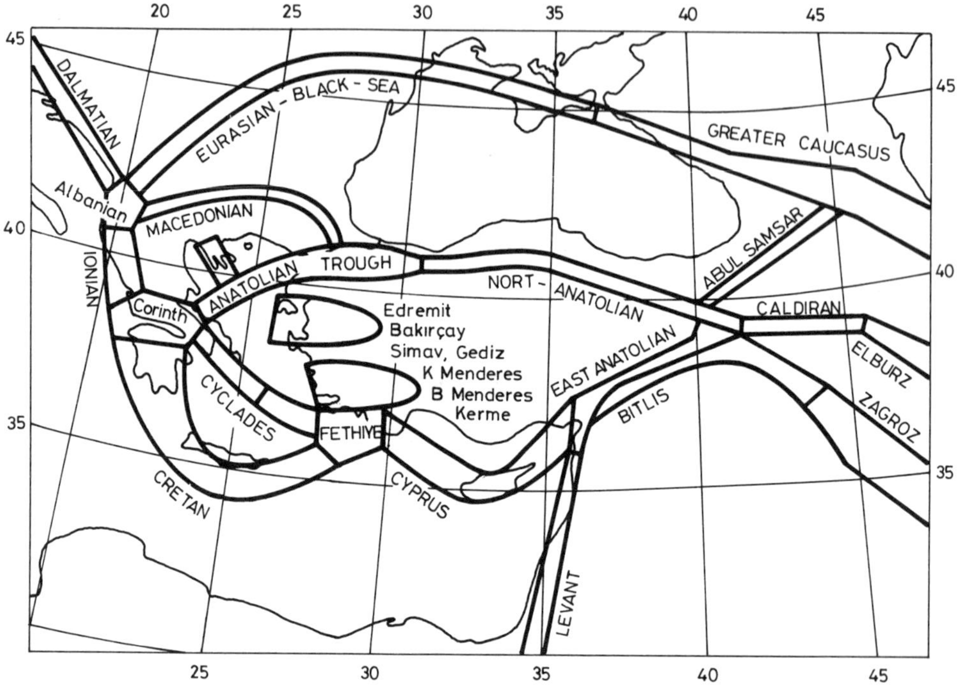

Figure 4. Common Features of the Plate Seismotectonic models

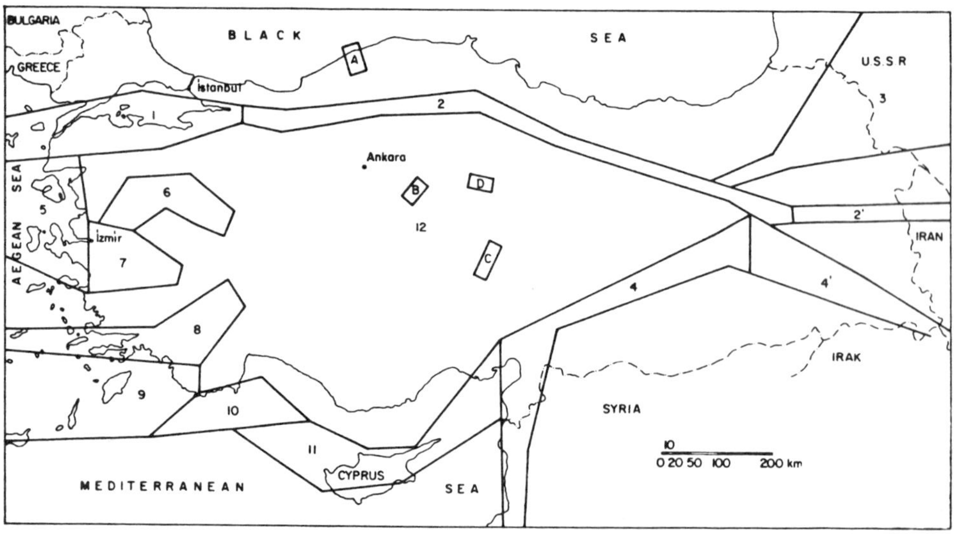

Figure 5. Seismic Regions Used in This Study

the basis of the seismological data.

East Anatolian Fault (Zone No. 4)

The EAF is an active left-lateral strike-slip fault which extends from Antakya to Karliova. The general trend of the fault is NE-SW. The fault zone is about 2-3 km wide and it consists of numerous parallel or subparallel, continuous and discontinuous fault traces. The EAF zone joins with the NAF zone in the east of Karliova (Seymen and Aydin, 1972 ; Arpat and Saroglu, 1972). The EAF joins with the Dead Sea fault system in the southwest, at the triple junction (Sengör and Yilmaz, 1981) near Maras.

Bitlis (Zone No. 4')

The Bitlis Zone, from which the North and East Anatolian faults originate west of a triple junction west of Lake Van (near Karliova) continues into the Zagros Crush Zone, the line of suturing between Arabia and Iran (Dewey and Sengör, 1979). The southeastern Anatolian thrust zone or the so-called "Bitlis suture zone" marks the boundary between the Arabian and Anatolian plates. The zone essentially consists of two thrust belts, the southern thrust belt forms the boundary between the Taurids and the border fold zone and it extends between south of Hakkari in the east and the Amanos mountains in the west (Ketin, 1966 ; Ilhan, 1971). The same belt extends westward to Cyprus (McKenzie, 1970) and eastward to join the Zagros Crush Zone (Dewey *et al.*, 1973). The thrust belt extends for a distance of 500 km in the east-west direction and its width ranges between 30 and 60 km. Northern thrust belt roughly extends in the northwest-southeast direction for a distance of 300 km. This belt has an average width of 50 km.

Sakiz and Midilli (Zone No. 5)

This zone, which extends from the south of the Gulf of Edremit to the south of the Izmir Peninsula, exhibits diffuse seismicity that can be associated with numerous small-scale grabens. The eastern boundary of this zone coincides with the north-south trending tectonic structures of the Aegean.

West Anatolian Graben Complex (Zones No. 6, 7, 8)

Western Anatolia is an area of intensive seismic activity. This activity is closely related to the east-west trending graben complexes which are bounded by high angle normal faults. McKenzie (1978) cites evidence to show that the dip of at least some of the normal faults becomes shallower at depth. The Sea of Marmara, Izmit-Sapanca, Gemlik-Iznik, Edremit, Bakirçay, Bergama, Simav, Gediz, Büyük Menderes, Küçük Menderes and Alasehir grabens are typical examples of this graben complex. The grabens close eastward and also terminate westward shortly after the western coast line of Anatolia.

Cyclades and Fethiye (Zones No. 9, 10)

In the southern Aegean, a Tertiary-Quaternary volcanic arc produces a prominent feature. It stretches from Thebes in the northwest across the southern part of the Cyclades towards the Bodrum peninsula in south-western Turkey. The eastward continuation of this arc has a higher seismic activity compared to its northwestern part. In the northwest and southeast the Cyclades arc terminates at the stress concentration areas referred to as Corinth and Fethiye, respectively.

Cyprus (Zone No. 11)

The southern boundary of the Anatolian plate runs in a large loop from the Gulf of Iskenderum, South of Cyprus and to Rhodes (McKenzie, 1972). The eastern Mediterranean lithosphere underthrusts the Eurasian lithosphere along the arcuate structure between the Ionian islands and Cyprus (Papazachos, 1974). The Cyprus zone terminates on the east at the EAF and/or the Bitlis suture zone.

Source zones A, B, C and D in figure 5 are defined to model the occurrences of infrequent but damaging events at Bartin, Kirsehir, Kayseri and Sivas. These are essentially point sources.

3.3. Relative frequencies

The recurrence relationship for earthquakes (Richter, 1958) :

(23) $$\log N = a + bM$$

where N is the number of the earthquakes above the magnitude M in a given region and within a given period and a and b are regression constants, has been extensively used in many seismicity studies. The coefficient a is a constant that is dependent on the location and time of the sample and b represents a constant that is characteristic of the region.

Earthquake catalogues are often biased due to incomplete reporting for smaller magnitude earthquakes in earlier periods. Thus, to derive the recurrence relationship for a region, one should decide between using a short sample that is complete in small events or a longer sample that is complete in larger events or a combination of the two data sets to complete the deficient data, thereby obtaining a homogeneous data set.

In this study, an artificially homogeneous data set was determined through the determination of the period over which the data in a given magnitude group are completely reported (Stepp, 1973).

In Table 1 we provide a list of the regression coefficients of the recurrence relationship for each seismic source zone based on the adjusted annual rates of occurrence. Zones No. 2' and 4' were given the same coefficients as those of zones No. 2 and 4, respectively.

TABLE 1. Regression Constants for the Zones

Zone No.	a	b
1	4.1	-0.78
2	3.6	-0.80
3	3.3	-0.75
4	4.3	-0.87
5	3.8	-0.78
6	5.2	-0.98
7	3.6	-0.78
8	4.7	-0.91
9	4.2	-0.76
10	4.4	-0.89
11	3.2	-0.72

3.4. Maximum magnitudes for zones

The analysis of the rupture length versus magnitude data has yielded the following expressions for the Turkish (excluding North Anatolian Faults events) and North Anatolian Fault earthquakes, respectively :

(24) $\log L = -3.9 + 0.8\ M$

(25) $\log L = -6.4 + 1.1\ M$

where L is the rupture length in km, and M is the surface wave magnitude.

In our work we used the fault rupture length procedure in connection with Equations (24) and (25) to determine the maximum magnitudes associated with the seismic sources encompassing major faults. The maximum historical earthquake procedure was applied to seismic sources where the tectonics is very complex and/or it is not possible to single out a prominent fault. In each case it has also been checked that the maximum magnitudes are in good conformity with the annual N = 0.001 intercept magnitude of the corresponding recurrence relationship. Table 2 summarizes the maximum surface wave magnitudes assigned to each seismic zone. Zones No. 2' and 4' were given the same maximum magnitudes as those of zones No. 2 and 4, respectively.

TABLE 2. Maximum Surface Wave Magnitudes for the Zones

Zone No.	Max. Mag.	Zone No.	Max. Mag.
1	8.0	7	7.5
2	8.0	8	8.0
3	7.5	9	8.0
4	8.0	10	7.5
5	7.5	11	7.5
6	8.0	Background	7.0

3.5. Attenuation

Seismic hazard studies commonly use attenuation relationships for peak ground acceleration (PGA) and the site intensity (I). The scarcity of the local strong-motion acceleration data in Turkey makes it necessary either to define the attenuation on the basis of local intensities or borrow the acceleration attenuation relationships of foreign data.

3.5.1. The PGA attenuation. The PGA is the most commonly used ground motion descriptor for the seismic hazard assessments. The seismic hazard in what follows is derived relative to the horizontal component of the PGA not necessarily because that it is the best parameter that describes the severity of the ground motion or governs the reponse of the structures, but because it has been accepted by other applications of similar nature.

Most of the existing attenuation relationships correlate the PGA with magnitude (M), distance (R) and with site soil conditions. The magnitude, by definition, is a band-limited measure of the ground displacement which necessitates an adherence to the definition used in the attenuation relationship in the compilation of the frequency-magnitude data and in assessment of the maximum magnitudes. It should be noted that, whereas most of the PGA attenuation relationships are based on the local magnitude (M_L), the magnitudes associated with the earthquakes in Turkey are a mixture of body wave, surface-wave and region-specific-local magnitudes.

The distance parameter (R) used in the attenuation relationship may be related to : (1) hypocenter, (2) epicenter, (3) center of energy release, (4) fault surface, and (5) fault trace. The definition of distance plays a great role especially in the near field and should be compatible with the regional characteristics. In regions like Turkey where most of the earthquakes are shallow and are associated with surface ruptures the use of the site to fault trace distance is physically justifiable.

In the type of attenuation relationships given by Equation (19) the PGA is correlated with M and R in a single equation. The majority of the data base of such relationships pertain to the medium field (i.e. 20-200 km) with limited near and far field data and no physical constraints are introduced in the regression analysis. Typically, ln (PGA) is proportional to 0.9 M with a standard deviation of about 0.7. Thus ± 1 standard deviation zones almost overlap the mean estimates for earthquakes of two magnitude units apart, thereby decreasing the significance of such attenuation relationships. Another type of attenuation relationship correlates the PGA with R for earthquakes of each given magnitude group. Such an approach is superior to the former because the magnitude does not explicitly enter into the regression analysis and the distance dependence of the PGA does not have to conform to the same shape for each magnitude class. Schnabel and Seed (1973) and Boore *et al.* (1978) have presented attenuation relationships of the second type.

The scarcity of the local strong-motion acceleration data in Turkey makes it necessary to borrow the already developed acceleration attenuation relationships based on foreign data. After detailed considerations and comparisons the attenuation relationships of Schnabel and Seed (1973) with near field corrections from Campbell (1981) were utilized for the seismic hazard maps developed in the present work.

The PGA attenuation relationships utilized in this study are strongly influenced by the western U.S.A. data. In order to provide a check on the applicability of these data to Turkish earthquakes, the available Turkish strong motion data (Ates and Bayülke, 1981) are compared with the western U.S.A. data after Joyner and Boore (1981) in Figure 6.

3.5.2 Intensity attenuation. Seismic intensities are evaluations based on the severity of the destructions. In countries like Turkey, where there is a lack of strong motion data, it is essential to understand the attenuation of intensities for regional hazard assessments. Yet, intensities do not easily lend themselves to the earthquake resistant structural design in a quantitative manner necessitating the use of various empirical equations to obtain the ground motion parameters from site intensities. Intensity attenuation relationships can be obtained regionally and thus are better predictors than the borrowed attenuation relationships, but the subsequent use of the empirical formulations to obtain the PGA from the intensity assessments may introduce greater uncertainties.

For the development of regional intensity attenuation relationships the isoseismal maps pertaining to the earthquakes within Turkey were compiled from various sources and fitted into a relationship of the form

(26) $$I_o - I = a_1 + a_2 \ln (R + R')$$

and

(27) $$M = a_3 + a_4 I$$

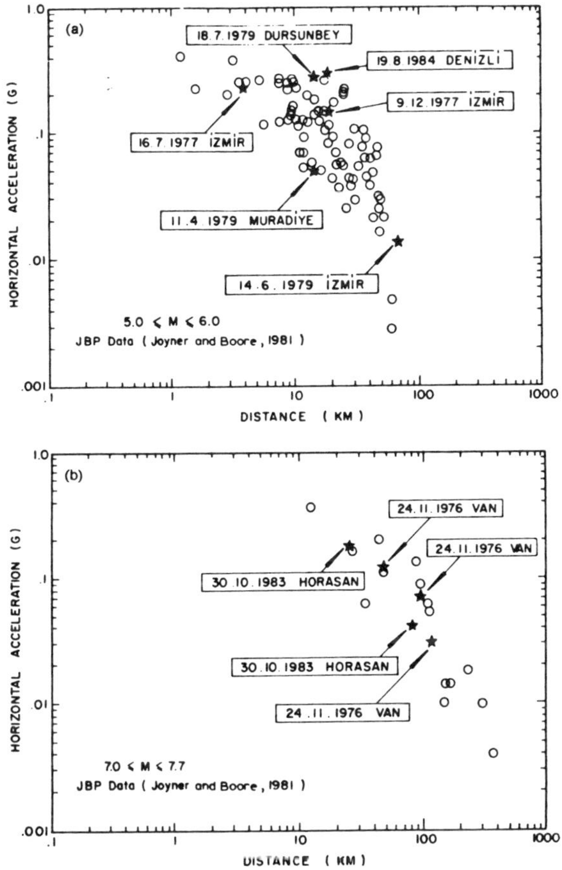

Figure 6. Comparison of Turkish and Western USA Data

yielding :

(28) $$I = a_5 + a_6M + a_7 \ln (R + R')$$

In these expressions a_1, a_2, a_3, and a_4 are regression coefficients, whose combination yields a_5, a_6, and a_7.

The value of R' has been found to be 7 km based on trial and error calculations to yield the best fit. For the epicentral distance R the equivalent circle method was used, where R is set equal to the radius of the circle centered at about the epicenter and with an area equal to that enclosed within the isoseismal.

In Table 3 the results of the regression analysis are provided for each source zone. This table does not contain entries for zones 3 and 10 because there exists insufficient data for them. In the calculations they were assumed to possess the general attenuation

TABLE 3. Intensity Attenuation Relationships

Zone No.	I - I (MSK)	M
1	-3.04 + 1.36 ln(R + 7)	-4.70 + 1.40 I
2	-2.90 + 1.59 ln(R + 7)	1.40 + 0.59 I
3		
4	-2.04 + 1.32 ln(R + 7)	1.25 + 0.66 I
5	-3.02 + 1.47 ln(R + 7)	8.00 - 0.40 I
6	-3.40 + 1.39 ln(R + 7)	2.00 + 0.55 I
7	-2/62 + 1.24 ln(R + 7)	0.53 + 0.83 I
8	-1.32 + 1.02 ln(R + 7)	6.77 - 0.20 I
9	-3.35 + 1.32 ln(R + 7)	-0.70 + I
10		
11	-2.01 + 1.11 ln(R + 7)	1.65 + 0.55 I
Entire Country	-2.68 + 1.38 ln(R + 7)	2.55 + 0.47 I

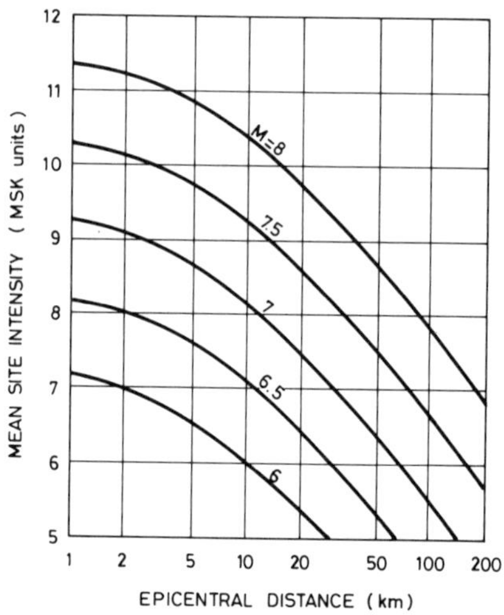

Figure 7. Intensity Attenuation Used in This Study

characteristics for the entire country.

The overall intensity attenuation relationship for whole Turkey has been found to be:

(29) $I_o - I = -2.68 + 1.38 \ln (R + 7)$

(30) $M = 2.55 + 0.47 I_o$

(31) $I = 2.74 + 2.13 M - 1.38 \ln (R + 7)$

(standard deviation = 0.7)

The last row in Table 3 gives the general relationships for Turkey as a whole. The mean site intensity versus epicentral distance curves for various magnitudes are plotted in Figure 7 using the relationship given by equation (31). This intensity attenuation relationship has been found to be reasonably accurate for all seismic source regions considered except for the strike-slip fault zones where the fault mechanism generates isoseismals of non-isotropic character thereby necessitating a different treatment.

For the North Anatolian Fault Zone the following intensity attenuation relationship based on the distance to the fault trace has been found to be compatible with the macroseismic observations and the fault-rupture model adopted for the seismic hazard assessment associated with this zone (Erdik and Oner, 1982) :

$$I = 3.92 + 2.08\, M - 0.98 \ln D \qquad (32)$$

where D is the shortest distance to the fault trace from the site in kilometers.

3.6. Results and discussion

We base our discussion on the two intensity maps for return periods of 225 and 475 years (Figures 8 and 9) in which for points in the vicinity of NAF Equation (32) was used ; for the remainder of the country equation (31) was assumed to hold. In Figure 10 we present the acceleration contours for a return period of 475 years ; these are based on a juxtaposition of the Schnabel and Seed (1973) and the Campbell (1981) attenuation relationships.

In closing we note the following general characteristics :

(1) The contour levels of the intensity hazard maps range from VI.5 to VIII.5. Below the VI level the ground shaking effects are generally controlled by small-magnitude events with unreliable recurrence rates and attenuation characteristics.

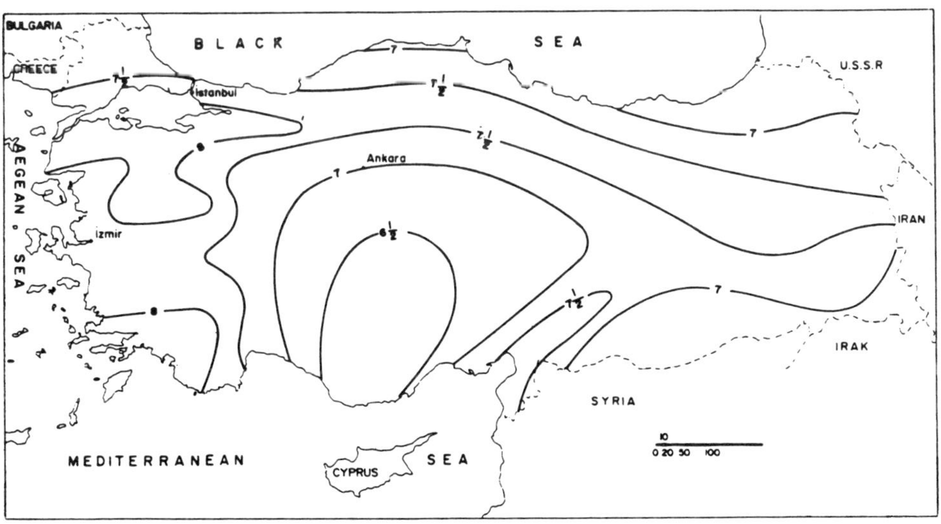

Figure 8. Intensity Contour Map for 225 Years

(2) The contour levels of the acceleration hazard map ranges from 5% to 60% g. Below the 5% level the ground shaking effects are generally controlled by small magnitude events with unreliable recurrence rates and attenuation characteristics.

(3) The highest intensity contours encompass the North Anatolian Fault Zone and the Aegean Grabens.

(4) The isolated regions of Bartin, Kayseri and Kirsehir have experienced earthquakes with rather large recurrence intervals compared to the other regions. The intensity and peak acceleration values associated with these isolated regions are indicative of their low reccurence interval rather than the size of the events that they have experienced.

(5) For important and/or critical structures for which very low probabilities should be considered these map can provide a preliminary idea of the relative hazard and can be utilized for preliminary site identification purposes.

(6) The peak ground acceleration attenuation relationship of Schnabel and Seed (1973) are for mean maximum accelerations in bedrock. With soil deposits of soft-and-medium-stiff sands and clays of appreciable depth the ground accelerations will be somewhat larger than those indicated on these hazard maps.

(7) In the delineation of the seismic sources the geological boundaries of the tectonic elements and their immediate vicinity have been considered. The seismic hazard near the source boundaries is strongly affected by the changes in the delineation of these boundaries.

(8) The recurrence relationship used in the preparation of these maps are based on artificially complemented data set that can account for the small magnitude data deficiencies in earlier reporting periods. The net effect of such a procedure is reflected in the increase of the slopes of the log-linear recurrence relationships and in the decrease of the relative number of earthquakes at the high magnitude end. The effect of such an approach on the hazard maps can be appreciable especially for high return periods.

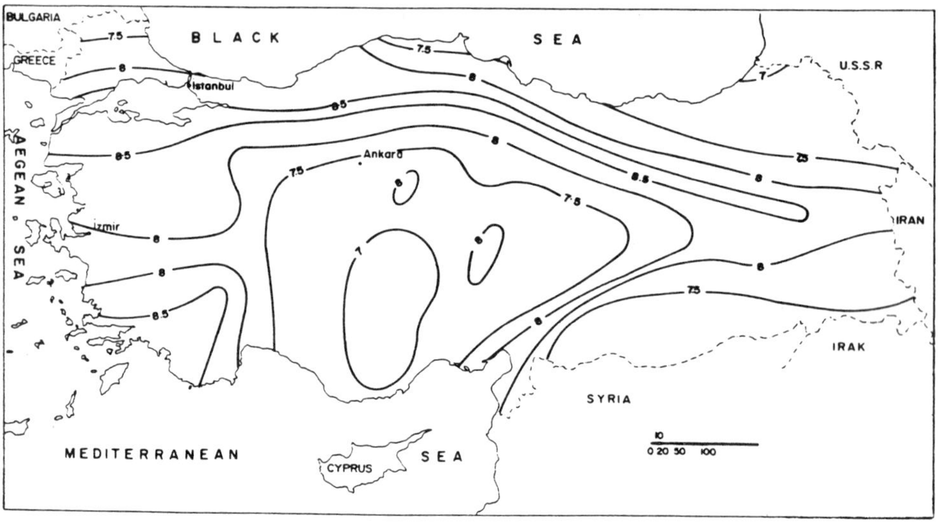

Figure 9. Intensity Contour Map for 475 Years

(9) The maximum magnitudes assigned to each seismic source zone have been based on physical interpretations. For lower magnitude zones, adopting higher maximum magnitudes can have pronounced effects on the iso-acceleration contours of these maps.

(10) The attenuation relationship utilized in these maps yield the mean peak horizontal ground acceleration with no statistical variations. Our findings on the same source zone configuration and recurrence relationships indicate that the inclusion of a standard deviation of ln PGA in the order of 0.62 does not appreciably affect the high amplitude ground motion levels near the tectonic elements but may have pronounced effects (reaching up to 50%) on the low-level ground motion regions.

The seismic hazard information conveyed here is based on the recurrence relationships given in Table 1 and represents extrapolations assuming that the rate of the seismic activity observed in the past 70 years will essentially repeat itself in the future. The seismic activity of the East Anatolian Fault (zone 4) has shown rather low activity in the past decade compared to its historic activity. However, the recent seismic activity of the East Anatolian Fault is comparable to that of the North Anatolian Fault. Thus, it may be reasonable to assume that the future seismic activity of the East Anatolian Fault will be similar to the past 70-year activity of the North Anatolian Fault.

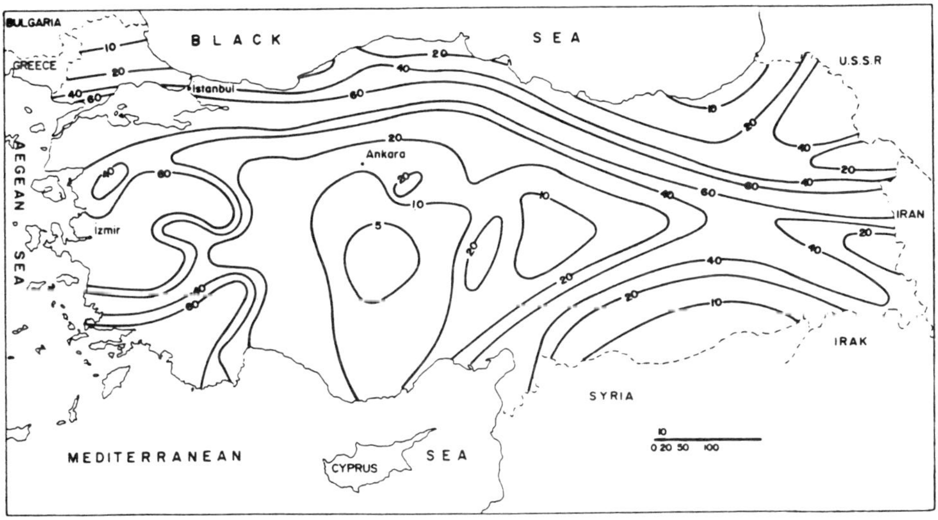

Figure 10. Acceleration Contour Map for 475 Years (% g)

REFERENCES

1. Allen, C.R., 'Active faulting in northern Turkey', Contrib. No. 1577, Div. Geol. Sci., Caltech, 1969, 32 pp.
2. Alptekin, O., 'Focal mechanisms of earthquakes in western Turkey and their tectonic implications,' Ph. D. Diss., New Mexico Instit. Min. Techn., 1973, 190 pp.
3. Alsan, E., L. Tezuçan, and M. Bath, 'An earthquake catalog for Turkey for the interval 1913-1970, 'Kandilli Observ., Istanbul, and Seismological Inst., Uppsala, 1975.
4. Arpat, E., F. and Saroglu, 'Dogu anadolu fayi ile ilgili bazi gozlemler ve düsünceler (Some observations and opinions regarding the East Anatolian Fault),' Bull. Min. Res. Expl. Instit. Turkey, **78**, 1972, pp. 44-50, (in Turkish).
5. Ates, R., and N. Bayülke, 'Strong motion records in Turkey,' Proc. of Strong Ground Motion Symp., UNESCO, RER/79/014 Balkan Project, 1981.
6. Bender, B., 'Modeling source zone boundary uncertainty in seismic hazard analysis,' Bull. Seism. Soc. Am., **76**, 1986, pp. 329 - 341.
7. Boore, D.M., A. Oliver, R. Page, and W.B. Joyner, 'Estimation of ground motion parameters', U.S. Geol. Surv., Open File Rep., No 78.509, 1978.
8. Campbell, K.W., 'Near source attenuation of peak horizontal acceleration,' Bull. Seism. Soc. Am., **71**, 1981, pp. 2039 - 2070.
9. Cornell, C.A., 'Engineering seismic risk analysis,' Bull. Seism. Soc.Am., **58**, 1968, pp. 1568 - 1606.
10. Cornell, C.A., and E.H. Vanmarcke, 'The major influences of seismic risk,' Proc.4th World Conf. Earthq. Engin., Santiago, Chile, Vol. I, 1969, pp. 69 - 83.
11. Cornell, C.A., and H. Merz, 'Seismic risk analysis of Boston,' J. Struct. Div., ASCE, **107**, 1975, pp. 2027 - 2043.
12. Der Kiureghian, A., and A.H. Ang, 'A fault rupture model for seismic risk analysis,' Bull. Seism. Soc. Am., **67**, 1977, pp. 1173 - 1194.
13. Dewey, J.F., W.C. Pitman, W.B.F. Ryan, and J. Bonnin, 'Plate tectonics and the evolution of the Alpine system,' Bull. Geol. Soc. Am., **84**, 1973, pp. 3137 - 3180.
14. Dewey, J.F., and A.M.C. Sengör,'Aegean and surrounding regions:complex multiplate and continuum in a convergent zone,' Bull. Geol. Soc. Am., **90**, 1979, 84-92.
15. Donovan, N.C. and A.E. Bornstein, 'Uncertainties in seismic risk procedures,' J. Geotechn. Eng. Div., ASCE, **104**, 1978, pp. 869 - 887.
16. Earthquake Engineering Research Institute (EERI), 'Evaluation of seismic hazards and decision making in earthquake resistant design,' Proc. EERI Seminar Reno, Nevada, 1983, 300 pp.
17. Erdik, M., and S. Öner, 'A rational approach for the probabilistic assessment of the seismic risk associated with the North Anatolian Fault,' *in* Multi-Disciplinary Approach to Earthquake Prediction, A. Isikara and M. Vogel (eds.), Vieweg, Braunschweig-Wiesbaden, 1982, pp. 115 - 127.
18. Erdik, M., V. Doyuran, N. Akkas, and P. Gülkan, 'A probabilistic assessment of the-seismic hazard in Turkey,' Tectonophys., **117**, 1985, pp. 295 - 344.
19. Esteva, L., 'Seismicity', *in* Seismic Risk and Engineering Decisions, C. Lomnitz and E. Rosenblueth (eds.), Elsevier, 1976.
20. Gülkan, P., and M.S. Yücemen, 'Seismic risk analysis for nuclear power plants,' METU J. Pure Appl. Sc., **10**, 1977, pp. 115 - 135.
21. Gülkan, P., and A. Gürpinar, 'Seismic hazard of northwestern Turkey,' Earthquake Engineering Research Institute Report No. 77-05, Middle East Technical University, Ankara, Turkey, June, 1977.
22. Gürpinar, A., M. Erdik, M. Oner, and M.S. Yucemen, 'Seismic risk analysis of northern Anatolia based on intensity attenuation,' Proc. U.S. Nat. Conf. Earthq. Eng., Stanford, Ca., 1977, pp. 72 - 81.

23. Joyner, W.B., and D.M. Boore, 'Peak horizontal acceleration and velocity from strong-motion records from the 1979 Imperial Valley, California, earthquake,' Bull. Seism. Soc. Am., **71**, 1981, pp. 2011 - 2039.
24. Ketin, I., 'Ueber die nordanatolische Horizontalverschiebung,' Bull. Min. Res. Expl. Instit. Turkey, **72**, 1969, pp. 1 - 28.
25. Le Pichon, X., 'Sea floor spreading and continental drift,' J. Geophys. Res., **73**, 1968, pp. 3661 - 3697.
26. McKenzie, D.P., 'Plate tectonics of the Mediterranean region,' Nature, **226**, 1970, pp. 229 - 243.
27. McKenzie, D.P., 'Active tectonics of the Mediterranean region,' Geoph. J. R. astr. Soc., **30**, 1972, pp. 109 - 185.
28. McKenzie, D.P., 'Active tectonics of the Alpine-Himalayan belt, the Aegean Sea and surrounding regions,' Geoph. J. R. astr. Soc., **55**, 1978, pp. 217 - 254
29. Merz, H., and C.A. Cornell, 'Seismic risk analysis based on a quadratic magnitude-frequency law,' Bull. Seism. Soc. Am., **63**, 1973, pp. 1999 - 2006.
30. Morelli, C., M. Pisoni, and C. Ganter, 'Geophysical studies in the Aegean Sea and in the eastern Mediterranean,' Boll. Geofis, **XVII (66)**, 1975, pp. 127 - 168.
31. Morgan, W.J., 'Rises, trenches, great faults and crustal blocks,' J. Geoph. Res., **73**, 1968, pp. 1959 - 1982.
32. Murphy, J.R., and L.J. O'Brian, 'The correlation of peak ground acceleration amplitude with seismic intensity and other physical parameters,' Bull. Seism. Soc. Am., **67**, 1977, pp. 877 - 915.
33. Nowroozi, A.A., 'Seismo-tectonics of the Persian Plateau, eastern Turkey, Caucasus, and Hindu-Kush regions,' Bull. Seism. Soc. Am., **61**, 1971, pp. 317 - 341.
34. OECD/Nuclear Energy Agency, 'Probabilistic methods in seismic risk assessment for nuclear power plants,' Proc. 2nd CSNI Specialist Meeting, CSNI Report No. 76, 1983, 491 pp.
35. Papazachos, B.C., 'Seismotectonics of the eastern Mediterranean area,' *in* Engineering Seismology and Earthquake Engineering, J. Solnes (ed.), Noordhoff, 1974,pp.1-32.
36. Papazachos, B.C., and D.E. Comninakis, 'Geophysical and tectonic features of the Aegean arc,' J. Geoph. Res., **76** , 1971, pp. 8517 - 8533.
37. Richter, C.F., Elementary Seismology, Freeman, 1958.
38. Schnabel, P.B., and H.B. Seed, 'Accelerations in rock for earthquakes in the western United States,' Bull. Seism. Soc. Am., **63**, 1973, pp. 501 - 516.
39. Sengör, A.M.C., and Y. Yilmaz, 'Tethyan evolution of Turkey : a plate tectonic approach,' Tectonophys., **75**, 1981, pp. 181 - 241.
40. Seymen, I., and A. Aydin, 'Bingöl deprem fayi ve bunun kuzey Anadolu fay zonu ile iliskisi (The Bingol earthquake fault and its correlation with the North Anatolian Fault zone),' Bull. Min. Res. Expl. Instit. Turkey, **76**, 1972, pp. 1 - 8, (in Turkish).
41. Shlien, S., and M.N. Toksöz, 'Frequency-magnitude statistics of earthquake occurrence,' Earthq. Notes Eastern Sect. Seism. Soc. Am., **41**, 1970, pp. 5 - 18.
42. Stepp, J.C., 'Analysis of completeness of the earthquake sample in the Puget Sound area,' *in* Contributions to Seismic Zoning, S.T. Handing (ed.), NOAA Tech. Rept. EERL 267-ESL 30, U.S. Dept. of Commerce, 1973.
43. TERA Corporation, 'Seismic hazard analysis : solicitation of expert opinion,' Washington, U.S. Nuclear Regulatory Commission Report No. NUREG/CR-1582, 1980, 200 pp.
44. Veneziano, D., 'Errors in probabilistic seismic hazard analysis,' Miscellaneous Paper S-73-1, U.S. Army Engineer Waterways Experiment Station, 1982.
45. Yücemen, M.S., 'Seismic risk analysis,' *in* State-of-the-Art in Earthquake Engineering, O. Ergünay and M. Erdik (eds.), Turkish National Committee on Earthquake Engineering, Ankara, 1981.

CHAPTER IV

SEISMOTECTONICS

GEODYNAMICS OF THE MEDITERRANEAN REGIONS

Jean BONNIN
Institut de Physique du Globe
Université Louis Pasteur
5, rue René Descartes
F - 67084 STRASBOURG Cedex

and

Jean-Louis OLIVET
IFREMER - Centre de Brest
B.P. 70
F - 29263 PLOUZANE Cedex

ABSTRACT. To describe the evolution of the Mediterranean domain in terms of plate kinematics, arguments from the region itself are lacking, because it is basically shrinking since 180 My, and most of the ancient oceanic parts have disappeared. One is forced to seek elements in other oceans where isochrons lines to be fitted are displayed, permitting reconstruction of relative palaeopositions of plates. The evolution of the North and Central Atlantic ocean is briefly reviewed, and the consequences on the Mediterranean domain are examined in terms of boundary conditions, which should be met by any model of evolution of the various blocks within it. The most recent relative motions are confronted with the present-day seismic activity : it appears that the concept of plate boundaries no longer prevails for most of the regions ; rather one should consider it as a large zone where deformation is widely distributed.

Foreword : What is discussed below is mainly the result of a long-lasting and active cooperation within a group of researchers, who have been occasionally with the same institution but are now dispersed in different laboratories. They still work together from time to time. They are : Jean-Marie AUZENDE (geologist), Paul BEUZART (Geophysicist), Jean BONNIN (geophysicist), Jean GOSLIN (geophysicist), Jean-Louis OLIVET (geologist), Patrick UNTERNHER (geologist).

*   *
*

It is not the intention of the authors to make a review of the many so-called models which have been designed to explain the tectonic history of various Mediterranean features. Those models are easy to look at in the literature, and it is doubtful that they are individually reliable enough to be integrated in a coherent view of the domain.

J. Bonnin et al. (eds.), Seismic Hazard in Mediterranean Regions, 257–281.

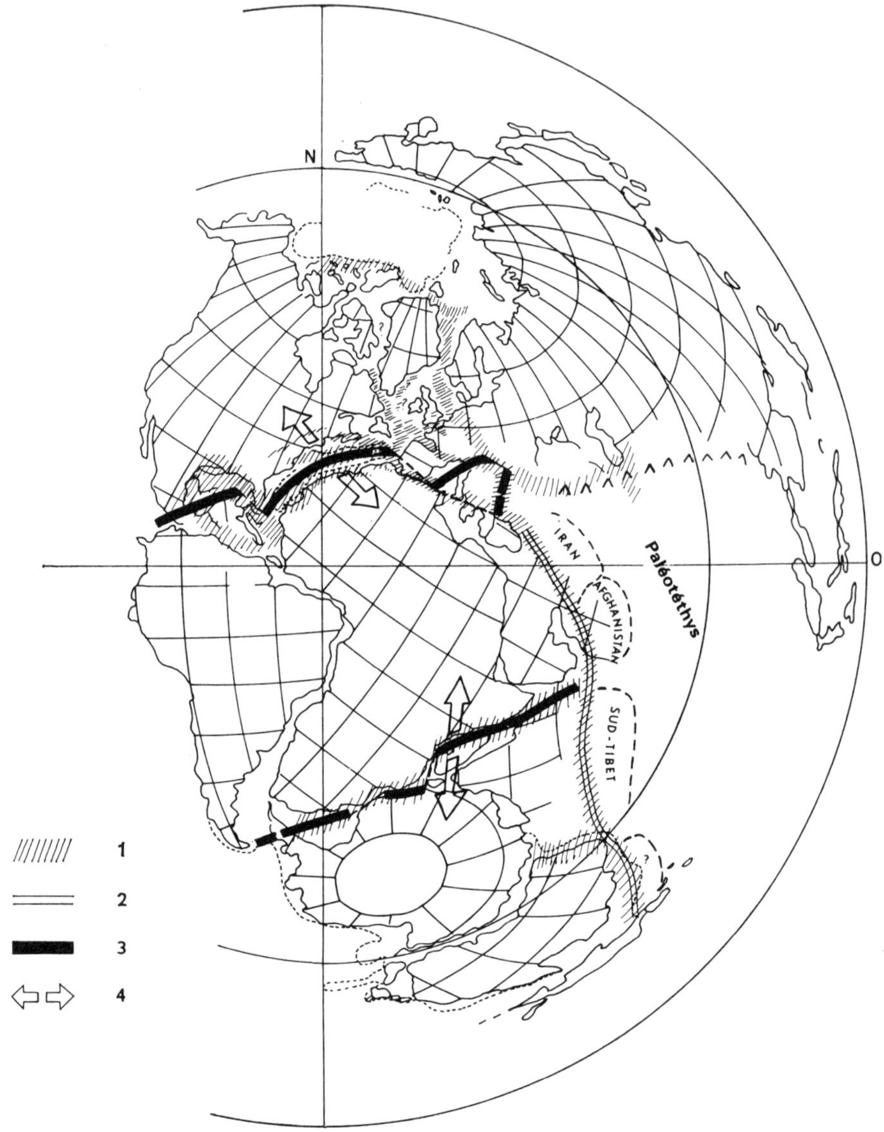

Fig. 1 : Pangaea in lower-Jurassic times (from Olivet *et al.*, 1984 ; modified from Smith and Briden, 1977). Hatched bands have experienced stretching during the uppermost Palaeozoic - lower Mesozoic period. Solid lines show zones where breaking will occur.

This contribution aims at placing some sort of boundary conditions for the evolution of the Mediterranean domain, which in turn will possibly give some insights into geographic distribution of seismic energy release in the region under consideration. The reasoning is as follows : since the very famous work of ISACKS *et al.* (1968) and BARAZANGI and DORMAN (1969), il is well known that the arrangement of seismic events delineates the different lithospheric plates, which are supposed to behave as rigid pieces of a spherical shell (*) ; conversely, if we are able to unravel the plate tectonic jigsaw puzzle in the Mediterranean regions, then we will be hope fully in the position of better understanding the seismic activity there, or at least of comparing the observed activity whith what could be expected.

Pangaea (lower Jurassic)

A simple glance at fig. 1 gives a rough idea of how things were arranged before creation of the Mediterranean. An angle-shaped domain referred to as the "Palaeotethys", of which coastlines are not precisely known, was most probably of oceanic nature, with its apex located where the western Mediterranean basins are presently. The large continental mass, called Pangaea, broke up in mid-Jurassic times into two main blocks, the southern one splitting soon into two others. At that time, the Palaeotethys began to shrink. Very little is known on that large ancient ocean because it has disappeared in subduction zones, except for limited remnants, most of them completely squeezed within the Alpine mountain chains (**). Arguments for deciphering its history have indeed vanished as it has been subducted beneath the southern Eurasian continental margin.

Forced are we, as direct clues do not exist any longer, to look for indirect arguments elsewhere to reconstruct the shrinking history of Palaeotethys and the development of the present Tethys (many authors have discussed this type of problem ; for a synthetic view on the Atlantic, see e.g. OLIVET *et al.*, 1984).

Fitting isochron lines

Arguments are usually found in oceans which have been spreading as early as possible, with no episod of shrinking. This is the case of the North and Central Atlantic, which have been produced by the drifting away from North America of respectively Eurasia and Africa. In both cases, relative motions have been regular enough to permit reconstruction of the relative positions of the plates with reasonable confidence at different epochs. Once one is able to restore North America and Africa on the one hand, and North America and Eurasia on the other hand in their relative positions at a given time, then the relative position of Africa and Eurasia is know at the same time.

(*) Incidentally, it is worth emphasizing here how useful has been the World Wide Standardized Seismographic Network (WWSSN) (which started operations in 1962 ; Murphy, 1966), for the computation of reliable focal solutions and the relatively homogeneous precision in location of events on a global scale by Lamont's seismologists.

(**) Whether the eastern Mediterranean basin is a piece of the Palaeothetys ocean is still the matter of debate ; the question will be solved once the original of southern Turkish blocks will be ascertained.

The procedure is thus to identify isochron lines (i.e. fossil traces of the common plate boundary at the time considered) on each plate of each pair (North America/Africa, North America/Eurasia), fit the isochron lines pair by pair, deduce the relative position of Africa with respect to Eurasia (or *vice-versa*). Then we have the contours of the Mediterranean domains at the time considered, as far as the limits between the deformed (Mediterranean) domain and the stable parts of the plates are possible to identify.

Most often, the isochron lines used are magnetic anomalies formed at the axes of the mid-oceanic ridges (except for the isochron which marks the initial splitting of the large blocks and which is unfortunately very often ill-defined). In the case of the Atlantic ocean, the following magnetic anomalies have been used (labelling of the magnetic anomalies comes from HEIRTZLER *et al.*, 1968 ; LARSON and PITMAN, 1972 ; LARSON and HILDE, 1975) :

13	approximately	36	My	(limit Eocene/Oligocene)
24	"	53	My	(limit Palaeocene/Eocene)
33	"	78	My	(Campanian)
(34)	"	86	My	(Coniacian-lower Santonian)
J (M0-M3)	"	112	My	(lowermost Aptian)
M 22	"	140	My	(Kimeridgian)

Thus, the relative positions of blocks are known only for a few discrete steps of the continuous history of the oceans. However, there is no indication of any major change in the spreading well within the intervals between the identified anomalies : all major changes appear to have occurred at (about) the time of one of the steps which have been worked out. Interpolating between these positions will yield the relative motions of plates during the corresponding time span.

Relative motions of Africa and Eurasia with respect to North America

When proceeding as described above, one is led to fit isochron lines from the youngest to the oldest : the ocean is progressively closed ; and it is important to check the consistency of the individual reconstructions when put all together. However, once confidence is gained on the reconstructions and their consistency, then it has more geological significance to look at the evolution as time goes naturally, i.e. opening the oceans progressively. The final results are shown on fig. 2 to 9 and a few brief comments follow.

In fig. 2, the initial fit (relative positions of the blocks just before very beginning of spreading) is shown : it has not been obtained by fitting any isochrons as none are available ; but simply by closing the gaps as much as possible, but taking much care of consistency with the directions of initial opening which will prevail soon, and avoiding unjustified overlaps. Most of the authors have an initial fit quite different from the one shown here. The main difference is that pre-existing, relatively important basins are featured here, while they are not by many other authors who assume usually a tighter fit of the continental margins.

At the time of anomaly M 22 (ca. 140 My) (Fig. 3), the Central Atlantic has formed by drifting of Africa away from North America ; spreading has been somewhat asymmetrical. Just before that time, Iberia has possibly started to rotate in such a way that no significant opening of the Bay of Biscay has occurred ; this remains to be ascertained.

At about 112 My (anomaly J = M0-M3) (fig. 4), Africa has continued moving

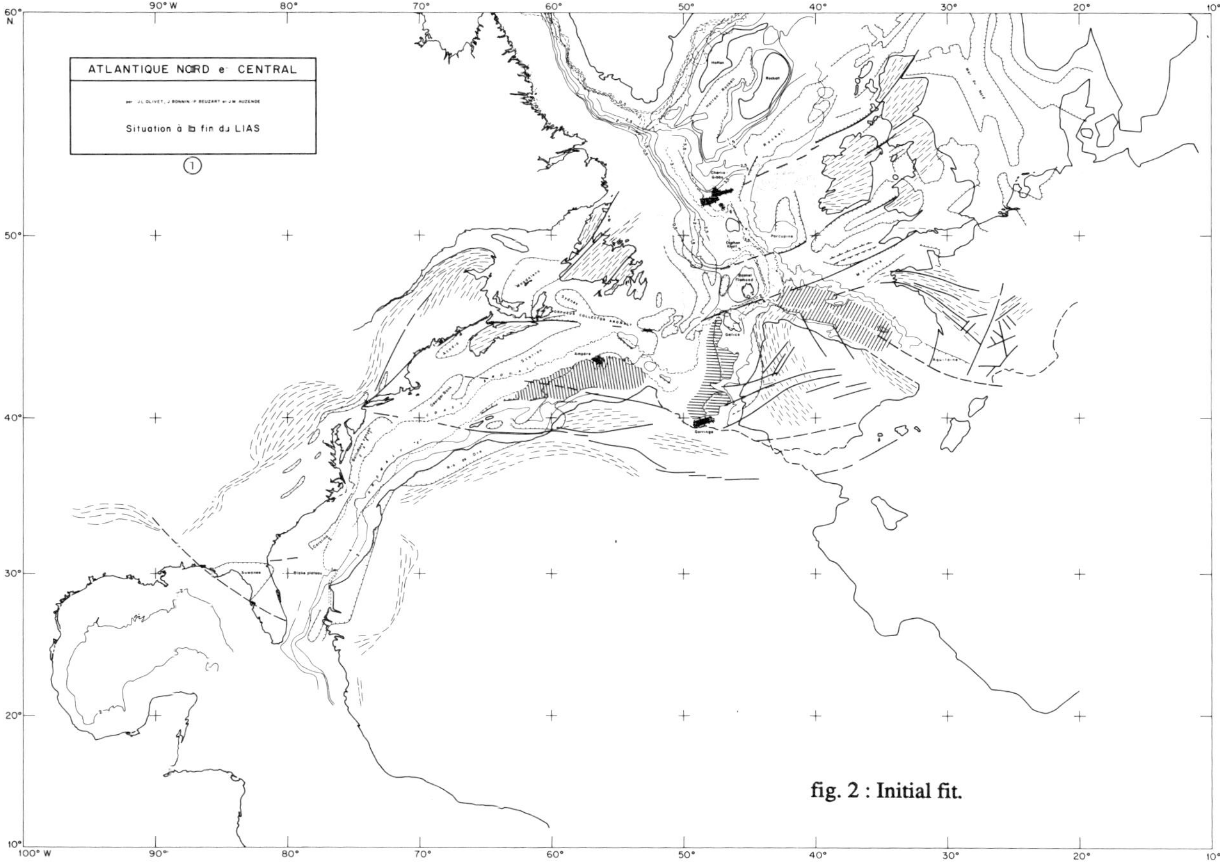

fig. 2 : Initial fit.

Fig. 2 to 9 : Palaeopositions of Africa, Iberia and Eurasia with respect to North America fixed in its present position (from Olivet *et al.*, 1984).

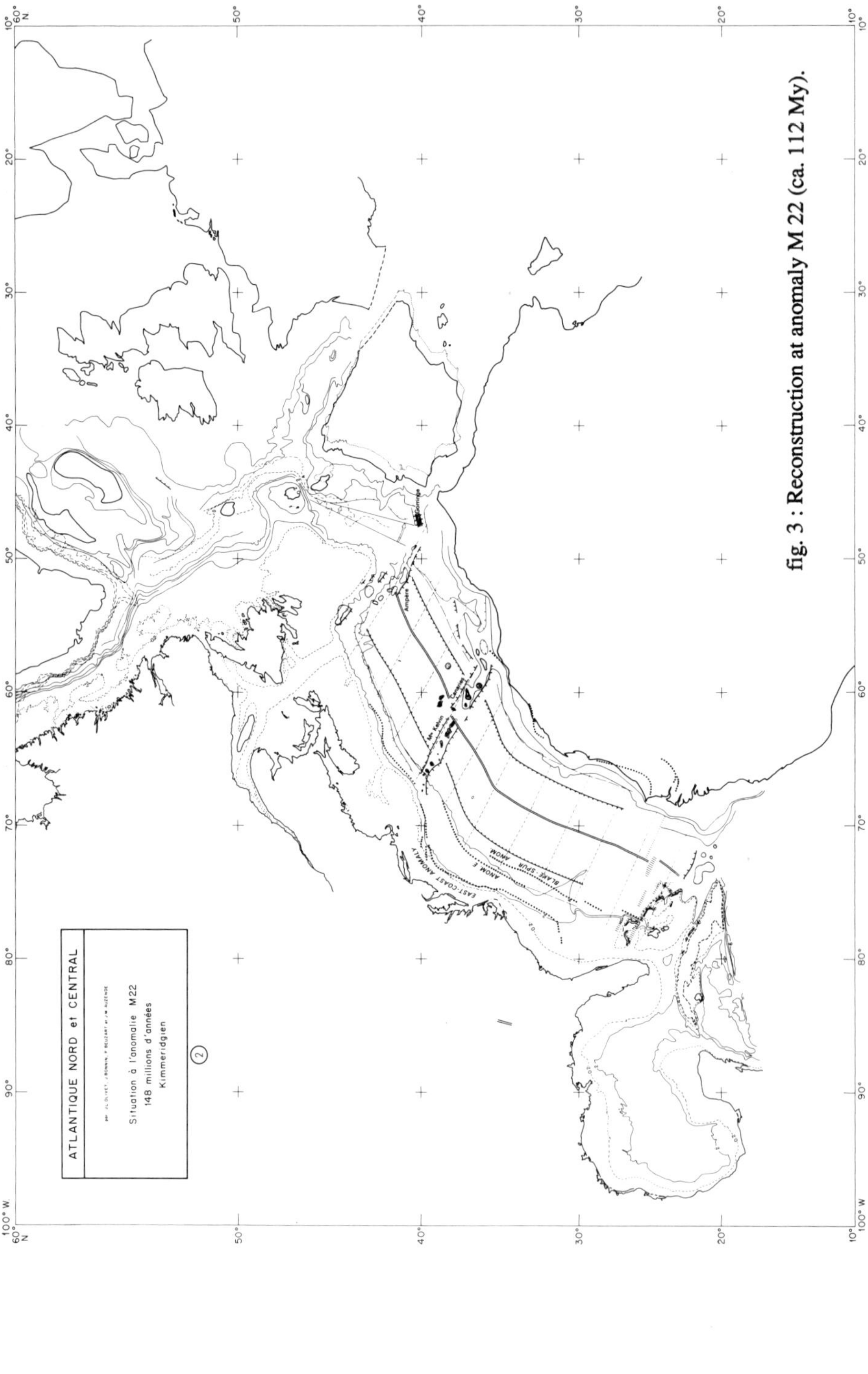

fig. 3 : Reconstruction at anomaly M 22 (ca. 112 My).

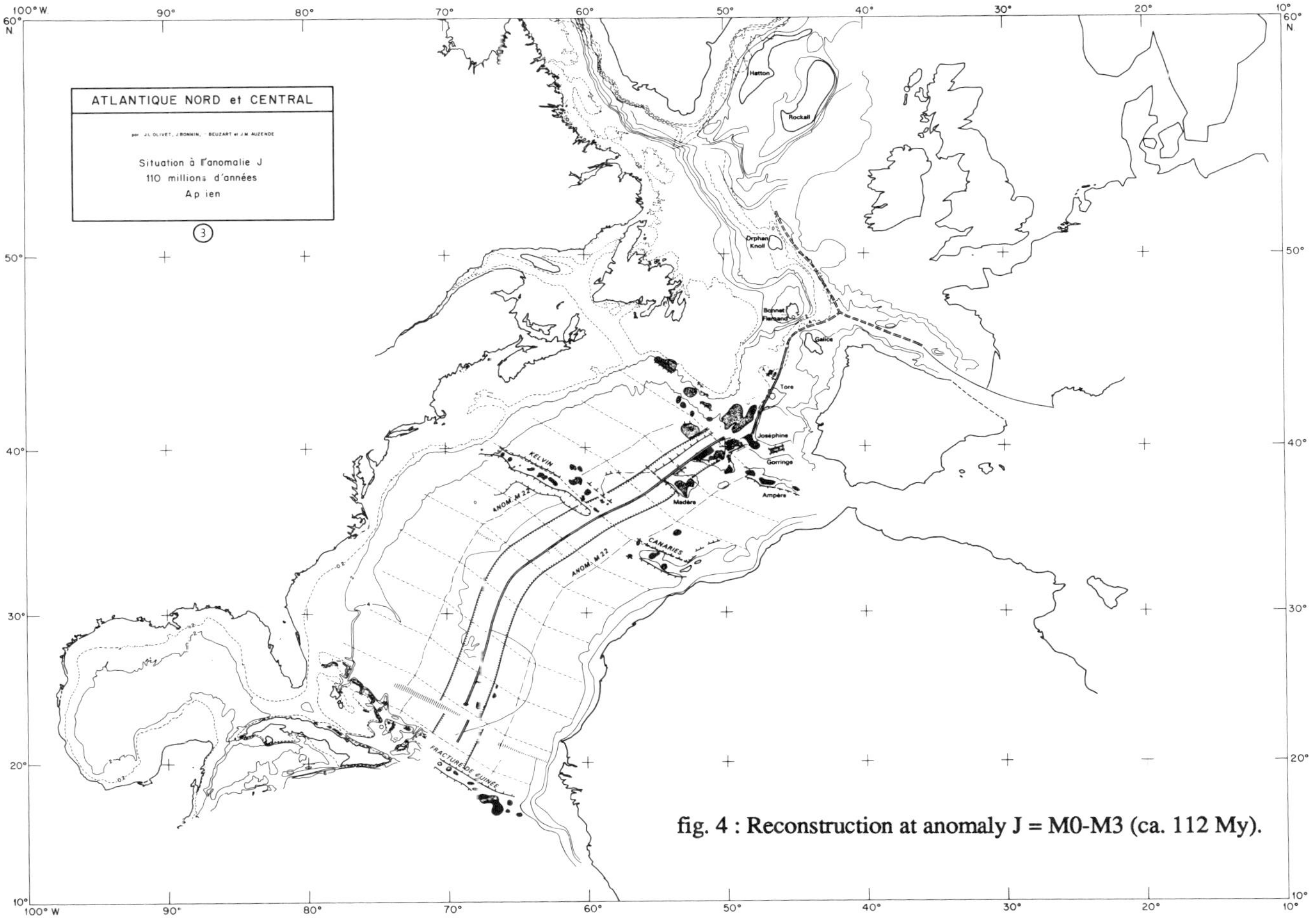

fig. 4 : Reconstruction at anomaly J = M0-M3 (ca. 112 My).

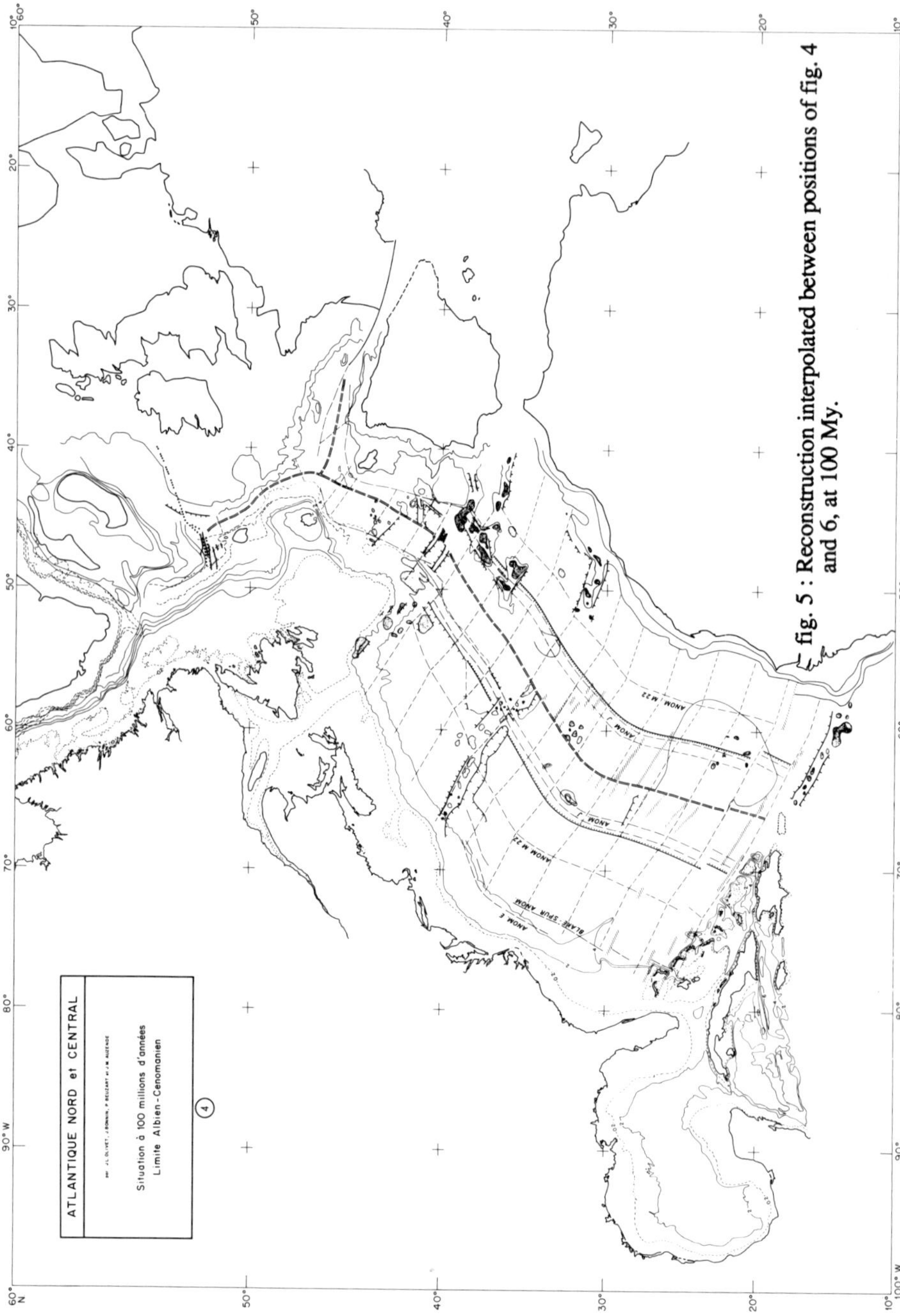

fig. 5 : Reconstruction interpolated between positions of fig. 4 and 6, at 100 My.

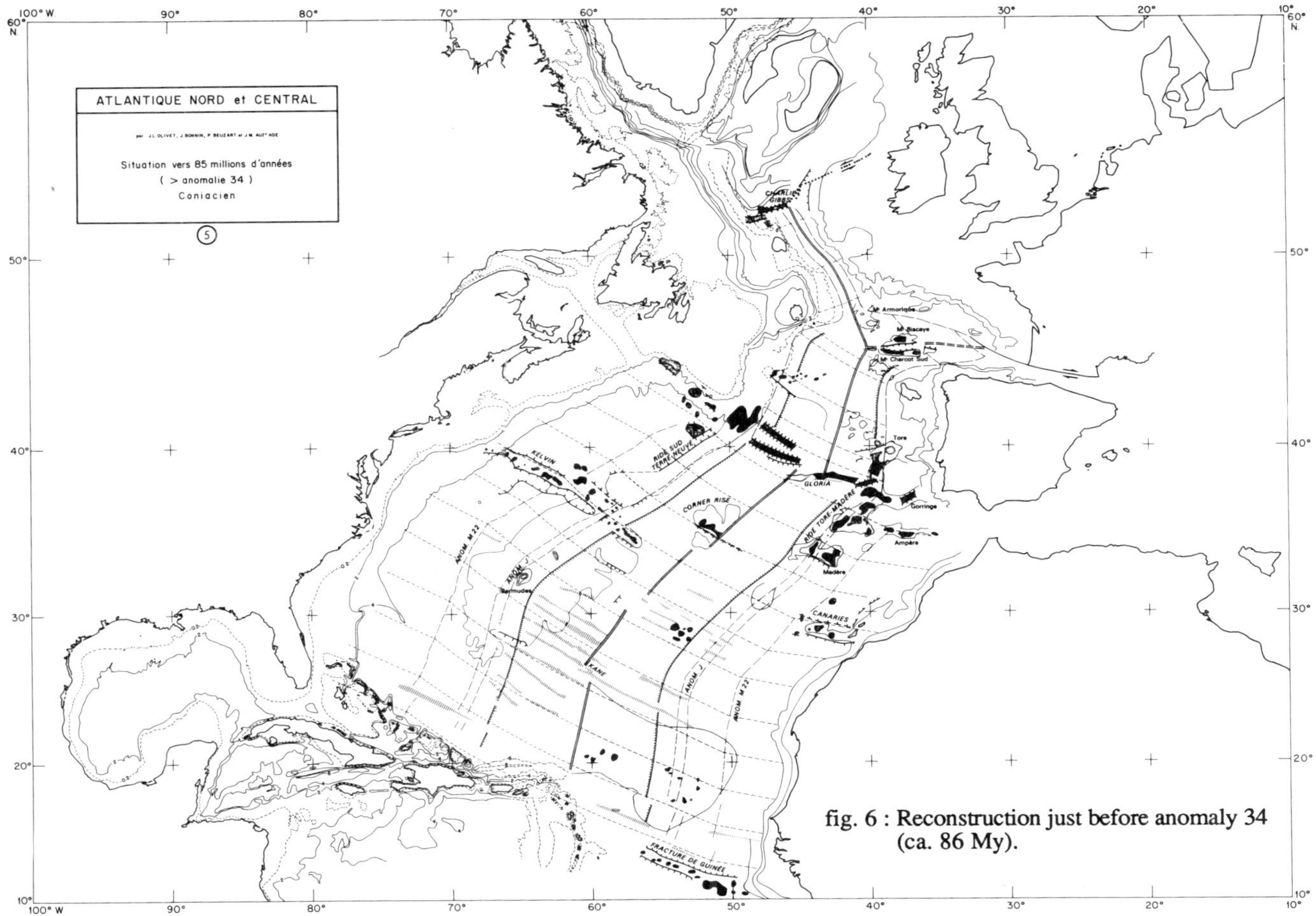

fig. 6 : Reconstruction just before anomaly 34 (ca. 86 My).

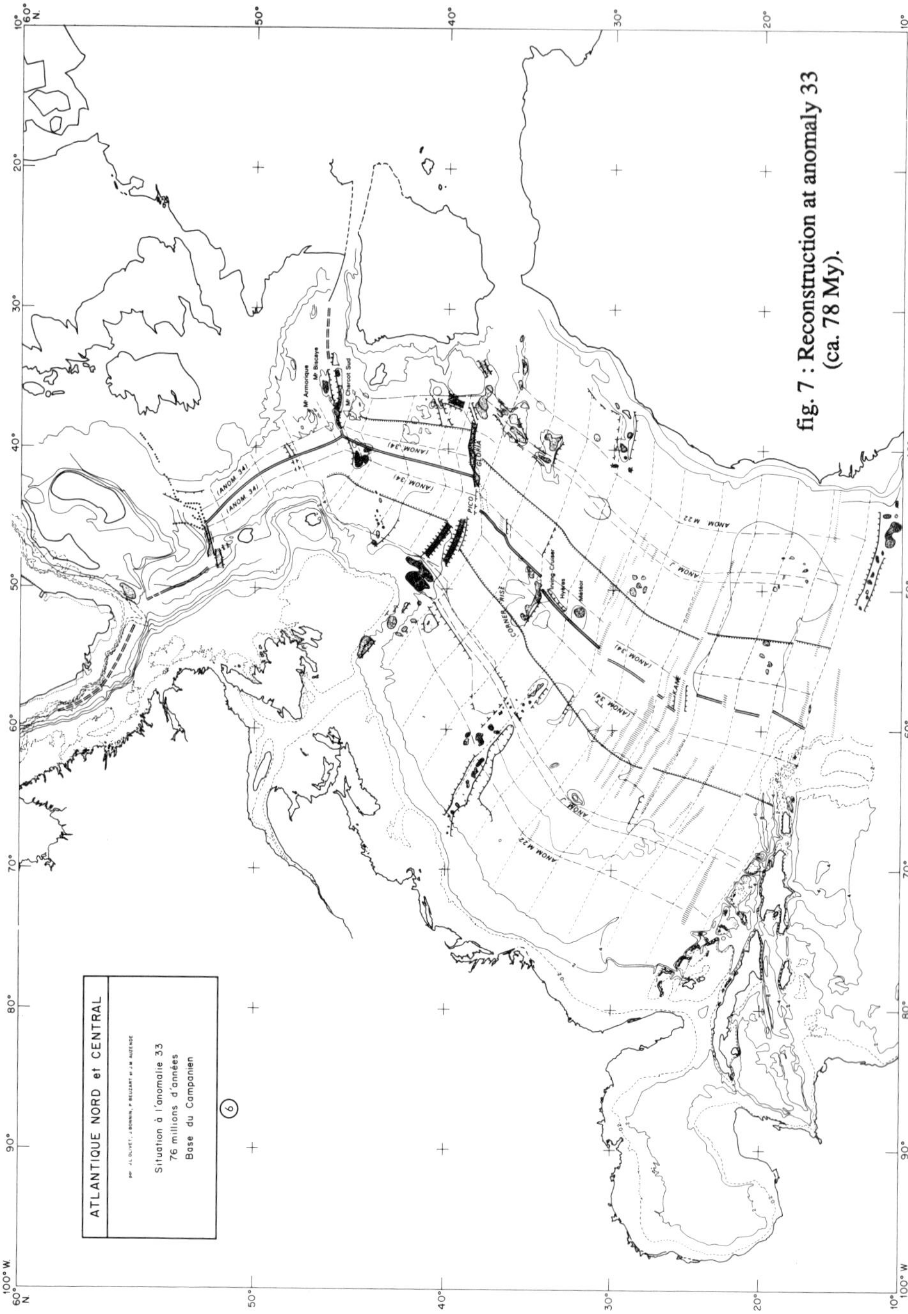

fig. 7 : Reconstruction at anomaly 33 (ca. 78 My).

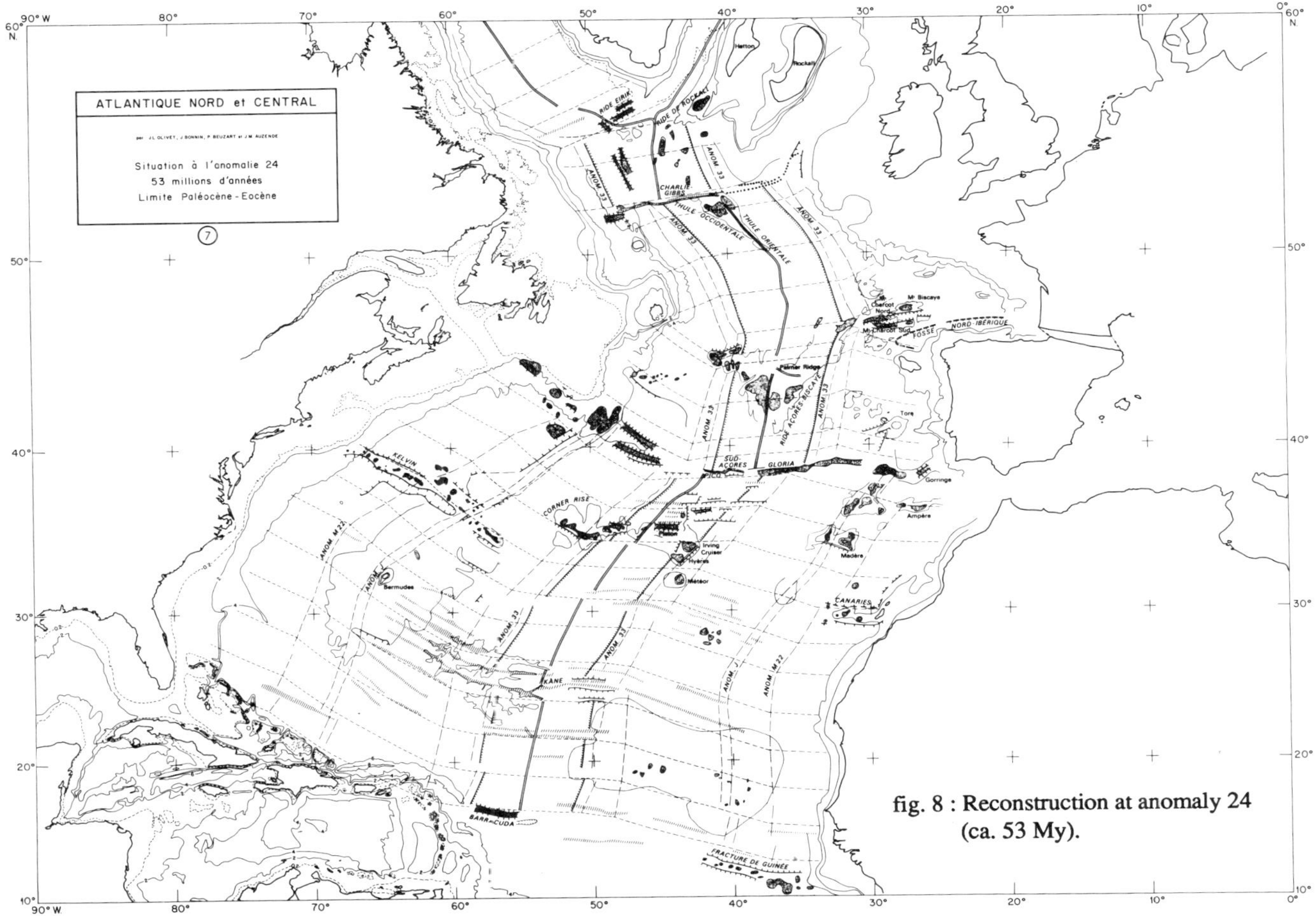

fig. 8 : Reconstruction at anomaly 24
(ca. 53 My).

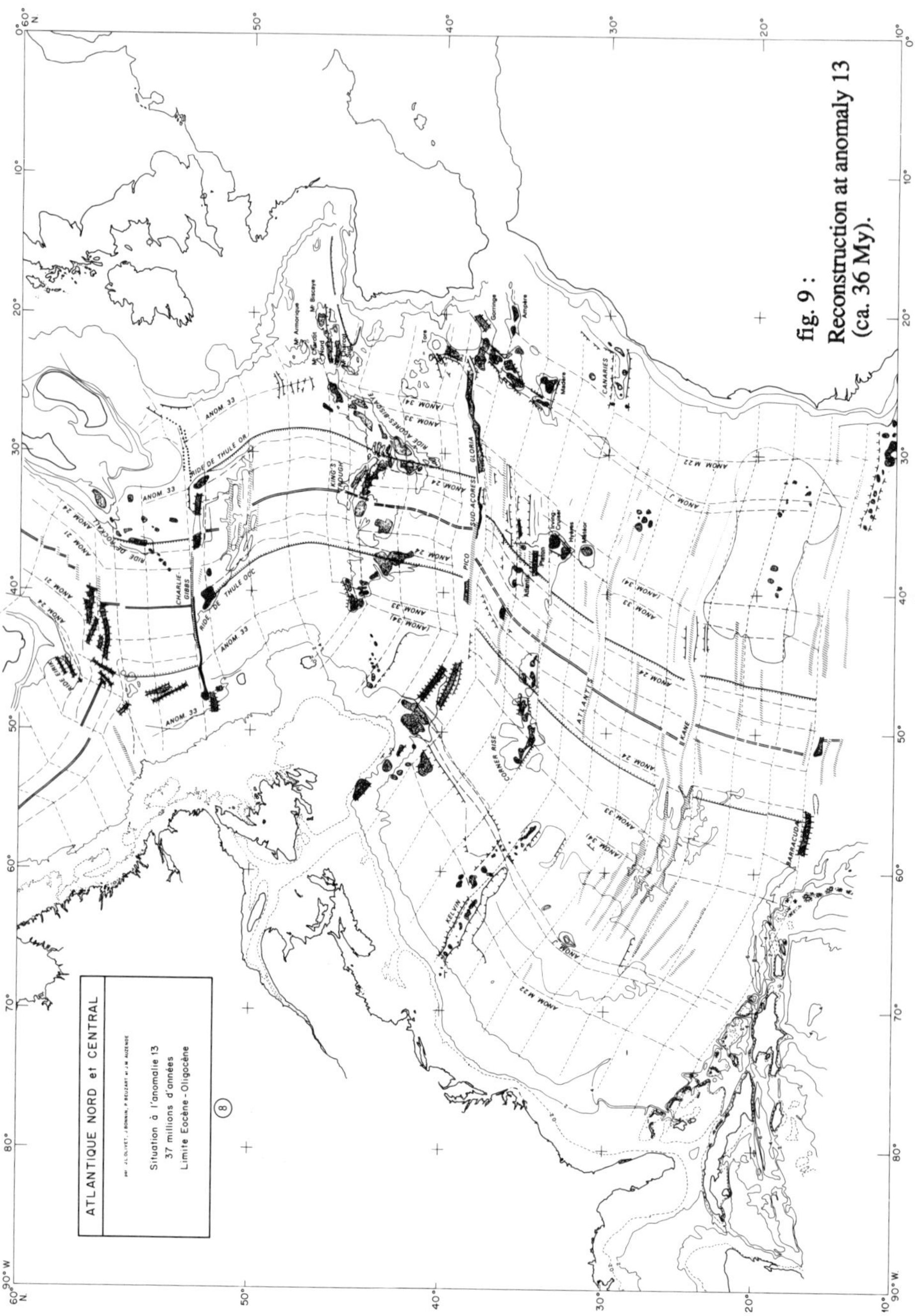

fig. 9 : Reconstruction at anomaly 13 (ca. 36 My).

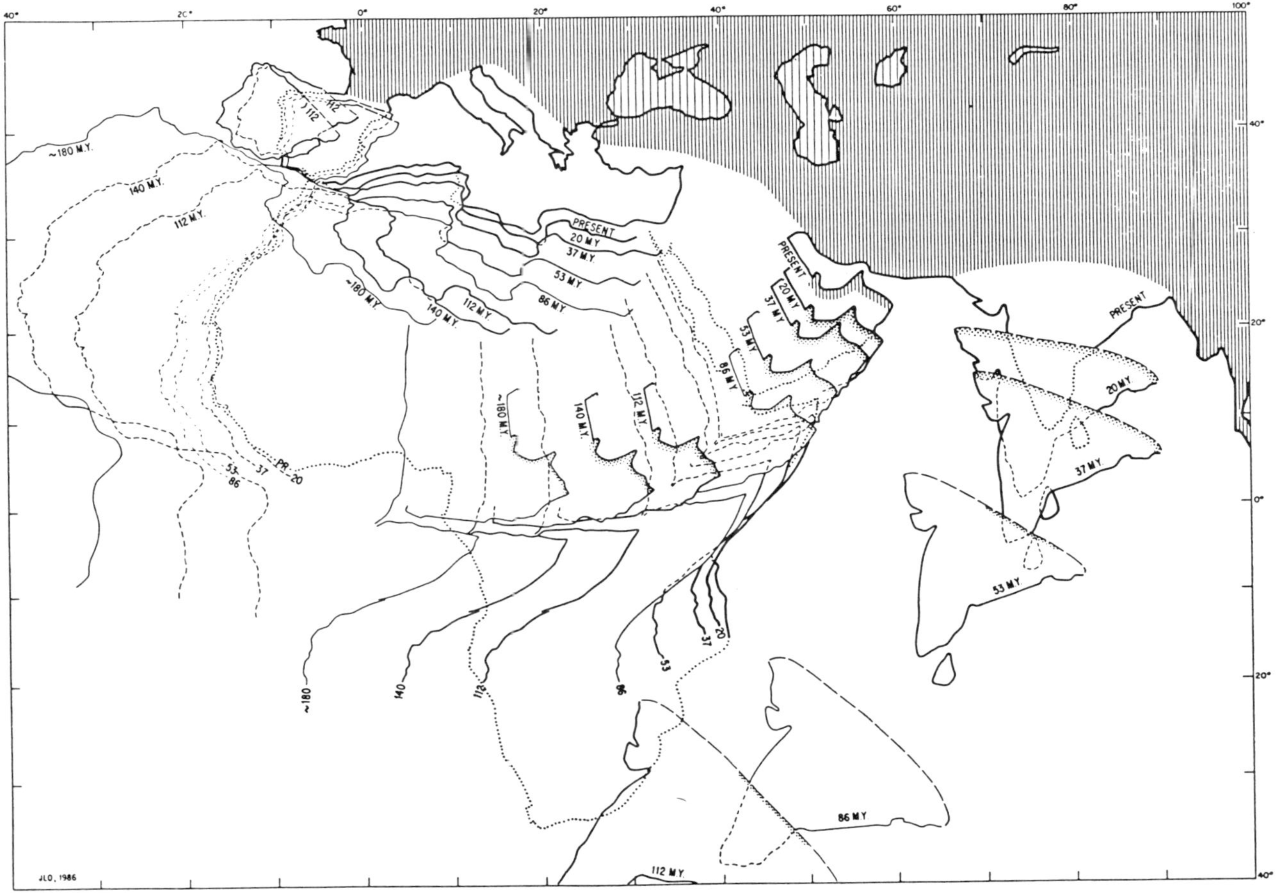

Fig. 10 : Successive palaeopositions of Africa, Iberia and India with respect to Eurasia fixed in its present position (from Olivet *et al.*, in preparation). See fig. 11 to 17.

away from North America, and has significantly dragged Iberia away from America. A fracture has propagated to the North and a branch has started the opening of the Bay of Biscay ; another branch could have initiated the very early stages of separation between Eurasia and America.

Fig. 6 shows the situation at about 86 My (what we denote by anomaly (34), the brackets indicating that it is taken a little before anomaly 34 *sensu stricto*). It is an important date for the history of spreading : in the Atlantic, the directions of spreading will be different from what they were before ; Eurasia will start to drift significantly away from America ; Iberia and Africa have sensibly completed their "longitudinal" motion relative to Eurasia. In the other oceans, India separates from Africa, and Australia from Antarctica, for examples. Fig. 5 does not correspond to any fitted isochron and has just been obtained by interpolation between the two controlled positions.

At about 78 My (anomaly 33) (fig. 7), the opening has propagated to the North in the Labrador sea, leaving Greenland temporarily attached to Eurasia. At the time of anomaly 24 (ca. 53 My) (fig. 8), the Norwegian-Greenland sea starts to open up ; this results in a triple junction to the South of the southern tip of Greenland as the Labrador sea continues spreading. At anomaly 13 time (ca. 36 My) (fig. 9), the relatively complex system becomes much simpler : the Labrador sea terminates spreading, attaching Greenland to North America ; Iberia is almost sutured to southern France : no significant relative motion (at the scale considered here) will occur now on.

Africa - (Iberia) - Eurasia relative motion

Changing the reference frame from North America to Eurasia in its present-day position, and summarizing for Africa, Iberia and Eurasia, lead to fig. 10, which has been drawn with the usual present-day coastlines for illustrative purposes. The main features displayed are :

i/ Africa has experienced two main episods in its motion relative to western Eurasia : a large scale strike slip motion from ca. 180 My up to ca. 86 My ; then on, the "longitudinal" part of the motion is almost nothing, and the global one is a counterclockwise rotation ;

ii/ looking at the longitude of Central Europe, the same overall picture still holds, but the converging motion of Africa toward Eurasia is much more marked ;

iii/ further to the East, the convergence has started since the beginning (first splitting of Africa apart from America), but it slowed down and changed its direction ca. 86 My ; one gains the impression that some sort of a collision at that time, in the eastern part of the system, has caused entire Eurasia to split away from America, resulting in a somewhat "residual" converging motion of Africa relative to Eurasia.

iv/Iberia has been dragged along by Africa until it collides with southern France (resulting in the well-known first Pyrenean orogenic phase) ca. 86 My ; it became practically sutured to France in Eocene times (the well-known second Pyrenean orogenic phase), due to the converging motion of Africa towards Eurasia.

Mediterranean domain

Looking at the Mediterranean domain at a slightly greater scale, and drawing only those contours which have not evolved considerably in the last 200 My, leads to the series of fig. 11 to 17.

The "boundary conditions" for the evolution of Mediterranean which were sought, are clearly shown : they leave plenty of room for speculations about the position of intermediate blocks in between stable Africa and Eurasia, such as : "inner zones" of northern Africa and southern Spain orogens, Corsica-Sardinia, Apulia, various Turkish blocks, and

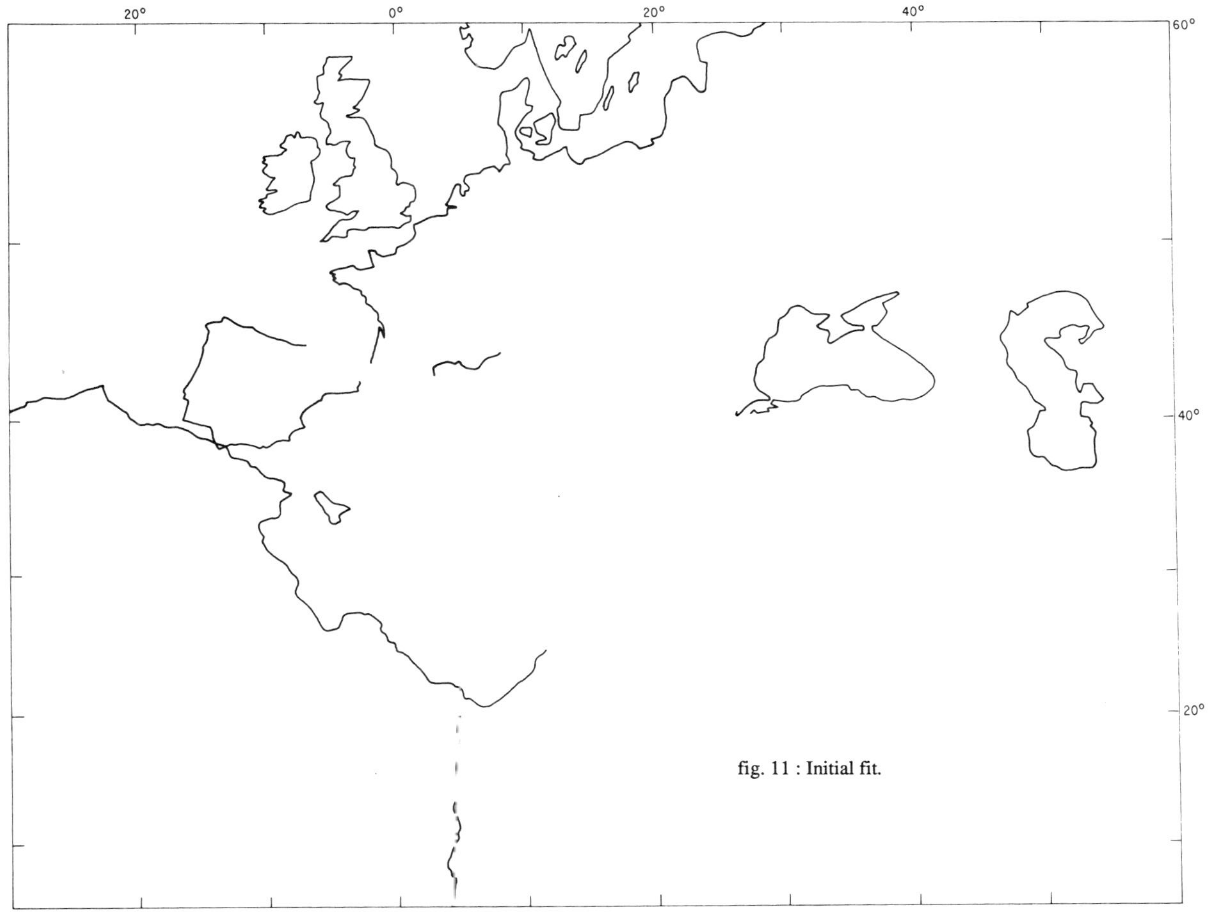

Fig. 11 to 17 : Palaeopositions of Africa and Iberia with respect to Eurasia fixed in its present position (from Olivet *et al.*, in preparation). Contours have not been drawn when they are known to have experienced complex geologic evolution.

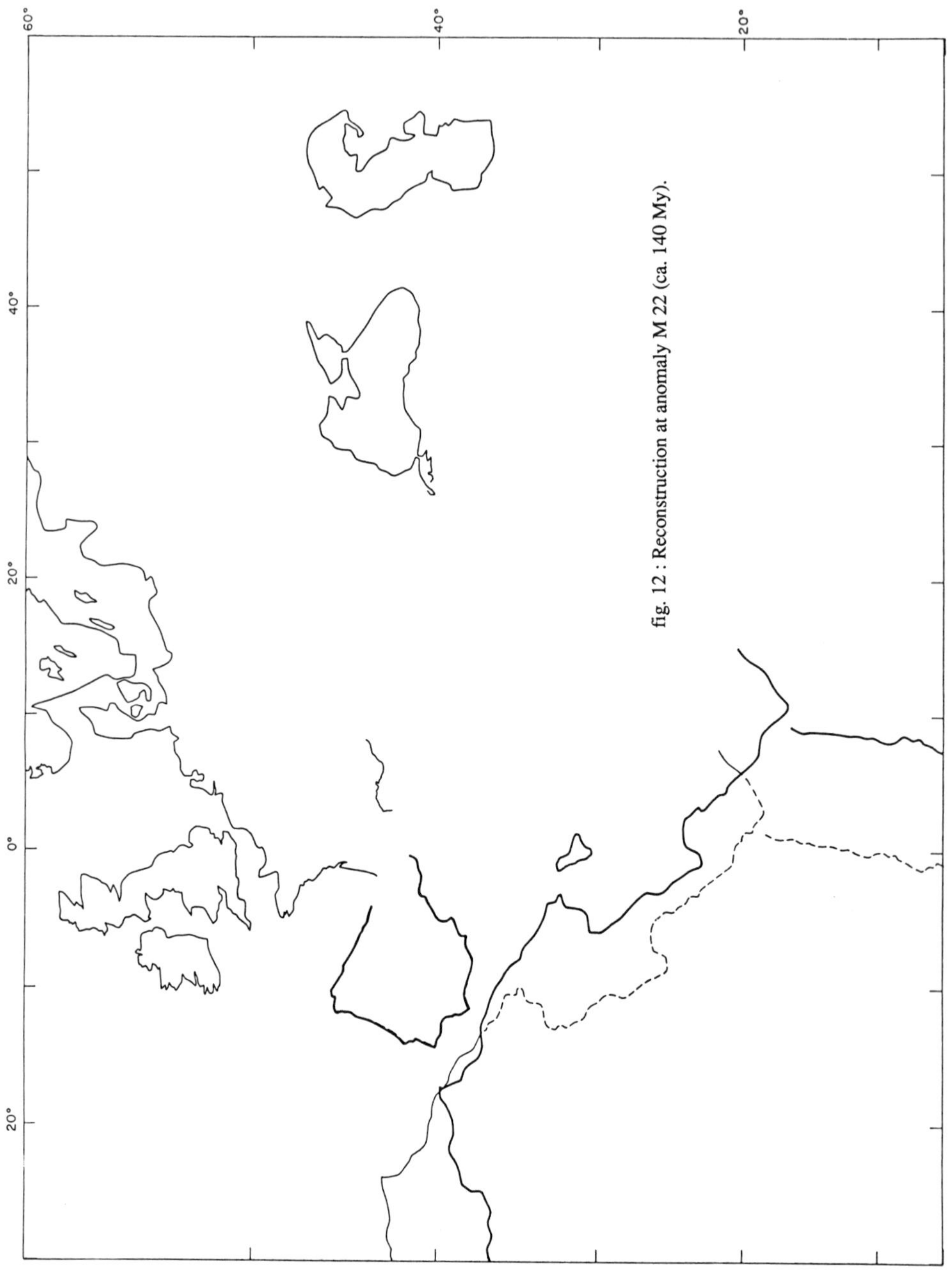

fig. 12 : Reconstruction at anomaly M 22 (ca. 140 My).

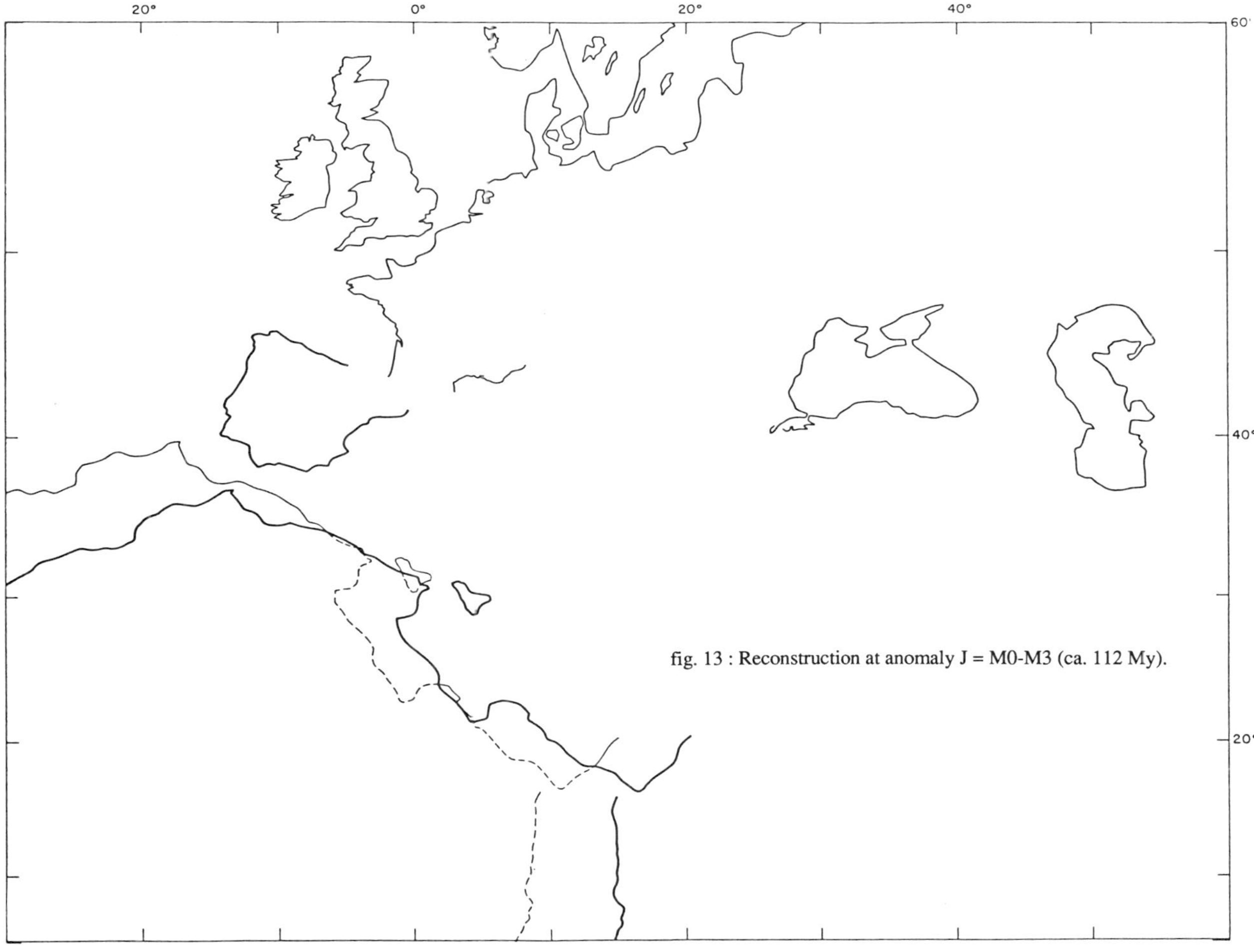

fig. 13 : Reconstruction at anomaly J = M0-M3 (ca. 112 My).

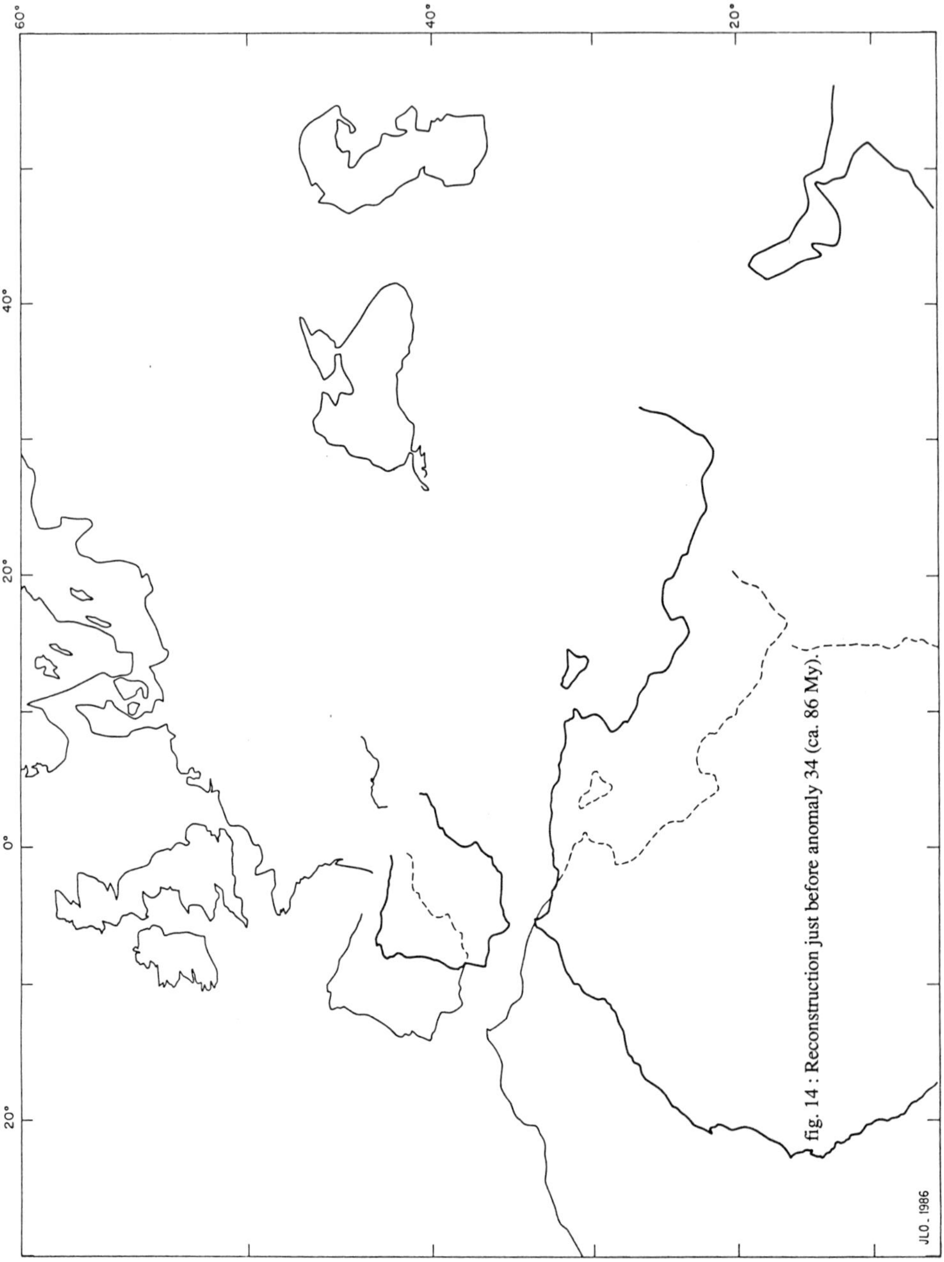

fig. 14 : Reconstruction just before anomaly 34 (ca. 86 My).

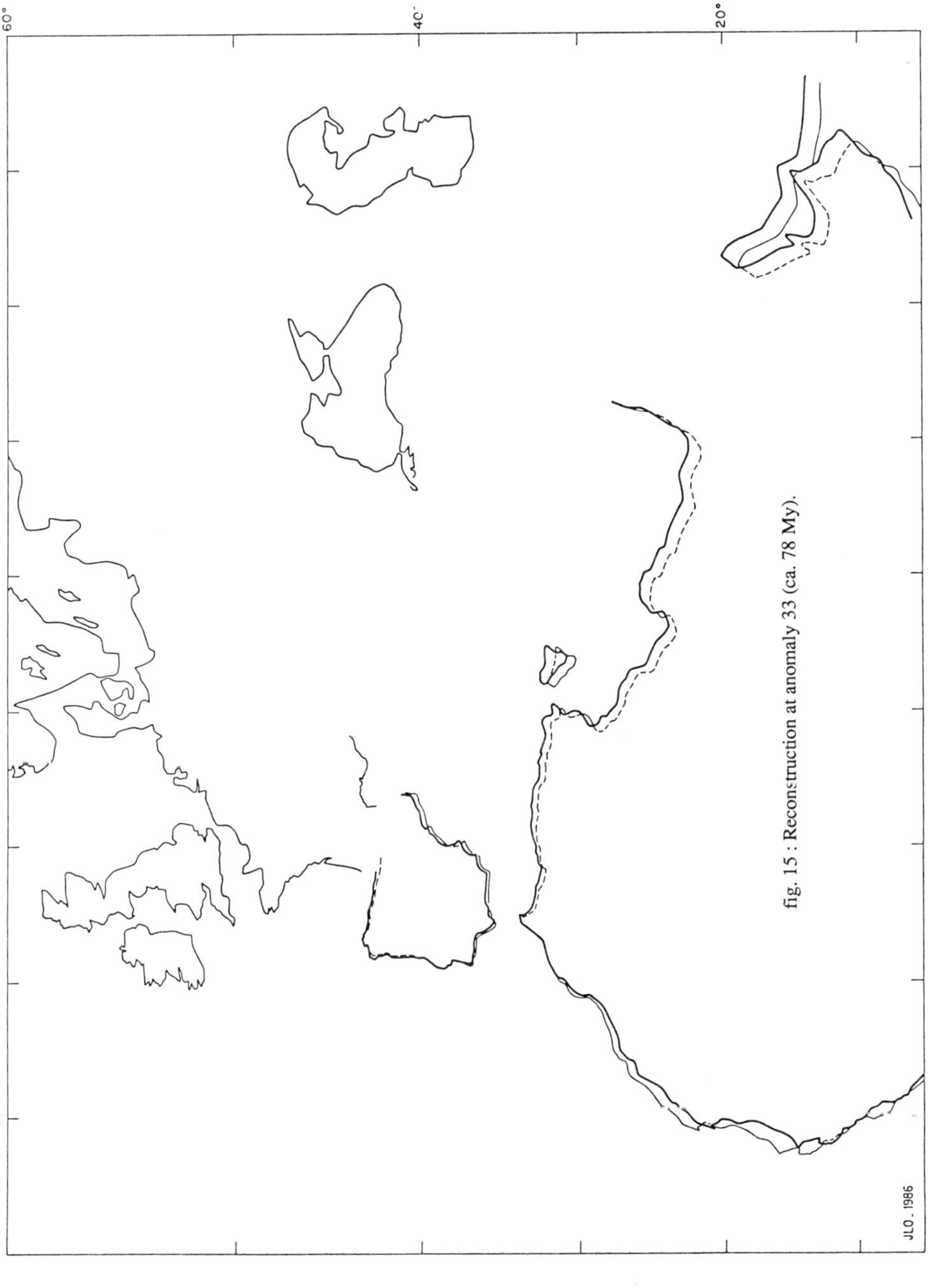

fig. 15 : Reconstruction at anomaly 33 (ca. 78 My).

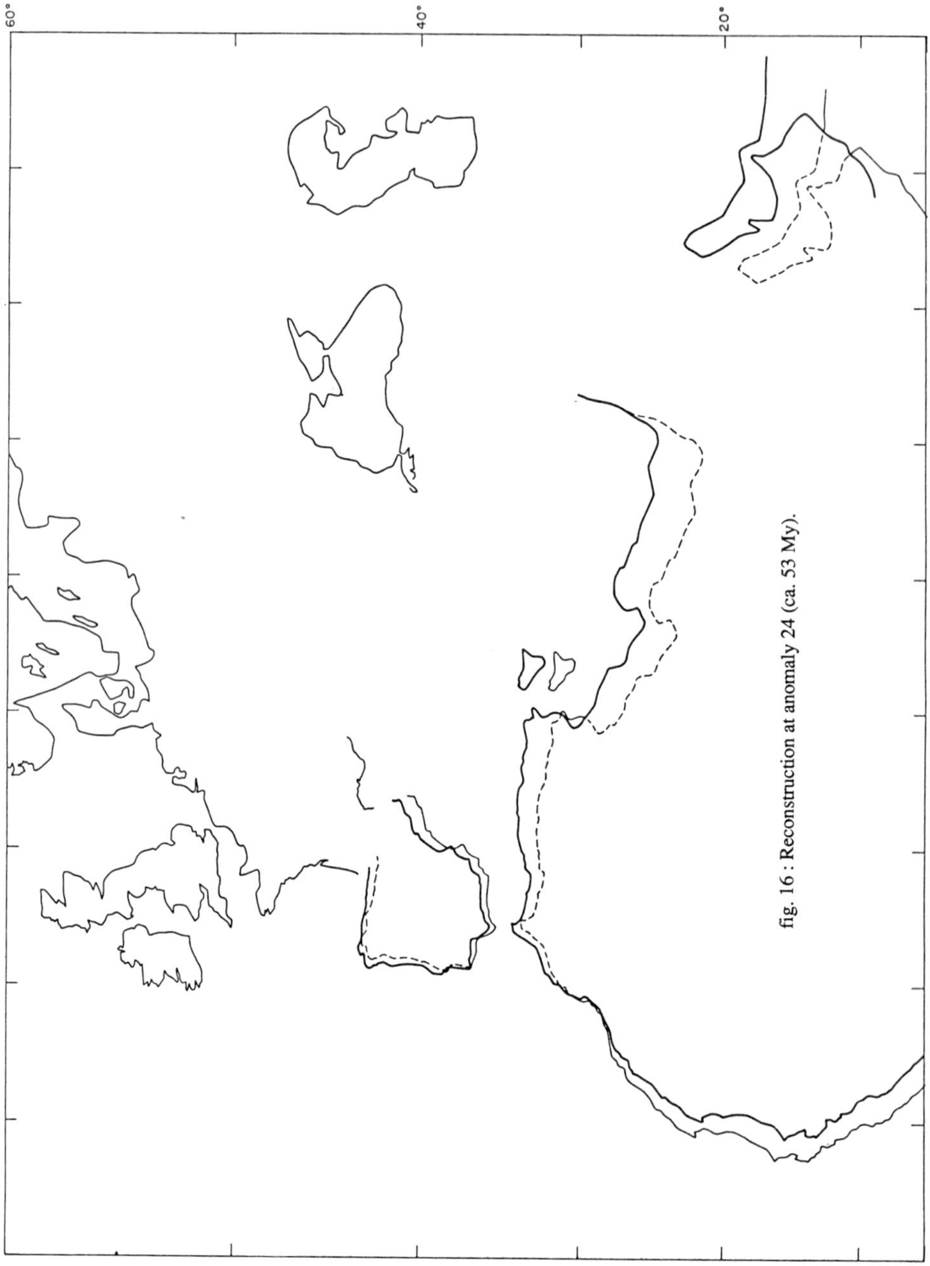

fig. 16 : Reconstruction at anomaly 24 (ca. 53 My).

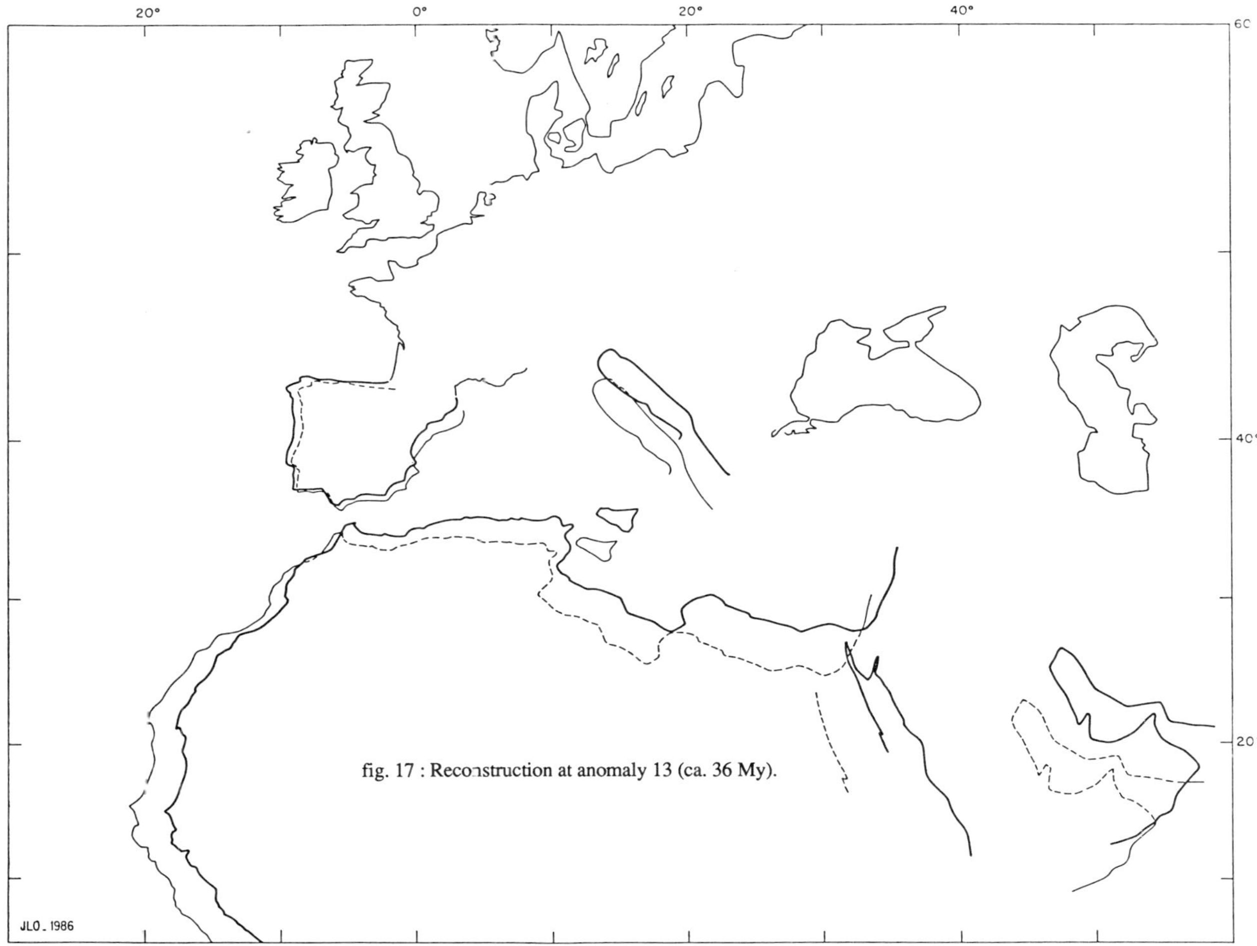

fig. 17 : Reconstruction at anomaly 13 (ca. 36 My).

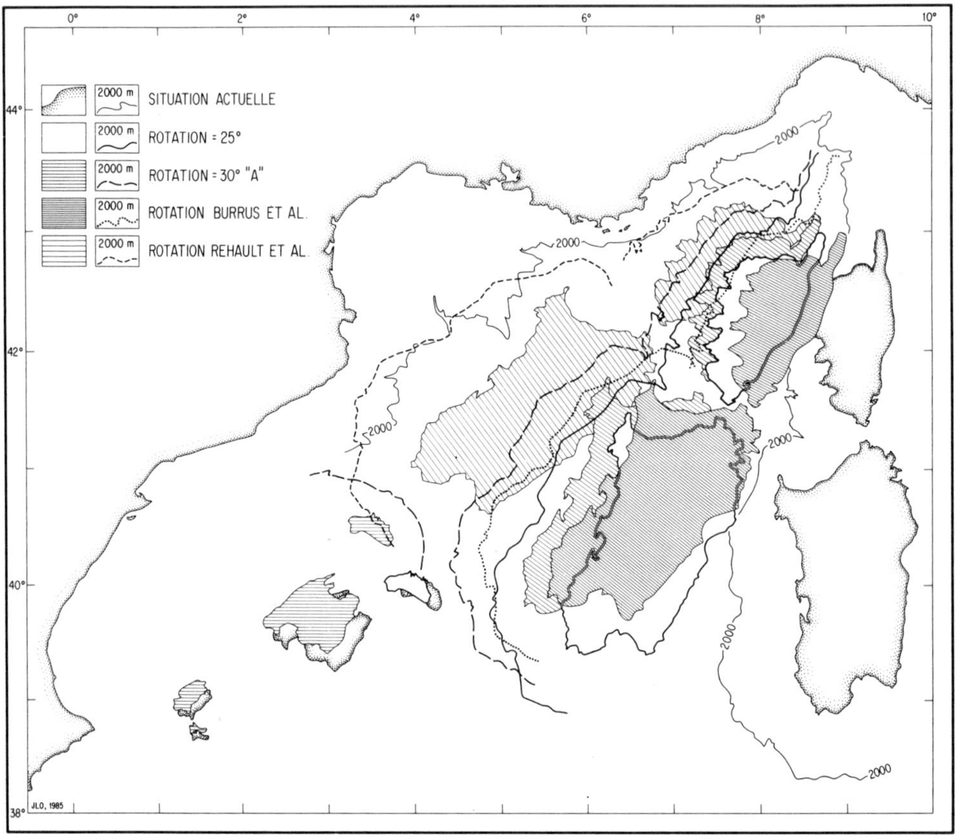

Fig. 18 : An example of small block reconstruction within the Mediterranean domain : the Corsica - Sardinia motion with respect to southern Europe in its present position (from Olivet *et al.*, in preparation).

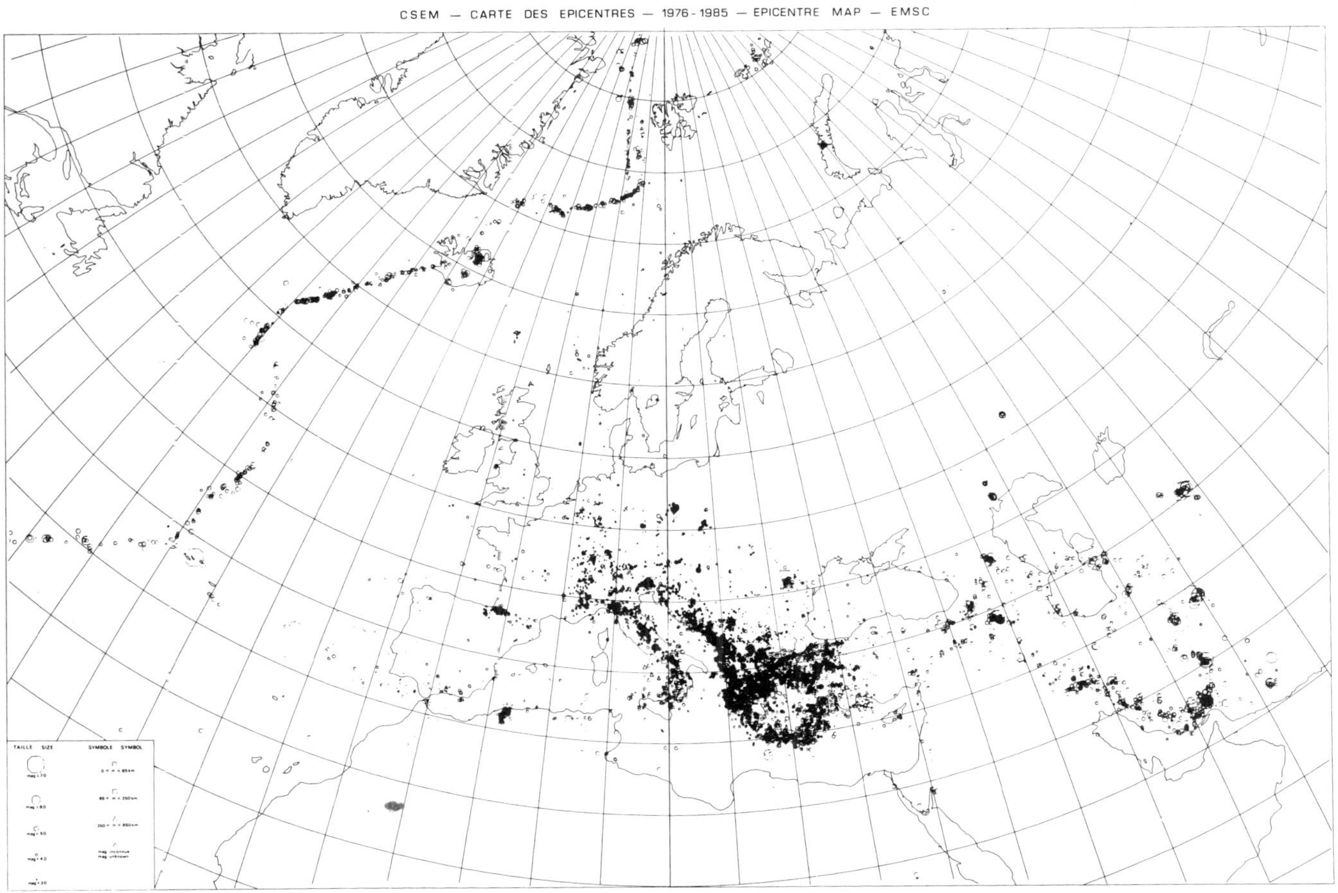

Fig. 19 : Epicentral determinations by the European-Mediterranean Seismological Centre during the period 1976-1985. There could be some inhomogeneity in the reporting of small events (magnitude less than 3.5) due to inhomogeneous coverage of seismographic stations and failures in data collection procedure.

many other smaller ones.

Many models and ideas have been put forward for the evolution of those inner parts of the Mediterranean. They will not be reviewed, neither cited ; just because they belong to a mode of reasoning which is completely different from the one which has been followed here : they rely mostly on identification of past block boundary environments on purely geologic grounds (ophiolites, margin-type deposits, volcanism, etc...) ; no isochron lines are recognized, and this is not surprising in a domain which is under shrinking since tens of millions of years.

In some special cases however, things can be worked in details when an ocean-like feature has been left behind the motion of blocks : this is the case of Corsica and Sardinia (fig. 18). But even in this favorable case, many uncertainties remain, as it is shown by the different positions of Corsica-Sardinia assumed by various authors.

Seismic activity in the Mediterranean regions

What precedes does not seem to lead straightforward to a study of the seismic activity in the Mediterranean regions. Nevertheless, the distance is not so large that it appears.

It is clear that the relative high level seismicity of the domain is due to the converging motion of the two large plates, Africa and Eurasia, with some smaller blocks trapped in between. There is not much oceanic space still to be consumed by well-developed subduction zones : as it could be expected in such a situation, seismic activity is scattered over a large zone, with oceanic-type and continental-type patches intermingled (fig. 19).

Plate tectonics calls for rigid plates and plate boundaries. Just in the Mediterranean regions, there is nothing like that, or at least nothing typical. Coming from the West, seismic activity defines very well the mid-Atlantic plate boundary ; the so-called Azores-Gibraltar ridge is approximately outlined by epicenters, even though the activity is somewhat strange (very large shocks (an event greater than magnitude 8 has occurred near Gorringe Bank, to the Southwest of Cape Sao Vicente), with little smaller magnitude activity). But approaching the longitude of Morocco and Portugal, the epicenters become scattered and no clear alignments appear : no sharp plate boundary is present in the Mediterranean regions ; with a possible exception : the Pyrenees mountain chain. Moreover, from the longitude of the Gulf of Lion eastwards, the width of the zone seismically active is largely greater than 1000 km, even though special concentrations of shocks are associated with recent and well-known compressive tectonic features, such as the Apennines, the Dinarides. A case should be made for Greece were a very highly active zone encircles a much calmer area : the southern branch is clearly associated with the Hellenic arc, while the northern one joins very active regions of Albania-southern Yugoslavia - Central Greece and northeastern Asian Turkey through the northern Aegean sea.

However, those active zones do not compare at all with clear plate boundary examplified by the mid-Atlantic ridge. Farther East, seismic activity affects large zones in Turkey with a tendency to concentrate on relatively narrow alignments such as the Jordan rift, the East Anatolian fault, and (not as clear) the North Anatolian fault ; it spreads out again when approaching Caucasus and northern Iran.

It is worth noticing that plate boundaries are observed at the axes of the mid-oceanic ridges in the Atlantic, and between Africa and Eurasia as far as the oceanic parts of the plates are considered, and to a lesser extent between Iberia and southern France. These are the places were seismic activity displays a well-defined concentration. But further East, where it implies much larger zones, nothing like well-marked, long-lasting plate boundaries has been identified and has left traces today. It does not mean that no plate boundary has ever existed in what has been in existence in the past : they have completely disappeared due to the shrinking of the entire zone.

* *
*

The direct implications for seismicity and seismic hazard studies are twofold:

i/ the Mediterranean regions should be regarded from a point of view quite different from the one used in other well investigated regions, such as California : there, the plate boundary is clearly expressed even though it is wider that a simple, unique fault ;

ii/ the entire Mediterranean regions are more or less undergoing deformation, on a width of one to several thousands of kilometres : they should be regarded as a continuum under deformation rather than as a set of micro-plates separated by plate boundaries. And this casts some gloom on the hopes that measuring, through space techniques, the converging rate of Africa and Eurasia will help predicting earthquakes in the Mediterranean regions.

BIBLIOGRAPHY

Barazangi, M., and Dorman, J. World seismicity maps compiled from ESSA, Coast and Geodetic Survey, epicenter data, 1961-1967. Bull. Seism. Soc. Am., **59,** p. 369-380 ; 1969.

Burrus J. Contribution to a geodynamic synthesis of the Provençal basin (north-western Mediterranean). Mar. Geol., **55,** p. 247-269 ; 1984.

Heirtzler, J.R., Dickson, G.O., Herron, E.M., Pitman, W.C. III, and Le Pichon, X. Marine magnetic anomalies, geomagnetic field reversals, and motions of the ocean floor and continents. J. Geophys. Res., **73,** p. 2119-2136 ; 1968.

Isacks, B., Oliver, J., and Sykes, L.R. Seismology and the new global tectonics. J. Geophys. Res., **73,** p. 5855-5899 ; 1968.

Larson, R.T., and Hilde, T.W.C. A revised time scale of magnetic reversals for the early Cretaceous and late Jurassic. J. Geophys. Res., **80,** p. 2586-2594 ; 1975.

Larson, R.L., and Pitman, W.C. III. World-wide correlation of Mesozoic magnetic anomalies, and its implications. Geol. Soc. Am. Bull., **83,** p. 3645-3662 ; 1972.

Murphy, L.M. World-wide seismic network, *in* ESSA Symposium on Earthquake Prediction. Proceedings, U. S. Gov. Print. Off., Washington D.C., p. 53-56 ; 1966.

Olivet, J.L., Bonnin, J., Beuzart, P., and Auzende, J.M. Cinématique de l'Atlantique Nord et Central. Publ. Cent. Nat. Expl. Océans, Rap. Sci. Tech. n° 54, 108 pp + maps ; 1984.

Olivet, J.L., Unternehr, P., Bonnin, J., Beuzart, P., and Goslin, J. Atlas of plate tectonic reconstructions of the Atlantic ocean and surroundings. In preparation.

Rehault, J.P., Mascle, J., and Boillot, G. Evolution géodynamique de la Méditerranée depuis l'Oligocène. Mem. Soc. Geol. It., 27, p. 85-96 ; 1984.

Smith, A.G., and Briden, N.T. Palaeocontinental maps for the Mesozoic and Cenozoic. Cambridge Univ. Press., 320 pp.

RECENT AND PRESENT TECTONICS IN THE MEDITERRANEAN REGION

H. PHILIP
Laboratoire de Géologie Structurale, U.S.T.L.
F-34060 Montpellier Cedex

ABSTRACT. In the Mediterranean region, tectonics are ruled by the convergence of the African plate with respect to the Eurosian plate. This convergence is absorbed through different mechanisms. Subduction and collision are dominant but other mechanisms help keep the convergence active. This is the case of various zones which illustrate the passage from oceanic subduction to continental collision. No simple and unique scheme can be found ; each case (Tunisia, Betic Cordillera, Italian peninsula) corresponds to particular evolutions which depend mainly on boundary conditions.

INTRODUCTION. Seismicity and present day deformations are an instantaneous picture (at geology scale) of the tectonic evolution of a region. Then tectonic history gives us the logical explanations to understand present day tectonic deformations associated or not to seismicity.

The Mediterranean region corresponds to a quite complex plate boundary between Eurasia and Africa. This domain is a wide mobile zone made up of different blocks trapped between two large relatively indeformable continental plates.

To understand this picture and to define the main recent tectonic characteristics of Mediterranean region it is necessary to point out the main stages of the relative motions between Africa and Eurasia since 200 My (Bonnin and Olivet, this volume ; Dercourt *et al.*, 1986).

In the framework of the Atlantic oceanic opening, two epochs of relative motion between Africa and Eurasia could be distinguished (Fig. 1) (Dewey *et al.*, 1973 ; Le Pichon *et al.*, 1977 ; Tapponnier, 1977 ; Olivet *et al.*, 1984).

From 190 to 80 My the relative motion between Africa and Eurasia corresponds to a sinistral movement (Fig. 2A). This movement induced the openning of numerous oceanic or thinned continental crust basins in the Mediterranean area. For example (Fig. 3), the Mediterranean region in Aptian times (110 My) presents a complex juxtaposition of numerous small basins and continental blocks. This picture constrasts very well with the large oceanic domain of the Tethys in the Far East.

At about 70 My (Fig. 2B), the relative displacement vector of Eurasia and Africa plates changes ; it becomes roughly N-S to NNW-SSE, and has remained relatively unchanged until present time.

One by one, from North to South, all the basins created during the Mesozoic time were absorbed under subduction, collision or any other mechanisms, depending on the different areas and epochs considered (fig. 4) (Dercourt *et al.*, 1984). The North-South convergence resulting from this last relative displacement, has also led to the opening of new oceanic basins (Algero-Provençal, Alboran basins ; Tyrrhenian sea;

J. Bonnin et al. (eds.), Seismic Hazard in Mediterranean Regions, 283–300.

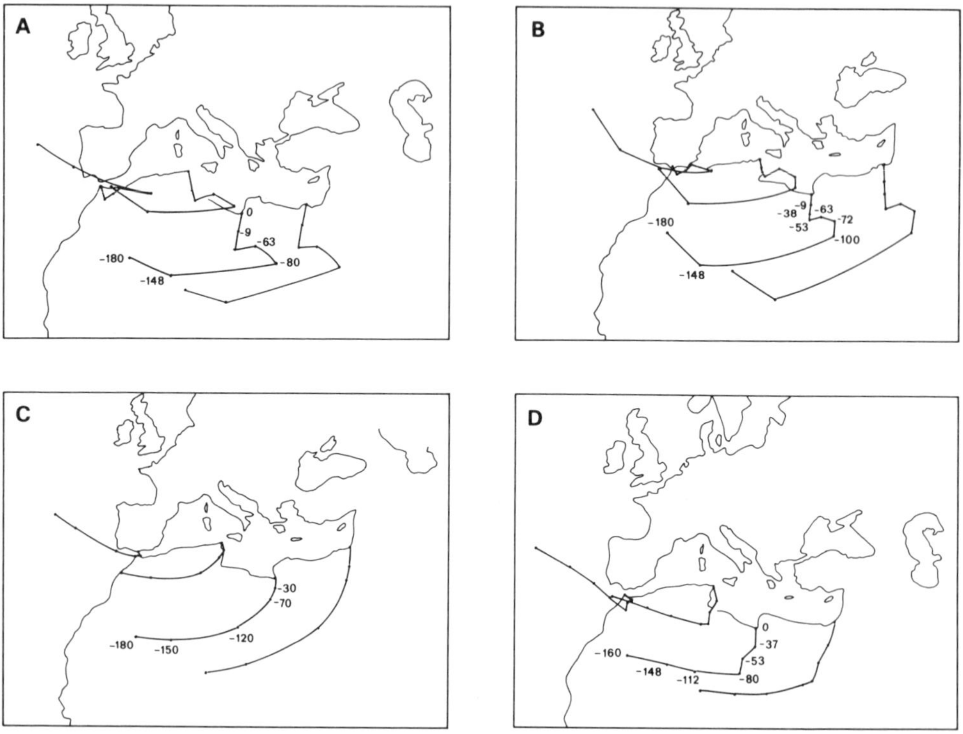

Fig.1: Relative motions of Africa with respect to Eurasia according to : A - Dewey *et al.*, 1973 ; B - Le Pichon *et al.*, 1977 ; C - Tapponnier, 1977 ; D - Olivet *et al.*, 1984.

Panonian basin...). But each individual geodynamic stage was of short duration regardless of the block sizes that were affected (Fig. 5 et 6).

Various geodynamical contexts can be present during the same period ; for example at the end of Miocene we can observe :

1. The Apennines and the Calabrian arc are under subduction since Tortonian times.
2. Subduction in the Aegean arc just begins.
3. Collision starts in the Maghreb chain and the Betic Cordillera.
4. Collision existing in the Alps since Cretaceous times, is still active.

METHODS FOR STRESS ANALYSIS

One hypothesis in tectonics is to propose that each geodynamical stage corresponds to a space distribution of various types of deformation. Then, reciprocally, the tectonic evolution permits to determine the geodynamic changes that occur in a region.

So the main problem is to define :

1. The type of deformation in "each" point of the region.
2. The space distribution of each type of deformation.

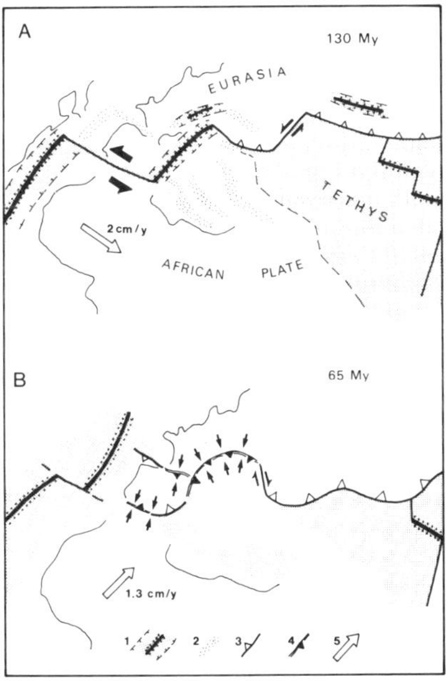

Fig.2: Simplified plate scheme during the two main stages of Atlantic ocean opening (modified from Dercourt *et al.*, 1986).
1 : oceanic accretion ; 2 : thinned continental crust ; 3 : oceanic subduction ; 4 : continental collision ; 5 : relative motion of Africa with respect to Eurasia.

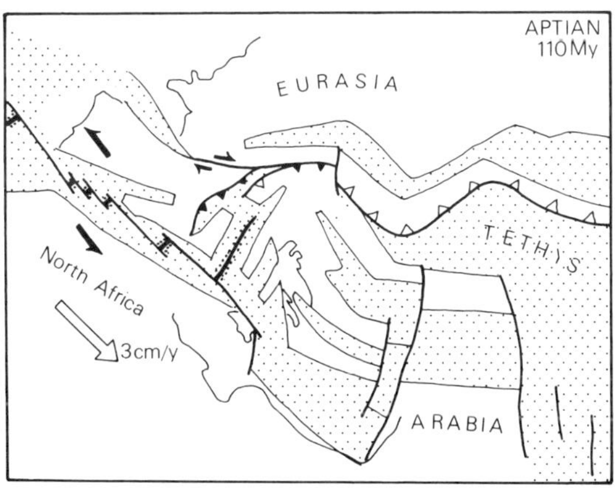

Fig.3: Mediterranean region during Aptian time. Dotted pattern indicates oceanic and continental basins ; open triangles : subduction ; closed triangles : continental collisions ; double dotted line : accretion (modified from Dercourt *et al.*, 1986).

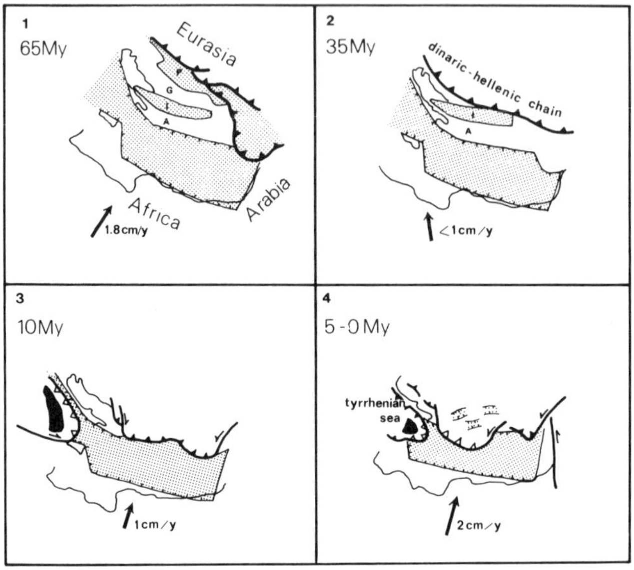

Fig.4: Southward progressive closure of the Mesozoic basins in the eastern Mediterranean region, from upper Cretaceous to present time.

3. The chronology of the different deformation states.

Before examining these points in some key sectors of the Mediterranean region under various geodynamic conditions, it is necessary to define briefly the different types of elementary superficial (i.e. brittle) deformations.

The seven basic types of deformation in brittle tectonics are specified by the orientation and the shape of the tectonic stress tensors (Phillip, 1987). The three first cases (Fig. 7) correspond to planar deformation defined first by Anderson (1951). These three cases have identical tensor form, characterized by an intermediate stress axis $\sigma 2 = \sigma 1 + \sigma 3/2$. Each case is specified by the orientation of stress axis and, of course, by the corresponding type of faulting (normal, reverse, strike slip).

In fact, these cases can seldom observed in nature. More often a combination of various types of faulting is observed, such as, for example, reverse fault and strike slip, normal and strike slip faulting (Fig. 8). Finally, the last two cases (constrictive and radial extension) correspond to specific conditions; constrictive deformation is quite rare and generally related to local conditions such as a "corner effect". Radial extension in much more frequent and could be expected in the back arc region of subduction zones.

The characterization of tectonic deformations and their evolution with time can be approached in different manners. To complement classical structural analysis, microtectonic analysis can be used. These methods rely on microstructural observations (fault planes, microfaults, tension gashes, stylolites,...) and usually aid the recognition of the types of deformation affecting a region, as well as their chronology. Appropriate

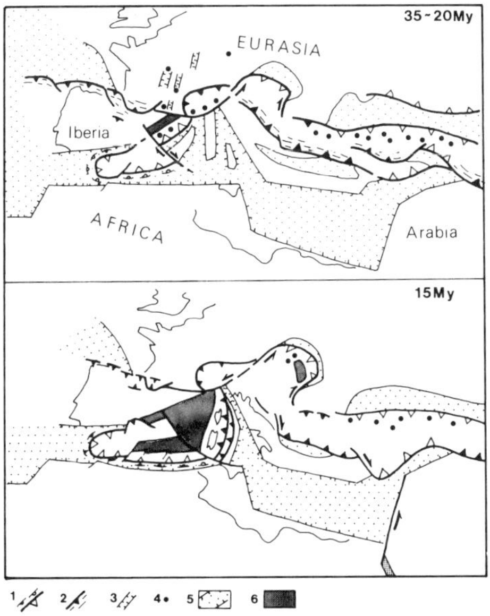

Fig.5 et 6 : Simplified tectonic scheme of the Mediterranean basin during Cenozoic time.
1 : Subduction ; 2 : Continental collision ; 3 : Grabens ; 4 : Volcanism ; 5 : Mesozoic oceanic or continental basins ; 6 : Cenozoic oceanic basins.

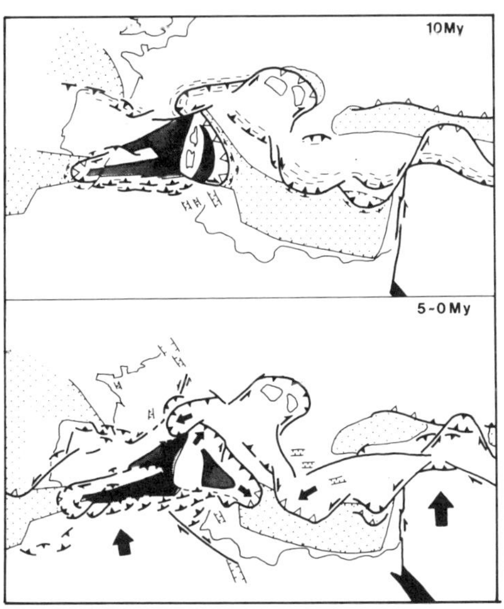

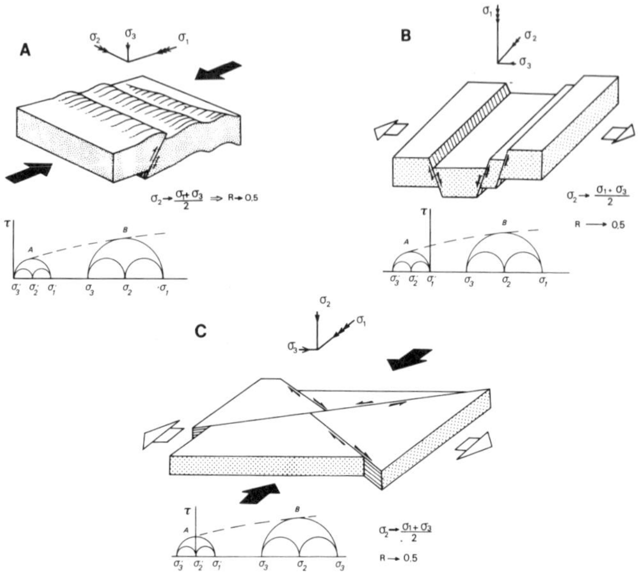

Fig.7: The three cases of planar deformation.
A : Pure reverse ; B : Pure normal ; C : Pure strike-slip. Schematic Mohr circles are given for deformations produced (A) under shallow conditions, and (B) under deeper conditions.

methods (Carey, 1976 ; Armijo and Cisternas, 1978 ; Etchecopar *et al.*, 1981) give us the orientation and the shape of the stress tensor.

Finally, earthquake focal mechanisms and in situ measurements of stress can be used to define the present day state of stress of a region (Vasseur *et al.*, 1983 ; Paquin et Froidevaux, 1980).

SUBDUCTION AND COLLISION

In the two relatively simple cases of subduction and collision, we can define from present day tectonic analysis differents sectors characterized by distinct stress conditions. Good examples are North of Europe and the Alps for collision (Fig. 10), and the Aegean arc for subduction.

Deformations related to collision (Fig. 9) correspond to compressive tectonics with folds and reverse faulting more or less associated with strike-slip faults. On the other hand, distensive tectonics, with normal faults, associated generally with strike-slip faulting, are also present in regions which are peripheral to orogenic zones. The distribution of these deformations related to collision shows that the maximum horizontal stress (σ1 for the first case and σ2 for the second case) retains in both cases the same direction which is also the direction of plate convergence.

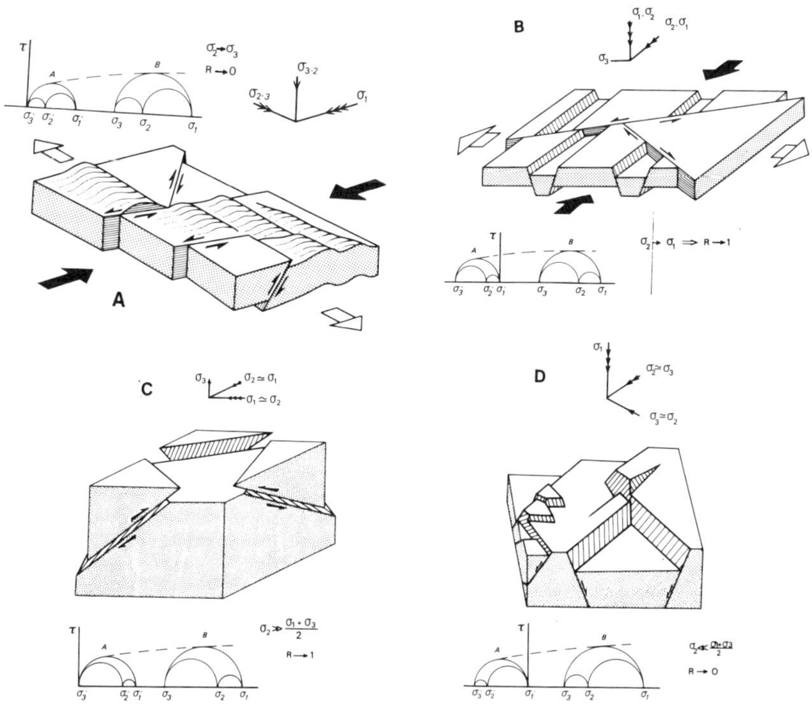

Fig.8: A : Deformation combining strike slip and reverse faulting. B: Deformation combining strike slip and normal faulting. C : Constrictive deformation. D : Radial extension deformation.

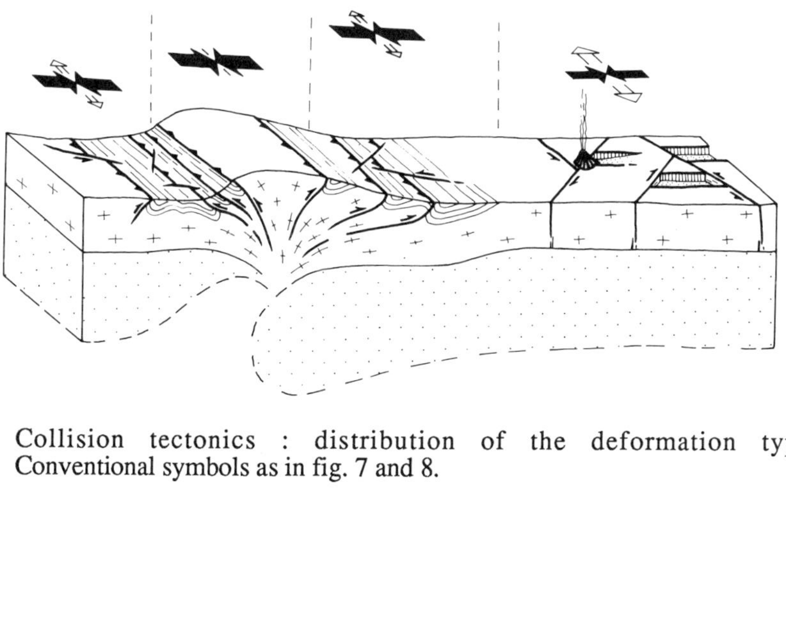

Fig.9: Collision tectonics : distribution of the deformation types. Conventional symbols as in fig. 7 and 8.

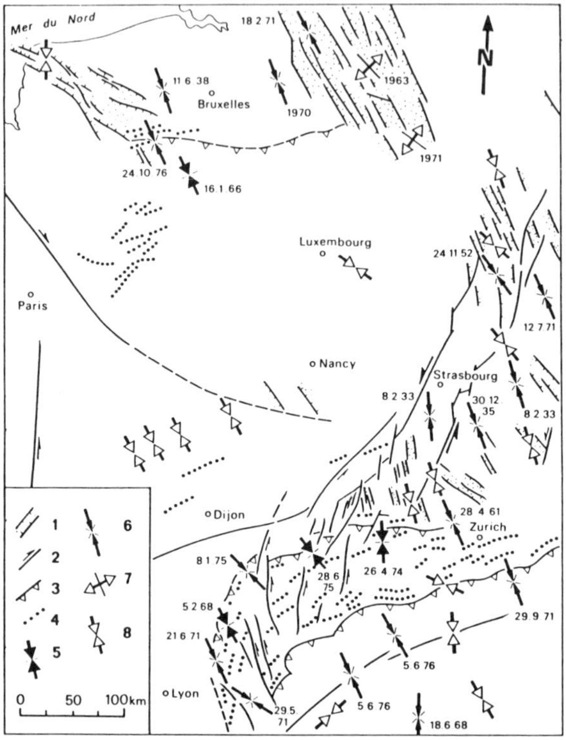

Fig.10: Neotectonic and present-day tectonics from Swiss Alps to the North Sea (Philip, 1987) 1 : Plio-Quaternay grabens ; 2 : Strike slip faults ; 3 : Reverse faults ; 4 : Fold axes ; 5 : P axis of earthquake focal mechanisms for reverse faults ; 6 : Same, but for strike slip faults ; 7 : T axis of earthquake focal mechanisms for normal faults ; 8 : Direction of maximum stress axis from in situ stress measurements.

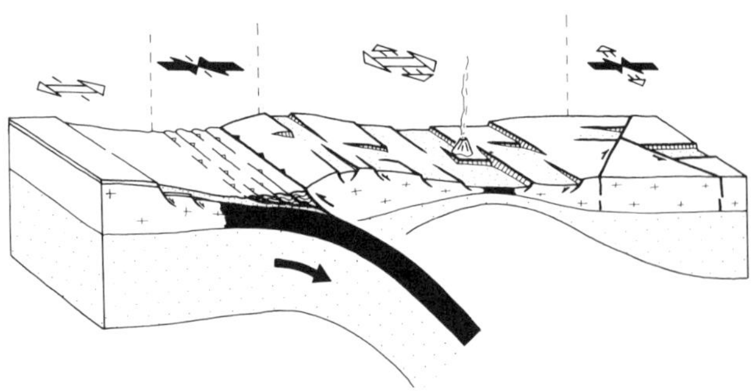

Fig.11: Oceanic subduction : distribution of the deformation types according to Aegean example. Conventional symbols as in fig. 7 and 8.

Tectonics related to oceanic subduction (Fig. 11) (Aegean arc for example) have been studied mostly from earthquake focal mechanisms (McKenzie, 1972). Different conditions can exist above subduction and might be related to the characteristics of the subduction (velocity, dip,...) (Nakamura and Uyeda, 1980 ; Mégard and Philip, 1976).

In the Aegean regions, compressive tectonics are continuous in the external domain of the arc and are related to the existence of an accretionary prism. In the back-arc regions, normal faulting characterizes quite exclusively the tectonics. The extension directions may be quite variable and microtectonic analysis of deformation reveals that horizontal stresses ($\sigma2$ and $\sigma3$) are generally very close (radial extention).

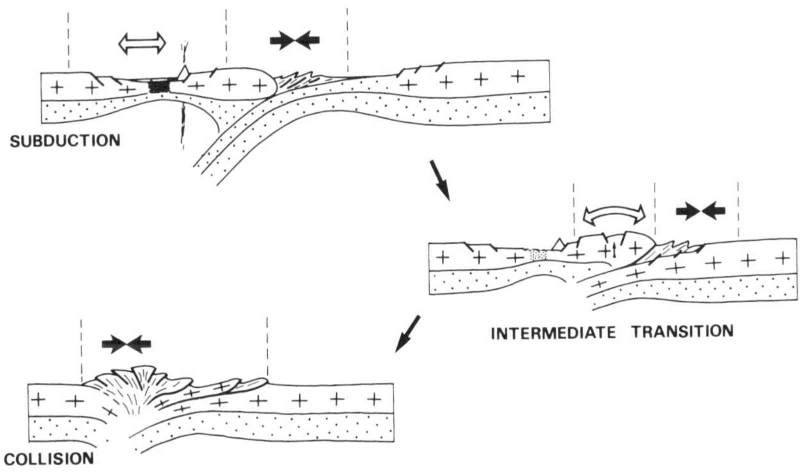

Fig.12: From oceanic subduction to collisional tectonics. This geodynamic evolution is illustrated by the case of Apennines and Calabrian arc.

INTERMEDIATE CASES

But we can consider intermediate cases that correspond to the progressive evolution from subduction to collision processes (Fig. 12).

The different possible cases of evolution are not so numerous because each one depends strongly on few parameters such as the nature (oceanic, continental) and the thickness of the different lithospheric plates which converge.

Such intermediate cases must be understood by new concepts such as "continental subduction" (Bousquet and Philip, 1986) (Fig. 13).

The duration of the different stages (tectonic phases) depend on the variation of the convergence rate, the size of each block and, mainly, the general evolution of the boundary conditions.

In the Mediterranean region, the Betic Cordillera, Tunisia and the Apenninic chain provide good examples of different intermediate cases of tectonic evolution (at different periods) from subduction to collision.

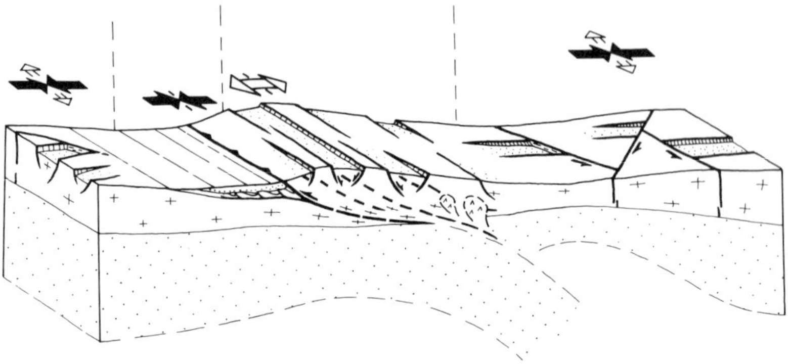

Fig.13: Intermediate transition from oceanic subduction to collision according to Apennine example.

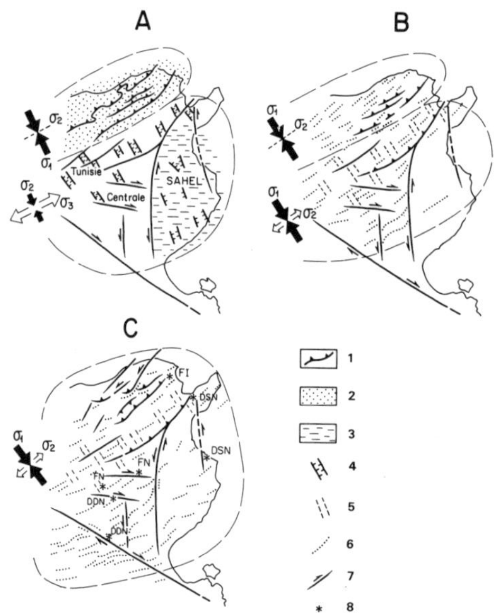

Fig.14: Schematic evolution of Tunisia from Aquitanian to Quaternary times.

A : Aquitanian to Tortonian time ; B : Tortonian to Messinian time ; C : Messinian to Villafranchian time.

1 : Reverse faults ; 2 : Allochtonous units ; 3 : Subsidence zones ; 4 : Grabens ; 5 : Grabens with weak activity ; 6 : Folds ; 7: Strike slip faults ; 8 : Tyrrhenian and upper Palaeolitic deformations : FN : Normal faults ; DDN : Dextral strikes ; FI : Reverse faulting and folds.

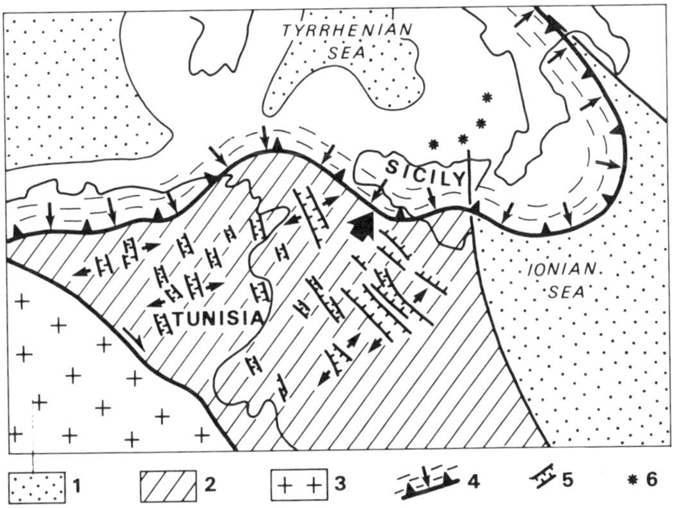

Fig.15: Schematic map of Tunisia and South Tyrrhenian zone during Miocene period.
1 : Oceanic or thin continental crust ; 2 : Central Tunisian and Pelagian block ; 3 : African craton ; 4 : Subduction ; 5 : Graben ; 6 : Eolian volcanism.

1. TUNISIA (Philip *et al.*, 1986)

From Aquitanian to Tortonian times, during the emplacement of allochtonous units related to subduction processes (Fig. 14A), extensive tectonics took place in the foreland (Central Tunisia). This extension, associated with large strike faults, may be related to the possibility for the Central Tunisian block to be ejected eastwards because of an active subduction in Sicily during the same epoch (Fig. 15). A compressional stage which began in the North of Tunisia (Fig. 14B) and moved progressively southwards until present days, (Fig. 14C) corresponds to the blocking of the subduction in Sicily and in Tunisia at the end of Miocene.

This second tectonic stage indicates clearly the beginning of continental collision in Tunisia.

2. BETIC CORDILLERA

Like in the North African ranges, subduction (continental one ?) stopped in the upper Miocene (Fig. 16A) and was followed by collision tectonics. This sudden change is connected to the closure of the foreland basin (Guadalquivir basin) in front of the nappes, except in the Gibraltar arc to the West where the Atlantic oceanic lithosphere remains.

During the upper Miocene and Pliocene (Fig. 16B), the internal domain of the Betic Cordillera was subjected to normal faulting associated with large strike-slip faults ;

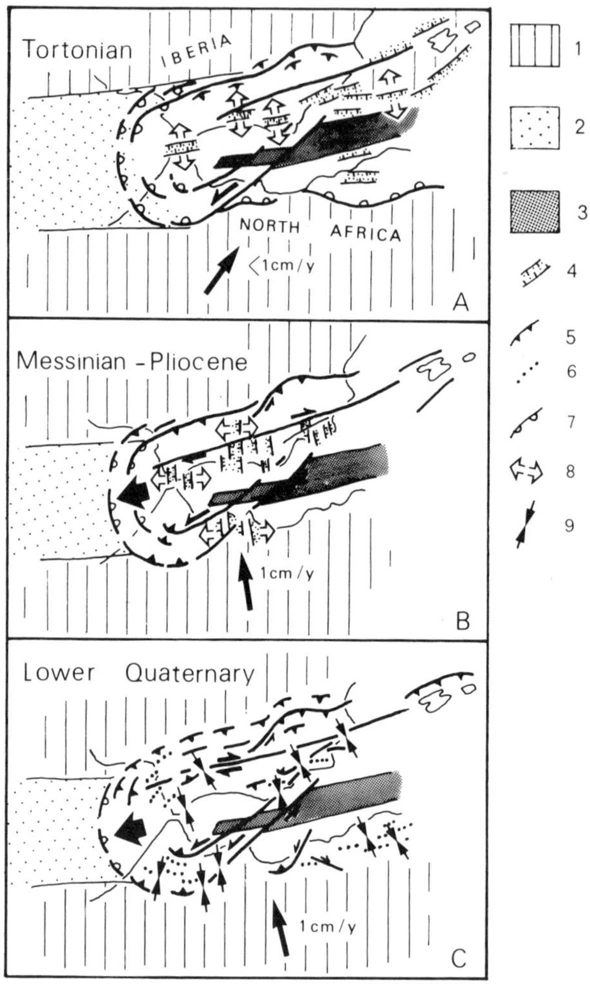

Fig.16: Tectonic evolution of the Betic Cordillera from Tortonian to lower Quaternary times.
1 : Continental crust of Africa and Iberia. 2 : Oceanic crust. 3: Intermediate crust. 4 : Graben. 5 : Reverse fault. 6 : Fold axis. 7 : Subduction front. 8 : Minimum horizontal stress direction ($\sigma 3$). 9 : Maximum horizontal stress direction ($\sigma 1$).

while, in the external domain compressive tectonics was still active. The ENE-WSW direction of extension in the internal domain was perpendicular to the direction of compressive stress in external domain. During this time alkaline volcanism replaces calc-alkaline volcanism associated to the previous stage of subduction.

At the end of the Pliocene this tectonics has been progressively replaced by an important N-S to NW-SE compression (Fig. 16C) which is still active in present time except in some (Bousquet and Philip, 1976) sectors (Granada basin, Mellilla region in Morocco) where normal faulting, identical to that of the Pliocene event, reappears.

During these two steps of collision, large strike slip faults play an important role in the region. The largest ones (the Cadix-Alicante right lateral strike-slip fault and the left lateral system of Carboneras - Nekor - Alhama - Totana faults) delimit a large wedge which was ejected westward due to the North-South convergence between Iberia and Africa. This was possible because thin Atlantic oceanic crust is still present to the West.

3. APENNINES AND THE TYRRHENIAN ARC (Philip, 1987 ; Bousquet and Philip, 1986)

From a geodynamical point of view, the Italian peninsula could be devided into two regions : the North-Central Apennines and the Calabrian arc (Fig. 17).

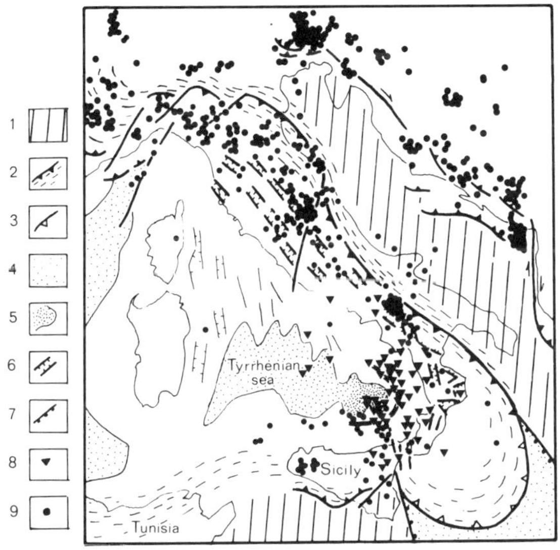

Fig.17:General scheme of recent tectonics and seismicity in the Apennines and Calabrian arc.
1 : Continental platform. 2 : Reverse fault and folding. 3: Front of the Calabrian "ridge". 4 : Oceanic or thinned crust of Cenozoic age. 5 : Plio-Quaternary oceanic crust. 6 : Graben. 7 : Mesozoic passive margin. 8 : Dip and intermediate earthquake focus. 9 : Shallow earthquake focus.

a) In the first region we can define four zones in any cross-section from East to West :

i) The Apulian platform which is a relatively undeformed Mesozoic carbonated platform over the continental crust.

ii) The peri-Apennine basin which corresponds to the last stage of a foreland flexural basin developed in front of allochtonous units (overturned towards the East). Oil industry seismic profiles indicate that the last displacement of these allochtonous units is quite recent, dating back only to the latest Miocene-early Pliocene in the Po plain. After the lastest emplacement, compression still continues until the present times.

iii) The Apennines. These mountain ranges were built by the piling up of low angle thrusts overturned towards the East, with the general structure of an accretionary prism. Normal faulting, parallel to the axis of the chain has been later superposed on the previous compressive structures.

iv) The Tyrrhenian basin corresponds to a back-arc marginal sea. This assertion is essentially based on the presence of a volcanic arc, of a rather thin oceanic lithosphere at the center of the basin, of high heat flow, of intermediate to deep seismicity in the Calabrian arc, and of distensive tectonics associated with the recent opening of the basin.

b) In the Calabrian arc we can define from SE to NW :

- The continuation to the South of the Apulian platform.
- The Ionian basin which corresponds to a thin continental crust covered with Mesozoic to Quaternary sediments.

The two previous zones are separated by a Mesozoic passive margin.

- The Calabrian "ridge" which covers the Ionian crust, and has a geological structure corresponding to an active accretionary prism. This ridge is characterized by compression which combines folds, thrust and strike faults, with South vergence sedimentary "décollements". The transition into the peri-Apennine basin to the North is continuous.
- Calabria : this region shows very similar features to those of the Apennines but its present day tectonic regime is closer to what is known in the Aegean arc during the Plio-Quaternary. The presence of an active volcanic arc and of a young oceanic crust (< 1 My) associated to the opening of the Tyrrhenian sea, are related to the subduction of the Ionian lithosphere underneath the Calabrian arc.

The characteristics of the seismicity permits also to individualize in the Italian peninsula the northern-central Apennines and the Calabrian Arc. In the first region the activity corresponds to shallow seismicity concentrated mainly on the axis of the Apennines and is associated to tensional tectonics (Irpinia and Perugia earthquakes, etc...). The eastern piedmont of the Apennines which has been the site of a weaker activity is associated to compressional tectonics (Parma earthquake). A very low level of seismicity characterizes the northern-central part of the Tyrrhenian basin and the western side of the Apennines.

The presence of an intermediate to deep seismicity in the Calabrian arc

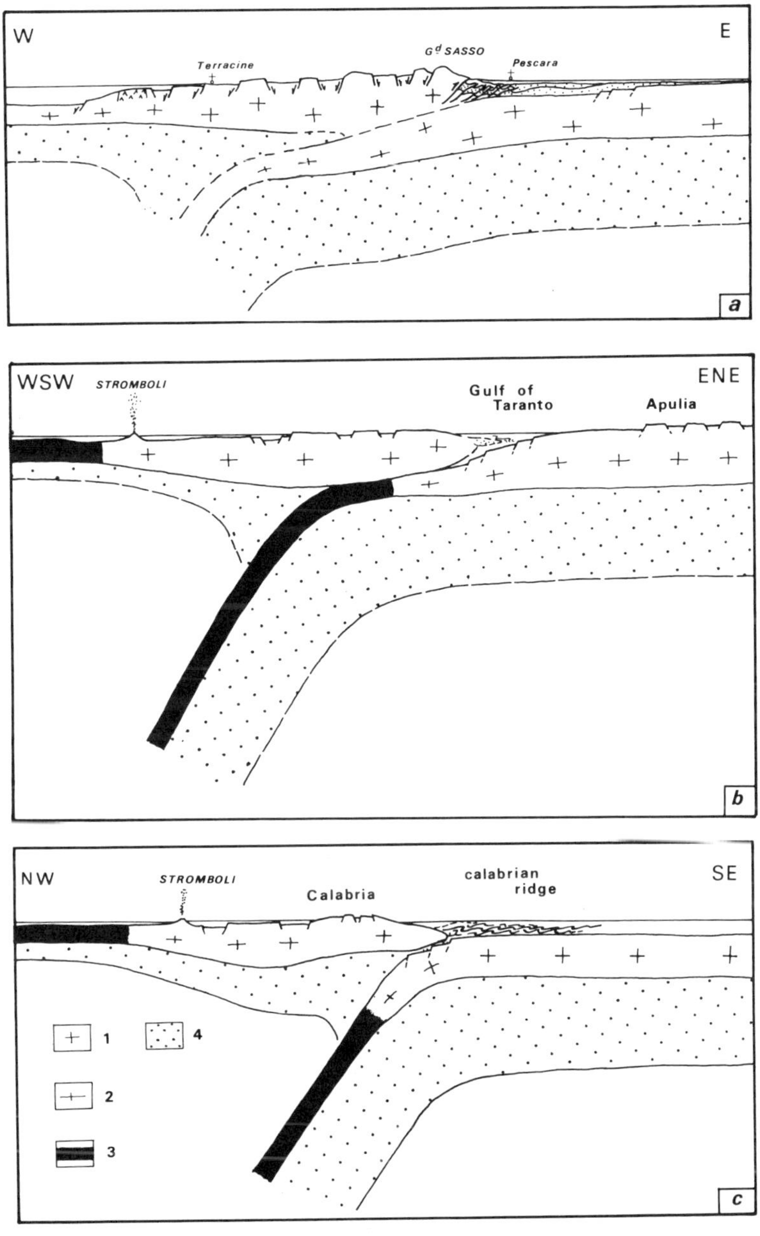

Fig.18: Lithospheric structure from (a) the Central Apennines to (b) and (c) the Tyrrhenian arc deduced from large scale profiling (see Philip, 1987). 1 : Continental crust. 2 : Thinned continental crust. 3 : Oceanic crust. 4 : Lithosphere.

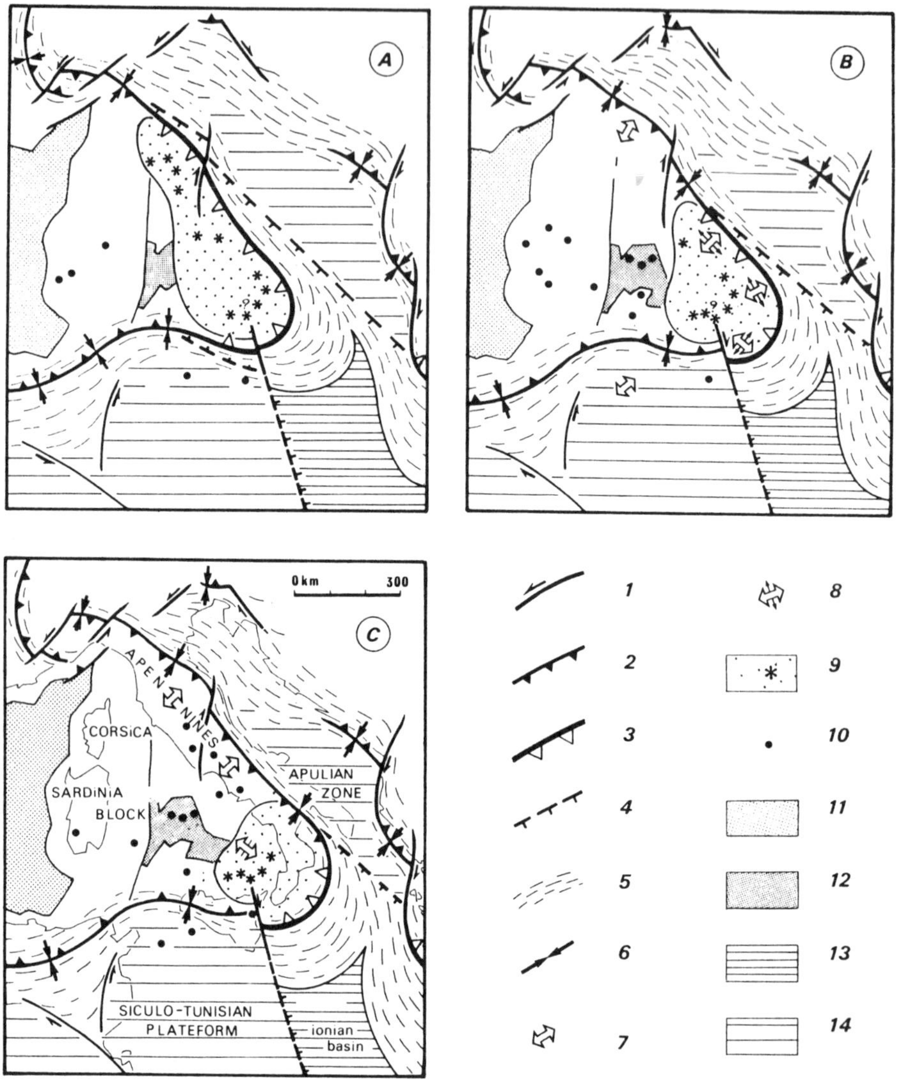

Fig.19: Plio-Quaternary tectonic evolution of the Apennines and of the Tyrrhenian arc. A. Early Pleistocene. Later Pleistocene-Calabrian. C. Latest Pleistocene to present day. 1 : Strike-slip. 2 : Major thrust. 3 : Subduction zone. 4: Apulian margin of Mesozoic age. 5 : Compression zones. 6: Convergence direction of "lithospheric blocks". 7 : Extension direction. 8 : Radial extension. 9 : Volcanic arc. 10 : Alkaline or tholeiitic volcanism. 11 : Algerian-Provençal oceanic basin. 12 : Tyrrhenian oceanic basin. 13 : Thinned crust of the Ionian basin. 14 : Mesozoic carbonate platforms over continental crust.

demonstrates an active subduction of the Ionian lithosphere underneath Calabria. Shallow (crustral) seismicity is also present in Calabria and in the southern Tyrrhenian basin, but weak seismic deformation in the accretionary prism.

CRUSTAL STRUCTURES (FIG. 18)

Long seismic profiles across the Apennines show a crustal doubling in the central part of the chain which is in agreement with a plunge of the Adriatic lithosphere underneath the Central Apennines. This crustal doubling continues towards the South where the Ionian lithosphere plunges underneath Calabria.

Regarding to geophysical and geological data it seems reasonable to assume that an oceanic or thinned continental lithosphere, a prolongation of the present Ionian sea crust, has been subducted under the Apennines since the latest Miocene-early Pliocene in a zone (from the Po Plain to Calabrian ridge) corresponding to the foreland basin of the Apennines (Fig. 19). This Ionian crust is bounded eastwards by a passive margin which limits the Apulian plateform.

The subduction ended earlier in the North than in the South because the Ionian basin was probably shorter in the North and was rapidly absorbed.

This progressive cessation of subduction is marked by magnatic evolution in the peri-Tyrrhenian volcanism, as well as by changes in tectonic regimes (Bousquet et Philip, 1986). In fact the Italian penninsula illustrates very well the passage from subduction to collision in a complex situation where North-South diachronous tectonic phenomenons are imbricated to West-East ones.

CONCLUSIONS

In the Mediterranean regions, neotectonics and seismic activity are ruled by the convergence of the African and the Eurasian plate along a NW-SE direction. Other mechanisms than subduction and collision help keep the convergence active and correspond to intermediate situation. These situations introduce new concepts as "intracontinental subduction" for example, which allow the reconciliation of neotectonic observations and geodynamics evolutionary models for different sectors of the Mediterranean regions during Mio-Plio-Quaternary times.

REFERENCES

Anderson E.M., 1951. The dynamic of faulting. Oliver and Boyd, Edinburgh. 203 pp.

Armijo R., and Cisternas A., 1978. Un problème inverse en microtectonique cassante. C. R. Acad. Sci., Paris, **287**, 5495-598.

Bousquet J.C., and Philip H., 1976. Observations microtectoniques sur la compression nord-sud quaternaire des Cordillères Bétiques orientales (Espagne méridionale - Arc de Gibraltar). Bull. Soc. Géol. Fr., **3**, 711-724.

Carey E., 1979. Recherche des directions principales de contraintes associées au jeu d'une population de failles. Rev. Géogr.

Phys. Géol. Dyn., **21** (1), 57-66.

Dercourt J., Zonenshain L.P., Ricou L.E., Kazmin V.G., Le Pichon X., Knipper A.L., Grandjacquet C., Sborshikov I.M., Geyssant J., Lepvrier C., Pechersky D.H., Boulin J., Sibuet J.C., Savostin L.A., Sorokhtin O., Westphal M., Bazhenow M.L., Lauer J.P., and Biju-Duval B., 1986. Geological evolution of the Tethys belt from the Atlantic to the Pamirs since the Lias. Tectonophysics, **123**, 214-315.

Dewey J.F., Pitmann W.C., Ryan W.B.F., and Bonnin J., 1973. Plate tectonics and evolution of the Alpine systeme. Geol. Soc. Am. Bull., **84**, 3137-3180.

Etchecopar A., Vasseur G., and Daignières M., 1981. An inverse problem in microtectonics for the determination of stress tensors from fault striation analysis. J. Struct. Geol., **3**, 51-55.

Le Pichon X., Sibuet J.C., and Francheteau J., 1977. The fit of the continents around the Atlantic ocean. Tectonophysics, **38**, 169-209.

McKenzie D.P., 1972. Active tectonics in the Mediterranean region. Geophys. J.R. astr. Soc., **30**, 109-185.

Mégard F., and Philip H., 1976. Plio-Quaternary tectono-magmatic zonation and plate tectonics in the Central Andes. Earth Planet. Sc. Let., **33**, 231-238.

Nakamura K., and Uyeda, 1980. Stress gradient in arc-back arc regions and plate subduction. J. Geophys. Res., **85**, 6419-6428.

Olivet J.L., Bonnin J., Beuzart P., and Auzende J.M., 1984. Cinématique de l'Atlantique Nord et Central. Rapports Scientifiques et Techniques, CNEXO, Paris,108 pp.

Paquin C., and Froidevaux C., 1980. Tectonic stresses in France. Rock Mechanics, **9**, 17-18.

Philip H., Andrieux J., Dlala M., Chihi L., and Ben Ayed N., 1986. Evolution tectonique mio-plio-quaternaire du fossé de Kasserine (Tunisie Centrale) : implications sur l'évolution géodynamique récente de la Tunisie. Bull. Soc. Géol. Fr., **4**, 559-568.

Philip H., 1987. Plio-Quaternary evolution of the stress field in Mediterranean zones of subduction and collision. Annales Geophysicae, **5B** (3), 301-320.

Tapponnier P., 1977. Evolution tectonique du système alpin en Méditerranée : poinçonnement et écrasement rigide-plastique. Bull. Soc. Géol. Fr., **3**, 437-460.

Vasseur G., Etchecopar A., and Philip H., 1983. Stress state inferred from multiple focal mechanisms. Annales Geophysicae, **1** (4-5), 291-298.

ACTIVE TECTONICS IN THE AEGEAN AND SURROUNDING AREA

B.C.Papazachos
Geophysical Laboratory,
University of Thessaloniki
Thessaloniki 54006
Greece

ABSTRACT. An attempt is made in this article to review the important information concerning active tectonics in the Aegean and surrounding area. Morphological and geophysical properties, distribution of earthquake foci, relation of shallow seismic activity to geology, seismicity patterns related to long and short term earthquake prediction and focal properties of earthquakes are reviewed. Geodynamic modes which have been proposed to interprete geophysical and other properties of this area are discussed.

1. INTRODUCTION

It is well known that the Aegean and surrounding area (34°N-43°N, 18°E-30°E), which includes Greece, the Aegean Sea, Albania, S.Yugoslavia, S.Bulgaria, W.Turkey and part of the North - Eastern Mediterranean, is seismically the most active region in the whole Mediterranean and in the whole West Eurasia. Almost every year at least one earthquake with magnitude $M \geq 6.5$ occurs in this relatively small region. This region is a part of the collision zone between the Eurasian and the African lithospheric plates but its present tectonic activity is much higher than in other regions of the same zone.

The seismic and other geophysical properties of this region are very similar to properties of other regions of the world of similar tectonic structure. It includes the Hellenic arc, and the Aegean Sea, which has the basic properties of the well known marginal Seas and is separated from the eastern Mediterranean by the Hellenic arc.

Due to its high seismicity and to the fact that special geophysical and geological properties, which are observed in much broader areas of similar tectonic structure, are concentrated in this relatively small region, Aegean and its surroundings have been extensively investigated during the last two decades and the results of this investigation is of general interest.

An attempt is made here to give basic information on the present tectonic activity of this region on the basis of seismological and o-

J. Bonnin et al. (eds.), Seismic Hazard in Mediterranean Regions, 301–331.

ther geophysical information.

A very brief description of the main geomorphological and geophysical properties is first made and after that basic information is given on the spatial distribution of the earthquakes, the seismicity patterns, the stress field and seismic faulting and on the proposed geodynamic models to interprete the seismic and other properties of the Aegean and surrounding area.

2. MORPHOLOGICAL AND GEOPHYSICAL PROPERTIES OF THE AEGEAN AREA

The most prominent morphological features of tectonic origin in the Aegean and surrounding area from south to north are: the Mediterranean ridge (or chain), the Hellenic trench, the Hellenic arc and the northern Aegean trough. Figure (1) shows the Hellenic trench, the main axis of the Alpine folding, the small Cretan trough between the sedimentary arc and the volcanic arc, the isodepth of 150Km of the earthquake foci which almost coincides with the volcanic arc, and the northern Aegean trough.

The M e d i t e r r a n e a n r i d g e is a submarine crustal swell which extends from the Ionian Sea to Cyprus and parallels the Hellenic trench and other tectonic features in the northern Mediterranean coast. This is not a mid-ocean ridge, but a compressional feature produced by the convergence of the African and Eurasian lithospheric plates (Rabinowitz and Ryan 1970, Comninakis and Papazachos 1972). Finetti (1976) investigated this tectonic feature in detail and suggested the name "east Mediterranean chain" instead of the name " Mediterranean ridge " .

The H e l l e n i c t r e n c h consists of a series of depressions with depth to about 5Km.It parallels the Hellenic arc and includes some linear trenches as are the Pliny and Strabo trenches southeast of Crete and the Ionian trench.

The H e l l e n i c a r c is formed by the outer sedimentary arc, a link between the Dinaric Alps and the Turkish Taurides, and the inner volcanic arc which parallels the sedimentary arc at a mean distance of about 120Km. The sedimentary arc consists of Palaeozoic to Tertiary rocks, while the volcanic arc consists of several volcanic islands and includes andesitic active volcanoes and solfatara fields. Between the sedimentary and the volcanic arc is the Cretan trough with water depth to about 2000m.

The most interesting feature of the northern Aegean is the n o r t h e r n A e g e a n t r o u g h with water depth to about 1500m. Its extension to northeast are probably the small depressions to the Marmara Sea.

The free air gravity anomalies are irregular and predominantly positive in the Aegean Sea, while the Bouguer anomalies are positive in the Aegean and eastern Greece and negative (-120to - 140mgal) in western Greece along the main mountain range (Allan and Morelli 1971, Makris 1973). Strong positive Bouguer anomalies (to +170mgal) have been observed in the Cretan trough, while the maximum value of these anomalies in the northern Aegean trough and in other places of the northern and central Aegean is about +50mgal. In the outer part of the Hellenic

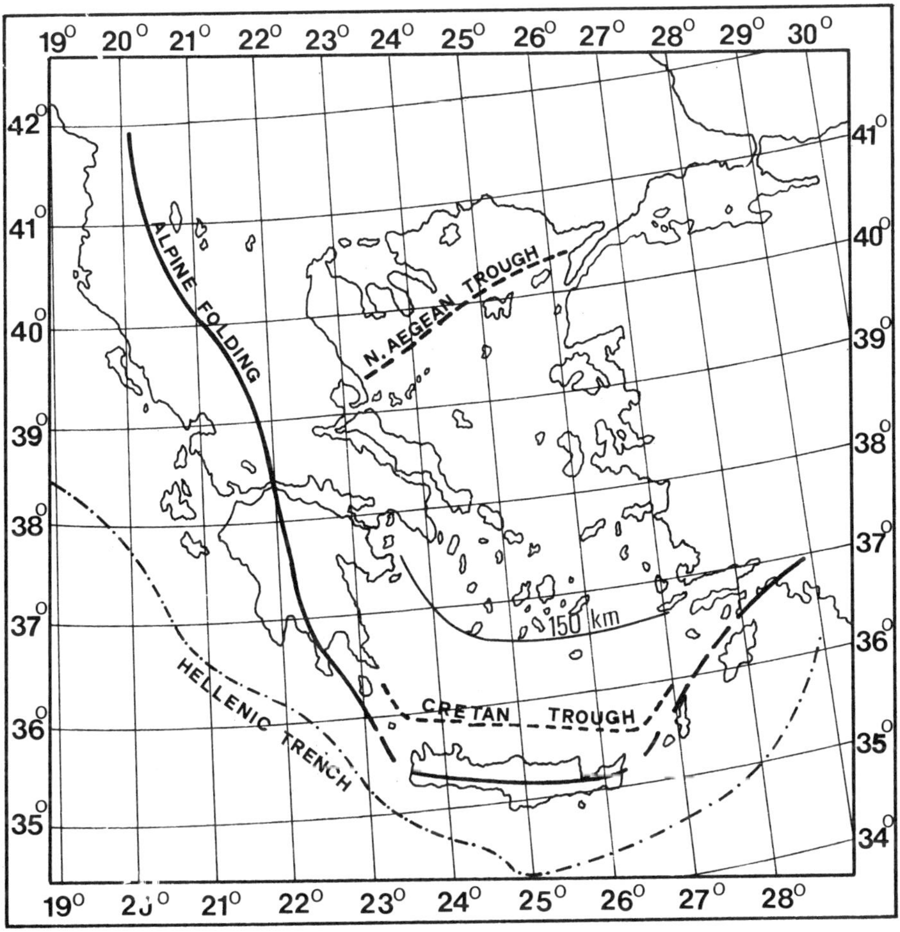

Fig. 1. The main morphological features of tectonic origin in the Aegean area.

arc (easterm Mediterranean) the free air anomalies are mainly negative (to -180mgal), while the Bouguer anomalies are positive (to +180mgal), but not so strongly positive as in the western Mediterranean or in oceanic areas in general (Morelli et al 1975).

Positive magnetic anomalies have been determined in several parts of the Aegean Sea. The strongest of these anomalies have been observed in the northern Aegean trough, along the volcanic arc of the southern Aegean and in the Cretan trough (Vogt and Higgs 1969, Makris 1973). Needham et al (1973) have determined positive magnetic anomalies of +250 gamma in the northern Aegean trough. The magnetic field is undisturbed in the Mediterranean Sea south of Crete (Vogt and Higgs 1969).

The heat flow is small ($\sim$ 0.7HFU) in the eastern Mediterranean south of the Hellenic arc (Erickson 1970, Morelli et al. 1975). In the Aegean Sea high heat flow ($\sim$ 2.1HFU) has been measured (Jongsma 1974). The highest value ($\sim$ 2.7HFU) has been observed in the volcanic arc of the southern Aegean but high values have been also found in the northern Aegean trough (to$\sim$2.5HFU).

An aseismic region, where the short period body waves are strongly attenuated, has been observed in the upper mantle above the Benioff zone of the southern Aegean (Papazachos and Comninakis 1971). It has been also observed, through studies of P wave residuals, that the wave velocity in this low Q asthenopheric region is smaller than the wave velocity in the lithospheric slab under the Benioff zone (Agarwal et al 1976, Gregersen 1977, Tassos and Papazachos 1985).

Crustal thickness between about 20 and 45Km and compressional velocities which characterise crystalline rocks (6.0-6.1Km/sec) have been determined in several parts of the Aegean area by seismic and other methods (Papazachos 1969, Makris 1973, Panagiotopoulos and Papazachos 1985, Panagiotopoulos et al 1985). These observations show that the crustal structure in the Aegean area is basically continental, in contrast to other marginal seas which are underlain by oceanic crust (Karig 1971). Large crustal thickness (40-47Km) has been found in the western part of the area, that is, along the Hellenides mountain range, while the crustal thickness becomes normal (28-32Km) in the whole eastern part of the area (eastern Greece, Aegean Sea, Western Turkey) except for the Cretan trough where a crustal thickness of the order of 20Km has been determined (Payo 1969, Makris 1973). The crustal thickness is of the order of 20Km in the eastern Mediterranean (Payo 1969, Papazachos 1969).

3. DISTRIBUTION OF EARTHQUAKE FOCI

One of the main contribution of Seismology to the new global tectonics is the determination of boundaries of lithospheric plates by determining the seismic zones which define these boundaries. Even in the cases of intraplate earthquakes the accurate determination of their spatial distribution is of primary significance for understanding the active tectonics because this distribution defines the fracture zones where present tectonic activity is manifested. On the other hand, one of the most important methods of determining the seismic hazard in an area is based on the accurate definition of the active zones in this area.Thus, accurate determination of the directions and slopes of the seismic zones in the Aegean and surrounding area will be of primary importance for any tectonic theory concerning this area and for the solution of seismic hazard problems as well.

Good knowledge of the spatial distribution of the seismic activity requires homogeneous and complete data in respect to the magnitude of the shocks and of acceptable accuracy in respect to their location (epicenter, focal depth) and magnitude. Furthermore, the sample of these data,must be as large as possible and must cover as large as possible time period. To fulfill this last condition it is necessary to increase the data sample by extending the considered time period to the past as much as possible. It is true that by including older earthquakes in the data we introduce errors because the foci of the older earthquakes are less accurately known. However, the accuracy of the determination of the earthquake foci depends on the earthquake magnitude too. The foci of the large shocks are better determined because of their larger macroseismic effect and of the larger number of the seismic stations by which these earthquakes were recorded. Therefore , we can reduce the errors considerably by dividing the time period considered in several intervals and properly choosing for each interval a lower limit of earthquake magnitude which will increase as the time increases to the past. The division of the whole time period in several time intervals and the determination of the lower limit of magnitude for each interval must be based on a carefull examination of the progress in the techniques used to determine earthquake foci as well as on the progress in the establishment of regional and global networks of seismic stations.

Epicenter maps and other information concerning distribution of earthquake foci which are given here are based on data which fulfill the conditions mentioned above in a satisfactory way.

3.1. Distribution of the Epicenters in the Mediterranean Area

Figure (2) shows an epicenter map of shallow ($h \leq 60$Km)earthquakes which occurred in the Mediterranean and surrounding area (Papazahos 1973, Comninakis and Papazachos 1978). Three symbols (circles) are used to denote earthquakes of three magnitude ranges and of corresponding time periods.

It is seen in this figure that there is not one or two lithospheric plates in this area but there are several aseismic blocks surrounded by active zones. One of the most important observations which can be made on the epicenter map of figure (2) is that the seismic activity in the Aegean and surrounding area (10^oE-30^oE, 34^oE-43^oE)is much higher than the activity in the rest of the whole Mediterranean zone which defines the boundary between the Eurasian and the African lithospheric plates. This indicates that the seismic activity in the Aegean and surrounding area cannot be attributed only to the fact that this area constitutes a part of the boundary of the two big lithospheric plates but additional local factors contribute much to this high seismic activity.

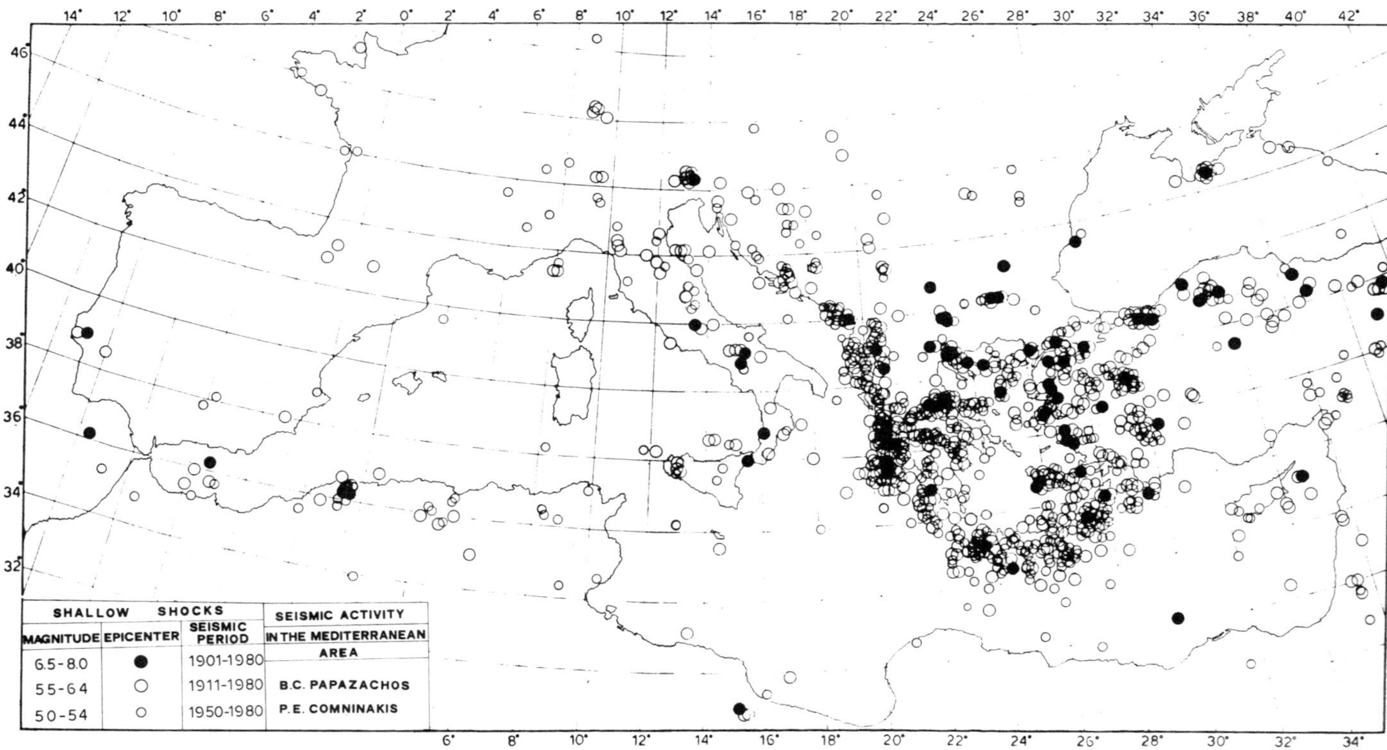

Fig. 2. Distribution of the epicenters of shallow earthquakes which occurred in the Mediterranean and surrounding area during the present century.

3.2. Distribution of the Epicenters of the Shallow Earthquakes in the Aegean and Surrounding Area

Figure (3) shows an epicenter map which concerns all known shallow shocks with M$\geq$ 6.0 which occurred in this area. Three sizes of open circles have been used to denote historical earthquakes (600B.C.-1900) of three magnitude ranges (6.0-6.4 , 6.5-6.9 , 7.0-7.8) and three sizes of black circles to denote the same three corresponding ranges of magnitudes for instrumental data (1901-1985). It is evident that the epicenters of the large shallow shocks form several seismic zones. The e x t e r n a l s e i s m i c z o n e s form a continuous large seismic belt along

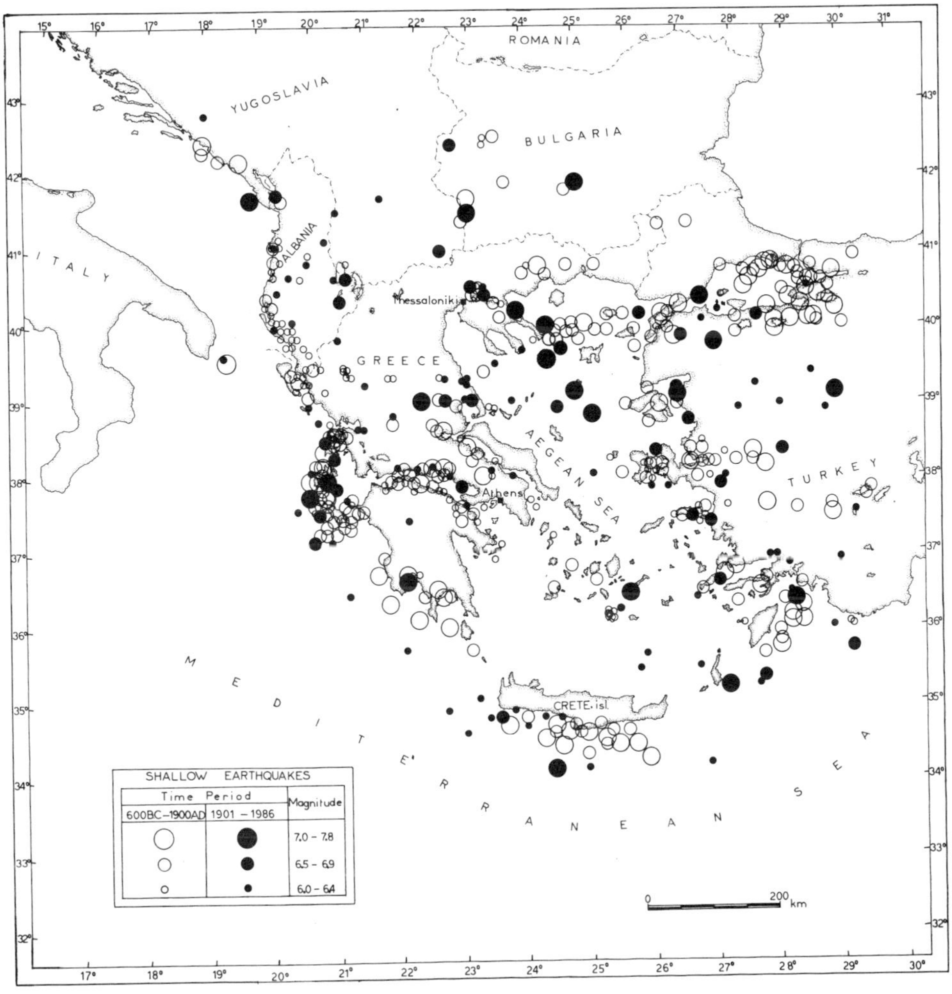

Fig.3. Epicenters of historical (open circles) and present century (black circles) earthquakes in the Aegean and surrounding area. Surface wave magnituedes are used.

the external (convex) side of the Hellenic arc, and its extension along the western coast of central Greece, Albania and Yugoslavia. All other zones constitute the i n t e r n a l s e i s m i c z o n e s, which have a tendency to have an east-west direction and this tendency is very clear in the eastern part of the area.

3.3. Relation of the Shallow Seismic Activity with the Geology of the Area.

The area of Greece has been divided in several geological zones on the basis of several petrographic , stratigraphic, structural and palaeogeographic criteria. By comparing the seismic zones in figure (3) with the geological zones of figure (4) we can see some similarities in their trends. For example, the outer seismic zones in the western part of the area follow the Ionian geologic zone and a seismic zone almost coincides with the Servomacedonian geologic zone. However, there are striking differences between the trends of the geologic zones and the seismic zones as one can see by comparing the maps of figures (3) and (4).

Analogous deviations between geologic structures and seismic sources have also been observed in other regions of the world and have been attributed to several reasons. Such reasons are: changes in the tectonic regime of the region, seismic events at depth may not be directly and simply related to surface geology, the recorded seismic history of a region may not cover a long enough period to represent true secular seismicity, earthquakes may not reccur along pre-existing fracture planes, from geological evidence alone it is difficult to determine the recent desplacements on a fault (Allen et al 1965, Nowroozi 1971).

Hatzidimitriou and his colleagues (1985) observed a systematic decrease of the values of the parameter b of the frequency-magnitude relation for shallow eartqhuakes in the area of Greece. These values decrease systematically from the external seismic zones to the internal ones and are clearly related to the geological zones. The whole Aegean and surrounding area has been divided in three regions A,B,C where the values of the parameter b are 1.03, 0.84 and 0.60, respectively (fig.4). These differences in the value of b, which indicate differences in the mechanical properties of the seismic regions A,B,C must represent differences in the structure and the tectonic evolution of the geologic zones which belong to the three seismic regions. In region A the material must have a higher degree of heterogeneity than the region B,which in turn must have a higher degree of heterogeneity than region C. This high degree of heterogeneity in region A could be due to the strong tectonization of the area, with a large density of recent fracturing of the rocks. On the other hand, region C must be the most homogenous and the less fractured region. These observations are in accordance with the geological conditions in the area of Greece.

Therefore, the available data for Aegean and surrounding area indicate that the directions of the seismic fracture zones in this area show little correlation with geological trends. On the other hand, the spatial distribution of the parameter b of the Gutenberg-Richter relation , calculated from data covering a time period considerably larger

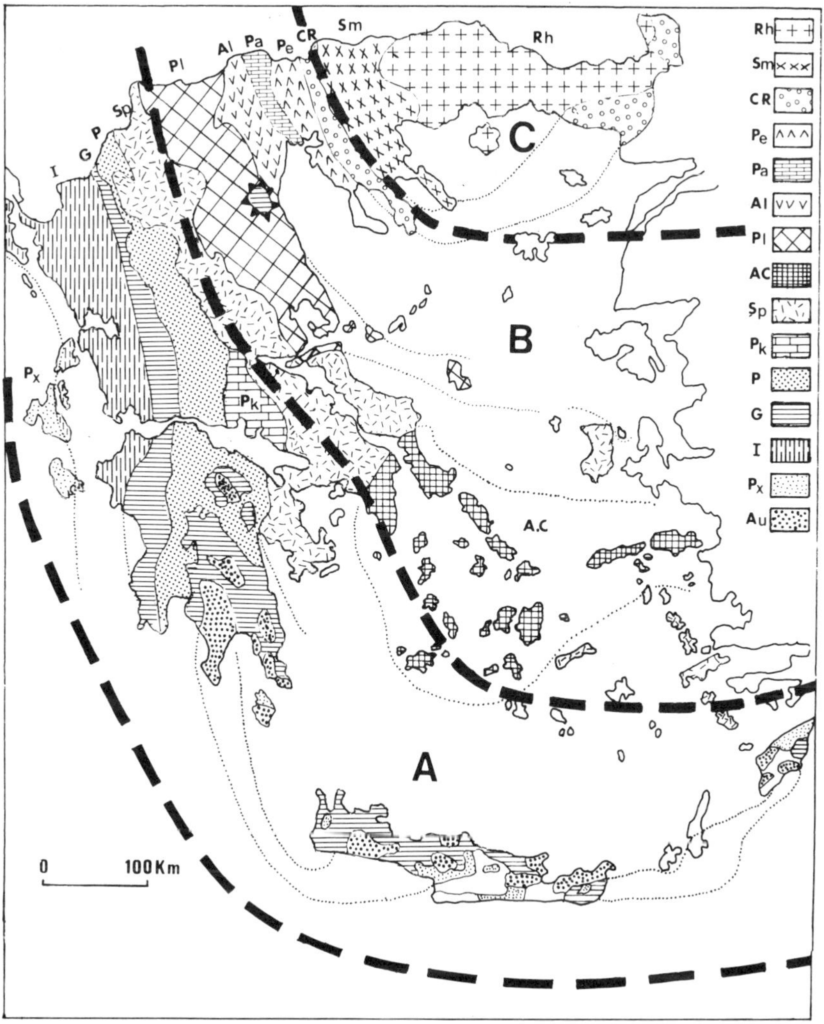

Fig. 4. The value of the seismic parameter b of the frequency-magnitude relation is 1.03 in region A, 0.84 in region B and 0.60 in region C (Hatzidimitriou et al.1985).

than the return period of the large earthquakes, shows a good correlation with the geological trends.

3.4. Distribution of the Intermediate Focal Depth Earthquakes

The spatial distribution of the foci of the intermediate focal depth

($70Km \leqslant h \leqslant 180Km$) earthquakes in the Aegean and surrounding area is of much interest because the existence or not of a dipping Benioff seismic zone defines basic properties of the deep tectonics in this area and because the strongest and most destructive earthquakes (with $M \sim 8.0$) in this region are of intermediate focal depth.

Comninakis and Papazachos (1980) and Panagiotopoulos and his colleagues (1984) used the most accurate recent data which concern the focal coordinates of the intermediate focal depth earthquakes in the Aegean and surrounding area to study the spatial distribution of these earthquakes. The upper part of figure (5) shows the epicenters of all intemediate focal depth earthquakes which occurred during the period 1964-1984 in the Aegean region and which were recorded by 26 or more stations (Panagiotopoulos et al 1984). Six symbols have been used to denote two ranges of magnitudes (3.5 - 4.4) , (4.5-5.8) and three ranges of focal depths (70-100Km), (101-130Km), (131-183Km). It is clear from this figure that the most extensive deep seismic activity in the Aegean and surrounding area is associated with the Hellenic arc in the Southern Aegean. A systematic increase of the focal depths, from the outer part of the Hellenic arc to the inner part of the arc is clearly observed. All earthquakes with focal depths greater than 100Km are located in the inner (concave) part of the arc, while only earthquakes with smaller focal depths are located at the outer (convex) part of the arc. The lower part of figure (5) shows projections of the seismic foci of the intermediate focal depth earthquakes onto the sections AA´,BB´,CC´, DD´, EE´, FE´which are perpendicular to the arc. To make these projections, all data shown in the upper part of figure (5) have been used. The straight lines represent the linear approximation to the data in the least squares´ sense. All six sections show that the seismic zone of the intermediate focal depth earthquakes dips under the Hellenic arc. The dipping angles are 36°, 37°, 32°, 33°, 38° and 38° for sections AA´, BB´, CC´, DD´,EE´ and FE´, respectively. These accurate data concerning the intermediate focal depth earthquakes in the southern Aegean leave no doubt that a well defined Benioff zone, which dips from the convex to the concave part of the Hellenic arc and has an amphitheatrical shape, is formed,as suggested by Papazachos and Comninakis (1969, 1971). At the eastern and western parts, the Benioff zone dips slightly steeper ($\sim 37^o$) than at the central part ($\sim 32^o$) of the arc.

The largest earthquakes in the whole Aegean and surrounding area are of intermediate focal depth and occur in the shallower part of the Benioff zone of the southern Aegean. Figure (6) shows the distribution of the epicenters of the known intermediate focal depth earthquakes with $M \geqslant 6.0$ which occurred in the Aegean and surrounding area during the last two centuries (Papazachos and Comninakis 1982). The isodepths of 150Km and 100Km (Papazachos and Comninakis 1971) are also shown. It is observed that all these strong shocks occur in the southern Aegean along the Hellenic arc. Eight earthquakes with $M \geqq 7.5$ occurred in this Benioff zone at a depth between 70Km and 100Km, while relatively weak ($M \geqslant 7.1$) earthquakes occur at larger ($h > 130Km$) focal depths.

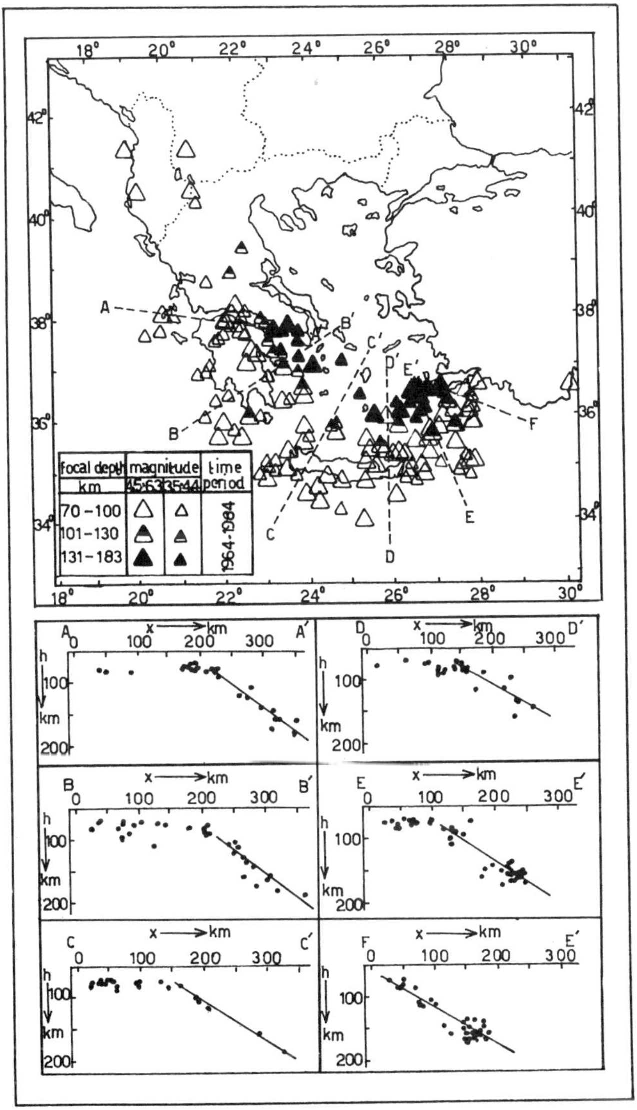

Fig.5. Distribution of epicenters of intermediate focal depth earthquakes (h ⩾ 70Km) which were recorded by 26 or more stations and occurred in the Aegean and surrounding area during the period 1964-1984 (upper part) and projections of their foci (lower part) onto the sections AA´, BB´, CC´, DD´, EE´, FE´ (Panagiotopoulos et al 1984).

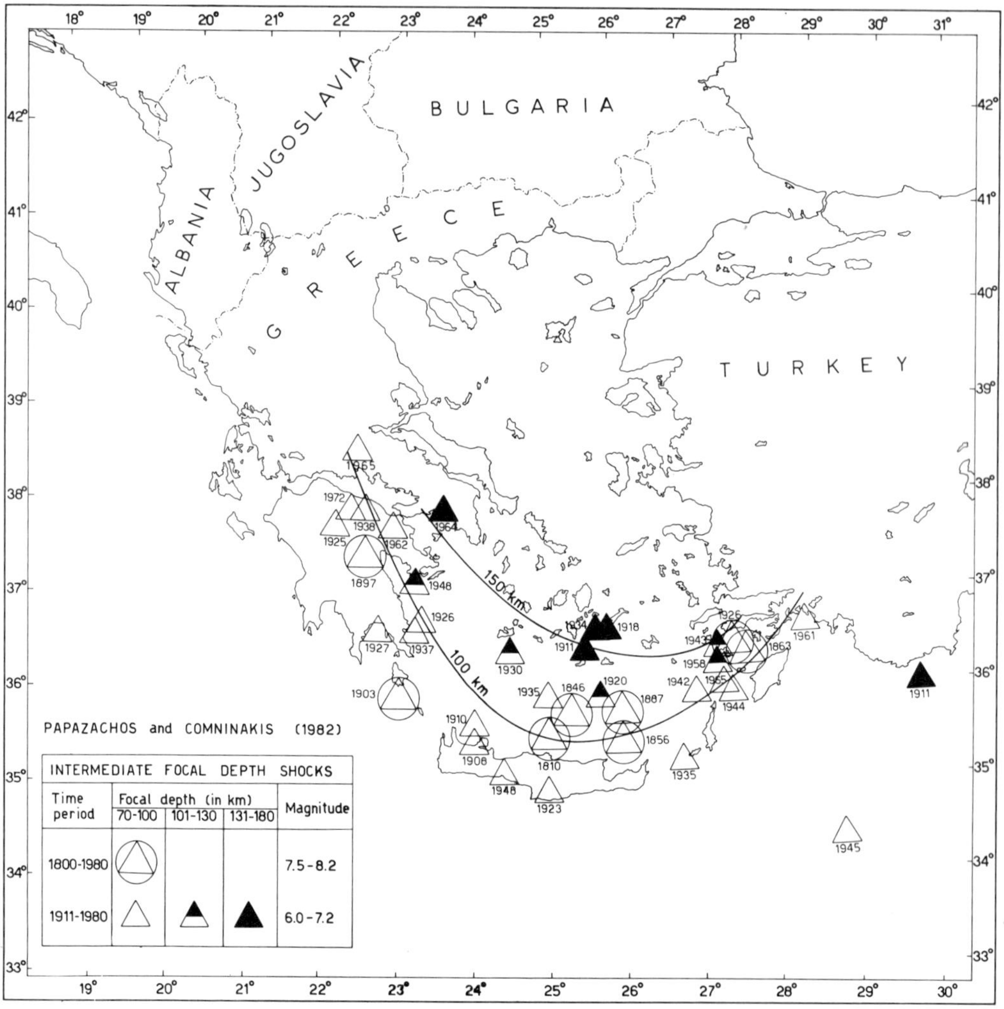

Fig. 6. Distribution of the epicenters of the intermediate focal depth earthquakes with $M \geq 6.0$ which occurred during the last two centuries (1800-1980) in the Aegean and surrounding area.

4. SEISMICITY PATTERNS

Investigation of seismicity patterns are of interest not only for earthquake prediction but for seismotectonic problems as well. It is known, for example, that seismic gaps in the seismogenic volumes of strong earthquakes (gaps of the first kind) have been mainly observed along plate boundaries. For these reasons several attempts have been made to identify such patterns in the Aegean area during the last ten years. Most of these seismicity patterns have been considered as possible long term precursors of large earthquakes but observations have been made to identify seismicity patterns as possible short term precursors as well. We present here some of the most interesting cases of such patterns.

4.1. Precursory Seismicity Patterns along the Hellenic Arc

Work on seismicity patterns along the convex side of the Hellenic arc (Ionian-Crete-Rhodos) has been performed by several authors. This work concerns mainly the seismicity quiescence, that is, the decrease of the frequency of the small earthquakes, in the seismogenic volume and surrounding area of strong earthquakes before the occurrence of these strong earthquakes.

Papazachos (1980) observed a quiescence period before the earthquake of August 12, 1953 (M = 7.2) in the Ionian island and before the October 7, 1947 (M = 7.0) south of Peloponnesus, while Wyss and Baer (1981) based on the seismicity drop observed in the region south of Peloponnesus-W.Crete concluded that an earthquake of magnitude 7.75±0.5 is expected in this region between 1980 and 1991.

Figure (7) shows the cumulative number of earthquakes with $M \geqslant 4.9$ as a function of time (Papazachos and Comninakis 1982) in the region south of Peloponnesus-West Crete. Four subperiods with different rates (number of shocks with $M \geqslant 4.9$ per year) are observed. These subperiods and rates are also shown in the same figure. The second subperiod (1932-1947) is a 14.4 years period of quiescence which preceded the earthquake of October 6, 1947 with M = 7.0 . Another quiescence period started in 1967 and it continues till now, which indicates that a strong earthquake is expected in this region as suggested previously by Wyss and Baer (1981).

Papazachos and Comninakis (1982) observed that the difference, D_1, between the magnitude of the main shock and that of the largest aftershock ($D_1 = M_o - M_1$) varies with time in an harmonic way with a period of 29 years in the western part of the Hellenic arc (Ionian-South Pelononnesus-W.Crete) as well as in the southeastern part of the arc (Crete-Karpathos-Rhodos) and that the largest earthquakes in both zones are associated with low values of D_1 (fig. 8). Based on these observations, Papazachos and Comninakis (1982) concluded that earthquakes with $M \geqslant 7.0$ are expected between 1983 and 1987. Furthermore, Papadimitriou and Papazachos (1985) defined two seismic gaps of the second kind (gaps in the low magnitude seismicity) in the Ionian islands (fig. 9) and concluded that earthquakes with magnitudes between 6.5 and 7.0 are expected in the areas of these two gaps. The occurrence of the earthquake of January 17, 1983 (M = 7.0) proved that the breaking out of strong seismic activity

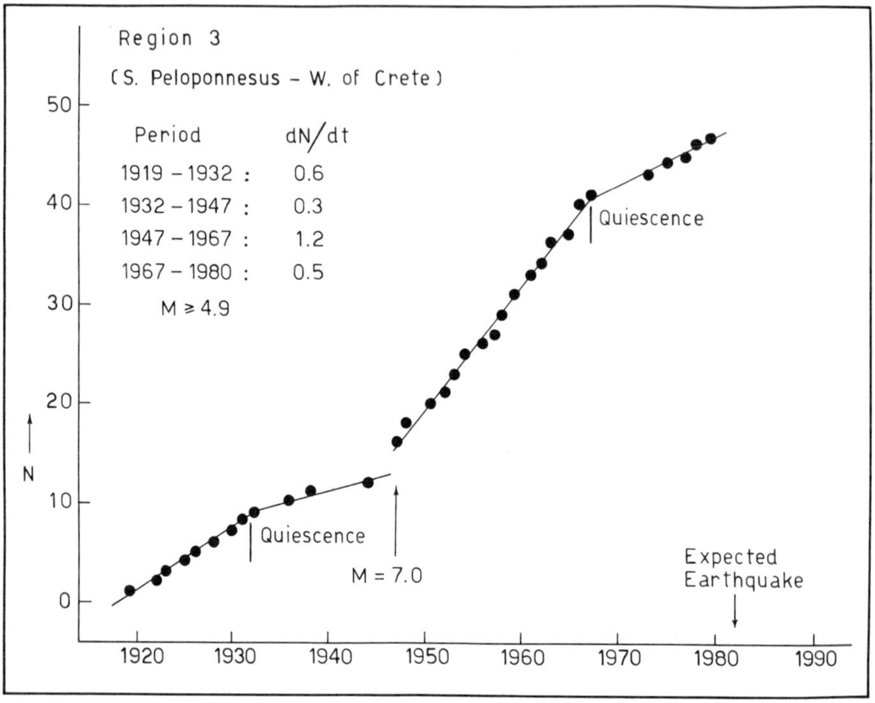

Fig. 7. Cumulative number of shocks as a function of time for the region south of Peloponnesus-West Crete (Papazachos and Comninakis 1982).

in 1983 along the Hellenic arc predicted by Papazachos and Comninakis (1982) and the occurrence of a strong earthquake in the southern Ionian islands (south of Cephalonia-west of Zante) predicted by Papadimitriou and Papazachos (their paper was send for publication before January 17, 1983) were true. According to these authors, strong earthquakes are still expected in the northern part of the Ionian islands (Papadimitriou and Papazachos 1985) as well as in the rest part of the arc (Wyss and Baer 1981, Papazachos and Comninakis 1982). These suggestions are strongly supported by new evidence (Papazachos et al 1986b).

A region which is at quiescence has been also observed in the inner (concave) part of the Hellenic arc but it concerns the intermediate focal depth seismicity. This region (west Peloponnesus-north of Crete-north of Rhodos) goes along the isodepth of 100Km shown in figure (6). Figure (10) shows the cumulative yearly number of the intermediate focal depth earthquakes of $M \geq 5.3$ (upper part) and of $M \geq 6.1$ (lower part since 1920. It is clear that this area is at quiescence since 1948, that is, for 38 years no earthquake of intermediate focal depth with magnitude larger than 6.1 occurred in this region (Papazachos et al 1985). It is known that in this region five great intermediate focal depth e-

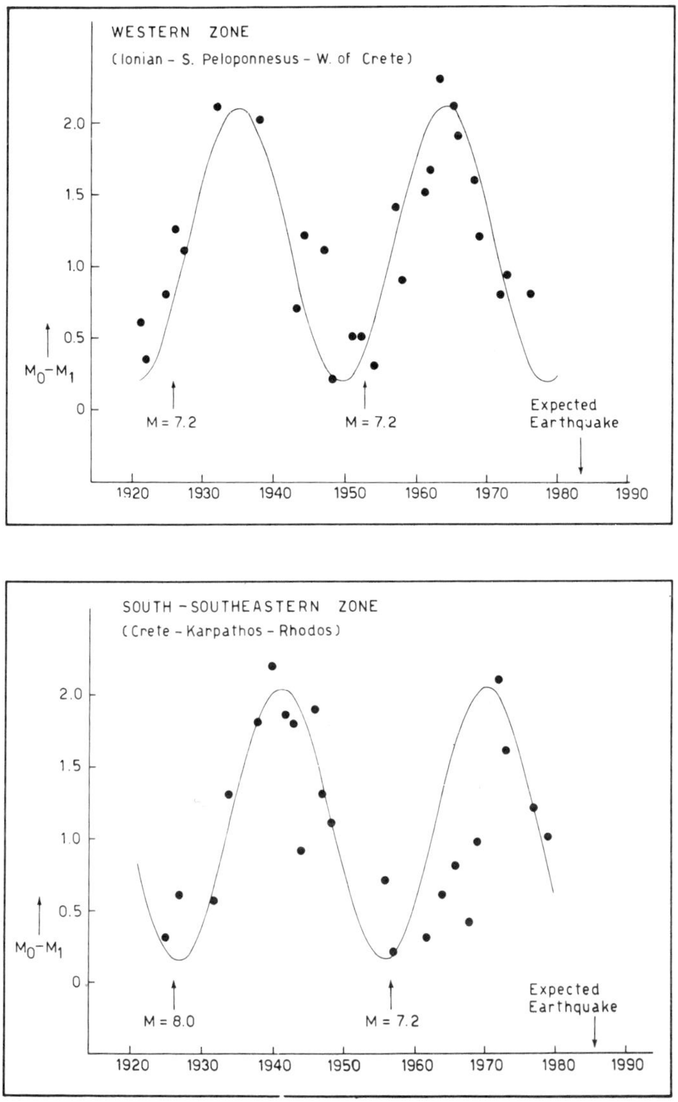

Fig. 8. Time variation of the difference in magnitude between the main shocks and the largest aftershocks in the western Hellenic arc (upper part) and in the south-southeastern Hellenic arc and times of the past and of the expected (predicted) strong shocks (Papazachos and Comninakis 1982). The occurrence of the Ionian islands earthquake of January 17, 1983 (M=7.0) proved that the prediction was correct for the western part of the arc.

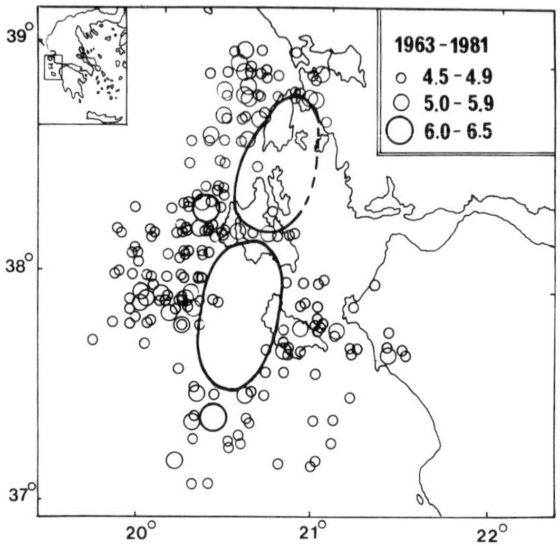

Fig. 9. The two seismic gaps in the Ionian islands (Papadimitriou and Papazachos 1985). The southern of these gaps (south of Cephalonia-west of Zante) was filled by the January 17, 1983 (M = 7.0) earthquake, while an earthquake of similar magnitude is expected in the northern gap (Leukas-north Cephalonia).

arthquakes with magnitude $M \geqslant 7.8$ occurred between 1810 and 1926 and no such earthquakes occurred in this region during the last 60 years. These observations are considered as strong evidence that great intermediate focal depth earthquakes are expected in this region during the next few decades (Comninakis and Papazachos 1980, Papazachos et al 1986b). This evidence and the information that great intermediate focal depth earthquakes, which have not till now affected modern high structures in Greece, generate relatively long period and very high amplitude seismic waves, lead to the conclusion that such earthquakes constitute a major seismic threat for the buildings and mainly for the high ones (hotels, etc) in the whole southern Greece during the next few decades (Papazachos et al 1985).

4.2. Migration Patterns in the Aegean Area

Work on the seismicity migration patterns in the Aegean and surrounding area has been performed by several authors. One of the most interesting cases of seismicity migration which is of importance for long term earthquake prediction is a nort-south-north migration of the epicenters of the large ($M \geqslant 7.0$) shallow earthtuqkes during the present century (Papadimitriou et al 1985). It has been shown that during the three successive time periods 1903-1932, 1933-1957 and 1958-1985, these earthquakes occurred in the northern, mainly in the

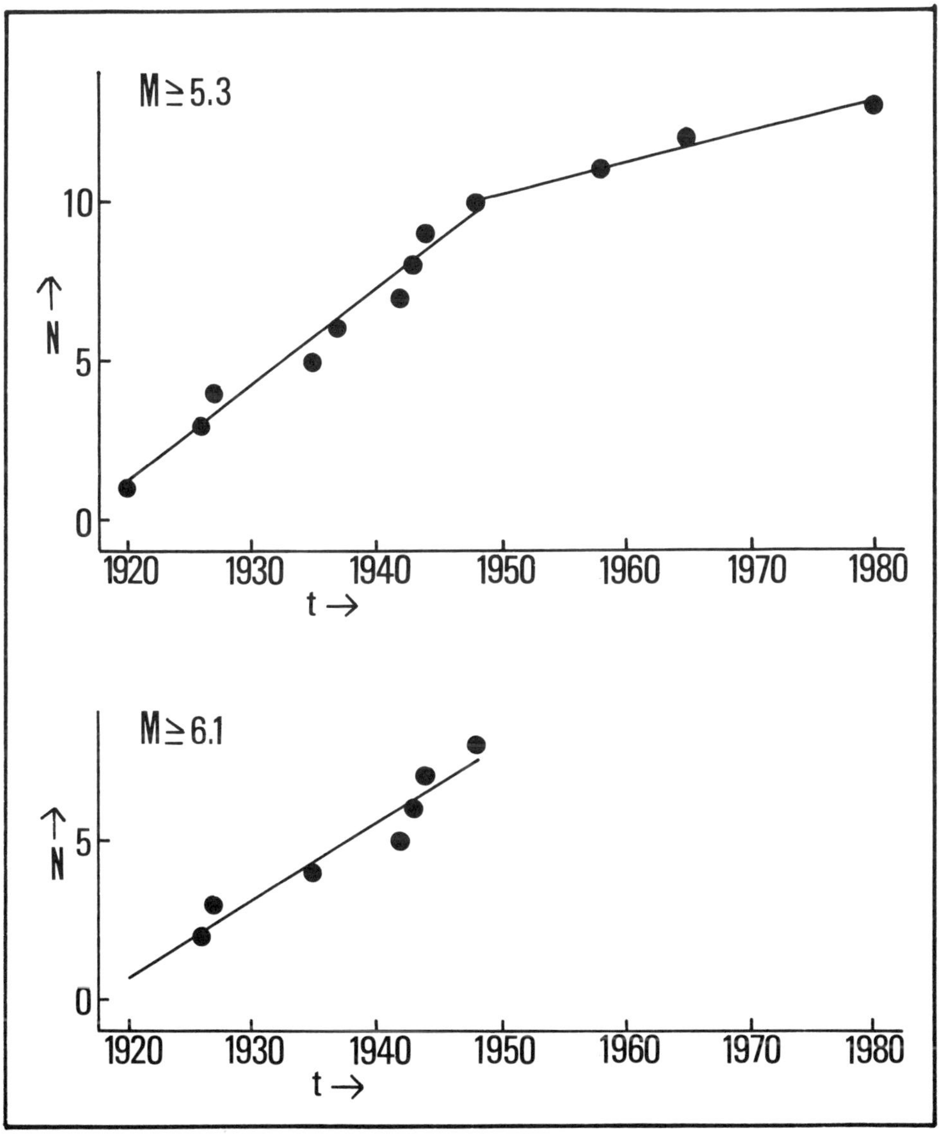

Fig. 10. Cumulative yearly number of intermediate focal depth earthquakes in the southern Aegean as a function of time (Papazachos et al 1985).

southern, and in the northern part of the Aegean area, respectively. This observation led to the conclusion that the seismic activity of the major shallow earthquakes will migrate to the southern Aegean during the next few years and that the earthquake of January 1983 (M = 7.0) in the Ionian islands is the first shock of this activity.

Figure (11) shows the epicenters of all shallow and intermediate focal depth earthquakes of M ≥ 6.5 which occurred in the Aegean and surrounding area during the period 1958-1985. It is observed that no shal-

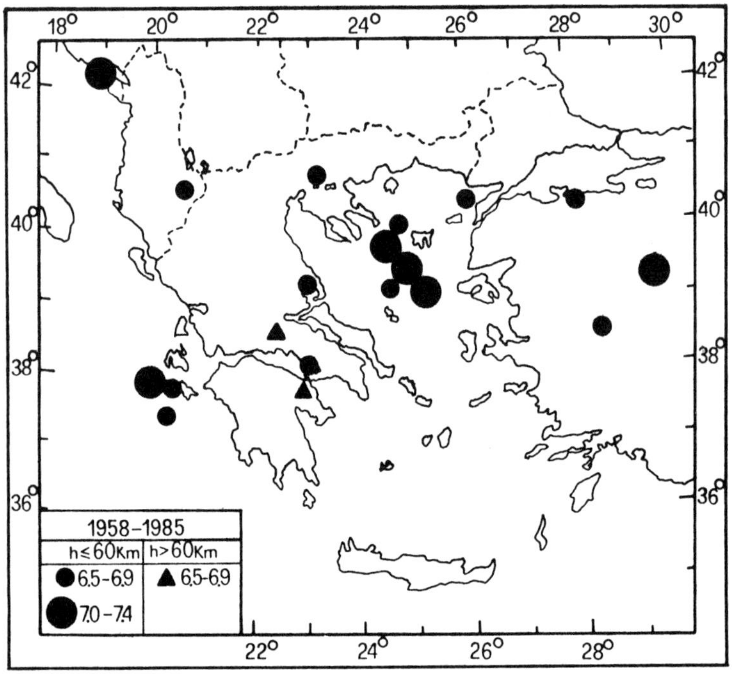

Fig. 11. Epicenters of shallow and intermediate focal depth strong (M≥ 6.5) earthquakes which occurred in the Aegean and surrounding area during the last 28 years (1958-1985). No such strong earthquake occurred during this time period in the southern part of this area.

low or intermediate focal depth earthquake with magnitude M ≥ 6.5 occurred in the southern part of this area during the last 28 years. For this reason, a period of high seismic activity is expected to break out soon during the next few years along the Hellenic arc with shallow earthquakes in the convex part of the arc (Ionian-south of Crete- south of Dodecanese) and with intermediate focal depth earthquakes in the concave part of the arc (east Peloponnesus-north of Crete-north of Rhodos) in agreement with independent evidence presented in the previous paragraph).

4.3. Space Time Distribution of the Shocks in Seismic Sequences

Work on the time, space and magnitude distribution of the shocks of seismic sequences in Greece has been performed by several authors. This work led to several conclusions which are of interest for the earthquake prediction.

Very recently, Karakaisis and his colleagues (1985) investigated in detail the space-time distribution of the shocks of all seismic sequences which are associated with main shocks of $M \geq 6.5$ and which occurred since 1978 in the Aegean and surrounding area. It was found that in three out of five cases the foreshocks were located very close to the epicenter of the main shock or were migrated to the epicenter of the main shock. The focus of the main shock can be located anywhere in the fault (near the middle or near the ends). In all cases the aftershock activity expanded to one or to both ends of the fault immediately after the occurrence of the main shock and reached these ends within a time of less than one day. In all cases, the largest aftershock occurred near the one end of the fault within a time period which varried from sequence to sequence between 15 minutes and 38 days after the occurrence of the main shocks. This observation shows that a continuous accurate determination of aftershock foci can lead to the prediction of the epicenters of the largest aftershocks which are usually large and produce damage comparable to that of the main shock.

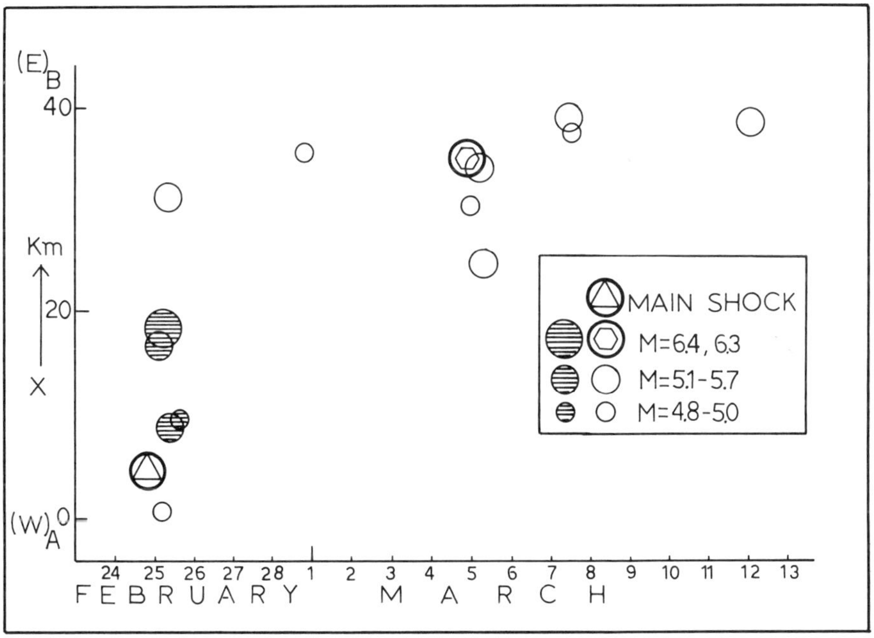

Fig.12. Plot of the distances, x, of the epicenters of the Alkionides gulf seismic sequence along the strike of the fault as a function of the time of occurrence of the corresponding shocks (Karakaisis et al 1985).

An example is given in figure (12) which shows a plot of the distances, x, of the epicenters for the Alkionides 1981 seismic sequence from the point A, located at the westernmost part of the fault , along a line parallel to the strike of the fault (east-west), versus the occurrence time of the aftershocks. This figure shows that after the occurrence of the main shock (M = 6.7) at the western part of the fault on February 24, the aftershock activity expanded in less than 24 hours to the eastern part of the fault where the second large aftershock (M = 6.3) occurred and produced extensive damage. The dashed circles in this figure are referred to earthquakes related to an antithetic fault which produced the largest aftershock (M = 6.4) of February 25.

5. FOCAL PROPERTIES OF EARTHQUAKES

The first attempts to determine focal properties of earthquakes in the Aegean and surrounding area started about thirty years ago and concern fault plane solutions of some strong earthquakes which occurred in the period 1953-1957. This work, which is based on the concept of the point source, was continued since then and recently has been supplemented by some attempts to get information on the dimensions of the seismic sources and on the process along the faults during the earthquake generation. However, most of the knowledge on the focal properties of the earthquakes in this area is based on fault plane solutions and for this reason some of the most interesting results coming from such reliable solutions are presented here.

Fault plane solutions for almost all major earthquakes ($M \geq 6.0$)which occurred in the Aegean and surrounding area during the last three decades have been determined (Papazachos and Delibasis 1969, Ritsema 1974 , McKenzie 1972, 1978, Papazachos et al 1983, 1984a,c, Scordilis et al 1985). However, only the fault plane solutions which are based on first onsets of long period instruments and concern mainly recent (after 1962) large ($M \geq 6.0$) earthquakes can be considered as reliable.

Papazachos and his colleagues (1984b, 1986a) examined all the available fault plane solutions of the shallow earthquakes and used the most reliable of these solutions to describe the basic properties of the stress field and seismic faulting in the Aegean area. Figure (13) shows the most reliable of these solutions with the known bubble symbols (Papazachos et al 1986a). The date of the corresponding earthquake is shown close to each symbol. Each bubble is separated by the projections of the fault plane and of the auxiliary plane in quadrants, two black and two white . The black quadrants are those with compressional first motions while the white ones are those with dilatational first motions. Thus, those symbols with black in the center of the bubble indicate thrust fault produced by horizontal compression (P axis horizontal) while those with white in the center indicate normal faulting produced by horizontal tension (T axis horizontal). Cases when all four quadrants meet at a point close to the centre of the bubble indicate strike-slip faulting.

The fault plane solutions presented in figure (13) show that thrust faulting is observed along the convex side of the Hellenic arc (Zante - S of Crete - S of Rhodos) except for the northwesternmost part of the arc

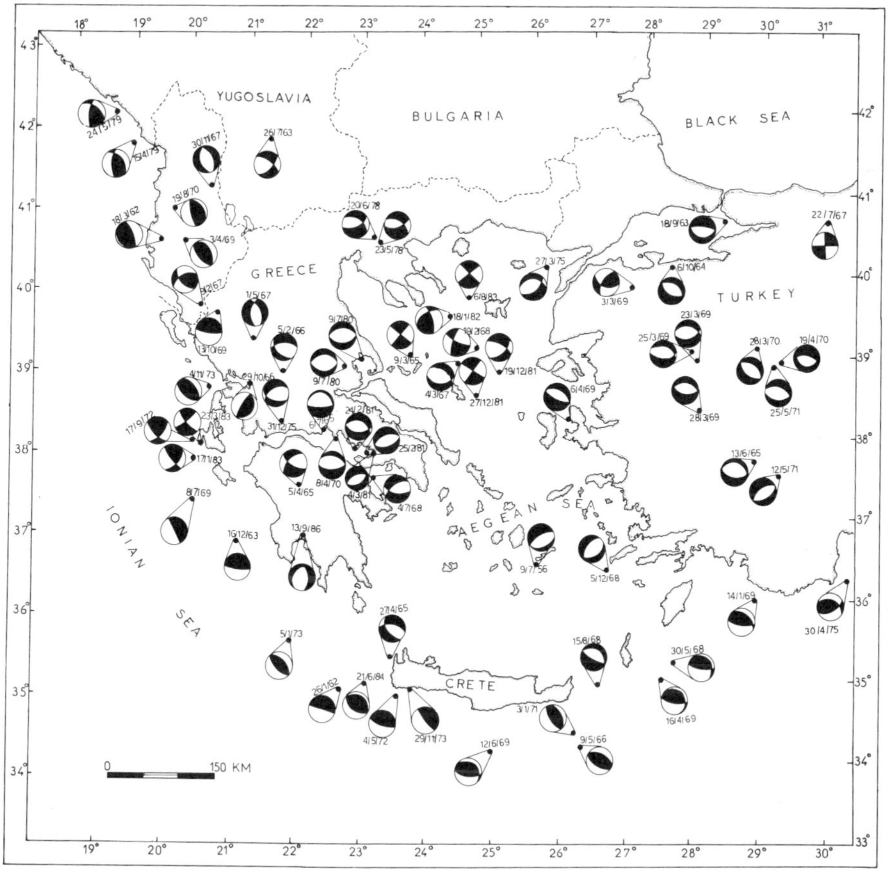

Fig. 13. The most reliable fault plane solutions in the Aegean and surrounding area (Papazachos et al 1986a).

where strike-slip faulting with thrust component is observed. The average trend of the slip vector is N 34°E and its average plunge is 19°. This is in full agreement with subduction of the eastern Mediterranean lithosphere (front part of the African lithospheric plate) under the Aegean lithosphere (front part of the Eurasian lithospheric plate). The strike-slip faulting in the northwesternmost part of the arc (Cephalonia) has been considered (Scordilis et al 1985) as dextral tranform fault, in agreement with the known relative motion of the Aegean and eastern Mediterranean.

The thrust faulting continues to the north in the land along the western coast of the central mainland of Greece and of Albania and Yugoslavia. In this case, however, there is no evidence for subduction (no

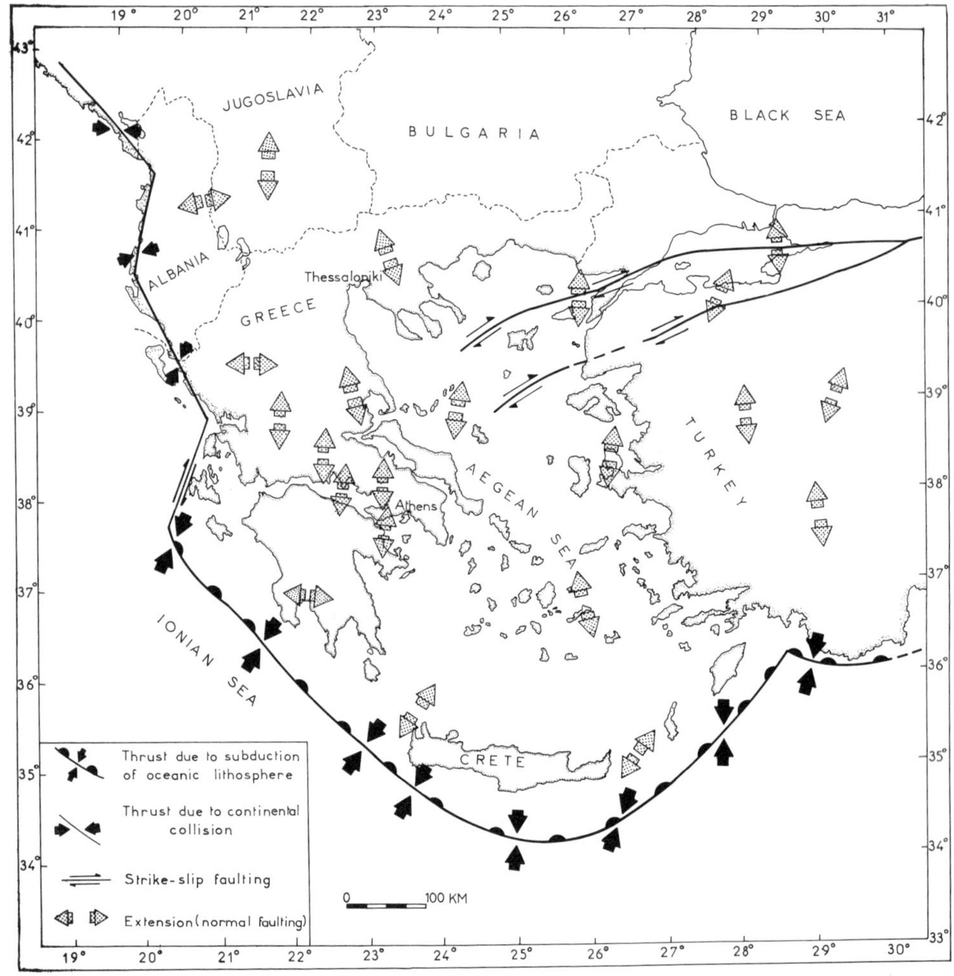

Fig. 14. The main features of active tectonics in the Aegean and surrounding area (Papazachos et al 1986a).

Benioff zone, etc) and collision between two continental lithospheres (Eurasian-Apulian) occurs.

Normal faulting is observed in the whole inner part of the Aegean area, from Crete in the south to central Bulgaria in the north and from eastern Albania and central Greece in the west to all western Turkey in the east. However, in the northwestern part of Turkey and in the northernmost part of the Aegean Sea strike-slip dextral faulting with thrust or normal component is also observed.

The tensional field (axis T) in the Aegean has an almost N-S trend but close to the boundary with the outer thrust zone seems to change direction and the axis T tends to become parallel to the P axis of the compressional field of the thrust zone.

The complication of the stress field along the line northwestern Anatolia-northern Aegean trough, where, in addition to normal faulting, strike-slip faulting with thrust or normal component is also observed , can be explained by accepting the opinion that the western part of the northern Anatolian strike-slip dextral fault is branched and extends up to the northern Aegean trough.

There are only few earthquakes of intermediate focal depth in this region for which reliable fault plane solutions exist and all these shocks have their foci in the western part of the southern Aegean Benioff zone. All these solutions show thrust faulting. Papazachos (1977) used the fault plane solutions of intermediate focal depth earthquakes, which concern three strong shocks which occurred close to the upper boundary of the western part of the Benioff zone, to show that these shocks occurred by thrusting parallel to the Benioff zone.

Figure (14) summarizes in a schematic way the properties of the active tectonics of the Aegean and surrounding area (Papazachos et al 1986a).

6. GEODYNAMIC MODELS

Several attempts have been made to interprete the seismic and other geophysical properties of the Aegean and surrounding area by models compatible with ideas of new global tectonics. These models can be divided in three general types. Those which assume external forces, that is, forces due to horizontal movements of adjacent lithospheric plates (westward movement of the Turkish plate, etc), those which assume internal forces in the Aegean (gravitational sliding) and those which assume hat the source of these forces is in the asthenosphere beneath the Aegean lithosphere and exert on the bottom of this lithosphere horizontal forces (convection currents, etc).

Mckenzie (1970,1972) suggested that the relative motion between several lithospheric plates is the main cause of the high seismic activity in the Mediterranean area. Except for the big lithospheric plates of Eurasia, Africa and Arabia, he considers four other relatively small plates. For the three of them he gave the names " Aegean plate ", " Turkish plate" and " Black Sea Plate " , while for the fourth , which is formed by the northern and central Greece and by the northern Aegean Sea, he gave no name but it is that which was called " Macedonian plate " by

Dewey and Sengör (1979). McKenzie (1972) believed that the fast motion of the Aegean and Turkish plates relative to Africa and Europe and the subduction along the Hellenic arc can account for most of the seismic activity in the area.

Although the existence of the Turkish plate and its westward movement is supported by the available seismic and other data, this does not hold for the Aegean and the Macedonia plates because these " plates" are crossed by seismic zones of high seismicity (fig. 3). On the other hand, new reliable fault plane solutions do not support the strike-slip faulting in central Greece expected by McKenzie´ s model.

Dewey and Sengör (1979) suggested that the westward movement of the Anatolian (Turkish) plate from the Bitlis suture zone in North Arabia to consume oceanic lithosphere in the Eastern Mediterranean, can explain most of the seismic and other properties of the area. The result of this westward motion was the generation of the northern Anatolia strike -slip dextral transform fault. In northwestern Anatolia, the fault is branched in two strands which bend southwesterly south of the western end of the Marmara Sea and reach northern Aegean. The northern strand enters the Marmara Sea and continues in the Saros and Sporades Troughs where it terminates while the southern strand passes just south of Bursa and terminates in the Skiros trough. The bending of these strands result to a locking orientation in the northern Aegean that obstracts the western motion of the Anatolian plate relative to Greece and Balkans. This locking leads to compression in northern Aegean (Papazachos 1976) with a combined result of strike-slip and thrust faulting and ripping of the Macedonia plate from the Black Sea plate and the creation of their extensional boundary . Furthermore, Dewey and Sengör (1979) believe that the western Anatolian graben system, situated on a very broad dome, may owe its origin to the locking geometry of the north Anatolian transform strands. The east-west compression relieved by north-south extension on a smaller scale into Northern Anatolia is consistent with such an interpretation.

It is interesting to note that new reliable fault plane solutions in the northern Aegean during the period 1981-1983 (Papazachos et al 1984c, Rocca et al 1985) supports the idea of branching of the northern Aegean and Skiros troughs. However, there are several problems with the model proposed by these authors. The north-south extension is not limited in the western Anatolia and in the boundary between Macedonia and Black Sea plate but it is widespreaded all over the Aegean and the surrounding area (fig. 13) . On the other hand, there is no Macedonian plate because Macedonia is crossed in a NW-SE direction by the Servomacedonia seismic zone (fig. 3) which has been one of the most active zones during the present century in the whole Balkan region.

Rotstein (1985) assumes that Anatolia and the Aegean are parts of one plate complex which cendergoes counter-clockwise rotation. At present, however, free and undisturbed rotation is possible only for the Anatolian block (excluding western Anatolia) where the motion is accommodated by subduction along the Cyprean arc. Due to a local obstruction caused by a " side arc collision " of the Hellenic arc with the Turkish plate near Rhodos, a separate Aegean block is created which is not separated from the Turkish plate and for this reason this block cannot move freely.

Under this condition the greatest extension is expected along the suture zone between the two blocks and especially in the northeastern part of the Aegean block where the motion relative to Anatolia must be greatest.

There are some problems with Rotstein´s model too. It assumes that the boundary between the Turkish and Arabian plate is of thrust type in contrast to McKenzie´ s (1972) model which assumes strike-slip faulting and this problem must to solved by new seismic and other data. On the other hand, according to this model a variation of the direction of the slip vector in accordance with the counter-clockwise rotation of the Aegean block is expected, but the slip vectors in the whole Aegean and surrounding area have an almost constant north-south trend (Papazachos et al 1984b).

Makris (1976, 1978) to interprete geophysical data assumed that at the center of the Aegean a lithological system of low compressional velocity and density ascending from the asthenosphere into the lithosphere caused an upward movement of the crust of several kilometers, spreading by gravitational sliding and gradual expansion from the inner zones outwards. He believes that the ascending of this lithological system probably initiated by subduction of oceanic crust but at present the interaction is of continental-continental collision. However, the well developed Benioff zone in the southern Aegean (fig. 5) and a large amount of other seismological data shows that subduction of oceanic crust is still continued.

Berckhemer (1976) favors also the theory of speading by gravitational sliding and he believes that this is a result of the elevation difference between the Aegean and Mediterranean. Mercier (1981) , however, believes that gravitational sliding is present but stresses at the convergence boundary is the dominant factor for the geodynamic process in this region.

The present author and his colleagues (Papazachos and Comninakis 1976, Papazachos 1977, Papazachos and Comninakis 1978) based on seismic and other geophysical data (Papazachos and Comninakis 1971) proposed a geodynamic model to interprete these data (fig. 15). This model consists of a lithospheric slab, called the Mediterranean slab and being part of the African lithospheric plate, a back-arc brocken lithosphere, which is the lithosphere of the broader Aegean area, and a low Q region between these lithospheres where convection takes place in a manner suggested by Sleep and Toksöz (1971). The convective cells exert horizontal forces at the bottom of the back-arc southern Aegean lilthosphere and force it to expand. Under the action of these forces the Aegean lithosphere breaks and hot material intrudes into it. This results in volcanic activity, high heat flow, magnetic anomalies, modification of crustal structure, subsidence of crustal blocks and generation of shallow earthquakes in the Aegean lithosphere by tensional mechanism. These convective cells also drag the southern Aegean lithosphere to the Hellenic arc. The extensional properties at relatively large distances from the descending slab to the north may be attributed to secondary convecion centers such as those expected by the theory suggested by Toksöz (1975). The probable remnants of an old lithospheric slab which are still assimilated in the mantle under the northern Aegean may also con-

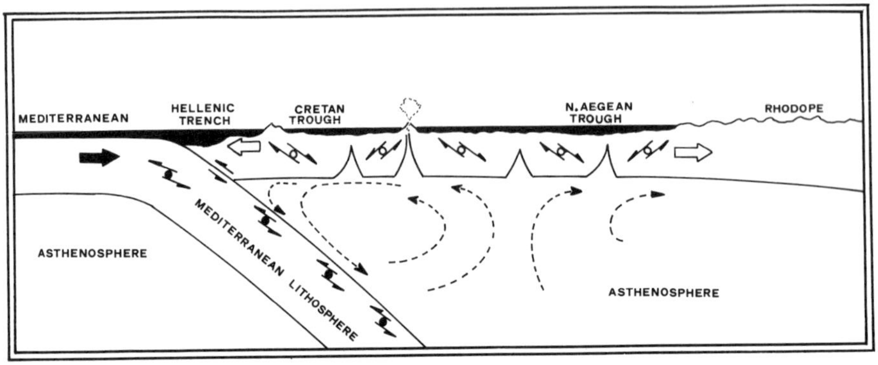

Fig. 15. Geodynamic model for the Aegean Area (Papazachos and Comninakis 1978).

tribute to this field and the northern Aegean trough can be considered as the remnant of an old back-arc marginal Sea (Papazachos and Papadopoulos 1977).

McKenzie (1978) based on fault plane solutions, Landsat photographs, information on the crustal structure of the region and on some theoretical considerations, concludes that the Aegean has been stretched by a factor of two since the Miocene and the sinking slab produced by subduction along the Hellenic arc may maintain the motion. McKenzie does not deny the gravitational spreading of the Aegean but he attributes its extension mainly to forces acting below the lithosphere although he accepts that the geometry and widespread nature of normal faulting is not easily explained.

Le Pichon and Angelier (1979) used seismic and other geophysical and geological information to estimate the relative motion occurring between the Hellenic arc and the adjacent sea floor, to evaluate the deformation in the Aegean area and to reconstruct the pattern of motion over the Eastern Mediterranean region for the last 13 m.y. They suggested that subduction along the Hellenic arc occurs by rotation of the sinking slab by 30° around a 40°N, 18°E pole during the last 13 m.y. They do not accept neither the idea that the strain pattern in the Aegean is due to forces acting at the bottom of the lithosphere nor the idea that the generalized extension within Aegean is due to the extrusion of Turkey to the west, but they agree that gravitational speading is responsible for the extensional field in the Aegean.

Although the model presented by Le Pichon and Angelier (1979) explains fairly well several geophysical and geological properties of the area, there are still some important problems regarding the interpretation of earthquake data. The model is based on the hypothesis that the slip vector in the eastern part of the convex side of the Hellenic arc turns more to the north than in the western part but the available evidence are very week. No earthquake of magnitude larger than 6.0 occurred in this eastern region (east of the 25° parallel) after the installation of the long period standardized stations (1962) and for this reason the available fault plane solutions are not very accurate. On the other hand, they assume that the length of the subducted slab is much shorter in the western part than in the eastern part but this does not agree with the distribution of the intermediate focal depth foci which shows that the Benioff zone is clearly symmetric (fig. 5).

From the above discussion we can conclude that although there is still disagreement between scientists about the source of the forces which maintain the present pattern of active tectonics in the Aegean and surrounding area, much progress has been made towards the solution of this problem.

The compressional field along the convex side of the Hellenic arc and its extension to the north (western Greece-western Albania) and the westward motion of Anatolia indicates that external horizontal forces must be exerted on the Aegean area by the adjacent lithospheres (Mediterranean, Apoulian, Turkish). Evidence for the existence of internal forces (gravitational sliding) are also strong (Berkhemer 1976, Le Pichon and Angelier 1979). The fact, however, that the seismic and other geophysical properties of the Hellenic arc and the Aegean Sea are very similar with such properties of well known island arcs and marginal seas, for which the source of the driving forces are considered to be under the lithosphere (Karig 1971, Sleep and Toksöz 1971) give additional support to the idea that such forces are the main factor which determines these properties in the Aegean and surrounding area too.

RERERENCES

Agarwal, N.K., Jacoby,W.R. and Berckhemer,H. "Teleseismic P-Wave Travel Time Residuals and Deep Structure of the Aegean Region" , Tectonophysics" , **31** , 33-57, 1976.

Allan,T.D. and Morelli,C. " A Geophysical Study of the Mediterranea Sea" Bull.Geof.Teor. Appl. , **13** , 99-142, 1971.

Allen,C.R., Amand,P.S., Richter,C.F. and Nordquist,J.M. " Relation between Seismicity and Geological Structure in the Southern California Region " , Bull. Seismol. Soc. Am. , **55**, 752-797, 1965.

Berckhemer,H. "Some Aspects of the Evolution of Marginal Seas Deduced from Observations in the Aegean Region",Proceedings of the XXV Congress and Plenary Assembly of ICSE, Split, 22-30 October ,303-314, 1976.

Comninakis,P.E. and Papazachos,B.C."Seismicity of the Eastern Mediterranean and Some Tectonic Features of the Mediterranean Ridge " , Bull. Geol. Soc. Am. , **82**, 1093-1101 , 1972.

Comninakis,P.E. and Papazachos,B.C. " A Catalogue of Earthquakes in the Mediterranean and the Surrounding Area for the Period 1901-1977 ", Publ. Geoph. Lab. Univ. Thessaloniki , **5**, 1-96, 1978.

Comninakis,P.E. and Papazachos,B.C. " Space and Time Distribution of the Intermediate Focal Depth Earthquakes in the Hellenic Arc " , Tectonophysics , **70**, 35-47, 1980.

Comninakis,P.E. and Papazachos,B.C. " A Catalogue of Earthquakes in Greece and Surrounding Area for the Period 1901-1985" , Publ.Geoph. Lab. Univ. Thessaloniki, **1** , 1-172, 1986.

Dewey,J.F. and Sengör,C. " Aegean and Surrounding Region: Complex Multiplate and Continuum Tectonics in a Convergent Zone " , Geol.Soc. Am. Bull. , **90**, 84-92 , 1979.

Erickson,A.J. " The Measurement and Interpretation of Heat Flow in the Mediterranean and Black Sea " , Ph. D. Thesis, Mass. Inst.Tech. , 1970.

Finetti,I. " Mediterranean Ridge: A Young Submerged Chain Associated with the Hellenic Arc " , Boll. Geof. Teor. ed Applic., **19**,31-65 , 1976.

Gregersen,S. " P-Wave Travel Time Residuals Caused by a Dipping Plate in the Aegean Arc in Greece " , Tectonophysics , **37**, 83-93, 1977.

Hatzidimitriou,P.M., Papadimitriou,E.E., Mountrakis,D.M. and Papazachos, B.C. " The Seismic Parameter b of the Frequency-Magnitude Relation and its Association with the Geological Zones in the Area of Greece " , Tectonophysics , **120**, 141-151, 1985.

Jongsma, D. " Heat Flow in the Aegean Area " , Geophys. Jour. Roy.Astr. Soc. , **37**, 337-346, 1974.

Karakaisis,G.F., Karacostas,B.G., Papadimitriou,E.E., Scordilis,E.M.and Papazachos, B.C. " Seismic Sequences in Greece Interpreted in Terms of the Barrier Model " , Nature , **315**, 212-214, 1985.

Karing,D.E. " Origin and Development of Marginal Basins in the Western Pasific " , Jour.Geophys.Res., **76**, 2542-2561, 1971.

Le Pichon,X. and Angelier,J. " The Hellenic Arc and Trench System:a Key to the Neotectonic Evolution of the Eastern Mediterranean Region", Tectonophysics, **60**, 1-42, 1979.

Makris,J. " Some Geophysical Aspects of the Evolution of the Hellenides", Bull. Geol. Soc. Greece, **10**, 206-213 , 1973.

Makris,J. " A Dynamic Model of the Hellenic Arc Deduced from Geophysical Data " , Tectonophysics , **36**, 339-346 , 1976.

Makris,J. " Some Geophysical Considerations on the Geodynamic Situation in Greece " , Tectonophysics , **46**, 251-268, 1978.

McKenzie,D.P. " The Plate Tectonics of the Mediterranean Region ", Nature , **226**, 239-243, 1970.

McKenzie,D.P. " Active Tectonics in the Mediterranean Region" , Geophys. Jour. Roy. Astr.Soc. , **30**, 109-185, 1972.

McKenzie,D.P. " Active Tectonics of the Alpine-Himalayan Belt: The Aegean Sea and Surrounding Region ", Geophys.Jour. Roy. Astr. Soc. , **55**, 217-254, 1978.

Mercier,J.L. " Extensional-Compressional Tectonics Associated with the Aegean Arc: Comparison with the Andern Cordillera of South Peru - North Bolivia" , Philos. Trans. R. Soc. London , **300**, 337- 355 , 1981.

Morelli,C., Pisani,M. and Gantar,C. " Geophysical Studies of the Aegean Sea and in the Mediterranean ", Boll. Geof. Teor. ed Applic. ,**18**, 127-167, 1975.

Needham,H.D., Le Pichon,C., Melguen.M., Pautot,G., Reanard,V., Avedic , F. and Carre,D. " North Aegean Sea Trough: 1972 Jean Charcot Crusie " , Bull. Geol. Soc. Greece , **10**, 152-153, 1973.

Nowroozi,A.A. " Seismotectonics of the Persian Plateau, Eastern Turkey, Caucasus and Hindu Kush Region " , Bull. Seismol. Soc. Am. , **61** , 317-341, 1971.

Papadimitriou,E.E. and Papazachos,B.C. " Evidence for Precursory Seismicity Patterns in the Ionian Islands (Greece) ", Earthq. Predic. Res. , **3** , 95-103, 1985.

Papadimitriou,E.E., Karacostas,B.G., Karakaisis,G.F. and Papazachos , B.C. " Space-Time Patterns of Seismicity in the Aegean and Surrounding Area " , Proc. 3rd Intern.Symp.Analys.Seismicity Seismic Risk, Liblice, Czechoslovakia, June 1985, 1-5, 1985.

Panagiotopoulos,D.G., Scordilis,E.M., Hatzidimitriou,P.M., Rocca, A.C. and Papazachos,B.C. " Further Evidence on the Deep Tectonis of the Aegean and Eastern Mediterranean Area " , 19th Conference of the Europ. Seismol. Commission, Moscow, October 1984 , 1-14, 1984.

Panagiotopoulos,D.G. and Papazachos,B.C. " Travel Times of Pn Waves in the Aegean and Surrounding Area " , Geophys. J.R. Astr.Soc. , **80**, 165-176, 1985.

Panagiotopoulos,D.G., Hatzidimitriou,P.M., Karakaisis,G.F., Papadimitriou, E.E. and Papazachos,B.C. " Travel Time Residuals in Southeastern Europe " , Pure and Appl. Geophys., **75**, 47-55, 1969.

Papazachos,B.C. " Phase Velocities of Rayleigh Waves in the Southeastern Europe and Eastern Mediterranean Sea ", Pure and Appl. Geophys.,**75**, 47-55, 1969.

Papazachos,B.C. " Distribution of Seismic Foci in the Mediterranean and Surrounding Area and its Tectonic Implication ", Geophys.Jour.Roy. Astr.Soc., **33**, 419-428, 1973.

Papazachos,B.C. " Seismotectonics of the Northern Aegean Area " ,Tectonophysics , **33**, 199-209, 1976.

Papazachos,B.C. " A Lithospheric Model to Interprete Focal Properties of Intermediate and Shallow Shocks in Central Greece ", Pure and Appl. Geophys. , **115,** 655-666, 1977.

Papazachos,B.C. " Seismicity Rates and Long-Term Earthquake Prediction in the Aegean Area ", Quat. Geod., **3,** 171-190, 1980.

Papazachos,B.C. and Comninakis,P.E. " Geophysical Features of the Greek Island Arc and Eastern Mediterranean Ridge " , C.R.Séances de la Conférence Réunie á Madrid , **16** , 74-75, 1969.

Papazachos,B.C. and Delibasis,N.D. " Tectonic Stress Field and Seismic Faulting in the Area of Greece ", Tectonophysics ,**7,** 231-255,1969.

Papazachos,B.C. and Comninakis,P.E. " Geophysical and Tectonic Features of the Aegean Arc " , Jour. Geophys. Res. , **76,** 8517 - 8533 , 1971.

Papazachos,B.C. and Comninakis,P.E. " Modes of Lithospheric Interaction in the Aegean Area " , Proceedings of the XXV Congress and Plenary Assembly of ICSE, Split, 22-30 October, 319-332 , 1976.

Papazachos,B.C. and Papadopoulos,G.A. " Deep Tectonics and Associated Ore Deposits in the Aegean Area " , Proc. VI Coll. Geol. Aegean Region, Athens 1977 , **3,** 1071-1080, 1977.

Papazachos,B.C. and Comninakis,P.E. " Geotectonic Significance of the Deep Seismic Zones in the Aegean Area " , Sec. Intern. Scient.Conf., Thera and the Aegean World, Santorini, Aug. 1978,121-129, 1978.

Papazachos,B.C. and Comninakis,P.E. " Long-Term Earthquake Prediction in the Hellenic Trench-Arc System ", Tectonophysics , **86,** 3- 16 , 1982.

Papazachos,B.C., Panagiotopoulos,D.G., Tsapanos,T.M., Mountrakis,D.M., Dimopoulos,G.C. " A study of the 1980 Summer seismic Sequence in the Magnesia Region of Central Greece " , Geophys.Jour. Roy. Astr. Soc. , **75,** 155-168, 1983.

Papazachos,B.C., Comninakis,P.E., Papadimitriou,E.E. and Scordilis,E.M. " Properties fo the February-March 1981 Seismic Sequence in the Alkionides Gulf of Central Greece " , Annal. Geophysic. , **2,** 537 - 544, 1984a.

Papazachos,B.C., Kiratzi,A.A., Hatzidimitriou,P.M. and Rocca,A.C." Seismic Faults in the Aegean Area " , Tectonophysics ,106, 71 - 85 , 1984b.

Papazachos,B.C., Kiratzi,A.A., Voidomatis,Ph., and Papaioannou,C. " A Study of the December 1981-January 1982 Seismic Activity in Northern Aegean Sea " , Boll. Geof. Teor. Appl., **26,** 101-113, 1984c.

Papazachos,B.C., Papadimitriou,E.E., Karacostas, B.G. and Karakaisis , G.F. " Long Term Prediction of Great Intermediate Depth Earathquakes in Greece " , 12th Seminar of Earthquake Engineering , European Association on Earthquake Engineering, Halkidiki , September 1985 , 1-12, 1985.

Papazachos,B.C., Kiratzi,A.A., Hatzidimitriou,P.M. and Karacostas,B.G . " Seismotectonic Properties of the Aegean Area that Restrict Valid Geodynamic Models " , 2nd Wegener Conference, Dionysos, Greece 14-16 May 1986 , 1-16, 1986a.

Papazachos.B.C., Papadimitriou,E.E., Kiratzi,A.A., Papaioannou,Ch.A. , and Karakaisis,G.F. " Probabilities of Occurrence of Large Earthqukes in the Aegean and Surrounding Area during the Period 1986 -

2006 " , Publ.Geoph.Lab. Univ. Thessaloniki, **4**, 1-22, 1986b.

Payo,G. " Crustal Structure of the Mediterranean Sea, Part II, Phase Velocity and Travel Times " , Bull. Seismol.Soc. Am. , **59**, 23-42 , 1969.

Rabinowitz, P.D. and Ryan,W.B.F. " Gravity Anomalies and Crustal Shortening in the Eastern Mediterranean " , Tectonophysics , **5-6**, 585 - 608, 1970.

Ritsema,A.R. " The Earthquake Mechanism of the Balkan Region ", Royal Netherl. Meteor. Inst., Scient. Rep. Nr. 74-4, 1-36, 1974.

Rocca,A.C., Karakaisis,G.F., Karacostas,B.G., Kiratzi,A.A., Scordilis , E.M. and Papazachos,B.C. " Further Evidence on the Strike-slip Faulting of the Northern Aegean Trough Based on Properties of the August-November 1983 Seismic Sequence " , Boll. Geof.Teor. ed Appl., **106**, 101-109, 1985.

Rotstein,Y. " Tectonics of the Aegean Block: Rotation, Side Arc Collision and Crustal Extension " , Tectonophysics , **117**, 117-137, 1985.

Scordilis,E.M., Karakaisis,G.F., Karacostas,B.C., Panagiotopoulos,D.G. and Papazachos,B.C. " Evidence for Transform Faulting in the Ionian Sea: The Cephalonia Island Earthquake Sequence of 1983 " , Pageoph , **123**, 387-397, 1985.

Sleep, N. and Taksöz,M.N. " Evolution of Marginal Basins " , Nature , **233** , 548-550, 1971.

Tassos,S.T., and Papazachos,B.C. " High and Low Velocity Zones, Attenuation and Intermediate Depth Seismicity in the South Aegean " , Proc. 3rd Intern. Symb.Analys.Seismicity Seism.Risk,Liblice, Choslovakia, June 17-22, 1985, 1-13, 1985.

Toksöz,M.N. " The Subduction of the Lithosphere " , American Scientist, November 1975, 89-98, 1975.

Vogt,R., and Higgs,R.H. " An Aeromagnetic Survey of the Eastern Mediterranean Sea and Its Interpretation " , Earth Planet.Sci.Letters, **5**, 439-448, 1969.

Wyss,M., and Baer,M. " Seismic Quiescence in the Western Hellenic Arc May Foreshadow Large Earthquakes " , Nature , **289**, 785-787, 1981.

PALEOSEISMIC STUDY ON THE EL ASNAM (ALGERIA) THRUST FAULT

M. Meghraoui*
Laboratoire de Géologie Historique, Université de Paris sud, Orsay, France, and C.R.A.A.G., Bouzareah, Algeria.

*Present address: Laboratoire de Tectonique, Institut de Physique du Globe, 4 place Jussieu, 75252 Paris.

ABSTRACT. Geological analysis of eight trench exposures across the El Asnam thrust fault (reactivated during the 10 October 1980 earthquake with M=7.3) shows various tectono-sedimentary structures related to past seismic activity. The trenches cut across: 1) The main 1980 fault traces, particularly at the southeastern flank of Sara El Maarouf anticline; 2) A secondary normal fault (extrados fault) observed at the top of the anticline. Moreover, earthquake-induced flooding seems to be constant in the area and the last one occurred just after the 1980 event. A stratigraphic log in the flood area displays 6 flood horizons representing the past seismic records with a magnitude M≥7. Buried fault scarps with cumulative vertical displacements and buried secondary normal faults are present in the exposures. C14 dating allowed us to calculate an average uplift rate of 0.6 mm/y during the late Holocene time and to determine an event distribution/time. The Holocene seismic activity is then characterized by sequences of earthquakes alternating with periods of quiescence. During seismic clusters the recurrence interval of large earthquakes (M≥7) will vary between 300 and 500 years.

1. INTRODUCTION

Among the numerous geologic methods to investigate an active fault zone, trenching techniques provide reliable data, particularly on the past seismic activity. A seismogenic fault is able to produce surface ruptures with lateral and/or vertical offsets large enough to be recognized at field. When repeated, large seismic events with surface ruptures are registered within the young quaternary deposits and the faulting process produces cumulative displacements. Hence, the paleoseismic history of a fault can be determined from the study of young deposits. Trench exposures through the El Asnam thrust fault display cumulative offsets with buried fault scarps, buried subsidiary normal faults and secondary effects (sand-blows, flood deposits) that indicate an intense past seismic activity.

J. Bonnin et al. (eds.), Seismic Hazard in Mediterranean Regions, 333–346.

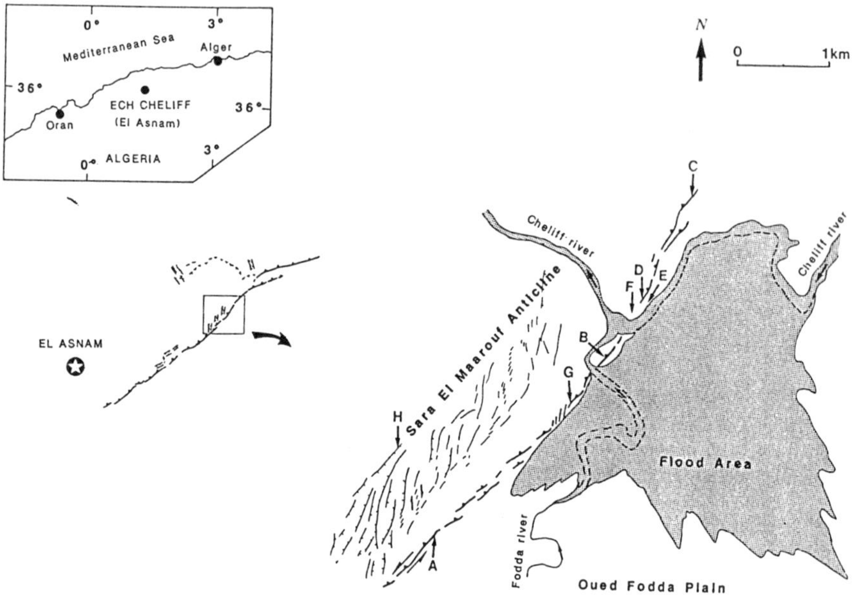

Figure 1. Situation of the study area. The city of El Asnam is now named Ech Cheliff. Surface ruptures were mapped soon after the 10 October 1980 earthquake (M=7.3)(Philip & Meghraoui, 1983). Flood area in March 1983 extends over 25 km2, it becomes completely dry during the summer time. Trench survey took place during March 1983 and June-July 1985 and trench locations are indicated with letters A to H (Meghraoui et al, 1987)

The El Asnam active zone has been the site of numerous studies after the 10.10.1980 earthquake with M=7.3 (Ouyed et al, 1981; King & Vita-Finzi, 1981; Yielding et al, 1981; Cisternas et al, 1982; Ruegg et al, 1982; Philip & Meghraoui, 1983; Meghraoui et al, 1986). Seismologic as well as tectonic results conform to the existence of a thrust fault, striking N 060, with a dip of 50° to the northwest and with 36 km long (fig 1). Detailed analysis of surface ruptures shows vertical motion (average amount=2m), and left lateral offset (1,21m), while remarkable subsidiary normal faults have been observed on the upthrown block (Philip & Meghraoui, 1983). The central segment of the thrust fault is related to Sara El Maarouf active fold (fig 1) where different quaternary terraces have been uplifted. Indeed, tilted and uplifted young quaternary deposits in this zone indicate a recent tectonic activity in close relationship with previous earthquakes (King & Vita-Finzi, 1981). The trench exposures that cross the 1980 surface ruptures (fig 1) display deformations that affect Holocene deposits (last 10 000 years). The past seismic events have been recorded in these deposits, it is then possible to characterize them by looking for ruptures associated to prehistorical or historical earthquakes.

The aim of the present work is three-fold: (1) to present the trenching survey and trench locations which are important to get the maximum reliable geological informations; (2) to study past seismic reconstruction from some examples of buried fault scarps, buried subsidiary normal faults and secondary effects, together with the correlation between trenches; (3) to propose slip rate values and recurrence intervals of large earthquakes in the area of El Asnam.

2. TRENCHING SURVEY

The trenching procedure is a heavy and costly operation that requires a careful planning in order to find the most appropriate location.

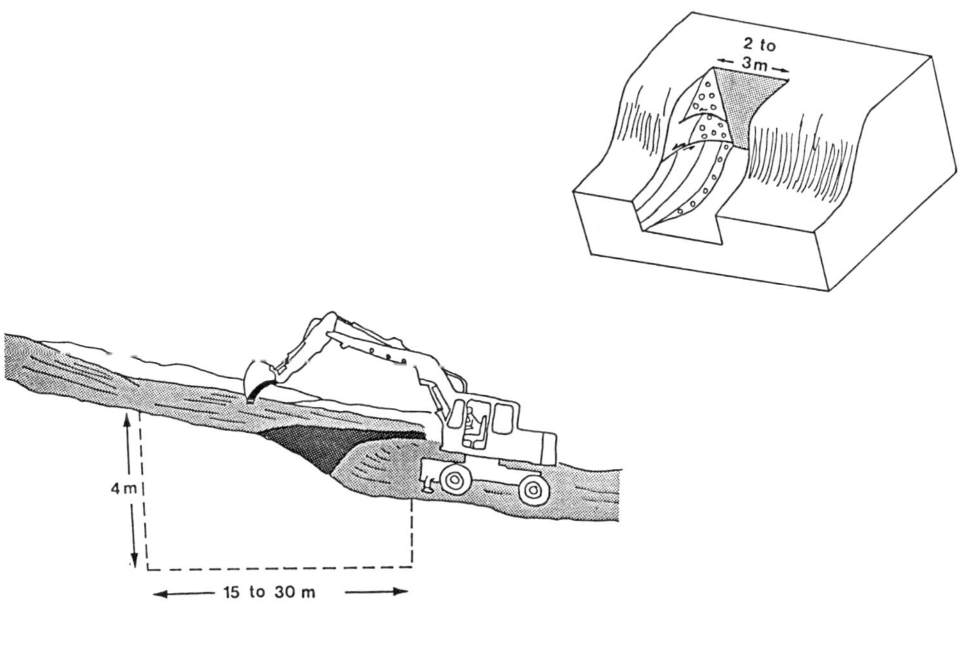

Figure 2. Schematic illustration showing sizes of trench-exposures and the material (digger) used for trenching operation.

Choice of trench locations must follows the Quaternary geology techniques mainly based on aerial photograph studies coupled with local geological investigations which are made up of the following points: (1) young fault scarps definition and study of recent tectonic movements (stream offsets, terrace deformation); (2) Pleistocene and Holocene deposits (slope deposits, colluvial wedges, alluvial fan, caliche), and soils development analysis; (3) faulting process (existence of subsidiary faults, fault plane geometry) and its secondary effects (bending moment fault, landslides and liquefaction). How rich would be the tectono-sedimentary informations in trenches is the necessary question before trenching. Length, width and depth of the trench depend on the fault scarp offset, topographic conditions, geological environments (nature of the rocks and sediments) and first order observations into the trench.

Figure 2 shows a field appreciation of an excavation, its exposures and the equipment commonly used for trenching (digger, backhoe), as indicated the trench sizes are generaly about 2 to 3m wide, 3 to 4m deep and 15 to 30m long. Cleaning of trench walls is necessary, it needs brushes, long knifes (particularly for soft sediments), shovel etc... . In order to locate every detail point, a grid (1m square mesh size) is set up to cover as much surface of the exposure as possible. Dateable materials carefully sampled are composed of bulk soil with charcoal fragments, peat, shells, organic matter and sometimes pottery pieces. Sizes of the samples depend on the used C14 dating technique.

2. OTHER EXAMPLES

Trenching techniques have been extensively used particularly on the San Andreas fault zone, western United States, (Sieh; 1978, 1984) and across the median tectonic line faults, Japan, (Okada; 1981, 1983). Sieh (1978) pointed out the main geologic indicators of past events which are: (1) the sedimentary deposits and their relationships to the main fault; (2) the subsidiary structures (small faults) that exist within the young deposits; (3) sand-blows and the liquefaction effects relative to an earthquake. Pallet Creek segment of the San Andreas fault is one of the remarkable places for a paleoseismic study. Indeed, the vertical and the right lateral slip components on the fault, that affect the young deposits that fill an ancient river gorge (Pallett Creek), are registered within the sediments. Sieh (1984) defined 12 seismic events during the last 2000 years, the recurrence interval of large seismic events being around 145 ±8 years. Trench investigations across the Atotsugawa fault, Japan, (Okada, 1983), Ventura fault, California, (Yeats, 1982), and El Asnam fault, Algeria, (Meghraoui et al; 1985, 1987) set the problem of the near-surface geometry of a thrust fault and indicate its complex paleoseismicity.

3. PAST SEISMIC RECONSTRUCTION

Buried fault scarps with cumulative vertical movements are often present in trenches that cut the 1980 surface ruptures at El Asnam. Slope and colluvial wedge deposits, paleosoils development, thickening of the downthrown compartment near the fault (photo 1) together with observed fault offsets are the main arguments to reconstruct the past seismic activity. Only part of the data and their interpretations will

Photo 1. View of a trench exposure (trench A, fig 1) where the arrows show the trace of the thrust fault. Scale (under the left arrow) indicates 10 cm. One can notice thickening of colluvial wedge sediments in the downthrown block (gravels and sands on left side of the fault trace) and faulted white calcareous gravels in the upthrown block (Meghraoui et al, 1985 and 1987).

be presented in this paper, more details on this trench study can be found in the following references : Meghraoui et al (1985, 1987).

3.1. Paleoseismic Indentification

Figure 3 shows exposure C (see figure 1 for location) that cut across alluvial fan deposits that lie on southeastern flank of the Sara El Maarouf anticline. Thickening of the downthrown block is due to erosion underwent by the uplifted block. Cumulative movements on the fault are the result of 4 stages (fig 4): (1) figure 4-1 indicates a pre-seismic situation where the active fault is buried and does not affect units b and c; (2) the next stage (fig 4-2) is post-seismic, the fault scarp has undergone an erosional process. The thickness of unit d is representative of the vertical amount of the fault deformation; (3) figure 4-3 is the second post-seismic situation; the same scenario is repeated and units e and f correspond to the fault scarp degradation that occurs soon after a large seismic event; (4) the vertical offset corresponding to the 1980 seismic event was about 0.25 m and since that period the new scarp has been partly eroded (fig 4-4).

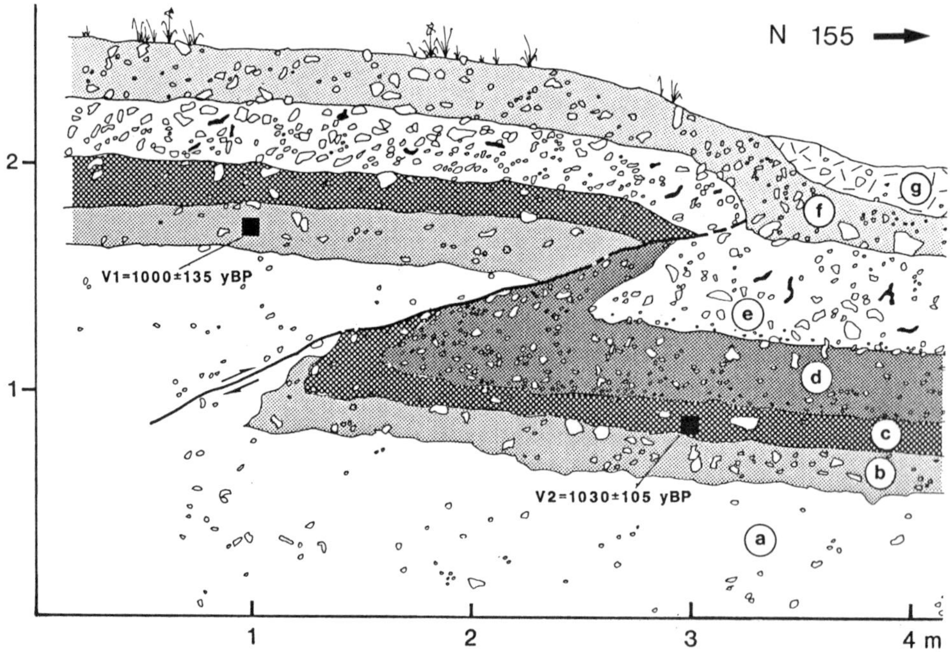

Figure 3. Exposure C showing a thrust fault affecting units (mostly gravels) a, b, c, d, e and f. The vertical offset at that place, during the 1980 event was about 0.25 m. Repeated coseismic displacements indicate a cumulative dip slip offsets of 2.35 m since 1030 ±105 years BP (i.e. Before Present),(Meghraoui et al, 1985).

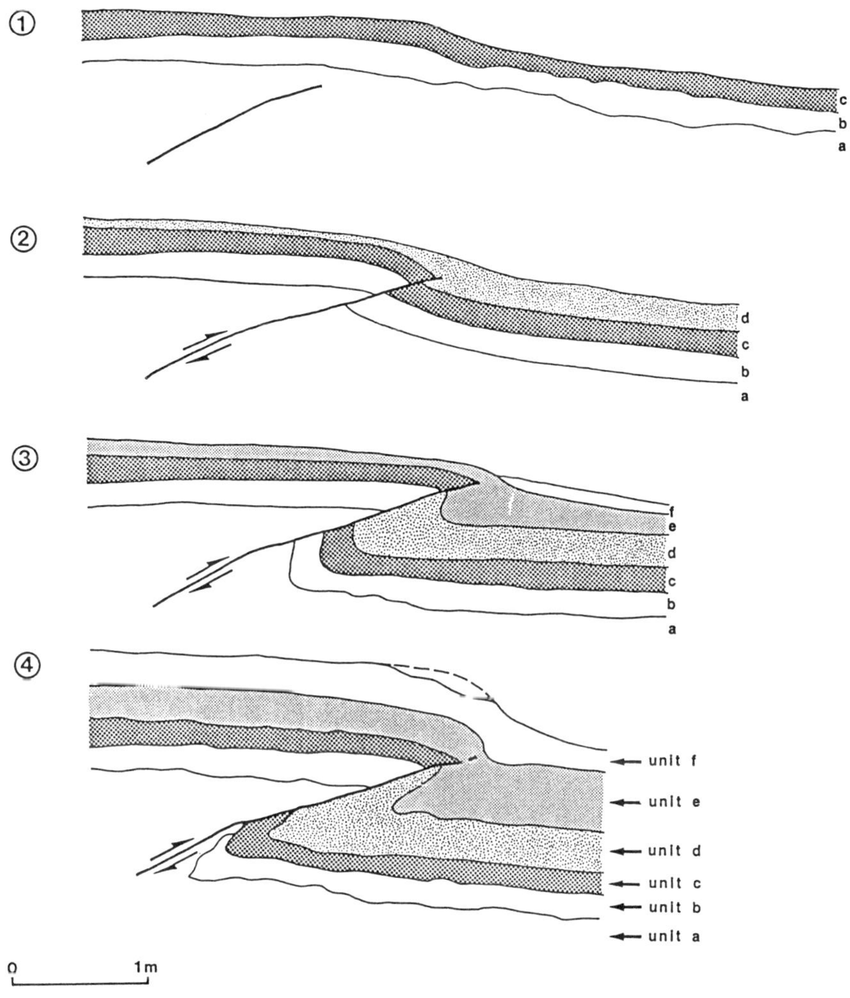

Figure 4. Reconstruction inferred from exposure C (fig 3), it shows 4 stages of the tectono-sedimentary processes. Thickening of the downthrown block is due to fault scarp erosion. After each coseismic movement, thickness of colluvial wedge deposits is proportional to the vertical offset. Three coseismic displacements took place after unit c, i.e. after 1030 ±105 y BP (Meghraoui et al, 1985).

The proposed reconstruction of figure 4 illustrates the tectono-sedimentary processes that explain the final state of figure 3. Three seismic events occurred then since the deposition of unit c, the last one being the 1980 earthquake. Dating of unit c corresponds to 1030 ±103 y BP, consequently the three seismic event took place during the last 1000 y.

3.2. Numerical Values

The total dip slip on the fault is about 2.35 m (fig 3), it is the offset between unit c in the upthrown and downthrown blocks. The rupture process is accompanied with a syncline fold that can be seen near the fault plane in the downthrown block (fig 3). Variations of unit thicknesses may exist, we consider in that case its mean value. The total dip slip component TO of a given fault can be divided in a shortening component of movement Sh and a vertical component of movement Vm (fig 5). Displacement measurements taken from figure 3

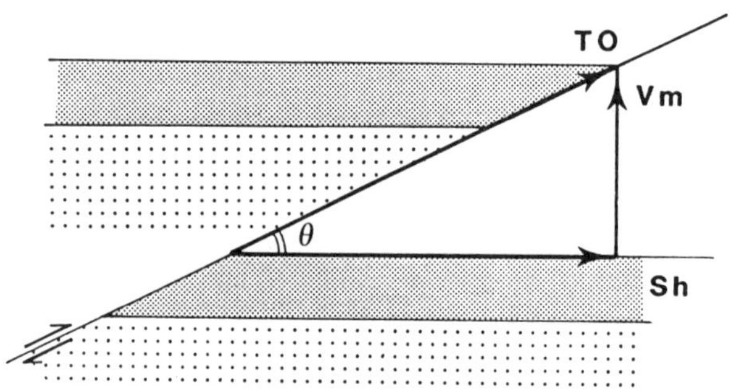

Figure 5. Example of a thrust fault movement. TO is the dip slip component, whereas Sh and Vm are the shortening and Vertical components on the fault respectively .

give Sh=2.20 m and Vm=0.82 m, they represent cumulative amount of movements along the thrust fault. One can notice that value of the cumulative vertical movements (Vm=0.82 m) is about three times the vertical motion of the 1980 seismic event (0.25 m) measured at that place (Philip & Meghraoui, 1983). Consequently, the cumulative vertical movement (including the 1980 displacement) near trench C might be the result of 3 seismic events. This observation is in good agreement with the proposed reconstruction described above (fig 4), and the calculated uplift rate is about 0.8 mm/y.

Correlations with other trenches are important in order to verify whether these values have a local signification or can be applied to an entire fault segment. The trench E (figures 1 and 6) was dug to

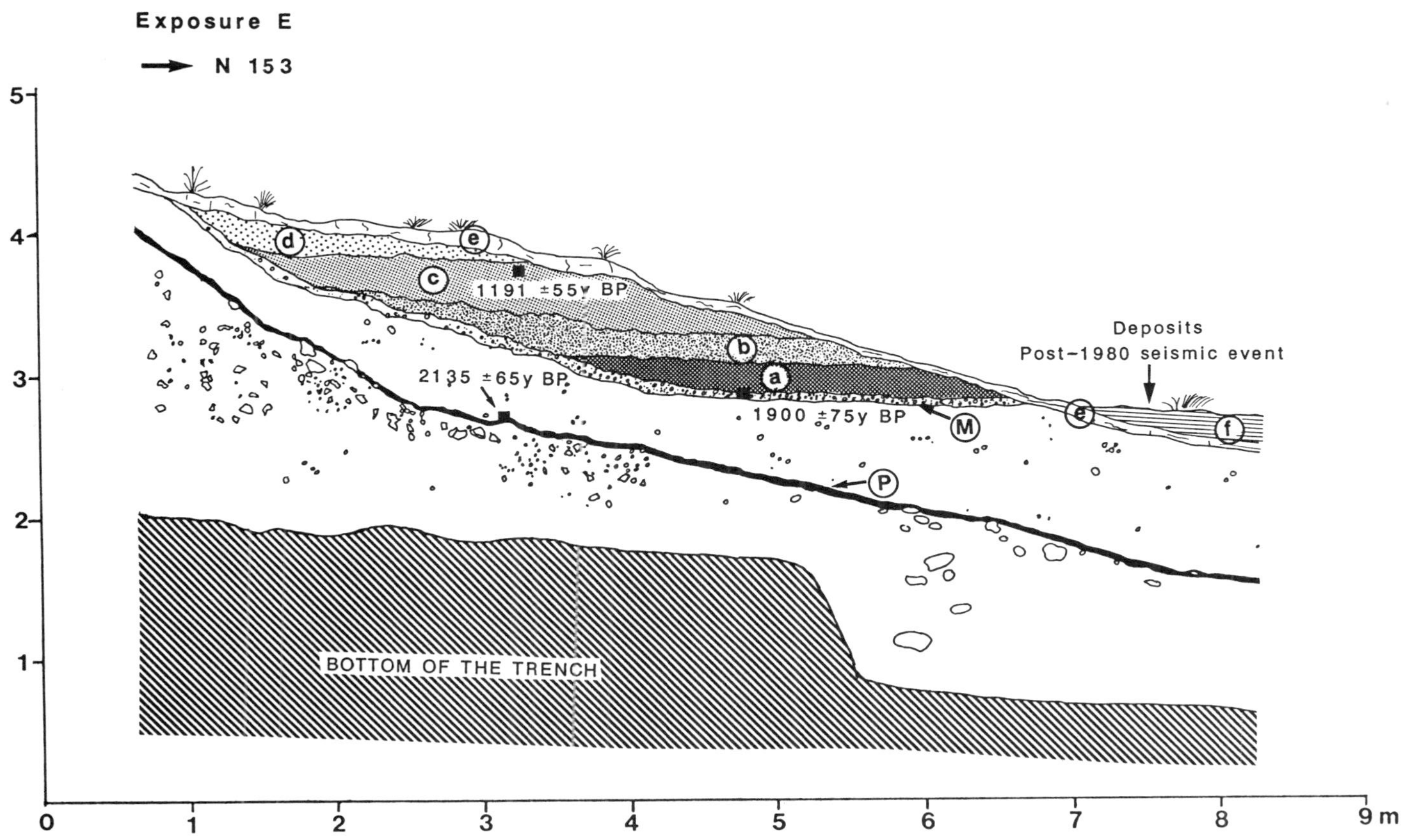

Figure 6. Exposure E. Units a, b, c and d which are comparable to the post-1980 unit f, have been uplifted by thrust fault movements. The fault rupture tends to flatten and does not reach the surface at that place (Meghraoui et al, 1987).

study uplifted young Cheliff river terraces. These river deposits are situated at the foot of the hillslope that forms the southeastern flank of the Sara El Maarouf anticline. Units a, b, c and d of trench E (fig 6) represent terrace deposits identical to the post-1980 deposit f, they are composed with sandy silt layers. Units c and d are the most recent deposits, they have an age equal to, or younger than 1191 ±55 y BP (fig 6). These observations lead us to associate uplifted river deposits to previous seismic events. Since their deposition and compared to level unit f, units c and d underwent 0.85 m of vertical motion during the last 1191 years. Uplift rate values of the Late Holocene time calculated from trench C (0.8 mm/y) and trench E (0.7 mm/y), are consistent. It is important to notice that measured offsets on the fault represent only part of the deformation and ruptures on other branches are not considered; calculated uplift rates are therefore lower bound values.

3.3. Flood Deposits

One of the remarkable secondary effects that occurred just after the 1980 seismic event is the flooding of the Oued Fodda plain (fig 1). Uplift motion of the northwestern block of the El Asnam fault (Ruegg et al, 1982; Philip & Meghraoui, 1983) and the intersection of the fault scarp with the Cheliff river stopped the Cheliff and Fodda river flows inducing the flood (fig 1). A thick sedimentation (0.40m of silty sand deposits) took place after the 1980 event. Furthermore, we assume that the earthquake-induced flood, when repeated, will form a sedimentary succession which reflects the large seismic activity (M≥7).

Detailed stratigraphic analysis of trench B reveals the existence of 8 flood deposits labelled a, b, c, d, e, f, g and h in figure 7, and alternating with paleosoils (Meghraoui et al, 1985). This sedimentary succession is affected by minor normal faults and sand-blows; one can also notice bending of layers which indicates the surface impact of the thrust fault. It is often difficult to obtain reliable tectonic data from soft sediments, particularly when a superficial water level is present. Nevertheless, buried grabens and normal faults under unit c represent evidences of a past seismic activity. The present and ancient flood deposits are remarkably similar and among the eight flood layers, units a and c have been correlated to tectonic structures observed in other trenches (Meghraoui et al, 1985; Meghraoui et al 1987). Units d and e belongs to the same flood event f. Because correlations with tectonic data are not established yet, large seismic events associated with flood deposits b, f, g and h are considered as probable events. Unit c characterizes a seismic event dated between 3810 ±90 y BP and 4170 ±70 y BP, whereas unit a correspond to a seismic event defined between 1945 ±65 y BP and 2070 ±80 y BP.

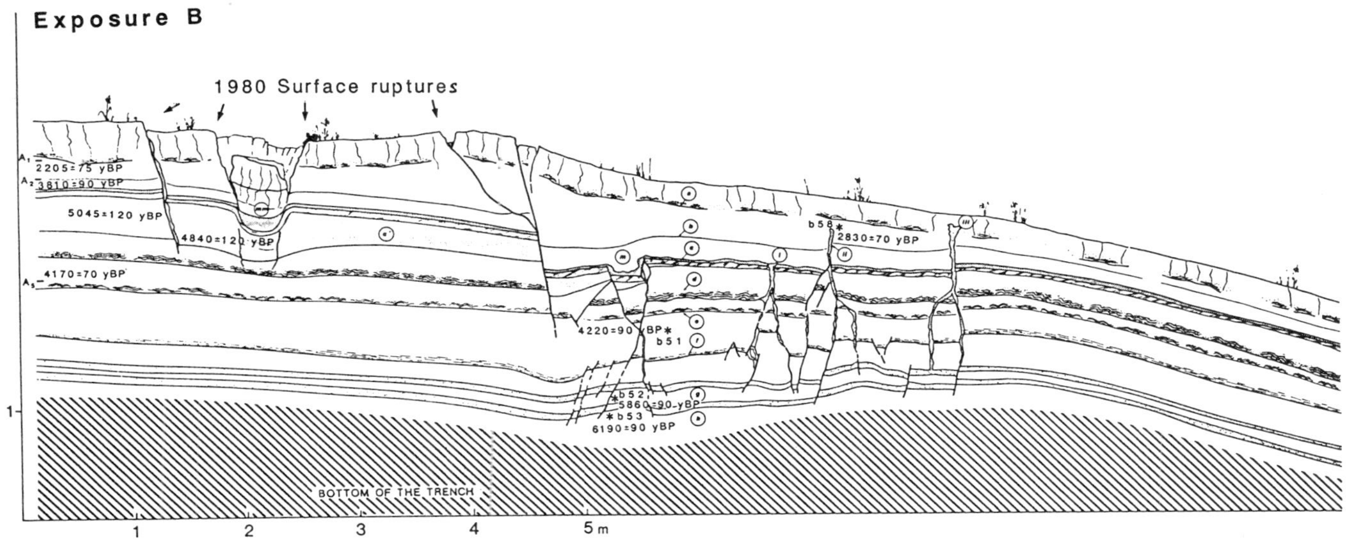

Figure 7. North to the flood area (fig 1), exposure B crosses the 1980 fault scarp. This exposure displays six earthquake-induced flood deposits labelled a, b, c, f, g and h. Subsidiary normal faults and grabens are covered by unit c, whereas bending of layers reflects the thrust fault impact at the surface (Meghraoui et al, 1985).

4. SLIP RATES AND RECURRENCE INTERVALS

Geological analysis of 8 trenches across the El Asnam thrust fault reveals the existence of past large seismic events comparable to the one of the 10.10.1980 (M=7.3). Tectono-sedimentary structures observed in exposures show faulting events associated with previous large earthquakes. Fault-throw measurements and dating allow us to estimate a slip rate and a recurrence interval of large earthquakes. Two periods of the seismic activity during the Late Holocene time can be defined from the diagram of figure 8. It shows the alternance of quiet periods (around 1500 y BP and 3500 y BP) with clustering of large earthquakes (M≥7).

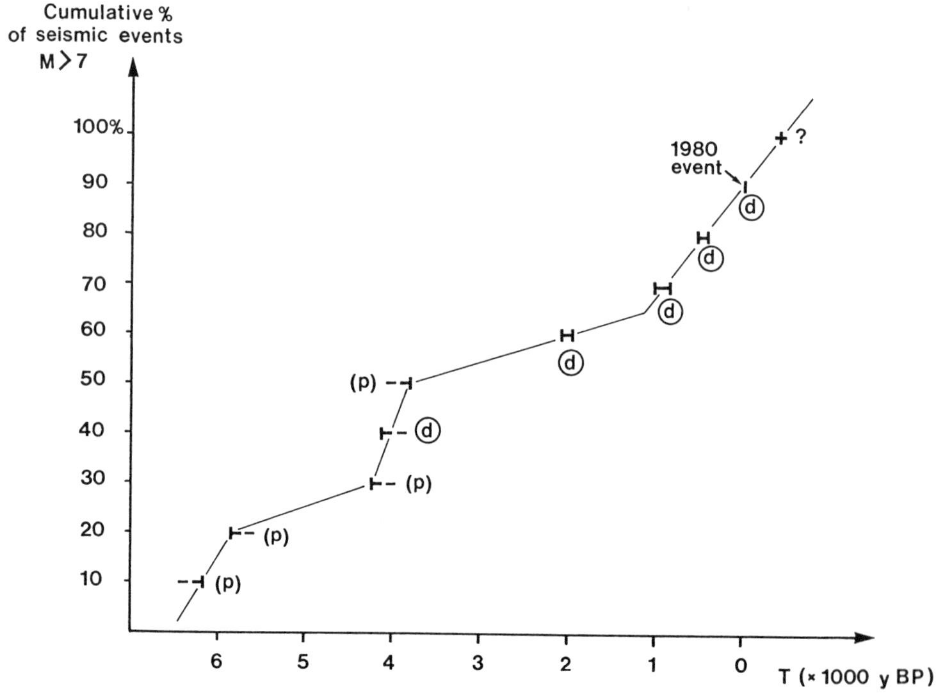

Figure 8. Paleoseismic events/time distribution (Meghraoui et al, 1987) pointing out the alternance between seismic clusters (around 4000 y BP and during the last 1000 years) and periods of quiescence. p is probable and d is well defined.

We define nine seismic events that occurred during the last 6500 years, and among them 5 are well defined whereas 4 are probable. Average recurrence time of large earthquakes is about 720 years, during clusters of seismic events this average time will be however reduced to 300 to 500 years.

Slip rates calculated from observed displacements (dip slip) will vary between 0.6 mm/y, for the Holocene time, and 2.35 mm/y during seismic clusters. Uplift rates also vary between 0.4 mm/y and 0.8 mm/y with an average value of 0.6 mm/y; taking into account the topographic offset of the Sara El Maarouf active fold (about 150 m), the El Asnam thrust fault activity and its related active fold is taking place since a period between 0.183 and 0.326 Million years (Meghraoui et al, 1985, 1987).

ACKOWLEDGMENTS. The author would like to thank the local authorities of Ech Cheliff (formerly El Asnam) for trenching permission and the C.R.A.A.G. (Bouzareah, Alger) colleagues, particularly Dr H. Benhallou, for field supports. A. Cisternas and H. Philip helped with advices and discussions. F. Albarede and R. Jaegy provided the C14 datations. This research program was financed by the C.R.A.A.G. (Algeria) and I.N.S.U. (France).

5. REFERENCES

Cisternas, A., J. Dorel, and R. Gaulon, 'Models of the complex source of the El Asnam earthquake', Bull. Seismol. Soc. Am., 72 (6), 2245-2266, 1982.

King, G. C. P., and C. Vita-Finzi, 'Active folding in the Algerian earthquake of 10 October 1980', Nature, 292 (5818), 22-26, 1981.

Meghraoui, M., A. Cisternas, and H. Philip, 'Seismotectonics of the lower Cheliff basin: structural background of the El Asnam (Algeria) earthquake', Tectonics, 6, n°7, 1-17, 1986.

Meghraoui, M., H. Philip, F. Albarede, and A. Cisternas, 'Trench investigations through the trace of the El Asnam thrust fault: evidence for paleoseismicity', Bull. Seismol. Soc. Am., in press, (submitted in 1985).

Meghraoui, M., R. Jaegy, K. Lamalli, and F. Albarede, 'Late Holocene earthquake sequences on the El Asnam thrust fault', submitted 1987.

Okada, A., M. Ando, and T. Tsukuda, 'Trenches, Late Holocene displacement and seismicity of the Shikano fault associated with the 1943 Tottori earthquake', Disaster Prevention Research Inst. Annuals, Kyoto Univ., 24B-1, 105-126, 1981.

Okada, A., 'Trenching excavation at the Atotsugawa fault (Central Japan)', Rep. Coordinating Committee for Earthquake Prediction, 30, 376-381, 1983.

Ouyed, M., M. Meghraoui, A. Cisternas, A. Deschamps, J. Dorel, J. Frechet, R. Gaulon, D. Hatzfeld, and H. Philip, 'Seismotectonics of the El Asnam earthquake', Nature, 292, 26-31, 1981.

Philip, H., and M. Meghraoui, 'Structural analysis and interpretation of the surface deformations of the El Asnam earthquake of October 10, 1980', Tectonics, 2, n°1, 17-49, 1983.

Ruegg, J. C., M. Kasser, A. Tarantola, J. C. Lepine, and B. Chouikrat, 'Deformations associated with the El Asnam earthquake of October 10, 1980: Geodetic determination of vertical and horizontal movements', Bull. Seismol. Soc. Am., 72, 2227-2244, 1982.

Sieh, K., 'Pre-historic large earthquake produced by slip on the San Andreas fault at Pallett Creek, California', Journal of Geophysical Research, 83, n°B8, 3907-3939, 1978a.

- , 'Lateral offsets and revised dates of large pre-historic earthquakes at Pallett Creek, southern California', Journal of Geophysical Research, 89, 7641-7670, 1984.

Yeats, R. S., 'Low shake faults of the Ventura basin, California', in J.D. Cooper (Compiler), guidebook, Neotectonics in southern California., 78th annual meeting of the Cordilleran section of the Geological Society of America, Araheim, California, April 19-21, 3-15, 1982.

Yielding, G., J. A. Jackson, G. C. P. King, H. Sinvhal, C. Vita-Finzi, and R. M. Wood, 'Relations between surface deformation, fault geometry, seismicity and rupture characteristics during the El Asnam (Algeria) earthquake of 10 October 1980, Earth. Planet. Sci. Lett., 56, 287-304, 1981.

ACCURATE EPICENTERS LOCATION WITH A LARGE NETWORK EXAMPLE OF THE 1984/1985 REMIREMONT SEQUENCE

J. L. Plantet, Y. Cansi
LDG
BP 12
91624 Bruyeres le Chatel
France

ABSTRACT. Seismic location with only permanent stations has often a poor precision because station geometry is not well adapted to the epicenters. In this study two methods are presented (P.M.E.L. and Doublets) which improve location accuracy. These two methods applied to the Remiremont crisis allow a precise description of the space-time distribution of the complete sequence of earthquakes.

1. INTRODUCTION

The accuracy of location is essentially depending on:

-the seismic network geometry with respect to epicentral area
-our knowledge of the structure along propagation
-the accuracy of arrival times determination

The network geometry has the most significant influence on the accuracy of epicenter determination because it affects also the two other factors : if epicenter distances are small, seismic wave on-set are well defined and lateral heterogeneities are less important.

So, the most reliable locations are obtained with seismic networks that cover adequately area to control. But these networks need installation and maintenance expenses which are not justified except for high seismicity areas(California, Japan or Arette in France).

For less active regions,one less expensive solution might be to set up,as soon as possible after a large earthquake, a mobile network on the concerned area in order to record aftershocks. But this method does not allow to obtain any information on the beginning of the crisis. Consequently, it appears to be often necessary to use the seismic data recorded by permanent stations whose situation with respect to epicentral area is not always optimum.

The aim of this study is essentially to present two methods using in an appropriate way the data recorded by such stations to control any seismic crisis.

-The Progressive Multiple Event Location(P.M.E.L Pavlis and al. 1983) takes into account the lateral heterogeneities in the propagation structure and leads to an accurate location for major quakes of the

J. Bonnin et al. (eds.), Seismic Hazard in Mediterranean Regions, 347–358.

crisis (absolute situation of the epicentral zone).

-The Doublets method improves the on-set time evaluation and leads to an accurate relative location which might allow a consistent description of the fault geometry.

These two methods have been applied to the Remiremont crisis (France December 1984). The aftershocks locations we obtained are compared to those determined by the mobile network runned by the Institut de Physique du Globe of Strasbourg (Haessler and al, 1985).

2. THE REMIREMONT CRISIS

One medium size earthquake was felt on december, 29 1984, in the Remiremont region (Vosges) (Io=VI near by Eloyes). This quake (Ml=4.8), the most important since the one of 1682 occuring in the same area (Io=VII-IX) ,followed an important and irregular precursory seismic activity starting on december 11 and was followed by many aftershocks (more than three hundred quakes have been recorded by the Haudompre station) up to february 1985.

As soon as december 30,1984 the Institut de Physique du Globe of Strasbourg set up a telemetered seismic network with 8 stations . The well adapted network geometry (epicentral distances less than 20km and one station very close to the epicenters) leads to incertitudes less than 250m on horizontal coordinates and less than 500m on the depth estimate (Haessler and al. 1985).

The faultplane solution of the main shock and aftershocks (Haessler and al. 1985), and the frequency magnitude relationship are in good agreement with those of the Rhinegraben and Black Forest (Bonjer and al. 1984).

In this study, we do not use the data of the temporary network, but the locations they lead to, will be taken as a reference to test the validity and accuracy of our methods using only permanent stations.

3. ABSOLUTE LOCATION OF THE EPICENTRAL AREA

3.1. P.M.E.L. Method

When epicentral distances are large, waves propagate through structures which might disturb the propagation. The arrival time of one wave is a complex function of both epicenter (Ho,L,M,P) and station (li,mi) coordinates:

$$hi = Ho + F(li,mi,L,M,P) \quad (1)$$

The inversion by block's method (Aki and al. 1977,Grasso and al. 1983) gives simultaneously location and blocks structure. But this method needs a dense network.

If only locations determination is carried on for quakes along a fault of small size in comparison with epicentral distances, equation (1) becomes simpler by introduction of:

-A mean propagation model represented by a propagation function depending only on epicentral distance.

-Propagation anomalies si, for each station, representing deviations between the correct structure and the mean model :

$$hi = Ho + F(Di) + si \qquad (2)$$

If the computation of a mean model is rather easy (for example from well localized events) the determination of anomalies is more complex.

The easiest way is an iterative procedure as

-a) Location of a set of quakes by equation (2) (fixing si=0 at the first iteration).

-b) Evaluation of station anomalies by computing the mean of travel time residuals at each station.

Unfortunately, by this method, computed anomalies are biased by the locations : during the location process (a) residuals are constrained to be minimized with a given norm(usually L2 norm from leastsquares). As a result, anomalies depend on the set of stations of each quake.

The P.M.E.L method (Pavlis and al. 1983) allows to become free of that problem by correcting the residues for each quake. This is done by a transformation that takes care of the stations set involved in the location for each shock. Beside that, the process is similar to the previous one with a depending station set correction for residuals between the a) and b) phases.

3.2. Application to the Remiremont crisis

Only 6 major earthquakes of the crisis (Ml>3.9) have been selected to compute the propagation anomalies for their clear seismic signals recorded on 38 european stations , but epicentral distances and azimuths are not the best (fig 1.). Arrival times for different phases (Pn,Pg and Sg) are read with an accuracy of 0.05s. Anomalies obtained by the P.M.E.L method and a rough mean model (fig 2.)do not show azimuthal variation for Pg and Sg waves. Nevertheless, the anomalies for refracted Pn waves are systematically positive for alpine stations (fig 1.) which might be associated with a crust thickening.

By taking into account these anomalies in the location process, the size of the epicentral area is decreased (fig 3.). It defines a north-south fault which dimensions are similar to the ones obtained by the temporary network of IPG of Strasbourg.

The absolute location is greatly improved, but it remains a bias of the order of 1.5km. This bias, which is not suppressed by any mean model change, is only explained by the too large S-E heterogeneities (Alps) not compensated by stations on the opposite side (N-W).

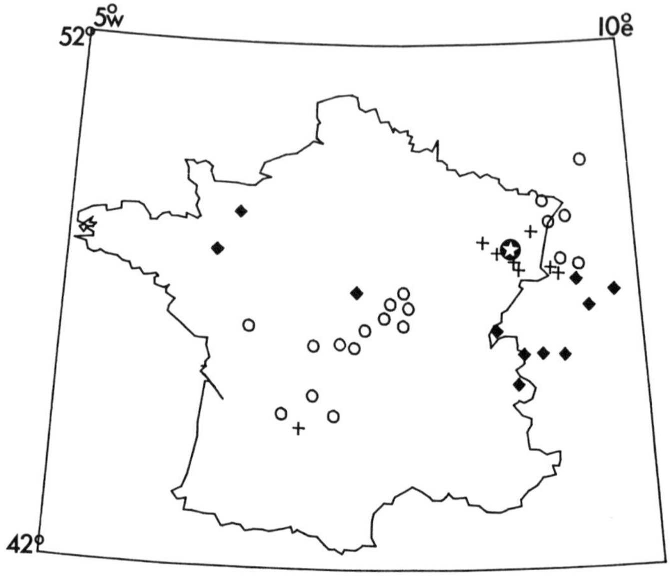

Fig. 1 The permanent seismological network used in the P.M.E.L method for location of the epicenter area of Remiremont ✪. The sign of Pn anomaly is shown by a circle if positive or a losange if negative.

Depth km	V_P km/sec	V_s km/sec
0.		
	3.00	1.73
0.9		
	6.03	3.56
25.9		
	8.16	4.65

Fig. 2 Velocity model used in P.M.E.L method

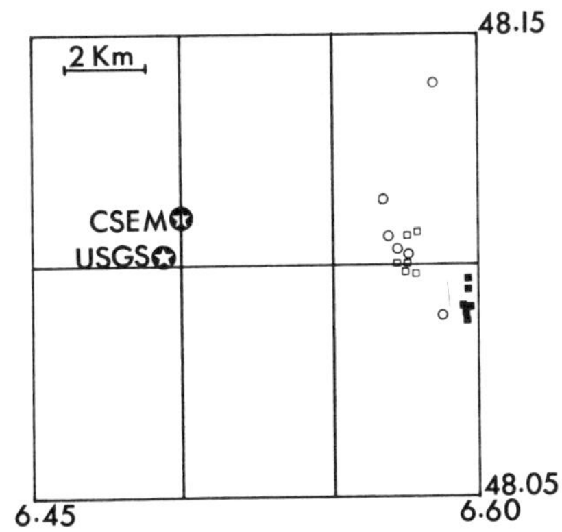

Fig. 3 Location of the six major quakes of Remiremont crisis. Comparison between large network location without anomaly (O), with anomaly (□) and the local network location(■).The main shock epicenter found by USGS and CSEM is also shown.

4. RELATIVE LOCATION OF EARTHQUAKES

By the use of the P.M.E.L method,large shocks have been localized with a good accuracy. The study is carried on with weak shocks locations.

In order to uncouple the location from propagation problems, the locations are computed with respect to one reference earthquake(absolute location might nevertheless be obtained by a shift using the P.M.E.L method results).

Because weakest quakes are only recorded by few stations, the location accuracy is strongly affected by the measurement accuracy.

4.1. "Doublets" method

The accuracy of on-set times measurement is limited by the definition of signal beginning, disturbed by the seismic noise. And , even without seismic noise, the measurement incertitude on digitalized data cannot be smaller than the sampling period.

Nevertheless, the evaluation of the time difference between two similar signals is independant of these two limitations:

-The processing is applied on the complete phase duration and therefor the noise is less influent.

-The phase spectrum difference versus frequency allows to obtain time differences with an error much smaller than the sampling period.

Data processing methods based on Wiener filters and coherency (Poupinet and al. 1984, Fremont 1984, Frechet 1985) allow to measure time differences with an incertitude lower than one millisecond with a signal sampling rate of 100s/sec, this in case of very similar signals on large time duration(coherency> 0.95 in a large frequency band).

In the case of a seimic crisis where epicenters are distributed along a fault of some kilometers long and where quake magnitudes range on several units, correlation coefficients between two signals recorded on the same station fall off very rapidly when the time duration increases. Consequently, the application of methods based on the phase spectrum difference measurement is very instable. More robust methods are needed in that case. thus, for example the interpolation of the cross-correlation curve around its maximum gives less accurate time differences (3 to 5 milliseconds of incertitude on perfectly similar signals sampled each 20 milliseconds) but is still applicable to very short duration (8 samples) and poorly coherent(coherency> 0.7) signals.

4.2. Application to Remiremont crisis

4.2.1. Data

In order to keep the measurement accuracy of the "Doublets" method (some milliseconds) we need to use the data from a centralized network with one unique clock. These data have been recorded by the LDG permanent network. Among the 28 available stations only 11 stations having recorded the seimic data with a sufficient signal to noise ratio have been selected (fig 4.). Epicentral distances range from 20 to 390km. The azimuthal pattern is only sampled by 3 points(55, 150 and 240 degrees).

Location is performed with Pn, Pg and Sg waves(fig 4.). Only quakes with Ml>1.7 are giving useful seimic signals at least in 3 stations. These lead to a study limited to 78 shocks (23 precursors and 54 aftershocks of the main shock).

For the smallest quakes of this set(Ml< 3.0) only Pg and Sg waves recorded by the three nearest stations(d< 100km) are available. For larger shocks, additionnal stations, as the Morvan stations, record Pn, Pg and Sg phases.

4.2.2 Measurement accuracy

Depending on the signals likeness, the correlation curve interpolation leads to incertitudes of 5 to 20 milliseconds (fig. 5 and fig. 6) with an estimated mean value of 10 milliseconds(half the sampling period).

This reading incertitude corresponds to a theoretical relative location error of:

a) +/-100 meters in the three dimensions for earthquakes localized with Pn, Pg and Sg phases on the 11 stations.

b) +/-250 meters on horizontal coordinates and +/- 1100 meters on the depth in the case of Pg and Sg on the three closest stations. As the depth variation of the shocks is less than the depth error the depth for shocks with Ml<3.0 can not be computed and is set to the mean focal depth(7.0km).

4.2.3. Location

The swarm of earthquakes localized by this method defines a north-south accident (fig 7.) of 2.1km length, and of 0.26km width at the most, which is of the same order of magnitude as theoretical location incertitudes.

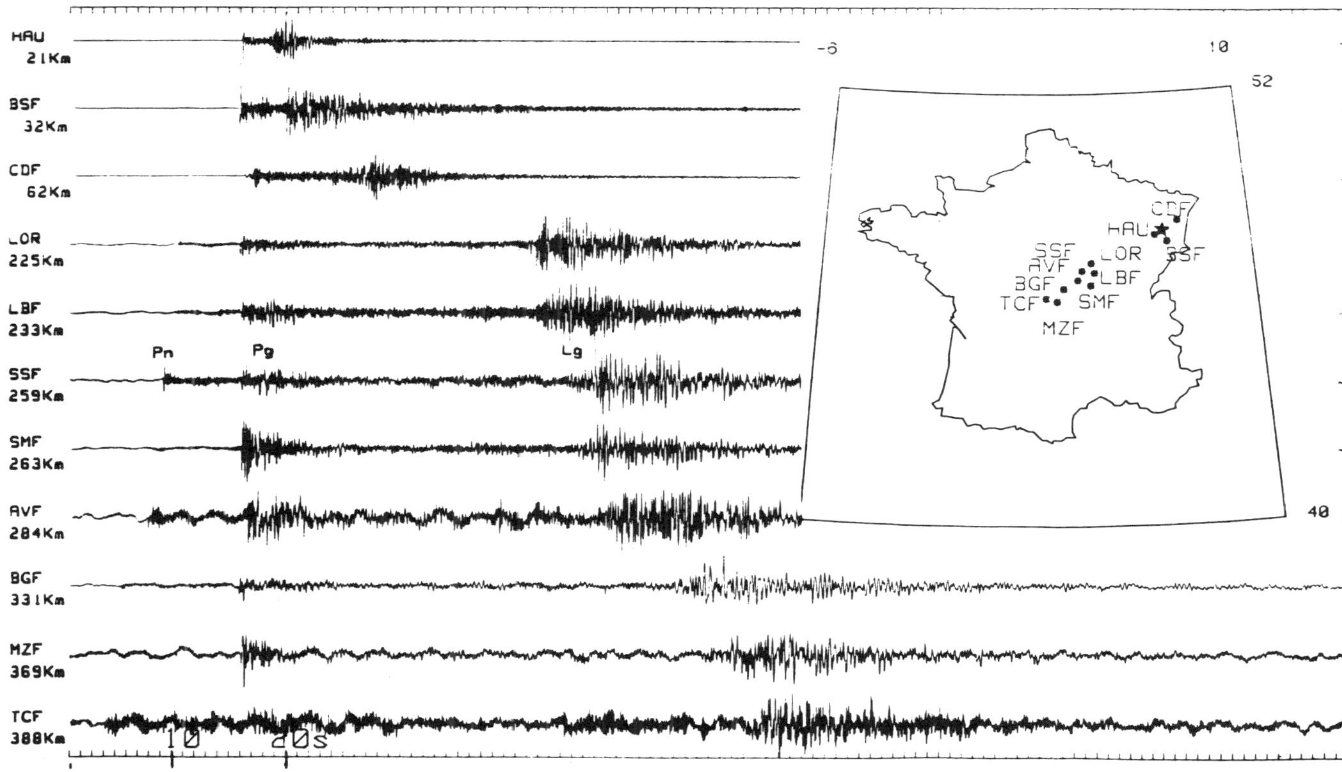

Fig. 4 Permanent LDG stations used in the relative hypocenter determination. Signals correspond to an aftershock (Ml=3.1).

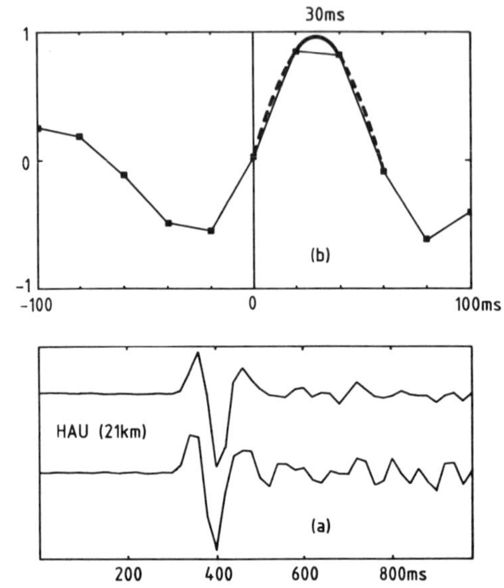

Fig. 5 Example of a doublets recorded at 21km (a) and the correlation curve with interpolation near its maximum (b).

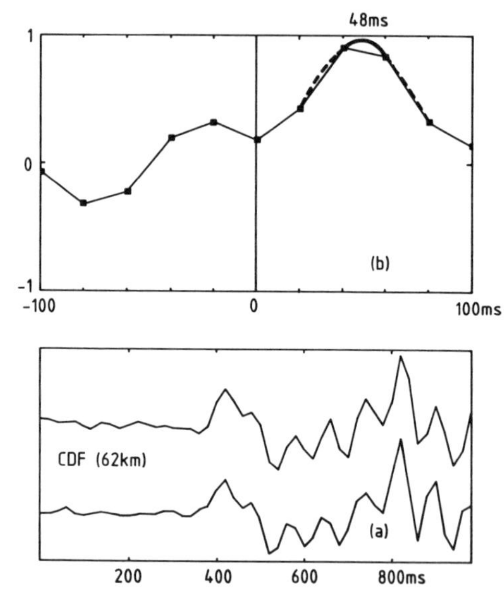

Fig. 6 Example of a doublets recorded at 62km (a) and the correlation curve with interpolation near its maximum (b).

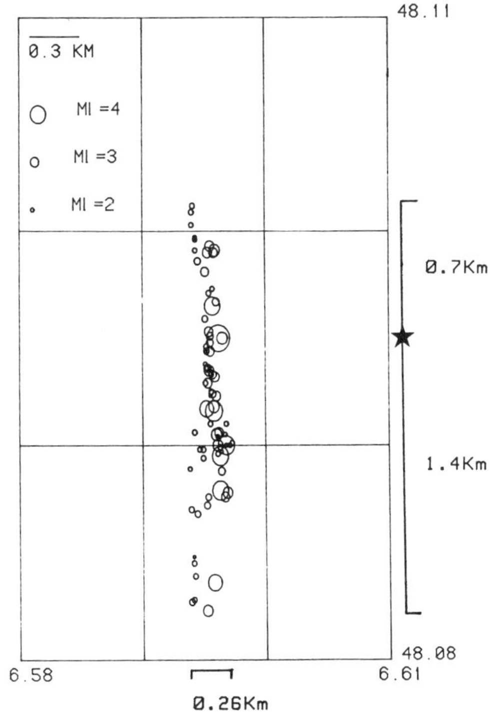

Fig. 7 Spatial distribution of the Remiremont swarm obtained by the large network. The main shock epicenter is shown on right side.

Depth determinations, when significant, are between 6.3km and 7.5km with a maximum of cases around 7.0km.

Thus, with only data recorded between 20 and 390km of epicentral distances, we obtain results similar to those given by the IPG temporary network with epicentral distances less than 20km.

For the common set of events(aftershocks), the pattern of location obtained by the large permanent network and the local array is the same(fig 8.). Event by event, the differences between these two locations (and this without taking into account the 1.5km bias due to absolute location by P.M.E.L method) gather inside an ellipse with its great-axis smaller than 500 meters and an orientation compatible with the theoretical error ellipse (fig 8.)

4.2.4 Space-Time distribution

The use of permanent stations gives the opportunity to follow the whole crisis, and the location accuracy allows to observe the seismicity evolution versus time.

Because of a north-south faulting orientation and concentrated depth around 7km, the shock latitude is taken as "Space" variable. The main shock origine time is taken as history origine time. To take into

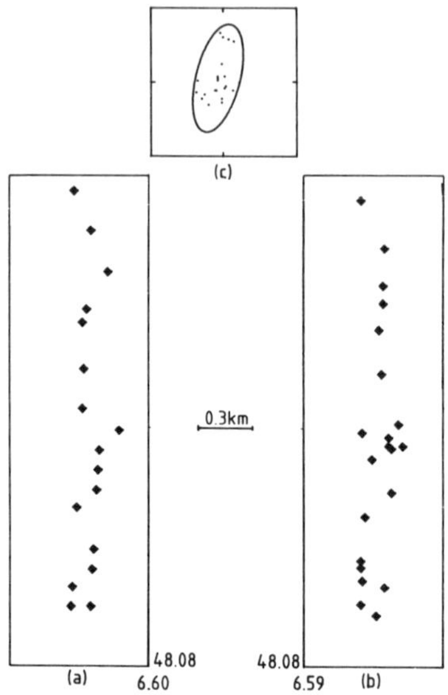

Fig. 8 Comparison between aftershocks location with local network (a) and large network (b). The event difference is shown in (c).

account the exponential variation of time duration between two succesive aftershocks, the "Time" variable is defined by:

$$\mathrm{sign}(t-Ho)*\mathrm{Log}(\ t-Ho\ )$$

"Space" variations versus "Time"(fig. 9) show out five different phases for the Remiremont crisis, three of which concerning the fore-shocks period.

a)Phase A: from -18 days to -3 days
During this phase, the 15 located foreshocks are taking place in the middle part of the fault. Both Ml=4.1 quakes are localized at the same place.

b)Phase B: from -3 days to -70 minutes
During this period, no quake of Ml>1.7 is observed

c)Phase C: from -70 minutes to 0
After 3 days of calm, the seimicity starts again 70mn before the main shock. The 8 localized foreshocks(the largest reaching Ml=3.0) are concentrated on the main shock area.

d)Phase D: from 0 to 2 days
The active zone is increasing around the main shock to reach a length of 1.5km. The extension is more progressive southward than northward.

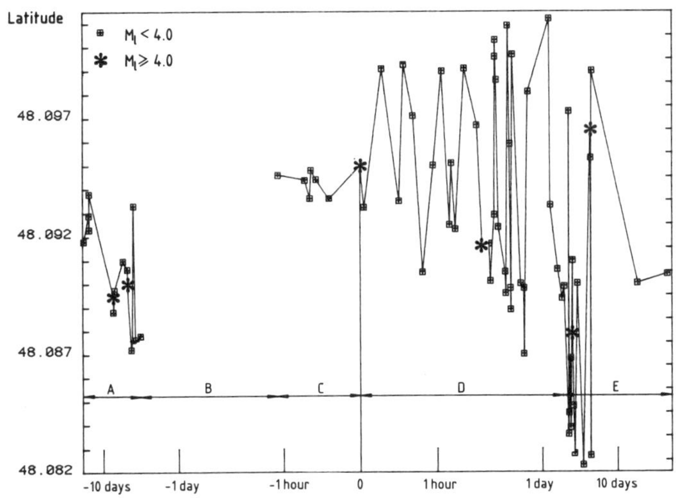

Fig. 9 Time-spatial distribution for the Remiremont crisis. Time scale is logarithmic and origin corresponds to the main event origin time. The spatial variable is the latitude of the shocks.

e)Phase E: from 2 days to the end of the crisis

At the beginning of this phase, seismic activity is increasing with a sudden faulting growth towards south. The swarm is reaching its maximum size of 2.1km. The crisis ends up with 2 quakes localized in the first foreshocks area.

5. CONCLUSION

The use of adapted methods applied on a large aperture network has allowed us to obtain locations as good as those calculated from a local network.

The P.M.E.L method (Pavlis and al. 1983) which takes into account the propagation anomalies leads to the location of an epicentral area with an incertitude less than 2km in the case of the Remiremont crisis.

Signals correlation gives a measurement of arrival time differences with errors clearly smaller than the sampling period. This allows a relative relocation of quakes belonging to Remiremont crisis with an incertitude of some hundreds of meters.

On the other hand, the seismic data from permanent stations enable us to observe the complete crisis evolution. Thus, the space versus time distribution for foreshocks shows up three distinct phases within the period of time previous to the main shock (december 29). In the same way the two different roles of the Northern and the Southern parts of the fault are clearly shown by the aftershocks sequence.

REFERENCES

AKI K., CHRISTOFFERSON A. and HUSEBYE E.S. 'Determination of the three dimensional seismic structure of the lithosphere'
J. Geophys. Res., 82, 277-296, 1977

BONJER K.P., GELBKE C., GILG B., ROULAND D., MAYER-ROSA D., and MASSINON B. 'Seismicity and dynamics of the upper Rhinegraben'
J. Geophys., 55, 1-12, 1984

FRECHET J. 'Sismogenese et doublets sismiques'
These d'Etat, Univ de Grenoble, 1985

FREMONT M.J. 'Mesure de variation temporelle de paramètres de la croute terrestre et d'effets de source par traitement de doublets de seismes'
These 3 Cycle, Univ de Grenoble, 1984

HAESSLER H. and HOANG-TRONG P. 'La crise de Remiremont (Vosges) de décembre 1984. Implications tectoniques régionales'
C. R. Acad Sc Paris, 1985

GRASSO J.R., CUER M., and PASCAL G. 'Use of two inverse techniques Application to a local structure in the New-Hebrides Island Arc'
Geophys. J.R. astr. Soc., 75, 437-472, 1983

PAVLIS G.L., and BOOKER J.R. 'Progressive multiple event location (PMEL)'
Bull. Seism. Soc. Am., 73, 1753-1777, 1983

POUPINET G., ELLSWORTH W.L. and FRECHET J. 'Monitoring velocity variations in the crust using earthquakes doublets:an application to the Calaveras fault, California'
J. Geophys. Res., 89, 5719-5731, 1984

THE REMIREMONT (VOSGES) SEISMIC CRISIS OF DECEMBER 1984 AND REGIONAL TECTONIC IMPLICATIONS*

Henri Haessler
Pho Hoang-Trong
Institut de Physique du Globe
5, rue R. Descartes
67084 Strasbourg Cedex
France

ABSTRACT. The seismic crisis of Remiremont (Vosges) is related to a submeridian quasi vertical tectonic accident. Most of the aftershock focal mechanisms correspond to a left lateral strike slip on a mid-crustal fault. The similarity of focal mechanism in the Swabian Jura (West-Germany) and the Remiremont area, 180 km apart, supports the idea that a uniform regional stress field controls active faulting in the southern Rhine Graben and the surrounding areas.

1. INTRODUCTION

The Remiremont earthquake which occurred on December 29, 1984 at 11:02 UTC, close to the village of Eloyes between Epinal and Remiremont (Vosges, France), is the most important shock in this region since the damaging earthquake of 1682. It was widely felt in the surrounding area, apparently more strongly on the cristalline basement as suggested by the south and east broadening of isoseismals IV and V (fig.1).

Ten years of instrumental monitoring for the Rhine Graben area (Bonjer et al., 1984) shows that the western flank of the Vosges Mountain exhibits a low seismic activity marked by a microearthquake swarm (M<4) in 1974, north of Epinal.

The Remiremont crisis was recorded by the permanent seismic stations of the Rhine Graben network radiolinked to the Geophysical Institute of Strasbourg. In order to improve hypocenter location and focal mechanism studies, a temporary telemetered network was installed in the epicentral area. The time distibution of events is characterized by foreshocks prior to the main shock (fig.2).

This premonitory activity corresponds to the type 2 of Mogi's classification of earthquake sequences and could be explained by a moderate fracturation of the focal medium (Mogi,1963).

---

* This paper resumes the essentials of a note presented in french at the Academy of Sciences in Paris, C.R.Acad.Sc.Paris, t300, 1985, 671-675.

J. Bonnin et al. (eds.), Seismic Hazard in Mediterranean Regions, 359–365.

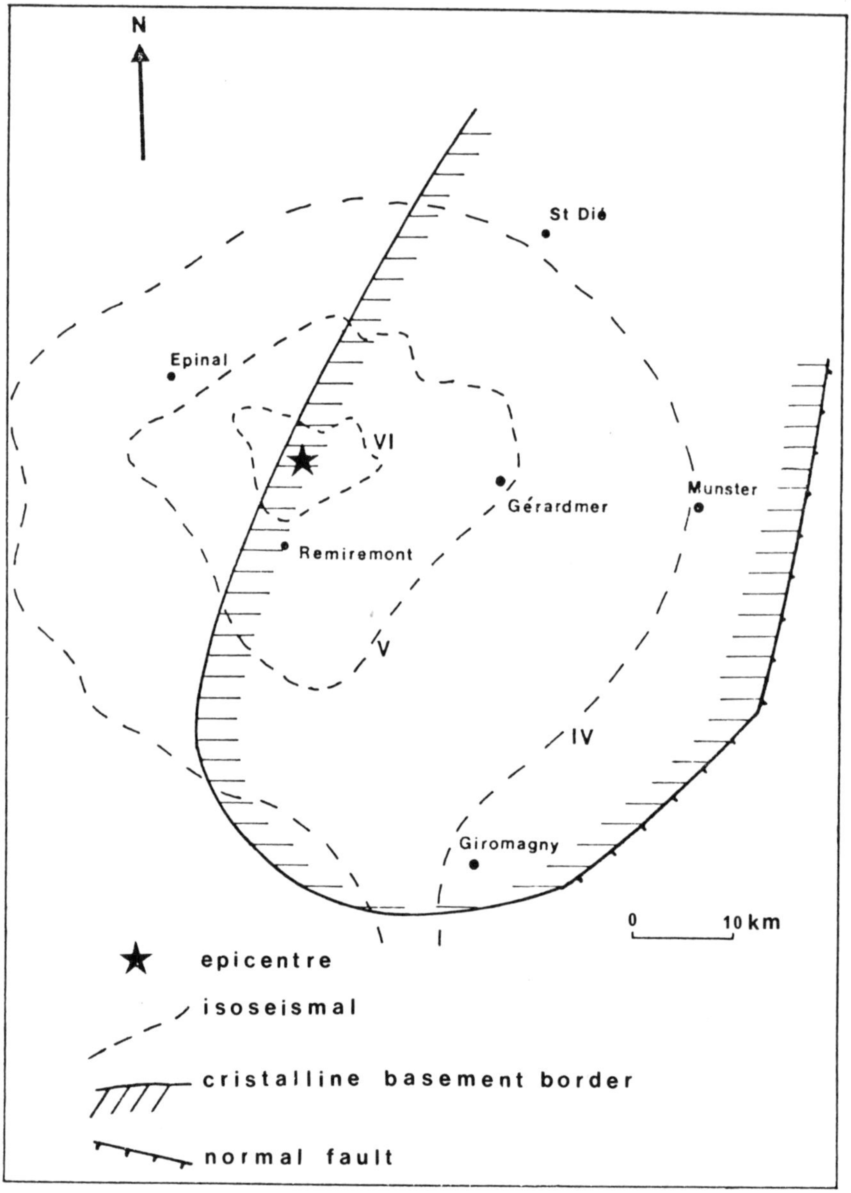

Figure 1. - Macroseismic map from J. Lambert (BRGM). Contours of outcropping Hercynian basement rocks and main normal faults.

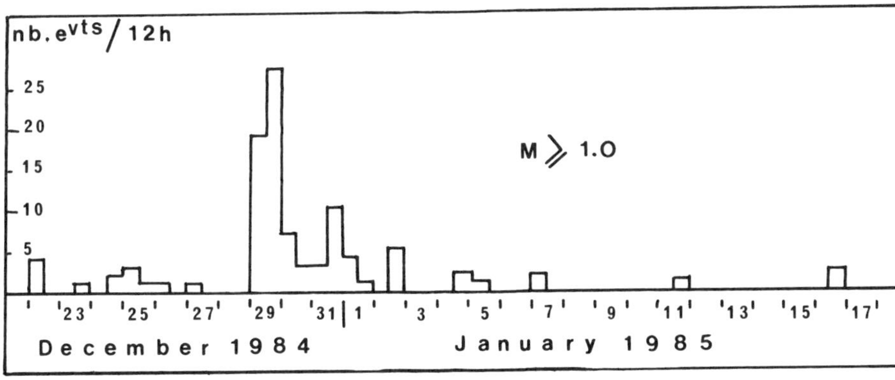

Figure 2. - Temporal evolution of earthquakes: events with magnitudes greater or equal to 1.0 detected by the regional seismological network.

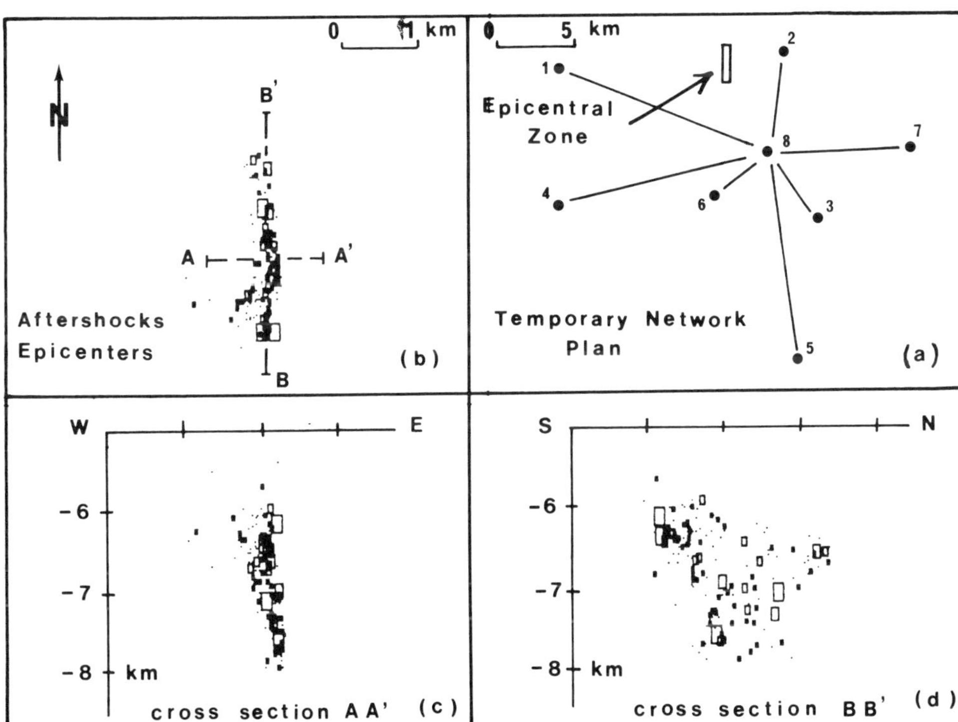

fig. 3. - The temporary seismological network and the spatial distribution of aftershocks.

2. HYPOCENTRAL LOCATION

For an homogeneous determination of hypocentral parameters we have only used travel times data of P an S waves from the local network. The epicentral distances of the 8 stations are less than 20 km. Some of them are close to or smaller than the focal depth (fig.3a). This configuration is almost optimal and leads to a precision of 250m for the horizontal coordinates and 500m for the depth. We used a half space velocity model with a mean velocity Vp adjusted to 6.0 km/s and a Vp/Vs ratio of 3 which fits very well the data (fig.4).

The epicentral zone, which describes a NS narrow band of 3km long, is located at 1 km westward of the village of Eloyes (fig.5a) on the left bank of the Moselle river. The

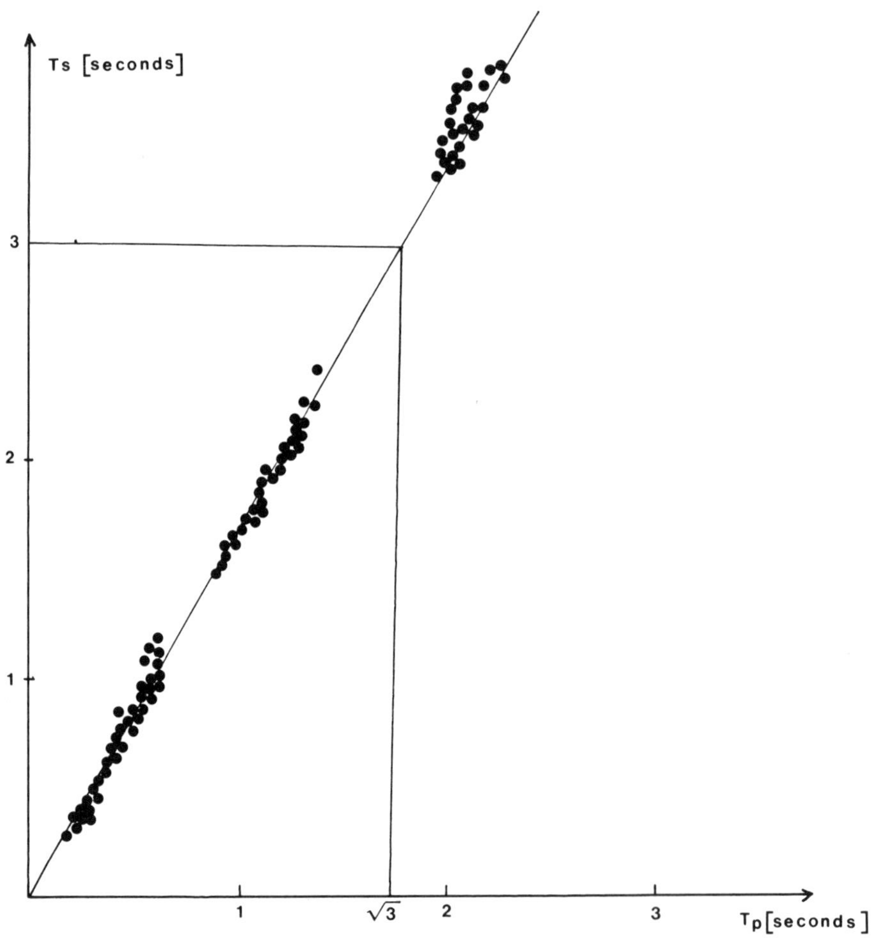

fig. 4. - S waves travel times versus P waves travel times, in seconds, obtained by G.Wittlinger using a Wadati composite method.The slope is very close to the theoretical value 3 .

aftershock foci are distributed along a vertical plane striking N03E and ranging from 6 to 8 km depth (fig.3b,3c,3d). The epicenter of the main shock obtained by relative location is inside this cluster (fig.5a) which defines a remarkable vertical rupture plane for the main event.

Aerial photographic pictures of this region show two tectonical lineaments intersecting in the epicentral zone. The first lineament has a NS direction along the aftershocks, the second one, starting at the northern tip of the epicenters' zone, strikes SE-NW from Jarmenil to Epinal. The course of the Moselle river follows successively these two lineaments with a sharp curve at their junction in the epicentral zone.

3. FOCAL MECHANISM

Figure 5b depicts P-waves first motion polarities on an equal area projection of the lower focal hemisphere for the main shock. All the data are read on seismograms for which epicentral distances are less than 5°. The Pn waves data set, distributed on a circle, exhibits a clear polarity change within the Swiss seismic network (azimuth about 180°) and the nodal

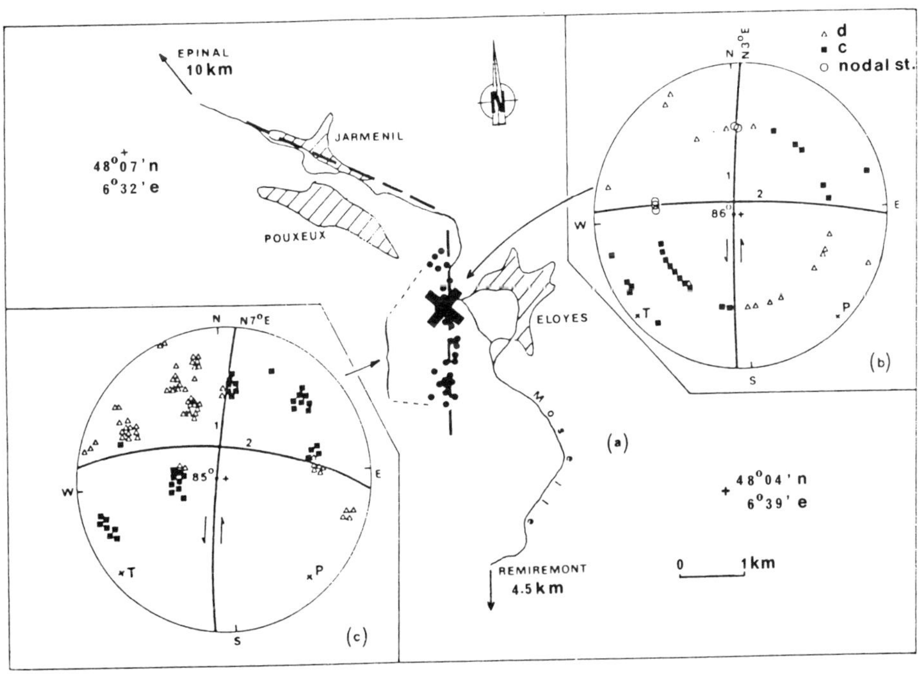

fig. 5. - The epicentral area (a) and nodal solutions of the main event (b) and of aftershocks (c). Dashed lines: Lineations shown by aerial photographies. Black cross: epicenter of the main shock.

character of the LDG stations of Normandie (azimut about 270°). The nodal planes are thus well constrained.

A composite P-wave first motion plot of aftershock data presents a very good internal coherence and a nodal solution similar to the one of the main shock (fig.5c). A focal solution of the same kind was found for the strongest event of the 1974 crisis North of Epinal (Bonjer et al, 1984). The source mechanism of the Remiremont earthquakes corresponds to a pure left lateral strike slip along a NS vertical plane.

4. TECTONICS AND CONCLUDING REMARKS

The similarities between the Remiremont earthquake and the 1978 Swabian Jura earthquake, which occurred on the eastern flank of the Rhine Graben, is amazing: this event, located symetrically of Remiremont with respect to the Graben axis, yields the same focal mechanism and aftershock characteristics (Haessler et al., 1980). These features are also found in the graben itself (Rouland et al., 1980). The similarity of fault pattern suggests that the same stress field would act in the Rhine Graben and its uplifted shoulders (fig.6). The Rhine Graben was formed as an extensional rift valley from mid-Eocene to lower Miocene, probably along preexistent zones of weakness in the hercynian basement (Illies, 1965). Since the middle Tertiary, the maximum horizontal compressive stress had undergone an anticlockwise rotation of about 60°; its present NW-SE direction is oblique to the graben axis which reacts by a sinistral shear (Illies et al., 1981). Off the graben, the

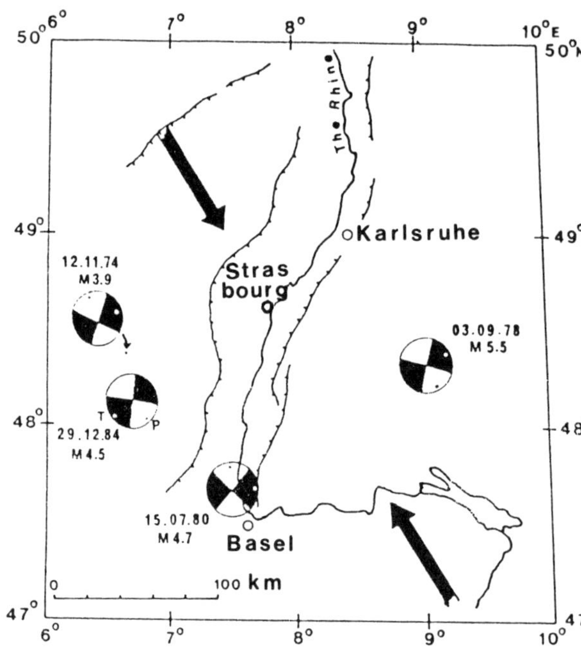

fig. 6. - General map of the Upper Rhine Graben area and Fault plane solutions of the main seismic events. White quadrants: compression. Black quadrants: dilatation. White dots: T axes. Black dots: P axes. Black arrows: mean direction of the maximum compressive stress.

active faults of the Swabian Jura and Epinal-Remiremont regions display the same movement, following the same Rhenish direction which is representative of the general orientation of crustal weaknesses.

The SE-NW maximum compressive stress which prevails in all the NW Alps foreland up to North Sea (Ahorner, 1975) is roughly perpendicular to the axis of the northern Alps. So, that stress pattern could reflect either the collision between African and Eurasian lithospheric slabs or a topographic stress due to the differences in altitude between the Alps and their northern foreland (Illies et al., 1981).

REFERENCES

Ahorner, L., 1975. 'Present-day stress field and seismotectonic block movements along major fault zones in Central Europe,*Tectonophysics*, *29*, 233-249.

Bonjer, K.P., Gelbke, C., Gilg, B., Rouland, D., Mayer-Rosa, D. and Massinon, B., 1984. 'Seismicity and dynamics of the Upper Rhine Graben, *J.Geophys.*, *55*, 1-12.

Haessler, H., Hoang-Trong, P., Schick, R., Schneider, G. and Strobach, K., 1980. 'The September 3, 1978, Swabian Jura earthquake,*Tectonophysics*, *68*, 1-14.

Illies, H., 1965. 'Bauplan und Baugeschichte des Oberrheingrabens,*Oberrh. Geol. Abh.*, *14*, 1-54.

Illies, H., Baumann, H., and Hoffers, B., 1981. 'Stress pattern and strain release in the alpine foreland,*Tectonophysics*, *71*, 157-172.

Mogi, K., 1963. 'Some discussions on aftershocks, foreshocks and earthquake Swarms - the fracture of a semi-infinite body caused by an inner stress origin and its relation to earthquake phenomena, *Bull. Earth. Res. Inst.*, *41*, 615-658.

Rouland, D., Haessler, H., Bonjer, K.P., Gilg, B., Mayer-Rosa, D. and Pavoni, N., 1980. 'The Sierentz Southern-Rhinegraben earthquake of July 15, 1980: Preliminary Results, *Proc. E.S.G.-E.G.S. Gen. Assembly, Budapest*, 441-446.

CHAPTER V

HISTORICAL STUDIES

SISMICITE HISTORIQUE : AMBIGUITES SISMOLOGIQUES

J. Vogt
1 rue du Dr. Woehrlin
67000 Strasbourg
France

R E S U M E. *A priori* , la complémentarité de la sismologie instrumentale et d'une sismologie historique critique et fiable peut paraître évidente dès lors qu'elle est une condition *sine qua non* de démarches essentielles, fondamentales ou appliquées : typologie de séismes et de crises sismiques, discussion sismo-tectonique, appréciation du risque sismique au sens probabiliste du terme, génie sismique, diagnostic des risques indirects. En fait, si la sismologie historique a connu une remarquable renaissance, c'est d'une manière paradoxale qu'un fossé la sépare encore trop souvent de la sismologie instrumentale. Ce fossé traduit des problèmes psycho-sociologiques, rapidement survolés, en même temps que sont évoqués quelques "clichés" dont pâtit la mise en oeuvre de la sismologie historique et le rôle néfaste de catalogues hâtifs. Les uns et les autres contribuent à une large méconnaissance de la richesse et de la variété des sources, mises en oeuvre d'une manière critique, dans un esprit d'interdisciplinarité, avec le souci de l'appréciation de la fiabilité. A la lumière des recherches et de publications récentes sont esquissés les apports de l'archéologie, de l'épigraphie, des textes historiques publiés, des archives, de la presse, avec les problèmes d'exégèse posés en particulier par les contextes économiques, sociaux et politiques, le vocabulaire, la toponymie, la chronologie. Avec quelque détail, sont examinés les modes d'expression adaptés à l'état des connaissances, laissant la porte ouverte à l'intégration d'autres éléments et à de nouvelles interprétations, aux antipodes de présentations entières, affirmatives, d'allure définitive, qui répondent certes à un besoin de confort intellectuel, mais font grand tort à une sismologie historique bien comprise. Qu'il s'agisse d'intensités ou d'épicentres, l'accent est mis sur la *relativité*, maître-mot de la sismologie historique, avec *fiabilité* et *exégèse*. Ainsi se développe peu à peu un corps de doctrine interdisciplinaire.

Avertissement : Dans une large mesure, ce texte, français, est différent de la conférence faite par l'auteur en anglais. Cette dernière a cherché à mettre en relief quelques traits majeurs, à l'aide d'exemples particulièrement concrets et éloquents (1), en multipliant croquis et projections de documents. Le présent texte approfondit la discussion de l'arrière-plan et du contexte, en mettant l'accent sur des aspects méthodologiques, voire épistémologiques, en donnant parfois d'autres exemples.

1 PROBLEMES D'EPISTEMOLOGIE ET DE METHODES

L'accent sera mis sur les problèmes de méthode, souvent perdus de vue, faute de recul, posés par une informatisation dont la sismicité historique est d'ailleurs souvent victime. A cet égard seront présentées en quelque sorte des "thèses", compte tenu des enseignements de

J. Bonnin et al. (eds.), Seismic Hazard in Mediterranean Regions, 369–382.

récentes réunions. Est donnée une sélection bibliographique permettant de mieux saisir un "corps de doctrine" complexe en gestation.

Depuis des années est combattu le fossé artificiel qui s'est creusé entre la sismologie instrumentale et la macrosismologie et singulièrement la sismologie historique. Ce fossé ne subsiste que trop souvent dans les esprits, malgré la démonstration incessante de l'indispensable complémentarité des deux domaines, en particulier sous l'impulsion des pionniers de la renaissance de la sismologie historique, Ambraseys en tête. C'est la maîtrise de cette convergence, certes complexe, qui témoigne précisément d'une véritable culture scientifique, sans oeillères, avec un large recul, loin des travaux étroitement techniques et répétitifs.

Dans une large mesure, la renaissance de la sismologie historique et sa convergence avec la sismologie instrumentale sont dues au poids croissant des problèmes pratiques avec leurs exigences de fiabilité. Qu'il suffise d'un exemple : au cours de la discussion de lacunes de sismicité ("gaps") on ne s'interroge pas toujours au sujet de la valeur des catalogues mis en oeuvre pour apprécier la fréquence et les caractéristiques d'anciens séismes. A ma surprise, une étude sismologique du fonctionnement de la plaque des Caraïbes omet un élément majeur, à savoir le séisme survenu en 1842 dans la région du Cap Haïtien, de sorte qu'il serait possible de parler de "pseudo-gap". Peut-être faut-il incriminer l'oubli de ce séisme par un catalogue mondial. Pour une discussion académique, un tel oubli est sans conséquences majeures, dès lors qu'est jugée une logique interne, mais il ne pourrait être admis dans une discussion de risque sismique ou de génie sismique, dès lors que la prise de responsabilité accroît les exigences de rigueur et impose un examen approfondi de sismicité historique, à une échelle appropriée.

Il serait possible d'énumérer une foule d'autres obstacles à la prise de conscience de cette vision large. Insistons sur un aspect psychologique parmi d'autres. D'une manière étonnante, un rôle essentiel revient à la coupure chronologique parfaitement artificielle entre ère macrosismique et ère instrumentale, relevant l'une de quelque sombre Moyen-Age, l'autre d'une radieuse civilisation scientifique qui commencerait miraculeusement au début de ce siècle avec l'apparition d'instruments, en introduisant objectivité et précision. Vision manichéenne d'un autre âge, mais qui, à notre surprise, subsiste parfois. Il est volontiers fait fi d'une longue période au cours de laquelle la sismologie instrumentale accroît progressivement sa fiabilité, compte tenu de la conception des instruments, de la configuration des réseaux, de l'élaboration de modèles, de l'apparition et de l'évolution de notions telles que la magnitude. Longtemps les localisations instrumentales sont très approximatives, avec parfois une marge d'erreur de l'ordre de la centaine de km, à une échelle donnée, dans l'un ou l'autre domaine, de sorte que la démarche macrosismique est souvent bien plus précise. Prenons l'exemple de la Tunisie. Les cartes classiques montrent des alignements d'épicentres qu'il peut être tentant d'interpréter en termes sismotectoniques. En fait, cette représentation résulterait de localisations grossières, par degrès, de sorte qu'il serait possible de parler d'artefacts sismotectoniques. Dans une certaine mesure de telles remarques s'appliquent aussi à l'Irak.

N'oublions pas que certains pays, démunis de réseaux de surveillance à leur échelle, sont tributaires de données recueillies à distance. Même dans les pays les mieux équipés ne cesse de se poser un problème de fiabilité. Le catalogue instrumental de tel pays méditerranéen a le mérite de proposer trois degrès de fiabilité, ce qui est certes une condition *sine qua non* d'une discussion sismotectonique tant soit peu significative, dans un contexte caractérisé par une forte densité d'accidents susceptibles d'être sismogènes. En France, la confrontation des épicentres macrosismiques et instrumentaux de séismes alpins relativement récents a posé des problèmes délicats. Même des conditions favorables n'excluent pas des problèmes de réinterprétation d'enregistrements relativement récents (localisation, profondeur, mécanisme), sans parler de délicats problèmes d'étalonnage et d'exégèse d'enregistrements anciens.

Même dans le meilleur des cas, ne cesse de s'imposer la complémentarité des deux démarches, instrumentale et macrosismique, à la fois pour mieux saisir les caractéristiques d'un séisme dans son cadre sismo-tectonique, faciliter les discussions de risque sismique et permettre des travaux de génie sismique, démarches qui exigent, répétons-le, une bonne connaissance de l'arrière-plan et du contexte.

La projection dans le passé de notions instrumentales est d'un maniement particulièrement délicat. Elle peut conduire à des triomphes de la "pseudo-objectivité". Telle étude coréenne, base de travaux de génie sismique, propose des magnitudes dotées de deux décimales pour des séismes anciens mal connus dont il est malaisé d'apprécier l'intensité ! Toute carte de sismicité historique présentant des magnitudes d'une manière affirmative doit être considérée avec une extrême prudence dès lors qu'elle est susceptible d'être un document abstrait, insuffisamment fondé. Si elle peut flatter l'esprit en présentant les apparences de la rigueur scientifique, sa mise en oeuvre hâtive en matière de risque et de génie sismique peut conduire à des déconvenues.

Certes, la sismologie instrumentale éclaire souvent d'une manière remarquable d'anciens événements survenus dans un même cadre sismotectonique, à condition de prendre une foule de précautions. Mais il est non moins vrai que la sismologie historique à l'échelle du siècle, fondée sur des recherches approfondies et une discussion critique, est susceptible d'éclairer des séismes récents parfois discrets, en facilitant l'élaboration d'une typologie d'événements et surtout de crises, à condition de dominer les problèmes d'échelle. Un tel va-et-vient exclut la tyrannie du seuil ("threshold"), pesante, appauvrissante. Avec quelque angoisse est sans cesse posée la question du seuil d'intensité, voire, même en matière de sismicité historique, de magnitude, avec les inconvénients dont il vient d'être question. Cette manière de faire peut conduire à fermer des portes. Prenons un exemple simple : un séisme à première vue spécifique et mineur, d'après les catalogues classiques riches en erreurs, peut fort bien être rattaché, après examen critique, à un séisme notable dont la connaissance s'enrichit de la sorte. Pour notre part, nous avons procédé plus d'une fois au regroupement en séismes notables d'événements artificiellement émiettés, faute d'une confrontation sérieuse des sources. D'autre part, des séismes mineurs, facilement victimes d'un seuil, sont souvent déterminants pour la compréhension d'un ensemble sismo-tectonique, même s'ils n'ont aucune implication immédiate pour la discussion du risque sismique. Au demeurant, les partisans des seuils font parfois l'impasse sur des séismes notables trop mal connus pour être caractérisés facilement et informatisés d'une manière simple, non sans conséquences pour l'appréciation du contexte sismo-tectonique et du risque sismique. Bref, il convient de faire grand cas des séismes "significatifs" d'un domaine donné, même si cette notion est susceptible de troubler les esprits attachés aux apparences de la rigueur de la notion de seuil, c'est à dire à une pseudo-objectivité.

Dans cette perspective, abordons rapidement les effets indirects des séismes, liquéfaction, mouvements de terrain..., à propos desquels la sismologie, dont les possibilités sont certes limitées faute d'expérience de terrain, converge avec la géotechnique. Ici la démarche historique est cruciale, en faisant connaître des précédents, en les situant dans un large contexte géotechnique. A vrai dire, les géotechniciens méconnaissent souvent ses possibilités, non sans conséquences pour leurs analyses de risques. A notre stupéfaction, tel enseignement géotechnique dans tel pays méditerranéen de forte sismicité passe sur la liquéfaction, sous prétexte que le pays n'en fournit aucun exemple. Précisément une foule d'exemples anciens, modèles du genre, sont décrits avec précision par les contemporains, dès le XVIème siècle pour le moins. Ce cas extrême permet véritablement de parler d'inculture scientifique et technique, sans parler d'un problème de responsabilité (2). De même, la discussion des techniques de prévention souffre elle aussi d'un manque de recul et d'une ignorance des précédents (3).

Après ce survol de "problèmes relationnels", il convient d'insister sur quelques pratiques et points de vue susceptibles de limiter les possibilités de convergence et de complémentarité.

Pour commencer, écartons une pseudo-sismologie historique, celle qui revient à redécouvrir des catalogues oubliés au cours de quelques décennies et de les reprendre tels quels, sans esprit critique, sans recherches nouvelles. Pis, de tels catalogues, souvent répétitifs, sommaires, riches en erreurs et en oublis, sont parfois informatisés tels quels, en hâte, et dotés de la sorte d'une apparence scientifique qui peut leur assurer, aux yeux de certains, un caractère définitif, intangible, en justifiant fatalisme et dérobade, une fois de plus. C'est, véritablement, trahir la sismologie historique et tomber dans un piège.

Même des esprits parfaitement convaincus de l'apport essentiel de la sismologie historique à la sismologie proprement dite et à ses applications, qu'il s'agisse d'effets directs ou indirects, font parfois preuve d'un certain schématisme. A cet égard, certaines "spécifications" rigides, formulées *a priori*, témoignent d'une méconnaissance des réceptacles et de la nature des sources dont la recherche et la mise en oeuvre exigent au contraire une grande souplesse. Toutes choses égales, nous retrouvons ici la rigidité des seuils. De plus, elles sont susceptibles d'enfermer les spécialistes dans une "tour d'ivoire" alors qu'une incessante collaboration avec les "scientifiques" est le meilleur moyen de mener à bon terme une recherche de sismicité historique et d'en tirer le meilleur parti possible. Revient à l'esprit la notion de "va-et-vient".

Les mêmes esprits soulignent volontiers que la sismologie historique se limite à des séismes de plus en plus importants au fur et à mesure que l'on remonte dans le passé. En fait, cette "règle" souffre de nombreuses exceptions. Innombrables sont les sources qui font connaître des séismes mineurs et permettent en particulier, répétons-le, la reconstitution de crises. Ainsi est-ce d'une manière détaillée que la réexploitation de sources et de nouvelles découvertes permettent d'analyser les longues crises sismiques qui affectent la Souabe au milieu du XVIIe siècle et la moyenne vallée de la Durance en 1708 et en 1812, exemples parmi bien d'autres.

2 RICHESSE ET VARIETE DES SOURCES

La richesse, la souplesse et, à son échelle propre, le caractère multidisciplinaire de la sismologie historique sont illustrés par la variété des sources qu'elle met en oeuvre. A proprement parler, il importe de faire feu de tout bois en soumettant les sources, quelles qu'elles soient, à un examen aussi rigoureux que possible, compte-tenu du contexte historique, économique, sociologique, psychologique et technique de chaque époque, exigence qui vaut d'ailleurs aussi, à notre époque, pour les enquêtes macrosismiques. Pour le monde méditerranéen, ces démarches et ces préoccupations sont illustrées d'une manière remarquable par le volume qui regroupe les contributions aux "Rencontres d'Antibes" de 1983 (4).

Par l'analyse archéologique, les édifices sont une source de choix, qu'il s'agisse de l'Antiquité ou du Moyen Age. Pour la première, qu'il suffise de citer les travaux consacrés à Delphes et à Pompéi (reconstitution des effets du séisme de l'an 62 par l'analyse des réparations) (5). Pour le second, il convient de mettre en relief la belle analyse du séisme lombard de 1117, par un "épluchage" systématique des églises romanes (6). Au Proche-Orient, de telles recherches se succèdent à un rythme rapide.

L'épigraphie est étroitement associée à l'archéologie. Nombreuses sont, dans l'Antiquité, au Moyen Age et parfois au-delà, les inscriptions fixant le souvenir de séismes notables, soit d'une manière directe, soit indirectement, à propos de faveurs du pouvoir, de reconstructions. Cependant, ce n'est que depuis quelques années que les *Corpus* d'inscriptions antiques sont examinés d'une manière systématique et critique en matière de sismicité, certaine ou hypothétique (7). Pour le Moyen Age, citons encore les inscriptions relatives au séisme de 1117 (8). Dans le monde arabe, mes lectures m'ont fait connaître de telles inscriptions au Proche-Orient, pour l'une ou l'autre mosquée, mais j'ignore si elles font l'objet d'un recensement et d'une discussion critique.

Innombrables sont les textes historiques publiés qui renseignent au sujet de séismes, de l'Antiquité à nos jours. En général, c'est incidemment qu'il est fait état de tels événements. Souvent les index (si index il y a) ne tiennent pas compte de brèves mentions, au demeurant longtemps négligées par des historiens dédaignant des événements "terre à terre" (9). S'imposent donc des "épluchages" et des confrontations laborieuses, mais souvent fructueuses. Pour l'Antiquité et le Moyen Age, qu'il suffise de donner en exemple les mises au point critiques consacrées récemment par L. Ducat au séisme de Sparte en 464, dans un large contexte (10), par P.L. Gatier aux séismes de la fin du Vème et du début du VIème siècle en Asie Mineure (11) et par E. Guidoboni au séisme italien de 1222, en mettant l'accent sur la discussion des échos lointains, facilement mésinterprétés (12).

A elles seules, ces mêmes sources alimentent parfois des mises au point régionales. A la suite d'un mémorable séisme récent qu'il convenait d'éclairer, Ambraseys et Melville, spécialistes de la sismicité du Moyen-Orient, pionniers de la sismologie historique en général, ont consacré dans les meilleurs délais une vue d'ensemble à la sismicité du Yemen (13). Fort à propos, des recherches systématiques dans une catégorie de sources historiques m'ont permis de la compléter par la relation contemporaine détaillée, publiée à Lyon, d'un séisme yéménite (?) du XVIe siècle. Bel exemple de catalyse et d'engrenage.... De la même manière, le récent séisme guinéen, surprenant pour certains, a suscité chez les uns et les autres une foule de lectures. Il en résultera en particulier une remarquable mise au point, par le même Ambraseys, sur la sismicité historique, généralement méconnue, de l'ensemble de l'Ouest africain. Insistons encore, à propos de ces sources, sur l'importance que prend souvent le contexte politique, voire militaire. Ainsi, si nous disposons de quelques détails sur le séisme qui détruit Oran en 1790, c'est que le préside est alors assiégé et que le séisme contribue à la chute de la place, non sans retentissements dont témoigne en particulier la publication d'un poème de circonstance (14).

Innombrables, souvent insoupçonnées, sont les ressources des archives (terme désignant ici les sources historiques inédites) dont l'exploitation fait connaître des séismes mineurs perdus de vue et est souvent une condition *sine qua non* d'une meilleure connaissance de séismes notables, qu'il s'agisse d'épicentres, d'intensités, d'aire macrosismique, d'effets indirects, etc. Considérons d'abord les archives publiques, en mettant l'accent sur les entités mineures, petits états, provinces, villes, particulièrement riches. Ce sont elles qui livrent des renseignements précis, surtout grâce à des états de pertes, voire à des projets de reconstruction. A titre d'exemple, c'est en termes de microzonage que viennent d'être exploitées des expertises consacrées au séisme de Rimini en 1786 (15). Pour un séisme de la fin du XVIIIe siècle, un tel document vient d'être retrouvé pour Villefranche-de-Rouergue (16). A ce propos, il n'est pas inutile de relater le processus de recherche. Pour commencer, j'ai, pour ma part, relevé dans une publication de sources provinciales une allusion à un séisme inconnu, à première vue notable, sans retrouver l'original en un premier temps. Par la suite, J. Lambert reprit les recherches et, à partir de ce repère, décela dans les archives de Villefranche-de Rouergue un autre document, essentiel pour la compréhension de l'événement, mais longtemps perdu de vue. Au hasard, citons les archives paroissiales. D'une part, elles consignent les victimes de séismes : c'est le cas d'une vingtaine d'entre elles lors de l'éboulement d'un rocher provoqué au milieu du XVIIe siècle par un séisme notable de l'arrière pays de Nice. Bien plus, un registre paroissial alsacien mentionne une victime du séisme de Raguse ! Mais surtout, de tels registres sont parfois riches en mentions de faits divers parmi lesquels des séismes même modestes occupent une place de choix. Ainsi sont-ils une source essentielle pour la connaissance de la sismicité des côtes africaines ou du tremblement de terre bourguignon de 1783 (17). Citons les surprises des archives privées dont l'accès peut certes poser des problèmes : c'est à un Ingénieur des Mines particulièrement motivé que je dois un remarquable écho au sujet du séisme alpin de 1839. Notons encore les journaux de bord, si précieux pour la connaissance de l'ancienne sismicité de la ride atlantique. A vrai dire, il serait possible de consacrer aux sources d'archives pour la connaissance des tremblements de terre non seulement un inventaire très fourni, mais aussi une étude-pilote de méthodologie archivistique. Et que de richesse disparues en raison des innombrables pertes d'archives !.

Terminons par la presse, d'une extraordinaire richesse dès le milieu du XVIIe siècle. Dans une large mesure, la connaissance de la sismicité de la France et de ses confins est ainsi fondée sur les échos de différentes feuilles provinciales -nous venons de découvrir l'apport méconnu des "Affiches du Poitou" - et sourtout de la "Gazette de France", certes exploitée en un premier temps par Perrey, mais d'une manière sommaire, et dont la réexploitation a multiplié les surprises. Il est vrai que l'ancienne presse pose souvent des problèmes de conservation et d'accès au même titre que les archives. A titre d'exemple, il semble bien que l'unique collection des "Affiches de Touraine", dont il était permis d'espérer de précieux renseignements, ait péri dans l'incendie de la bibliothèque de Tours en 1940. Globalement, l'apport de la presse est remarquable au XIXe siècle, qu'il s'agisse des journaux de haut niveau (tels que "Times", avec son précieux index), de plusieurs journaux du Proche-Orient ou de l'innombrable presse locale, au cours de la deuxième moitié du siècle, avec ses irremplaçables précisions, en partie mises en oeuvre par les "macrosismologues" contemporains, mais qu'il reste à exploiter dans une large mesure. Bien entendu, il se trouve des esprits supérieurs pour dédaigner le recours à la presse. Ils perdent simplement de vue que la plupart des catalogues classiques, parfois intangibles à leurs yeux, sont précisément alimentés par la presse qui fait longtemps figure d'unique source. Aujourd'hui encore, ce sont souvent les échos de presse, à défaut d'enquêtes macrosismiques, qui donnent corps aux informations abstraites fournies par la sismologie instrumentale. En France, c'est la presse qui m'a permis de corriger des erreurs et de combler des lacunes surprenantes des enquêtes macrosismiques entreprises il y a une vingtaine d'années, dans un contexte de crise il est vrai...

Certes, chaque région, chaque époque pose des problèmes particuliers, avec un dosage de sources et de recherches adaptées à chaque cas. Mais le plus souvent s'impose un laborieux "ratissage", riche en déconvenues, mais dont émergent en général des apports substantiels, déterminants pour la compréhension de séismes apparemment bien connus, sans parler de la (re)découverte d'événements perdus de vue. Contrairement à certaines idées reçues, un tel "ratissage" n'est pas anarchique, au gré d'inspirations littéraires, mais obéit à une logique, certes complexe, compte-tenu d'une foule d'aléas et d'engrenages imprévisibles, mais avec le commun dénominateur du souci constant de la fiabilité. Il est vrai qu'une recherche de sismologie historique, avec ses problèmes d'exégèse, peut prendre du temps. Récemment, j'ai estimé qu'une révision sérieuse d'un grand ensemble du Proche-Orient, à partir des seuls matériaux d'accès facile, exigerait deux années d'un travail intensif, plutôt que de se faire succéder des catalogues à tendance affirmative, pauvres en arguments et en références, se reprenant plus ou moins les uns les autres, qui posent à eux seuls des problèmes d'exégèse et qui sont susceptibles d'être autant de pièges ? Affaire d'éthique, de perspective ! Au demeurant, la révision de sismicité historique de la France et de ses confins, révision dont l'étape majeure a pris deux ans, est considérée par le CEA et EDF comme l'apport essentiel du Projet de la Carte Sismotectonique de France, entrepris il y a dix ans.

3 PROBLEMES D'EXEGESE

Abordons maintenant les problèmes de méthode pour lesquels *exégèse* et *fiabilité* sont les maîtres-mots. D'abord, il est parfois nécessaire d'établir la réalité d'un séisme. Nombreuses sont les confusions avec des mouvements de terrain qui ne doivent rien à la sismicité. Un bon exemple d'une telle confusion, en 1907, vient d'être rappelé en Aragon par J.M. Lopez Marinas (18). A notre surprise, il a été fait grand cas, par un publiciste célèbre, d'un séisme destructeur en Maurienne au XIIIe siècle. En fait est reprise une confusion commise jadis par un chroniqueur britannique avec un célèbre glissement de terrain à l'échelle géologique. Cette confusion avait été dénoncée, fort à propos, il y a un siècle, par un pionner de la sismologie savoyarde, en vain, dès lors que sa démonstration a été "escamotée" par un catalogue alpin publié il y a un quart de siècle. Le moindre contrôle, qui s'impose à tout esprit critique, aurait permis de faire l'économie de cette "bévue" et d'échapper au risque d'alarmer inutilement et d'une manière parfaitement irresponsable des populations qui n'échappent pas toujours à la "psychose sismique" à la mode. En pareil cas, il est permis de parler d'un dévoiement de la sismicité historique. Si des ambiguïtés sont possibles pour les mouvements de terrain anciens, certaines confusions relativement récentes méritent d'être "épinglées" avec d'autant plus d'insistance que les contrôles sont faciles. Nous songeons à des explosions et à des ouragans. A notre surprise, l'ouragan qui ravage en 1896 les plateaux de Bourgogne s'est traduit sur une esquisse des intensités maximales de la France, publiée il y a une vingtaine d'années, par une aire d'intensités notables et des isoséistes parfaitement concentriques, non sans conséquences pour l'interprétation sismotectonique du fossé de la Saône et de ses bordures. Rappelons aussi la confusion d'un mistral qui sévit en 1799 à Avignon avec un séisme d'intensité élevé qui a failli fausser l'interprétation d'une partie de la vallée du Rhône. Ces deux exemples montrent une nouvelle fois à quel point la sismotectonique est à la merci d'une sismologie historique pratiquée à la légère.

Passons aux problèmes chronologiques. C'est la grande affaire ! Que de "doublets" pour la simple raison que les auteurs de catalogues perdent de vue que l'année se termine longtemps à Pâques, que la réforme calendaire de la fin du XVIe siècle se traduit par un décalage de dix jours, selon qu'elle est adoptée ou non. Ainsi des "dédoublements" de séismes se produisent entre régions protestantes et catholiques au cours de la crise souabe du milieu du XVIIe siècle ou encore pour le séisme de Remiremont en 1682. Un minimum d'esprit critique aurait permis de déceler ces "artefacts" par une simple confrontation des témoignages. Ces "doublets" sont aussi multipliés par un récent catalogue d'un pays du Moyen-Orient : c'est deux années de suite qu'apparaissent les mêmes événements, sans doute à la suite d'auteurs qui avaient quelque mal à maîtriser la chronologie islamique. Est-il nécessaire de souligner à quel point les discussions probabilistes sont susceptibles d'être compromises par des démarches hâtives qu'une sismicité historique bien comprise ne peut que rejeter ? Il reste que certains problèmes chronologiques sont d'une complexité telle que seuls des spécialistes chevronnés peuvent tenter de les résoudre. D'une manière significative, est parfois pratiquée la démarche inverse, à savoir l'étalonnage d'une chronologie par un séisme majeur, mais attention à ne pas "tourner en rond"!

Bien entendu, la toponymie multiplie les pièges. A tout hasard, relevons la confusion des Blauen de Forêt-Noire et du Jura Bâlois, à propos du séisme en 1356, du Munstertal des Vosges et de Forêt-Noire, de Muhlhausen d'Alsace et de Thuringe, de Verdun-sur-Meuse et de Verdun-sur-le-Doubs, sans parler d'une foule d'ambiguïtés au Moyen-Orient. L'auteur d'un article remarquable par ailleurs est tombé dans un piège classique, en faisant état, à propos d'un séisme du Moyen-Age, de Berne, alors qu'il s'agit de Bern/Vern/Vérone. Si certains problèmes sont ardus et parfois insolubles à première vue, au Moyen-Orient surtout, d'autres font figure de "faux problèmes" dont une élémentaire curiosité et le simple bon sens permettraient de faire l'économie. A l'émiettement (cf. *Supra*) s'ajoute l'amalgame. A cet égard, une présentation du séisme majeur survenu en 1531 au Portugal est exemplaire. A des confusions toponymiques (par exemple Santarem/Santander) s'ajoute l'amalgame avec un tremblement de terre survenu la même année en Andalousie orientale, ce qui conduit à une extension jusque dans la région d'Almeria...

Terminons par le vocabulaire, en prenant deux exemples empruntés à l'ancien français. Il est facilement perdu de vue que le terme "tremble-terre" a longtemps un sens très large, en englobant mouvements de terrain et même coups de vent (cf. *Supra*). Volontiers le terme "croller" est traduit, en contexte de sismicité, par "crouler", ce qui conduit à admettre un tremblement de terre destructeur. En fait, synonyme de "trembler", il correspond à des intensités modestes. Ainsi, c'est à tort qu'un tremblement de terre signalé à Metz au XVe siècle a été largement surévalué. A notre surprise et à notre plaisir, une paysanne comtoise nous a fait savoir, à l'occasion d'une enquête consacrée à un modeste séisme récent, que la terre avait "crollé" ! La consultation d'un Dictionnaire de l'Ancien Français n'est pas interdite au sismologue.. D'autres langues posent de tels problèmes... Est-il nécessaire d'ajouter que la connaissance de plusieurs langues est recommandée pour la plupart des recherches de sismicité historique, non pas pour résoudre personnellement les innombrables problèmes rencontrés pas à pas, mais pour être en mesure de les saisir, de les poser aussi clairement que possible, de les débroussailler et d'être en mesure de s'adresser en connaissance de cause à des spécialistes, historiens ou linguistes. En France, la révision de sismicité historique conduite à partir de 1975 a exigé le maniement d'une foule de sources espagnoles, catalanes, provençales, italiennes et allemandes - ancien allemand surtout, flamandes et anglaises, en raison de la localisation de nombreux événements sur les confins du pays ou dans les pays voisins. Bien entendu une certaine connaissance du latin est indispensable pour toute recherche remontant dans le temps. *A fortiori* de tels problèmes se posent-ils dans une grande partie du monde méditerranéen, avec ses sources syriaques, arabes, turques, grecques, hébreues, slaves, etc. (19).

La plupart des innombrables "bévues" qu'il est possible de relever et dont quelques exemples ont été donnés au hasard sont le fait d'esprits enfermés dans une tour d'ivoire "scientifique" ou plutôt "pseudo-scientifique", rétifs à une pluridisciplinarité critique et enrichissante privant parfois leur activité scientifique de bases solides. Paradoxalement, c'est au nom d'une étrange "déontologie" que les efforts de révision de sismicité historique ont parfois été combattus par des esprits plus soucieux de figer les connaissances que d'admettre leur évolution dans un esprit constructif. D'une manière préremptoire et autoritaire ne m'a-t-il pas été déclaré il y a une dizaine d'années : "Perrey a tout dit" ! Sans diminuer les mérites de ce pionnier du milieu du XIXe siècle, nous savons maintenant à quoi nous en tenir. Arrêtons cette incursion dans le passionnant domaine de la "sociologie scientifique", avec ses pesanteurs...(20)

4 MODES D'EXPRESSION

Passons aux moyens d'expression d'une sismicité historique qui doit être soucieuse, avant tout, de fiabilité. Pour d'anciens séismes, l'appréciation des intensités est souvent malaisée, quels que soient les critères mis en oeuvre. Souvent la nature des constructions est mal connue. D'ailleurs, les sources n'évoquent en général que les dégâts subis par des édifices notables, temples dans l'Antiquité, églises, chateaux et enceintes au Moyen Age, tandis que le comportement d'autres bâtiments nous échappe. D'une manière significative, le séisme de Bâle (1356) est connu pour l'essentiel par les dégâts subis par de nombreux chateaux, proches ou à quelque distance. Progressivement, les renseignements se font plus précis en portant de plus en plus sur d'autres constructions. Il est particulièrement difficile d'interpréter les informations psychologiques qui envahissent parfois les sources, qu'il s'agisse de l'Antiquité, du Moyen Age, du XVIe siècle et parfois bien au-delà. Ainsi convient-il de s'interroger au sujet du contexte de processions. Sont-elles faites à tout propos? Présentent-elles un caractère exceptionnel? Telles sont quelques raisons, parmi d'autres, pour lesquelles l'affirmation entière d'une intensité est souvent trompeuse. En matière de sismicité historique s'imposent souvent des formules exprimant une hésitation, fourchette d'intensité, signe > ou <, etc. Reflétant l'état des connaissances, elles laissent la porte ouverte à l'intégration de nouveaux éléments, même si elles irritent parfois des esprits férus de données "définitives", propres à une informatisation susceptible de les figer.

Ces remarques conduisent au problème de la localisation des épicentres, exercice parfois périlleux. Relevons d'abord le cas-limite que représente un repère unique, avec les problèmes d'informatisation qu'il pose. A vrai dire, il est souvent permis d'espérer qu'il ne le restera pas. A ce propos, rappelons les cas d'émiettement en notations ponctuelles attribuées à des séismes différents, en raison des habituels vices méthodologiques (cf. *Supra*). De tels émiettements ne peuvent conduire qu'à des spéculations hasardeuses. Heureusement des regroupements ouvrent d'autres perspectives pour la discussion des épicentres. Ne cesse de se poser le redoutable problème de la "fixation urbaine" d'informations clairsemées dont l'interprétation des particulièrement malaisée. En pareil cas, la recherche de repères ruraux est une tâche par excellence de la sismologie historique. A condition de s'écarter des sources classiques et d'accès faciles, qui alimentent les catalogues, cette quête est parfois fructueuse et peut être décisive pour formuler une hypothèse d'épicentre. En tout état de cause, il convient d'être circonspect à l'égard d'épicentres inspirés soit par un raisonnement par analogie sommaire, soit par un "coup de pouce" prétendument sismotectonique. A plusieurs reprises ont été relevés des épicentres localisés "là où ils devraient être", en vertu d'une utilisation simpliste de quelque carte géologique. Un bon exemple de "nourrissage" est donné par deux tremblements de terre du dernier quart du XIXe siècle, localisés à tort dans la région des îles anglo-normandes. A première vue, l'un d'eux a sans doute été localisé de la sorte, par Mourant, par analogie avec le séisme de 1927, connu instrumentalement. Quoiqu'il en soit, ces localisations seront reprises par les catalogues, en étoffant abusivement une "province sismotectonique". En effet, il suffit de consulter les informations et interprétations contemporaines pour se rendre compte que ces événements sont sans doute survenus en pleine Manche, entre Cotentin et Wight, en définissant un nouvel ensemble sismotectonique. Au demeurant, c'est au prix d'invraisemblables contorsions d'isoséistes que la localisation de l'un d'eux à Jersey a pu être longtemps maintenue, en dépit du bon sens.

Certes, la précaution est parfois prise de distinguer par la graphisme épicentres certains et incertains. L'expérience montre cependant que des lecteurs pressés ne tiennent guère compte de cette distinction essentielle. A cet égard, voici deux anecdotes significatives. D'une part, il m'a été demandé une fois, avec le plus grand sérieux, pourquoi tel épicentre incertain était "localisé" en tel point. D'autre part, c'est *in extremis* que j'ai pu intervenir et purger un texte destiné à une revue internationale d'alignements prétendument sismotectoniques utilisant aventureusement épicentres certains et incertains. S'imposent d'autres formules, plus "percutantes", mieux adaptées aussi à l'échelle des événements. C'est à ce point de vue que le recours à l'informatique est susceptible d'être fructueux.

Ne relèvent pas d'une sismologie historique bien comprise des croquis d'épicentres simplistes, "passe-partout", imposés au lecteur sans explication ni interrogation. Tel est paradoxalement le cas du fossé rhénan dont la sismicité historique est suffisamment maîtrisée pour justifier une représentation élaborée, critique. Loin de figer des épicentres, le propre de la "sismologie historique en marche" est d'en remettre en question un certain nombre, non sans conséquences sismotectoniques. Tel est le cas des séismes notables de 1504 et de 1692, localisés dans la région d'Aix-la-Chapelle/Liège, au terme d'une longue enquête dans les bibliothèques et archives d'Allemagne, de Belgique, des Pays-Bas et de France, alors que le premier avait été vaguement situé en Wesphalie par le catalogue de Sieberg, avec d'ailleurs une large sous-estimation des intensités, et que le second avait connu d'un catalogue à l'autre une longue promenade en Belgique, avec, par exemple, une localisation à Malines. C'est semble t-il à tort que le catalogue ibéro-maghrébien localise dans l'Atlantique l'épicentre d'un séisme destructeur survenu dans la région de Meknès-Fez peu après le tremblement de terre de Lisbonne. Une fois de plus, un raisonnement par analogie risque de conduire à une de ces erreurs dont il conviendrait de purger les listings hâtifs.

A vrai dire, la sismicité historique contribue à mettre en question une conception excessivement ponctuelle de l'épicentre macrosismique en faisant apparaître des "mosaiques épicentrales", caractérisées par la juxtaposition de différents degrés d'intensité, juxtaposition appelant une analyse serrée, qu'il s'agisse de conditions de site ou d'aspects proprement sismologiques. Pour prendre trois exemples, de tels problèmes se posent pour le séisme bordelais de 1759, le séisme du Bugey en 1822 et le séisme provençal de 1909 (dont il est question par ailleurs).

C'est d'une manière très inégale que d'anciens séismes se prêtent au tracé d'isoséistes et d'aires macrosismiques. Prenons l'exemple du séisme bourguignon de 1783. Pour cet événement notable à l'échelle régionale, fût-ce en raison de son intérêt sismotectonique, nous ne disposions que d'une vision sommaire, sans isoséites, faute d'un dépouillement convenable de riches sources, en grande partie à portée de main. Désormais, il est possible d'esquisser une aire pléistoséiste (VI -VII), un isoséiste V, des fragments d'un isoséiste IV, une partie de l'aire macrosismique et, pour le reste, une enveloppe, avec des hypothèses d'extension (21).

Cependant, nombreux sont les événements qui ne se prêtent pas à de telles reconstitutions, en raison de la dispersion des repères et des déficiences des sources. Néanmoins, il est souvent possible de faire connaître quelques traits tels que la décroissance des intensités dans l'une ou l'autre direction ou l'enveloppe dans laquelle s'inscrivent les repères, sans préjuger de l'ensemble de l'aire macrosismique. Ces deux informations peuvent certes être discordantes. Elles n'en sont que plus suggestives - je l'ai montré pour le séisme sans doute berrichon de 1522 et le séisme nord-marocain de 1624 - en faisant apparaître la probabilité d'aires méconnues de tel ordre d'intensité, en permettant de formuler des hypothèses d'épicentres et en multipliant les interrogations orientant de nouvelles recherches.

En outre, les enveloppes forment de véritables toiles d'araignées permettant de mieux donner corps, à l'échelle régionale, à une sismicité mal connue et difficilement interprétable individuellement.

Pour les aires macrosismiques, les révisions de sismicité historique conduisent en général à une extension, illustrée en particulier, répétons-le, par le processus de regroupement d'éléments émiettés. Pour des événements très anciens entrés dans un domaine quelque peu mythologique, se produit parfois une évolution inverse. En pareil cas, une extension excessive est parfois admise sur la foi de témoignages lointains relevant de l'ouï-dire et qui n'ont pas été passés au crible de la critique. Tel est le cas des tremblements de terre lombards de 1117 et 1222, déjà cités (22). Dans l'Antiquité, il est fait grand cas, dans un contexte de troubles, du séisme de 365, responsable d'un tsunami signalé entre Sicile et Egypte. Procédant par amalgame, les auteurs antiques attribuent à ce séisme une extension démesurée. L'événement vient d'être ramené à ces véritables proportions par F. Jacques et B. Bousquet (23). Bien plus, des historiens modernes, emportés par l'élan ou la mode ont "misé" sur ce même séisme pour rendre compte de discontinuités d'histoire urbaine, en Cyrénaïque et dans la région de Sétif (Algérie). En fait, d'autres historiens eurent vite fait de démontrer qu'il s'agissait d'un *deux ex machina* (24). A défaut d'une telle oeuvre

critique, des sismologues auraient pu être tentés d'accepter telles quelles les hypothèses discutables certes chevronnés.
Brûlons les étapes, changeons d'échelle. Parmi les vues d'ensemble, il convient de mettre l'accent sur les cartes d'intensité maximale connue, mises en oeuvre par des démarches déterministes. En France, une première tentative a été faite il y a un quart de siècle, à 1/5.000.000. Affirmative, cette carte est victime de problèmes de méthode - rappelons la "fixation urbaine" -, d'erreurs notables et d'oublis. Ainsi la représentation du Roussillon est-elle faussée par l'oubli de la crise catalane du XVe siècle, avec ses intensités très élevées. Au terme de deux années d'un intense travail de révision une nouvelle carte, à 1/1.000.000 sera préparée en 1977. De conception très différente, elle met l'accent sur les degrés de fiabilité et les perspectives d'évolution, généralement confirmées par des recherches ultérieures qui multiplient les précisions régionales sans bouleverser l'ensemble. Néanmoins, cette carte, datant d'une décennie, est susceptible de figer les connaissances, compte-tenu d'une propension naturelle de nombreux lecteurs à rechercher des éléments "définitifs", malgré les précautions oratoires multipliées à cet égard. S'impose une mise à jour, ce qui semble bien être un voeu pieux en raison du contexte économique. Insistons sur le fréquent oubli des effets des séismes notables survenus à quelque distance. Ainsi, la représentation des intensités maximales du Nord-Ouest de l'Algérie serait faussée par l'oubli des séismes de 1522 et de 1755. Personnellement, je considère comme un échec une récente carte des intensités maximales de l'Europe et de ses confins, soit pour une raison d'échelle, en raison de l'absence de repères ponctuels d'intensité élevée susceptibles d'intéresser une aire plus étendue, ou de lacunes qu'il aurait été possible de combler, par exemple au Maroc. D'autre part sont établies pour les besoins des démarches probabilistes des cartes de récurrence dont la fiabilité reflète par définition la qualité du patrimoine de sismicité historique et de son interprétation, sans parler des techniques mises en oeuvre. Encore que le travail de base puisse être éclipsé par la technique informatique, le moyen devenant facilement une fin, une telle carte est littéralement à la merci de toute évolution notable de la connaissance d'anciens séismes majeurs.
Il serait possible de poursuivre cette discussion en traitant en particulier des effets indirects, liquéfaction, éboulements (25), tsunamis...

5 UNE LEÇON DE RELATIVITE

Insistons pour terminer sur quelques points essentiels, à l'aide de quelques exemples. Exégèse et fiabilité, avons-nous dit. Il convient d'ajouter, répétons-le, *relativité*. A cet égard, certaines courbes d'activité sismique dans le temps sont éloquentes. Une telle courbe a été proposée pour l'Irak, avec un "creux" de plusieurs siècles. Alors qu'il est permis de songer en premier lieu à une explication historique et culturelle, cette courbe a été interprétée d'un point de vue étroitement sismologique, trompeur. L'ancienne sismicité du littoral iranien peut suggérer à première vue une migration de l'activité sismique, avec une activité notable à l'Ouest au Moyen Age, puis à l'Est, à partir du XVIe siècle. En fait, les informations relatives à l'Ouest seraient liées à la prospérité de deux ports tombés en décadence par la suite, tandis qu'à l'Est la fréquentation d'Ormuz se traduit par une floraison d'informations. Même une crise apparemment mineure au XVIe siècle est connue à Ormuz, grâce à la mention d'une procession organisée par les portugais... D'ailleurs, cette relativité est aussi illustrée par la réexploitation d'archives du Bureau Central Sismologique Français au cours d'un demi-siècle. Cette tâche a conduit à la découverte de vices de méthode, qui se traduisent en particulier par un remarquable déficit des intensités IV, et à une foule de réinterprétations qui sont loin de relever des préoccupations de "petit détail" de ceux qui "coupent les cheveux en quatre". Bien entendu, cette même relativité apparaît aussi au cours de récentes enquêtes macrosismiques, selon leur organisation, avec parfois des versions successives.

La sismologie historique est essentiellement évolutive. Innombrables sont les séismes anciens susceptibles de larges révisions, voire d'un bouleversement avec une foule d'implications, soit par une réinterprétation du patrimoine documentaire, soit par la découverte de nouveaux éléments dont les chances sont certes à la mesure de l'intensité des recherches dans leur ensemble. Encore faut-il une motivation. Elle peut être scientifique *sensu lato*, ce qui serait cependant l'exception, en raison de l'idée de la science *sensu stricto* se fait parfois de la sismologie historique ou encore de réflexes négatifs relevant du conformisme ou du confort intellectuel. Souvent, elle serait pratique, pour des raisons évoquées à plusieurs reprises. Ainsi l'enrichissement de la connaissance de la crise sismique auvergnate de la fin du XVe siècle est-elle liée à un projet de centrale nucléaire qui n'a d'ailleurs pas vu le jour. Plus souvent encore intervient la surprise ou le désarroi provoqués par un événement inhabituel que les catalogues ne permettaient guère d'envisager. A propos de sources historiques, viennent d'être cités, par ces "catalyseurs" les récents séismes du Yemen et de Guinée. Mais sans doute serait-il préférable de disposer de *Corpus* de sismicité historique, solide et évolutifs, sans être pris de court par de tels événements. Pour ne prendre qu'un exemple parmi bien d'autres, l'effet de surprise du séisme d'Agadir aurait été moindre si quelque attention avait été prêtée à un "précédent" du début du XVIIIe siècle, sans que nous reprenions pour autant à notre compte la formule simpliste de l'inéluctable retour des séismes destructeurs.

Quoiqu'il en soit, la sismologie historique, pluridisciplinaire par excellence à son échelle propre, étroitement associée à la sismologie instrumentale, est une école d'ouverture, de culture et de rigueur. A ce titre, elle peut certes faire figure de trouble-fête, mais c'est précisément sa vertu majeure. Il serait regrettable que les progrès d'un état d'esprit trop étroitement scientifique, état d'esprit dont les ravages sont omniprésents, puissent contribuer à freiner la renaissance de la sismologie historique, essentielle pour l'avenir et la crédibilité d'une sismologie bien comprise dans son ensemble. En entendant certains propos désobligeants (érudition littéraire, sclérose) on se met parfois à rêver de l'atmosphère dans laquelle évoluaient les esprits universels du XVIIIe siècle.

NOTES ET REFERENCES

1) A été faite l'économie de deux annexes: "Un séisme algérien mal connu, le 14 avril 1839", et "Difficulté d'une vue d'ensemble des anciens séismes majeurs intéressant Ibérie et Afrique du Nord".
2) A cet égard, quelques aspects méthodologiques sont développés dans une perspective interdisciplinaire par J. Vogt, "Effets indirects des séismes" (Cours de Sismologie, Centre Universitaire Européen de Ravello, 1985, à paraître).
3) A ce sujet, voir J. Vogt, 1986, "Tradition et méthodes de prévention au cours des siècles..., Moyen-Orient, Italie, Afrique du Nord, Antilles",
4) "Tremblements de terre, Histoire et Archéologie", *Actes des IVe Rencontres internationales d'Archéologie et d'Histoire d'Antibes* (1983), Valbonne 1984.
5) J.P. Adam, 1984, "Conséquences du séisme de l'an 62 à Pompéi" *Actes des IVe Rencontres internationales d'Archéologie et d'Histoire d'Antibes* (1983), Valbonne 1984.
6) E. Guidoboni, 1984, "3 janvier 1117 : le tremblement de terre du Moyen Age roman, aspects des sources" *Actes des IVe Rencontres internationales d'Archéologie et d'Histoire d'Antibes* (1983), Valbonne 1984.
7) Y. Burnand, 1984, "Terrae Motus : La documentation épigraphique sur les tremblements de terre dans l'Occident romain" *Actes des IVe Rencontres internationales d'Archéologie et d'Histoire d'Antibes* (1983), Valbonne 1984.
8) E. Guidoboni, 1984, art. cité (Note 6).
9) Les pièges des index sont illustrés par P. Samperi, "Messana", 1742 (Annales de Messine). Si deux séismes (1456 et 1493) apparaissent bien sous le vocable Terraemotus, à savoir "Terra Terraemotu concussa..." et "Terramotibus concuritur...", le tremblement de terre de 1638 échappe facilement à l'attention dès lors que l'index le signale en ces termes : "*Sabatto* palmarum terraemotus magnus factus...". Tel est peut être aussi le sort d'autres événements. En

revanche, soulignons la qualité de l'index d'un ancien répertoire des sources italiennes des archives espagnoles (I. Carini, 1884, "Gli archivi e le biblioteche di Spagna in rapporto alla storia d'Italia in generale", t. II, Palerme). Malgré l'adjectif "général", relevons la mention de documents relatifs aux séismes de Pouzzoles en 1538, de Sicile en 1693 (reconstruction d'églises) et à un tremblement de terre à Bagnorea. Ces qualités se retrouvent dans une récente Histoire de Sicile (D.M. Smith, 1968, "A History of Sicily...", Londres) dont l'index permet en particulier de repérer des informations au sujet du mémorable séisme de 1169. Mais que d'ouvrages faut-il parcourir d'un bout à l'autre faute d'index !

10) L. Ducat, 1984, "Le tremblement de terre de 464 et l'histoire de Sparte", *Actes des IVe Rencontres internationales d'Archéologie et d'Histoire d'Antibes (*1983), Valbonne 1984.

11) P.L. Gatier, 1984, "Tremblements du sol et frissons des hommes. Trois séismes en Orient sous Anastase" *Actes des IVe Rencontres internationales d'Archéologie et d'Histoire d'Antibes*(1983), Valbonne 1984.

12) E. Guidoboni, 1986, "The earthquake of December 25, 1222 : analysis of a myth", *Proc. Int. Symposium on Engineering Geology*, Bari.

13) N.N. Ambraseys, C.P. Melville, 1983, "Seismicity of Yemen", *Nature*, t. 303, N° 5915.

14) *Tragedia Relacion del Lamentable Terremoto...que la Divina Majestad se ha servido embiar a la Plaza de Oran...*, publ. contemporaine, sans lieu ni date.

15) G. Ferrari, 1984, "Les séismes de la côte de l'Adriatique du Nord au XVIIIe au XXe siècle : contribution des recherches historiques à un microzonage sismique de la ville de Rimini" *Actes des IVe Rencontres internationales d'Archéologie et d'Histoire d'Antibes* (1983), Valbonne 1984.

16) Résumé des travaux scientifiques et techniques du Service Géologique National (B.R.G.M.), 1984.

17) J. Vogt, 1985, "Le séisme bourguignon du 6 juillet 1783 : essai de sismicité historique". *Annales de Bourgogne*. Une foule d'autres séismes, rhénans surtout, ont fait et feront l'objet de telles mises au point.

18) J.M. Lopez Marinas, 1985, "Un dezlizamiento de ladere convertido en un terremoto...", *Revista de Geofisica.*

19) Ces problèmes sont évoqués par plusieurs articles de l'auteur, par exemple: J. Vogt, 1985, "Riesame di Sismi : Ricerche in Francia", *Quaderni Storici*, N° *spécial Terremoti e Storia*, J. Vogt, "Problèmes de méthode de sismicité historique, base des discussions de risque sismique", comm. au Symposium de Brigue, 1986, à paraître.

20) A ce sujet, voir : J. Vogt, 1981, "Problèmes de sismicité historique", *Géologues* (Union Française des Géologues), N° 57, J. Vogt, 1983, "Quelques coupures entre démarches et spécialités des sciences de la terre", *Géologues*, N° 64.

21) J. Vogt, 1985, "Le séisme bourguignon du 6 juillet 1783 : essai de sismicité historique". *Annales de Bourgogne.*

22) E. Guidoboni, 1984, 1986, art. cités (cf. notes 4 et 12).

23) F. Jacques, B. Bousquet, 1984, "Le cataclysme du 21 Juillet 365 : phénomène régional ou catastrophe cosmique ?" *Actes des IVe Rencontres internationales d'Archéologie et d'Histoire d'Antibes* (1983), Valbonne 1984.

24) Blanchard-Lemée, 1984, "Cuicul, le 21 Juillet 365", C. Lepelley, "L'Afrique du Nord et le séisme du 21 Juillet 365 : remarques méthodologiques et critiques". *Actes des IVe Rencontres internationales d'Archéologie et d'Histoire d'Antibes* (1983), Valbonne 1984.

25) A titre d'exemple, voir J. Vogt, 1984, "Mouvements de terrain associés aux séismes dans les Pyrénées", *Revue Géographique des Pyrénées et du Sud-Ouest,* J. Vogt, 1984, "Mouvements de terrain associés aux séismes en Afrique du Nord", *Méditerranée.*

ANNEXE: ASPECTS PSYCHOLOGIQUES DE LA SISMICITE HISTORIQUE ALGERIENNE, ECHOS DU XIXe SIECLE

Plusieurs notations faites au cours de la deuxième moitié du XIXe siècle au sujet de la connaissance de la sismicité algérienne présentent un grand intérêt, qu'il s'agisse de l'histoire des sciences ou des discussions de fond, sans parler d'autres aspects. Pour commencer, soulignons le *fatalisme* qui tient longtemps le haut du pavé, en alternant de temps à autre avec le triomphalisme.

En 1863, W. de Fonvielle, vulgarisateur scientifique, écrit : "Depuis un grand nombre de siècles les côtes barbaresques forment un espèce de *Terra incognita* où la séismologie n'a jamais pu pénétrer, quoique le sol porte des marques d'innombrables convulsions. A peine si l'histoire paraît avoir conservé le souvenir des cataclysmes qui ont couronné l'oeuvre commencée par la main des barbares et jeté au milieu des vastes plaines les débris de maison somptueuses et d'édifices majestueux...". Evoquant les travaux de Mallet et de Perrey, il écrit : "... tous deux déplorent aussi énergiquement l'un que l'autre la *négligence séculaire des populations africaines*..." (1). Romantisme, inculture, "hybris" pseudo-scientifique (qui, au demeurant ne cesse de fleurir)... N'insistons pas... Visiblement, W. de Fonvielle ne s'est posé aucune question au sujet des archives et des bibliothèques, sans parler de la tradition orale. Au sujet des archives, qu'il suffise de ce propos d'un érudit lucide : "... En Algérie... un indigène a été accusé d'hostilité envers la France et de connivence avec les insurgés, uniquement parce qu'on avait trouvé en sa possession... un registre dans lequel cet homme lettré tenait note des principaux événements dont il avait connaissance, tels que éclipses, *tremblements de terre*, ruptures de barrages et soulèvements de tribus..." (2). Heureusement ce précieux document sera publié... D'autre part, ont été évoquées rapidement les mesures préventives prises jadis en Algérie et dont l'oubli sera déploré par la suite (3). Mais voici un autre son de cloche, pour la sismicité contemporaine. A la suite du séisme de 1867. W. de Fonveille s'en prend à l'administration : "... nous ne croyons pas que les autorités locales aient pris des mesures bien énergiques pour jeter quelque lumière sur un des plus grands et des plus instructifs phénomènes qu'il soit donné à l'homme d'observer". Un résumé de ce propos fait état de "l'inertie des administrations algériennes". La presse est elle aussi prise à partie : "... généralement, les journaux d'Afrique font preuve de bien peu de discernement en matière scientifique" (4). En fait, la fréquentation des archives et de la presse montre que ces propos sont tout aussi excessifs que les précédents.

A propos du rapport consacré au séisme qui détruit Mélouza en 1855, nous avons relevé cette suggestion : "Cette dépêche peut donner un article pour le Moniteur Algérien". Au demeurant, de nouvelles recherches et des discussions serrées se prêtent à de substantielles mises au point. En témoigne la précision avec laquelle le Professeur Ambraseys a récemment présenté les caractéristiques du séisme majeur de 1856 (5). D'autres événements, signalés d'une manière sommaire par des catalogues répétitifs, se prêteraient à de telles mises au point. C'est l'affaire de volonté, d'esprit critique et de lucidité, loin du fatalisme que peut inspirer une mentalité trop étroitement scientifique.

1) W. de Fonvielle, 1863, "Tremblements de terre en Algérie", *Revue du Monde Colonial*, 2e série, t. VIII, 5e année.
2) M.O. Houdas, 1885, "Monographie de Méquinez", *Journal Asiatique*, février-avril.
3) J. Vogt, 1986, "Tradition et méthodes de prévention au cours des siècles...", *Premier Colloque National de Génie Parasismique*
4) "Chronique de la Science et de l'Industrie", *Presse Scientifique des Deux Mondes,* t. I., 1863.
5) N.N. Ambraseys, 1982, "The Seismicity of North Africa : the Earthquake of 1856 at Jijelli, Algeria", *Boll. Geof. Teorica ed Appl.*, t. 26.

THE PROVENCE EARTHQUAKE OF 11TH JUNE 1909 (FRANCE). A NEW ASSESSMENT OF NEAR FIELD EFFECTS

A. LEVRET, C. LOUP, X. GOULA
Commissariat à l'Energie Atomique - Institut de Protection et de Sûreté Nucléaire
B.P. 6, 92265 FONTENAY-AUX-ROSES
FRANCE

ABSTRACT. The earthquake which took place on June 11, 1909 in the vicinity of Lambesc in Provence was the largest one this century in France ($M \simeq 6$). The discovery of an unpublished document (Spiess 1914) containing a very detailed description of the effects at 809 points in the epicentral area (geological and topographical characteristics, description of the buildings and the damage incurred as well as the description of each seismic shock felt and the direction of ground motion), has made it possible to draw up an isoseismal map, to correlate local variations in intensity with the geological and topographical features, to determine a focus situated in the post-Triassic covering at a depth of less than 5 km, and to draw a map of the directions of the ground motion which will be used to identify the source mechanism, covered in another study.

1. INTRODUCTION

France is a country with only moderate seismic activity and destructive earthquakes are rare. One of the strongest in our history took place in the 20th Century in Provence, on June 11, 1909 at 9:16 p.m., killing 46 people and causing very extensive damage. Although several studies were carried out at the time (1 to 5), the synthesis published does not enable us to reinterpret the effects observed in the light of present-day knowledge.

The discovery of an unpublished document consisting of the very detailed investigations carried out by Commandant Spiess between 1909 and 1914 in the affected area, changes previously held views and enables a more precise conclusion to be drawn. This large file was found in the library of the French Geological Society. Commandant Spiess, who was officially instructed by the War Ministry on July 13, 1909 to study the effects of the earthquake in the affected area, spent a total of three months, spread over several visits, inspecting the area. Only a precis of this detailed investigation has been published so far (6).

J. Bonnin et al. (eds.), Seismic Hazard in Mediterranean Regions, 383–399.

2. SORTING THE INFORMATION

2.1. Composition of the File

With the exception of the letters which help to reconstruct the history of the document, of 189 questionnaires returned by the townships and the 27 questionnaires drawn up for the description of damage to particular buildings or factories, the file is made up essentially of the study itself. This study deals mainly with the area in which damage was suffered (about 350 square kilometers) covering 61 townships, and with very detailed descriptions of the damage noted both in towns and villages and in hamlets, chateaux and farms situated within the township. The following type of information is given : ground formations, damage in the locality and to specific buildings (town halls, schools, churches, chateaux, farms, factories, etc.) and to public works (railway lines, canals, roads, etc.), effects on the ground, on springs and wells, seismic noises and miscellaneous effects (flashes of light, smells of sulphurous gas). Sometimes information is given on the previous condition of a building or a sketch showing the way it faced, the direction of the shockwave and the position of the damage. The macroseismic intensity estimated on an 11-degree Mercalli scale modified by the author has been compared with other scales, including the MSK scale (Figure 1).

2.2. Intensity Estimations

Sorting this large file has given access to a large number of observations (reinterpreted on the MSK scale) in the damaged area : 207 townships and 602 hamlets with intensities of at least VI for the most part. A bargraph (Figure 2) shows the contribution of the Spiess file for each intensity class, compared with information supplied by Angot (1) and those mentioned in the "Sirène" computer file of historical seismicity of France (7). This type of bargraph which should show a regular increase in the percentages as the intensity decreases, that is as the area involved increases, shows a number of irregularities which enable us to draw the following conclusions :

- the data provided by Angot is insufficient and heterogeneous in the epicentral area but very complete in the far field ;
- the density of the points noted in the "Sirène" computer file is very irregular (too many points for intensities IX and VII and not enough for intensities lower than VI-VII) ;
- the data obtained from the Spiess report is the most coherent down to the intensity VI, corresponding to the damage area : further out, information is clearly lacking.

For the estimation of intensities it was considered that most of the provençal houses were of type A (cottages and field stone structures), because they almost all displayed construction weaknesses : inadequate foundations, badly-built walls, lack of bonding between front and gable walls, eaves which were too high and protruded

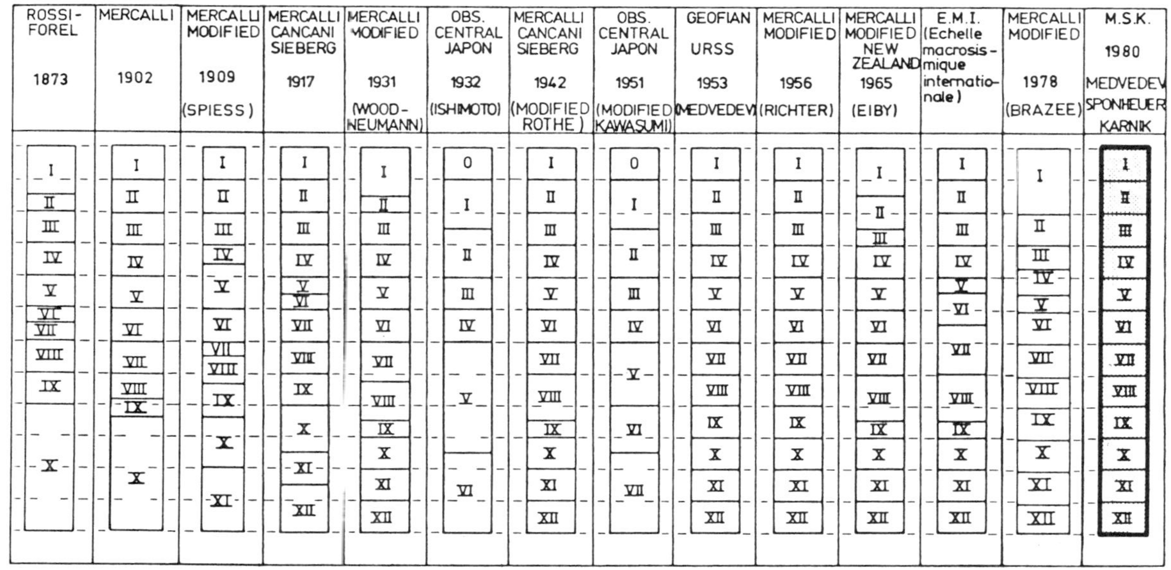

Fig. 1. Comparison between different intensity scales.

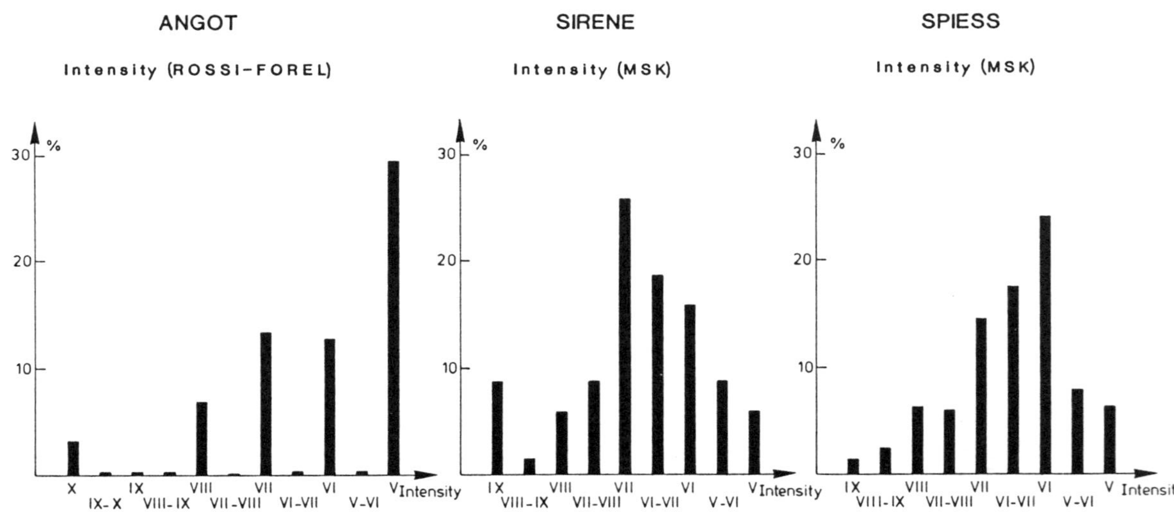

Fig. 2. Bargraphs of data according to intensity class showing the contribution of the Spiess data compared with information supplied by other sources.

too far, beams and floor joists not inserted far enough into the walls, lack of ironwork joining floors to walls and badly-designed roof timbers. Also, for the hamlets consisting of generally only one or two houses, the percentage of damage was obtained by counting the damage on each building. Consequently the intensity is more closely related to the condition of the building at the time of the earthquake and its estimation is thus less reliable. In the larger localities the intensity is usually a mean value expressing the overall damage in the locality, and often quoted in half-degrees (VIII-IX at Lambesc, St. Cannat, Vernègues). However, there are some exceptions in places with exceptional topographical features (for example, a village part of which is perched on a hillock) where the localisation and extent of the damage was so specific that two distinct intensities are quoted (Rognes : VII at the foot of the hillock and IX at the top in the Foussa quarter) (Figure 3). The intensity pattern in the epicentral area shows that three zones of maximum intensity IX existed at the western end of the Trévarese fault. Similarly, the isoseismal VIII is separated into two areas : one in the west including the townships of Lambesc, Rognes, St. Cannat and its extension towards the east to the south of the Trévaresse fault, and the other in the north-east, ahead of the fault, including the township of Puy Sainte Réparade.

3. INTERPRETATION OF THE MACROSEISMIC EFFECTS IN THE NEAR FIELD

3.1. Influence of the Local Geology

A detailed drawing of the isoseismals for intensities VI-VII and VIII (Figure 4) shows the influence of superficial formations having different geological characteristics. In order better to account for the specific geotechnical features of the ground, the isoseismals have been plotted on a simplified geological map, indicating 5 types of ground (Figure 4).

It was possible to quantify the specific response of different types of ground to the seismic motion by counting the number of points per intensity class and per type of ground. If we adopt as a starting hypothesis a homogeneous distribution of the various types of ground around the epicenter, we may attempt to do away with both the distribution of the intensities and the absolute area of each type of ground, by calculating first the percentage of points observed for each intensity, independent of the type of ground ; then, in each category of ground, the percentage of the distribution of each intensity, based on the percentages previously obtained is computed. The geological formations are grouped into 4 types of terrain in order of increasing cohesion : Quaternary, Mio-Pliocene, Oligocene and Secondary.

The bargraphs (Figure 5) depict the influence of the superficial geological layers on the observed intensity. In the formations having low cohesion, which are in the majority (64 % of the observations) and consist of sedimentary beds of Mio-Pliocene and Oligocene, the percentage of high intensities is larger than that of lower ones. This

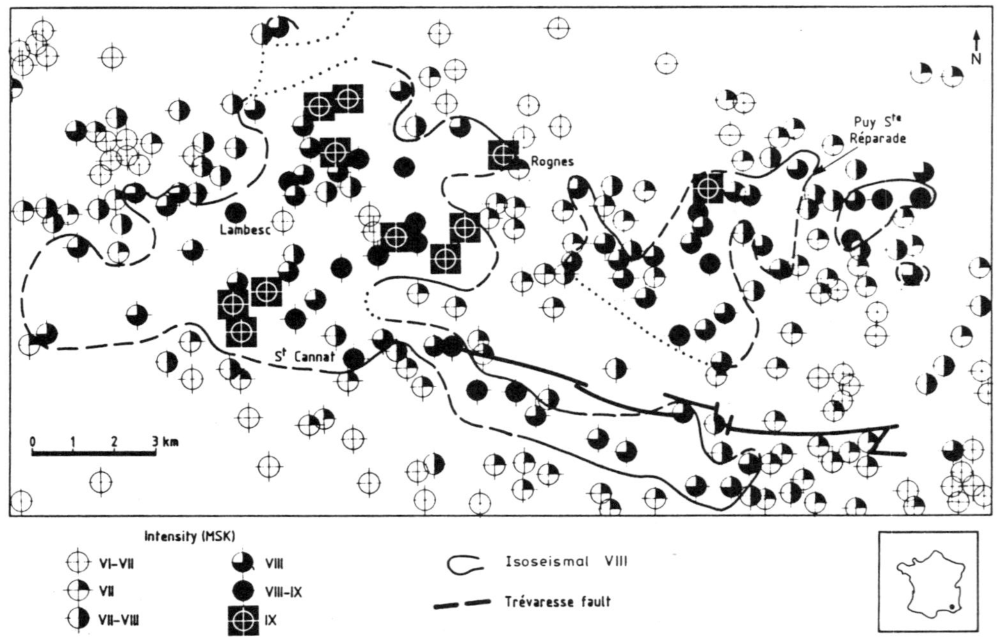

Fig. 3. Epicentral area of the June 11, 1909 Provence earthquake.

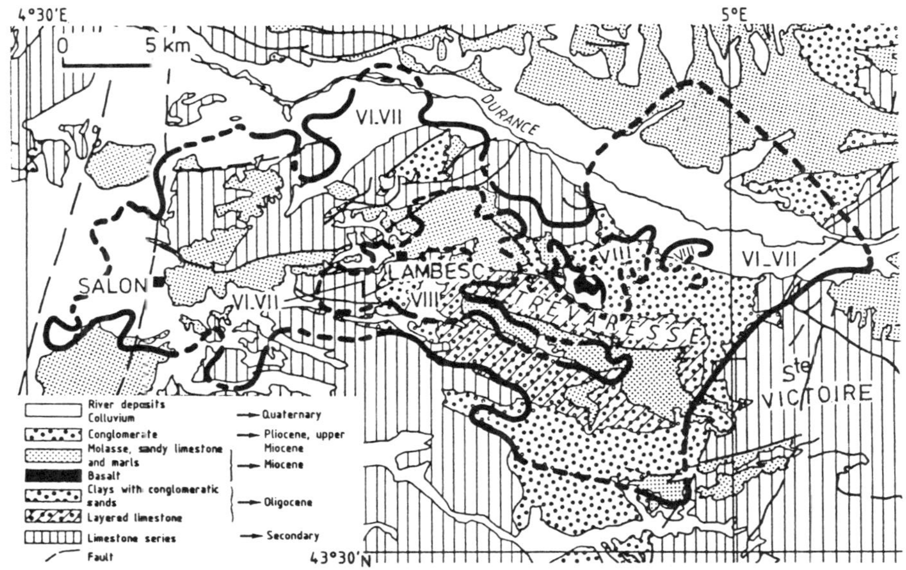

Fig. 4. Isoseismals for intensities VI-VII and VIII in the epicentral area on a geological map, showing the influence of superficial formations having different geological characteristics.

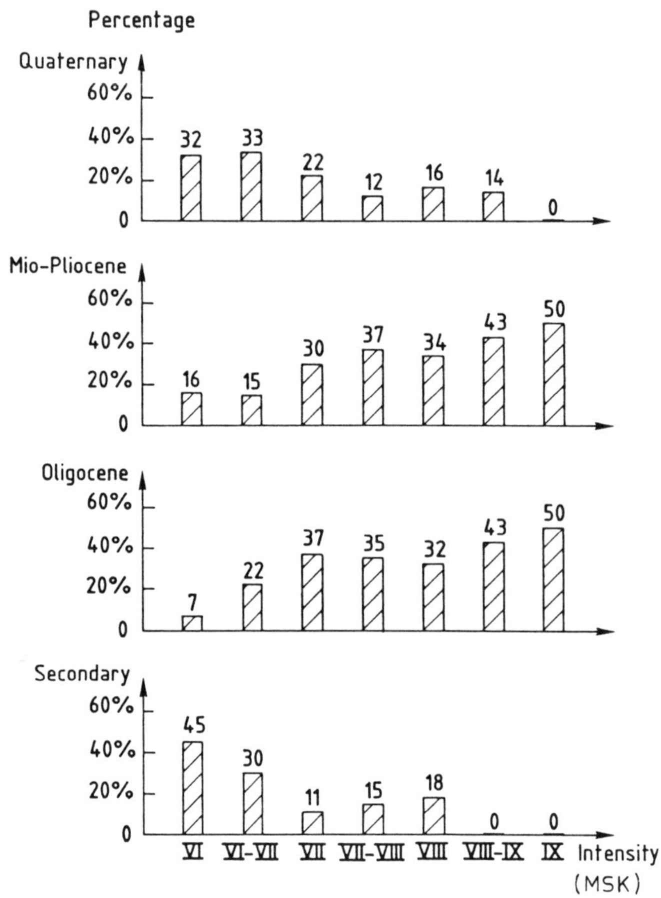

Fig. 5. Distribution of geological formations according to intensity class showing the influence of the superficial geological layers on the observed intensity.

tends to indicate that they have an amplifying role and would to some extent explain the groupings of intensities IX and the shape of the VIII isoseismal, although the geometry of the Trévaresse fault (Figure 4) probably also plays a part. On the other hand, the compact Secondary limestone, of which there is little (17 % of the observations) but which is distributed fairly regularly, seems to diminish the effects. This is particularly noticeable to the north, in the Costes range, in the south, on the Lançon-Provence-Coudoux plateau, in the east on the Ste. Victoire range, and in the center near Lambesc, where a small limestone outcrop exhibits intensities of VI and VII within a pleistoseismal zone with levels of VIII and IX. However, the results obtained on the Quaternary alluvions, which not widespread (18 % of the observations) and which are situated far from the pleistoseismal zone, do not indicate a damping effect : rather, judging from the isoseismal VI-VII (Figure 4) they would appear to have a slightly amplifying effect.

3.2. Influence of the Topography

At several individual points with greatly contrasting topography and sometimes ground type, effects were observed which were quite different from others in the same area : the intensities were greater than in neighbouring places (Table I).

All these examples, situated either at the top of a hillock or at the end of a steep ridge, or on the slopes of these escarpments exhibit an amplification of the effects. This is especially the case of Rognes and Vernègues, where we note a difference of two degrees of intensity between the old town, built on the slope, and the new town, built at the bottom of the hill. In the case of Vernègues (Figure 6), the old village, consisting of forty or so houses disposed like an amphitheatre on the south-west and south-east slopes of a ridge ending a long narrow ridge running north-west, was almost completely destroyed. On the other hand, the few houses situated in the west on less sloping ground did not collapse, but were seriously damaged. The high intensities observed in the old village (VIII-IX) were not caused by rockslide from the south-western part of the clift overhanging the houses, because these blocks did not reach the village. On the other hand, this amplification of the effects as compared with the new village (VI-VII) built lower down (90 meters height difference) on a milder slope or compared with the other hamlets situated at the base of the hill (intensities VI to VII) are probably due to a combination of factors :

Amplification due to resonance of the hillock as a whole and especially on the steeply sloping (nearly five per cent) and very sharply curving spur, where the most serious damage was sustained (VIII-IX).

Faulty building : age, poor foundations, front walls not bonded to the pinion walls (which was especially bad, for the front walls where were all facing south and thus practically at rightangles to the axis of the slope of the spur).

TABLE I : INFLUENCE OF THE TOPOGRAPHY ON INTENSITIES

PLACE	TOPOGRAPHY	NATURE OF GROUND	INTENSITY	NEIGHBOURING INTENSITY
Le Vieux Rognes (Le Foussa)	Southern and eastern slopes of a hillock	Miocene marine molasse	IX	VII
Vernègues	Southwestern and southeastern of a ridge ending in a long narrow crest trending NW-SE	Miocene marine molasse. At the summit : coarse shelly limestone ; at the base : soft sandstone and more or less marly sand	VIII-IX	VI-VII
Venelles	Hillock	Miocene limestone and marls	VIII	VI-VII
Cornillon-Confoux	Summit and end of a long ridge running NNW-SSE	Marine molasse	VII	VI
Eyguières	Upper part	Compact Neocomian limestone	VI	VI
	Lower part	Talus from neighbouring slopes	VII	
Mallemort	Part of the town is built on the top and slopes of a hillock	Eocene sand and Miocene molasse	VII	VI
Château de la Barben (La Barben)	Isolated steep rock	Compact Neocomian limestone	VIII	VII

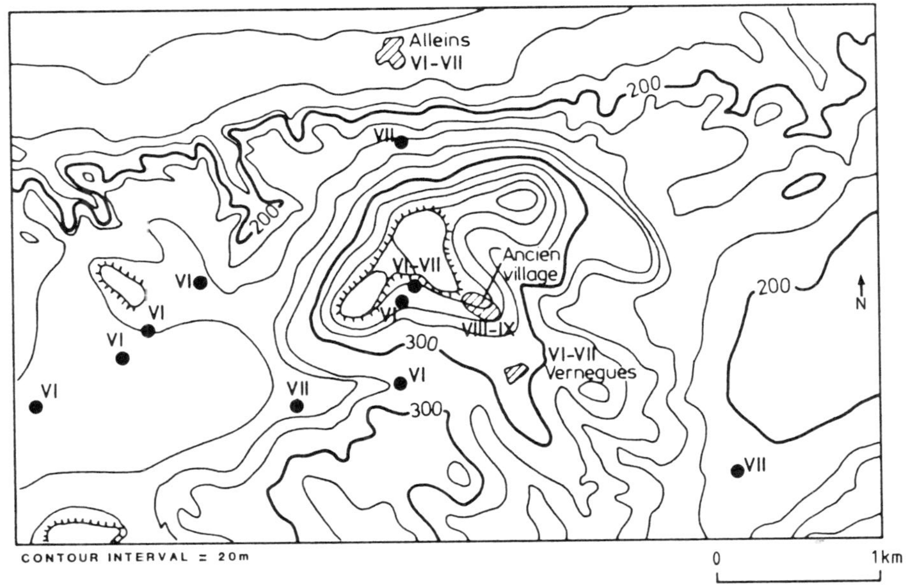

Fig. 6: Intensities observed near Vernègues on a topographical map showing an amplification of the effects between the bottom and the top of a steep ridge.

3.3. Directions of the Ground Motion

All the testimonies point to the existence of strong vertical motion in the epicentral zone : most of the witnesses had felt themselves lifted up, and some on the contrary said that they had felt their feet giving way. At Lambesc, the vertical motion seems to have been very strong, for boxes of coffee and sugar were emptied without being moved. At Salon de Provence, doors jumped off their hinges and candles fell from their candlesticks along with their little copper cups. On the other hand, no damage which could indicate the existence of surface cracking was noted on public works such as railway lines, irrigation canals or roads. Measurements of levels taken later (8) did not show any significant differences which could indicate the existence of vertical motions on the surface.

As to the horizontal direction of the seismic waves, a lot of information was gathered by Commandant Spiess. The directions established with varying degrees of certainty have been plotted on the background of a map with the areas of serious damage and the trace of the faults affecting the Mio-Pliocene areas (Figure 7). The most reliable directions were those deduced from the way objects fell (vases or ornemental balls which are quite common in Provençal houses).

The analysis of the directions observed reveals clear tendencies in various areas in relation to the Trévaresse fault. To the north of the fault, out of 32 observations, 60 % indicated a north-south direction and 30 % a north-west to south-east one. Hence the main component of the horizontal motion in this zone was in a direction slightly off north-south. The compartment south of the fault, less well documented (19 observations), shows a similar tendency, with 63 % of the observations giving a north-south direction, but a few observations (16 %) indicate a conjugate direction. At the western end of the fault, only 11 observations are divided between NW-SE (45 %) and NE-SW (36 %). In the north-west part of the fault which comprise 12 observations, 58 % of them indicated a NW-SE direction more or less parallel with the Trévaresse fault and 33 % an E-W direction. This distribution of vertical and horizontal motions displays an overall coherency, resulting from the mechanism which was probably at the origin of the break of the Trévaresse fault and which will be dealt with in another study.

4. INTERPRETATION OF THE OVERALL MACROSEISMICAL EFFECTS

In order to get a global view of the phenomenon and to study certain parameters connected with the wave propagation, it is necessary to consider the far field macroseismic effects. As Commandant Spiess's file deals almost exclusively with the damage area, it was necessary to supplement it with far-field information supplied by Angot (1) (intensity V or less). The only information accessible is in the form of intensities expressed on the 10-degree Rossi-Forel scale ; however,

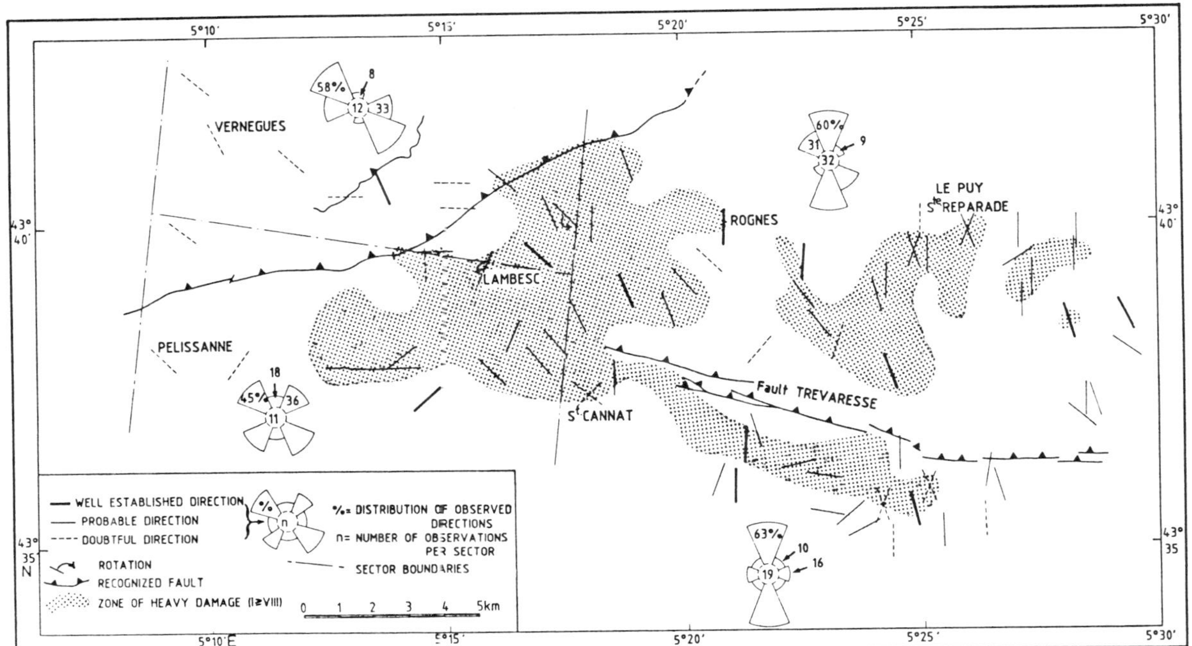

Fig. 7. Distribution of horizontal directions of ground motion in the epicentral area.

for these low levels the degrees are equivalent to those of the MSK scale (Figure 1). Considering the poor degree of accuracy obtained by this type of information, rough isoseismals have been drawn in far field and the detailed lines in the damage area have been smoothed after elimination as much as possible of the very localized effects so as to retain only the general isoseismal trends (Figure 8). This shows differences in wave propagation depending on the azimuth, clearly favouring the east-west direction, and a sudden change in the decreasing intensity versus distance in this direction : a rapid decrease up to the isoseismal VI, corresponding to the limit of most of Spiess's observations, and a slow decrease further out, where Angot's observations are reverted to. There is no doubt a problem here of coherency of information associated with variations in geometrical spreading factor, probably due to a change in the type of waves responsible for the effect observed : body waves in the near field (about the first 50 km), and surface waves after that. To the NNE, where the effects diminish very quickly (over 50 km), the rate of fall is regular.

To account for these observations, the simple Sponheuer-Koves-ligethy body-wave propagation relation was first applied (9), for a given azimuth on average isoseismal curve radii (Figure 9a). This relation is modified by a law of decreasing intensity I = Io - In, function of epicentral distance (Io and In being respectively the epicentral intensity and the intensity to the distance rn). The fitting is carried out by a least squares method entails solving a system of linear equations of the K order (number of parameters). The calculation is performed by successive iterations starting with an initial value of parameters and h (respectively, the absorption coefficient and the focal depth), until the solutions converge.

In a second stage, the model is implemented either considering two values of the geometrical spreading coefficient m (m = 1 in the near field, m = 0.5 beyond 50 kilometers, or by letting this coefficient vary freely in each zone : Figure 9b). In the second case, the number of points is too small (4 values) compared with the number of variables (3 values) for the results to be satisfactory. However, in all cases, a better fitting is noted. Whichever assumption is used, the values of focal depth obtained never exceed 5 km.

5. CONCLUSION

An analysis of Commandant Spiess's file enabled us to use the descriptions in 809 localities to compose a detailed picture of the epicentral area. This reflects the influence of three types of phenomenon on the damage incurred :

- the geometry of the Trévaresse fault probably responsible for the earthquake ;
- relative amplification of the effects in the Plio-Miocene and Oligocene sedimentary rocks as opposed to the Secondary limestones ;

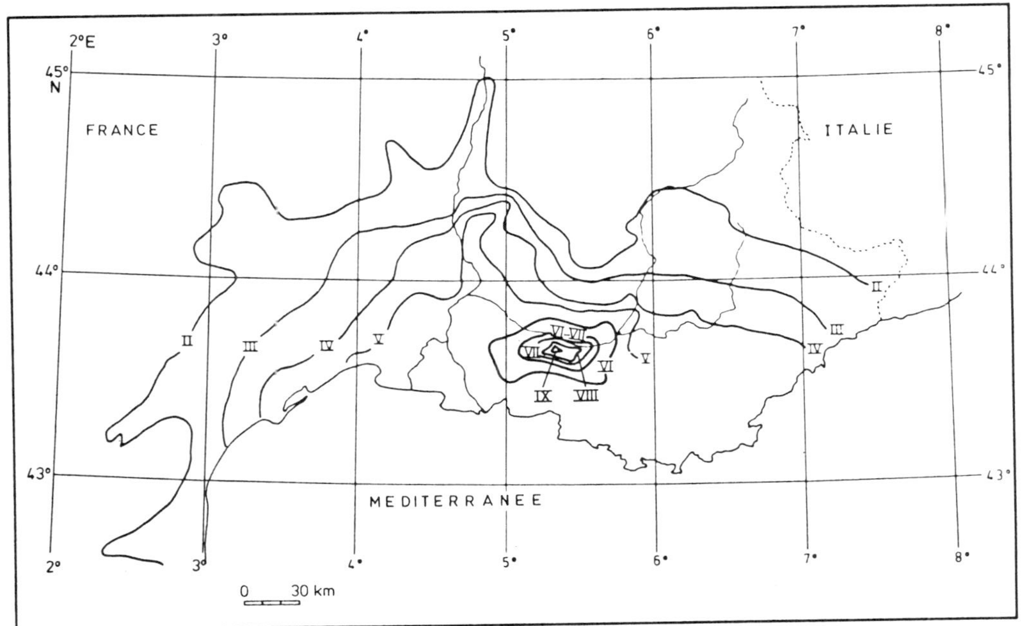

Fig. 8. Isoseismals of the June 11, 1909 Provence earthquake showing a clearly enhanced propagation in the east-west direction and a sudden change in the decreasing intensity versus distance in this direction.

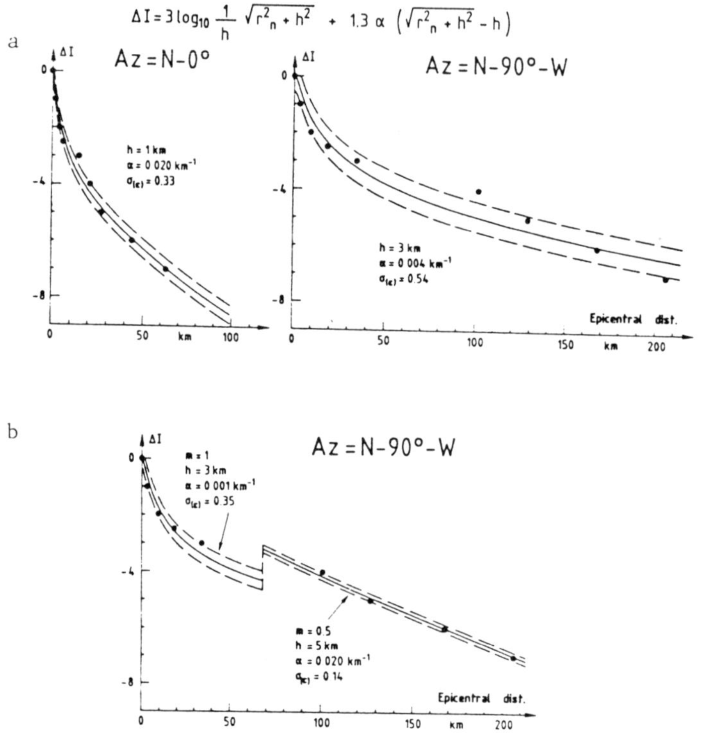

Fig. 9. Focal depth determination by fitting a law of decreasing intensity, function of epicentral distance. a : in two different azimuths. b : with two geometrical spreading factors.

- amplification of the effects on contrasting relief topography in conjunction with the nature of the foundations and the direction in which the buildings on the steep slopes faces.

Moreover, it is possible to establish from reliable testimonies the existence of a strong vertical motion, and the distribution of horizontal ground motions which will be used in another study to establish the source mechanism of the earthquake. The depths calculated from the macroseismic data place the focus within the first five kilometers of the Post-Triassic covering.

6. REFERENCES

1) - Angot A. (1913). "Le tremblement de terre de Provence". Annales du Bureau Central Météorologique de France, Paris. Année 1909 p. 37 à 93.

2) - Flammarion C. (1909). "Le tremblement de terre de Provence". Bulletin de la Société Astronomique de France, Paris. Juillet p. 297 à 312.

3) - Lemoine P. (1909). "Observations faites sur le tremblement de terre de Provence". Bulletin de la Société Philomatique, Série **X** tome **I**, No **3.**, Paris. p. 1 à 34.

4) - Repelin J., Laurent L. (1910), "Le tremblement de terre de Provence". Extrait du Bulletin de la Société de Géographie de Marseille. p. 3 à 22.

5) - Moreux Abbé (1909). "Les tremblements de terre". Ed. Jouve, Paris.

6) - Spiess Cdt (1926). "Note sur le tremblement de terre de Provence". Comptes rendus du Congrès des Sociétés Savantes, Paris. p. 173 à 195.

7) - Sirène (1984). "Fichier de sismicité historique de la France". BRGM/CEA/EDF.

8) - Lallemand C. (1911). "Rapport sur les changements du niveau du sol en Provence à la suite du tremblement de terre du 11 juin 1909. Comptes-rendus de l'Académie des Sciences. n° 152. p. 1560 à 1563.

9) - Sponheuer W. (1960). "Methoden zur herdliefenbestimmung in der makroseismik". Freiberger Forschungshefte Akademie Verlag, Berlin.